Computational Experiments

IEEE Press
445 Hoes Lane
Piscataway, NJ 08854

Computational Experiments

A Bridge Between Artificial Intelligence and Social Sciences

Xiao Xue
Tianjin University, China

Fei-Yue Wang
The University of Chinese Academy of Sciences
Beijing, China

IEEE Press
WILEY

Published by John Wiley & Sons, Inc., Hoboken, New Jersey.

Library of Congress Cataloging-in-Publication Data applied for:

Hardback ISBN: 9781394377459

Cover Design: Wiley
Cover Image: NeoLeo/Shutterstock

Set in 9.5/12.5pt STIXTwoText by Straive, Chennai, India

Contents

About the Authors

Xiao Xue received his Ph.D. degree in control theory and control engineering from the Institute of Automation, Chinese Academy of Sciences, Beijing, China, in 2007. Following that, he served as a postdoctoral researcher at the Department of Automation, Tsinghua University, Beijing, China, from 2008 to 2012. In 2013 and 2014, he worked as a Visiting Scholar at the School of Computer Science, Carnegie Mellon University, USA. Presently, he holds the position of Professor at the School of Computer Software, College of Intelligence and Computing, Tianjin University. His primary research interests include Agent-based Modeling, Computational Experiments, AI for Social Science (AI4SS), and service computing.

He actively participates in Technical Committees, including being a Senior Member of IEEE, and a Standing Committee Member of the Committee on Service Computing, the China Computer Federation (CCF). Additionally, he is a Deputy Director of the Committee on Computational Society and Social Intelligence, Chinese Association of Automation (CAA). Currently, he serves as the Deputy Director of the Tianjin Key Laboratory for Health Human Settlement Environment and Intelligent Technology. He also chairs the Young Experts in Services Computing (YESC) Committee. Furthermore, he holds editorial positions in prestigious academic journals. He is an Associate Editor of IEEE Transactions on Intelligent Vehicles; Associate Editor of International Journal of Crowd Science; Young Associate Editor of Complex System Modeling and Simulation.

Since the beginning of his career, he has made significant contributions to the academic community through his publications. He has authored 3 academic books and published over 60 articles in renowned international journals and conferences, including the esteemed IEEE Transactions series. His research output demonstrates his expertise and the impact of his work on the scientific community. In 2020, he published the book *Computational Experiments Method for Complex Systems*, which is the first monograph that systematically organizes the technical system of computational experiments method. His accomplishments have been recognized through various awards. In 2023, he received the IFAC Technical Committee award for outstanding achievements in Social Computing and Cyber Physical Social Systems. In 2024, he was the Winner of the 2024 Yang Jiachi Science and Technology Award.

Fei-Yue Wang received his Ph.D. degree in computer and systems engineering from the Rensselaer Polytechnic Institute, Troy, NY, USA, in 1990. He joined The University of Arizona in 1990 and became a Professor and the Director of the Robotics and Automation Laboratory and the Program in Advanced Research for Complex Systems. In 1999, he founded the Intelligent Control and Systems Engineering Center at the Institute of Automation, Chinese Academy of Sciences (CAS), Beijing, China, and in 2002, was appointed as the Director of the Key Laboratory of Complex Systems and Intelligence Science, CAS, and Vice President of Institute of Automation, CAS, in 2006. He founded CAS Center for Social Computing and Parallel Management in 2008 and became the State Specially Appointed Expert and the Founding Director of the State Key Laboratory for Management and Control of Complex Systems in 2011.

His current research focuses on methods and applications for parallel intelligence, social computing, and knowledge automation. He is a Fellow of INCOSE, IFAC, ASME, and AAAS. In 2007, he received the National Prize in Natural Sciences of China and numerous best papers awards from IEEE Transactions and became an Outstanding Scientist of ACM for his work in intelligent control and social computing. He was honored with the IEEE Intelligent Transportation Systems Outstanding Application and Research Awards in the years 2009, 2011, and 2015, respectively, and the IEEE Norbert Wiener Award in 2014. In 2024, he received IEEE CRFID Emily Sopensky Meritorious Service Award, IEEE SMC Lotfi A. Zadeh Pioneer Award, and 2025 IEEE Transportation Technologies Award. In 2025, he received IEEE ITSS Lifetime Achievement Award. Additionally, in 2021, he was selected as the IFAC Pavel J. Nowacki Distinguished Lecturer.

Since 1997, he has been serving as the General or Program Chair of over 30 IEEE, INFORMS, IFAC, ACM, and ASME conferences. He was the President of the IEEE ITS Society from 2005 to 2007, the IEEE Council of RFID from 2019 to 2021, the Chinese Association for Science and Technology, USA, in 2005, the American Zhu Kezhen Education Foundation from 2007 to 2008, the Vice President of the ACM China Council from 2010 to 2011, the Vice President and the Secretary General of the Chinese Association of Automation from 2008 to 2018, and the Vice President of IEEE Systems, Man, and Cybernetics Society from 2019 to 2021. He was the Founding Editor-in-Chief (EiC) of the International Journal of Intelligent Control and Systems from 1995 to 2000, IEEE ITS Magazine from 2006 to 2007, IEEE/CAA Journal of Automatica Sinica from 2014 to 2017, and China's Journal of Intelligent Science and Technology from 2019 to 2021. He was the EiC of the IEEE Intelligent Systems from 2009 to 2012, IEEE Transactions on Intelligent Transportation Systems from 2009 to 2015, IEEE Transactions on Computational Social Systems from 2017 to 2020, and the IEEE Transactions on Intelligent Vehicles from 2022 to 2024. Currently, he is the President of CAA's Supervision Council.

Preface

"Science is not merely a collection of facts, but a way of thinking"—Carl Sagan's insight resonates deeply in the era where artificial intelligence (AI) and social sciences converge. As algorithms weave into the fabric of social governance, economic operations, and public life, complex social systems present nonlinear dynamics, emergent characteristics, and inherent uncertainties that challenge traditional research paradigms. We can no longer rely solely on mathematical deduction to predict societal trajectories, nor can we afford large-scale, high-risk interventions in the real world to test hypotheses. Against this background, computational experiments have emerged as an important bridge—connecting the precision of AI with the depth of social sciences and opening a new window to decipher the complexities of human society.

Computational experiments are far more than a mere combination of technologies; they represent a paradigmatic innovation that integrates complexity science, computer simulation, statistics, and social theory. Guided by the core idea of "generative explanation," they construct artificial social models that embed the heterogeneity, bounded rationality, learning evolution, and interactive behaviors of micro-individuals. Through computational simulation, they replicate the evolutionary processes of macro-social systems, enabling mechanism exploration, future prediction, and intervention optimization. This "bottom-up" approach breaks the long-standing divide between macro and micro social research in social sciences, making the causal chain from individual behavior to collective emergence traceable and verifiable—turning the "black box" of social operations into a "transparent laboratory."

The birth of this book stems from our long-standing exploration of resolving the dilemmas in complex social system research. For years, we have witnessed AI's rapid advancement, offering powerful tools for social science, while social science's real-world demands provide rich scenarios for AI application. Yet, a systematic methodological framework to bridge these two fields has been lacking. Computational experiments fill this gap: they absorb AI's strengths in machine learning, multiagent systems, and causal inference, while rooting themselves in social science's profound insights into human behavior, social structures, and institutional evolution. The result is a complete research cycle of "modeling-experiment-analysis-optimization" that integrates theory with practice.

The book is structured into four parts, unfolding in a logical thread of "Theoretical Foundations–Methodological Framework–Key Technologies–Development Trends."

The first part, Theoretical Foundations, lays the conceptual and technical groundwork. It starts with the core concepts of complexity science, clarifies the limitations of traditional research methods in addressing complex social systems, and introduces the positioning of computational experiments as a fusion of agent-based modeling (ABM) and experimental logic. It then extends from physical system simulation to social system simulation, elaborating on the basic principles of similarity, system modeling, and experimental design, while exploring the core models and causal inference methods of social simulation.

The second part, Methodological Framework, constructs the overall architecture of computational experiments. It focuses on three core links—model docking between artificial societies and the real world, experimental scenario generation, and causal analysis—while presenting an independently developed multiagent computational experiment platform (MACE). This platform integrates theoretical and technical achievements, providing tool support for modeling, experiment design, causal analysis, and report generation.

The third part, Key Technologies, dissects the implementation steps of computational experiments. It systematically elaborates on artificial society construction, experimental system development, experimental design with generative explanation, experimental analysis with causal inference, and maturity evaluation. Each technical link is supplemented by practical logic and case demonstrations, ensuring operability and applicability.

The fourth part, Development Trends, looks toward the future of the field. It explores the integration of large language model (LLM)-based agents with social simulation, analyzes the trends of multiagent collaboration and computational experiment automation, and proposes core issues to be addressed in future development—including real-world computational modeling, virtual-world causal reasoning, and the real-world transfer of experimental laws.

The authors of this book have long been deeply engaged in the theoretical research and practical application of computational experiments, striving to strike a balance between theoretical rigor and practical feasibility: while elaborating on core concepts and methodological frameworks, we use classic cases and experimental details to help readers grasp the essence of computational experiments. It serves not only as a textbook for graduate students in computer science, systems science, sociology, economics, and public management but also as a reference for researchers, policymakers, and engineers in related fields.

As Alfred North Whitehead noted, "The progress of science depends on the invention of new concepts more than on the discovery of new facts." Computational experiments, as an emerging interdisciplinary field, are still evolving. With the continuous breakthroughs in AI and the deepening of social science research, its theoretical system and technical methods will surely grow and improve. We acknowledge that this book may contain omissions and imperfections, and we sincerely welcome criticisms and suggestions from readers. We also look forward to more scholars joining this field, jointly promoting the innovation and development of computational experiment methods, and contributing wisdom to solving complex social problems and advancing sustainable human development.

Finally, we extend our heartfelt gratitude to all experts, partners, and the editorial team of the press who supported this book. Special thanks go to the pioneering scholars and peers whose insights have inspired our research—we stand on the shoulders of giants to complete this work. May this book serve as a sincere link connecting AI and social sciences, illuminating new horizons and paths for research and practice in related fields.

January 10, 2026

Xiao Xue
Tianjin University
Fei-Yue Wang
The University of Chinese Academy of Sciences

1

Origin and Goal of Computational Experiments

This chapter clearly states that the core of computational experiments lies in "generative explanations." It studies social complexity from the bottom up, based on mathematical modeling of micro-level agents, using computer simulation techniques to derive the macro-evolutionary process of social systems and by designing experiments and causal inference methods to explore the underlying rules behind phenomena. Computational experiment method represents a higher level in computational simulation technology, offering a feasible way to analyze the behavior of complex systems and evaluate the effects of various decisions. It also bridges the gap left by traditional methods, which struggle or fail to experiment on complex systems.

1.1 Complexity Science

1.1.1 Basic Concepts of Complexity

Prediction is the oldest endeavor of mankind. From an individual's fate, including birth, aging, illness, and death, to the rise and fall of a country, all fall within this vast realm of prediction. Imagine if you could predict one's life, or if someone could predict the rise and fall of a country or even humanity, how significant that would be. Asimov's novel *Foundation* once depicted such a scenario. One Hari Seldon invented the science of "psychohistory" and was able to predict the rise and fall of the entire galactic empire. If we roughly classify the things in this world, we can divide them into two major categories: one is called "inanimate things," such as stars, solids, liquids, various molecules, atoms, and subatomic particles in the universe, and the other is "living things," including everything related to life, from various biological cell processes to neural activities in the brain to social and economic systems. For the first category of things, we have achieved highly successful predictions, while the second category is fraught with difficulties.

In the process of scientific development, scientists once viewed the universe as a mechanical clock, believing that as long as they understood the laws that govern the universe, they could infer the past and predict the future. Laplace once wrote: "If a wise man can analyze all the data of the natural world, he can put the movement of the largest celestial body and the smallest atom in the universe into a single formula. For such a wise man, nothing is uncertain, and the future is as clear as the past." The evolution of the planetary motion model is a classic example of this kind of deterministic prediction. At first, Ptolemy's cosmic model decomposed the movement of the sun, moon, and stars into a series of mutually nested circular motions; second, Copernicus proposed the heliocentric theory; third, Kepler turned circular orbits into elliptical orbits; and, finally, Newton transformed empirical theorems into a theoretical system and proposed Newtonian mechanics represented by differential equations. These theories can predict changes in various phenomena and can be tested by experiments. From the parabola of an apple falling to the ground to the movement of the sun, moon, and stars, they are all unified. This method of decomposing phenomena and abstracting the most general factors is physical thinking.

Soon, this deterministic thinking began to enter the microscopic world. Due to the uncertainty principle, these tiny particles have inherent uncertainty. If the motion of atoms is random, then why are the macroscopic substances they form, which have temperature, hardness, shape, and chemical properties, determined? Can we predict the macroscopic from the microscopic? This time, this is made possible by statistical physics, whose core idea is that although the movement of a large number of particles at the microscopic scale is random, the characteristics at the macroscopic scale obtained through statistical methods are stable and predictable. For example, we can calculate the magnetism of a magnet and the boiling point of water. The law of large numbers explains how random microscopic behavior yields predictable macroscopic phenomena. The key to the law of large numbers lies in the continuous addition of countless random and independent variables, which will result in a deterministic entity on a larger scale.

People used to be optimistic, thinking that they had understood the world and had given a perfect description of the world with mathematics. However, this optimism was short-lived. By the 20th century, with the rise of the Internet and big data, humanity's mathematical prediction ability has once again been unprecedentedly enhanced. People began to think further along the same lines as before: Does society itself also have first principles similar to Newton's laws, and can it be predicted using statistical physics? Such a desire is vividly described in Asimov's "Foundation": "Each member of human society is like a molecule in a gas. Their movement may be random and unpredictable, but the human society they form is highly predictable." Therefore, this science predicted the decline of the entire galactic empire, thus launching the entire plan to change human history. This is simply a reiteration of the view of statistical physics. Each person is random and unpredictable, but when a large number of people are considered collectively, the rise and fall of a country or a nation is highly predictable.

However, people soon discovered that the world is actually highly complex, especially in areas related to people or society, and the predictions made have largely gone unfulfilled. There is an interesting social experiment: "If the number of people in a group is fixed, everyone starts with 100 yuan, then plays a game, and each person randomly gives 1 yuan to the people around them; if this chain of funds is passed on in this way, then after a period of time, there will be a distribution of the funds of the people in the group; what kind of characteristics will this distribution take?" According to statistical physics, this process should be close to the physical diffusion process and thus result in a classical Boltzmann distribution. However, this is not the case. In fact, this distribution is much more extreme than the Boltzmann distribution, exhibiting a power-law distribution. Another typical example of prediction failure is the financial crises and stock market fluctuations. The essence of stock market prices is people's expectations of a company's total value in the future, which is determined by a large number of traders who engage in buying and selling. If these traders were like the particles that make up a material entity, behaving randomly but statistically predictable, then the market as a whole would conform to a classic stochastic differential equation. However, the reality goes against the original expectations of the geniuses. So many people entered the stock market to trade and compete, but few could have predicted the financial crisis in 2008 (Figure 1.1).

The German psychologist D. Dörner once raised a question: assuming that people have all the necessary intelligence, experience, and information conditions, why do they still make mistakes, sometimes even with disastrous consequences?

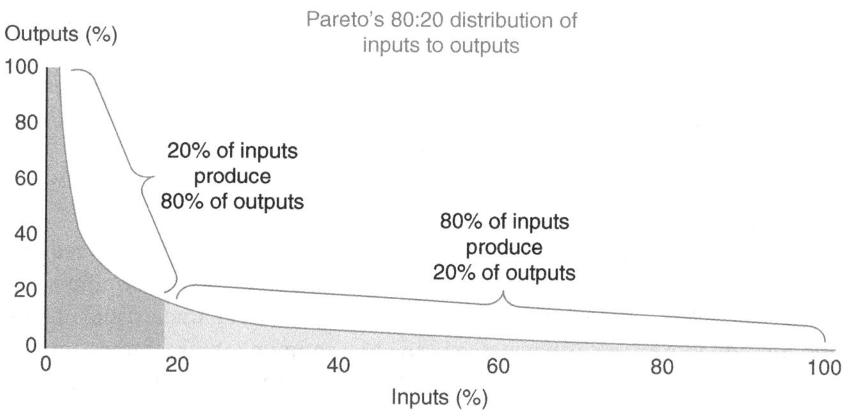

Figure 1.1 The cruel reality and the failure of statistical physics.

In his book entitled *The Logic of Failure*, he gave his own answer [1]: some of people's thought patterns are suitable for a simple world but have a disastrous impact on the complex world they live in. Derner found many examples, such as the planners of the Aswan Dam who only thought of the benefits of cheap electricity for Egypt, but did not realize that this would also interrupt the annual flood irrigation, which has maintained the fertility and prosperity of the Nile Valley for thousands of years. People reclaimed wasteland in the hope of increasing food production but failed to realize the need to protect the ecosystem, which ultimately led to land desertification and food failure. All these show that taking action before understanding all the interconnected factors in a complex system is bound to make big mistakes even if the original intent is good; small mistakes accumulate and eventually lead to serious mistakes; those who too often ignore the overall situation of the problem and only seek expedients in a local area can only be a drop in the bucket and will not help.

For human society, the independence assumption of statistical physics does not hold. The development of human society is closely related not only to its initial conditions but also to random events in its development process. A slight change in the initial conditions will lead to a huge deviation in the final results. In order to avoid failure to the greatest extent possible, it is necessary to understand the laws governing the operation of social complex systems. As J. Achenbach pointed out: "Science is gradually moving out of the field of easy-to-solve problems and beginning to touch on truly difficult problems. Science has reached a new turning point." Since the mid-1980s, some visionary scientists have begun to explore this new turning point, and a discipline known as the "science of the 21st century," complexity science, has emerged [2]. In April 1999, the American magazine *Science* published a special issue with the theme of "complex systems," reporting on complexity research in chemistry, biology, neurology, physical geography, climatology, economics, and other fields. As shown in Figure 1.2, complex systems have some common characteristics: a system exchanges matter, energy, and information with the external environment. Under the joint action of external stimuli and internal drives, the mutual influence between individuals cannot be ignored. System complexity will give rise to the following effects:

1.1.1.1 Nonlinear

The development of phenomena is not isolated, fragmented, or unconnected, but a unified whole that is interconnected, interactive, and mutually constrained. The characteristics of an organization do not depend only on the characteristics of individual components but also on how the individual components are interconnected to form an organization. The interaction makes it difficult to divide the system into parts and thus makes the overall characteristics difficult to predict, which is also called nonlinearity. The basic types of nonlinearity include negative feedback and

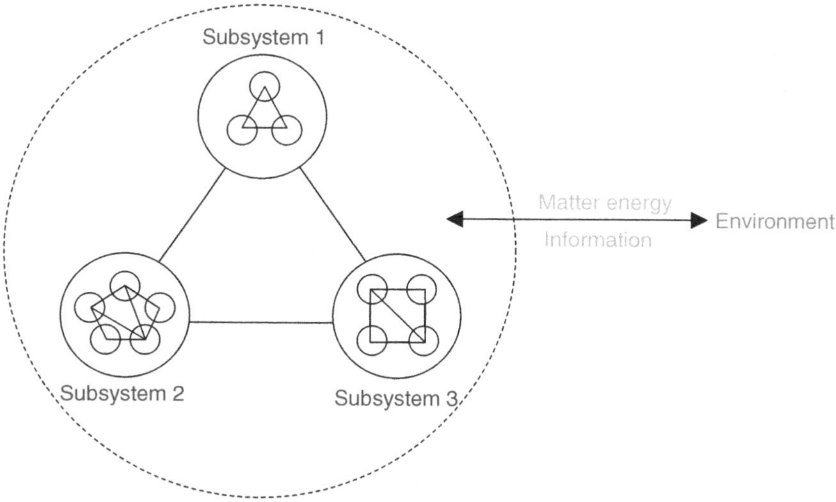

Figure 1.2 Schematic diagram of a complex system.

positive feedback, which make the result of the whole being more than or less than the sum of its parts, that is, $1 + 1 > 2$, or $1 + 1 < 2$, but basically not $1 + 1 = 2$. There are many such examples around us. A team composed of a group of excellent players may make mistakes repeatedly, while a rabble can succeed under a clear set of rules. Nonlinearity means opportunity for ordinary individuals. Due to continuous positive feedback, a single node can sometimes change the entire network, similar to how an ordinary individual sometimes suddenly changes the course of history.

1.1.1.2 Chaos

Weather forecasting has always been an important topic in agricultural production. Physicists once believed that as long as the fluid mechanics equations of gas motion can be solved, the weather one month later can be predicted through a large number of simulations. However, in the process of meteorologists trying to model the atmosphere, they discovered an incredible phenomenon: for the same data, if one more decimal place is retained, the calculated result is sunny; if one less decimal place is retained, the calculated result is rainy. This physical phenomenon is called "chaos." Chaos is caused by the increase in free dimensions in the system. The dynamic properties of the system are no longer confined to closed orbits, but open or unpredictable trajectories. The impact of slight changes in initial conditions on the future is far from known. Most nonlinear networks are in the conditional range where chaos occurs, which makes certainty gradually fade from our view. When the positive feedback in the nonlinear network reaches a certain level, it will transition from a relatively stable structure to chaos. This point on the edge of chaos is called the phase transition point, which is the foundation for the emergence of all complex structures (order).

1.1.1.3 Phase Transition and Emergence

Subsystems and elements within the system interact in various modes and exchange resources with the external environment, thus changing the internal structure and system behavior. After various external information, materials, and so on enter the system, the internal elements constantly collide and react with each other. After a certain period of amplification and development, new forms and characteristics will emerge as a whole, resulting in emergence. Emergence reflects the relationship between the macro and micro levels and is the macro effect of the micro behavior of complex systems. Phase change is the process of causing the entire system to transition from one phase to another through external variables. Taking the phase change from water to ice as an example, the controllable external variable is temperature. When the temperature is 0 °C, the system has the minimum degree of freedom and is in a uniformly ordered state; as the temperature rises, the degree of freedom of the disordered state gradually increases; at a certain temperature, disorder and order alternate, and this temperature is called the critical temperature. The so-called critical state is the state at the time of phase change, because this moment is the most distinctive. Emergence and phase transition points are also inextricably linked. Interested readers can refer to the theory of self-organized criticality.

1.1.1.4 Feedback Loop

When studying the evolution of a complex system, we often need to consider the impact of the current results on the system's output at the next moment. According to Jones's classification of the emergence mechanism of complex systems, emergence featuring a two-way feedback mechanism (feedforward and feedback) between the system as a whole and individual behavior is called second-order emergence, and the emergence with only a one-way feedback mechanism is called first-order emergence. As shown in Figure 1.3, all emergence in social complex systems belongs to second-order emergence: individual behavior generates new individual behavior due to system feedback; complex systems introduce feedback that affects individual behavior on the basis of the original first-order emergence, thus forming a two-way feedback loop within the system, and increasing the complexity of the system. Feedback is divided into positive feedback and negative feedback: negative feedback leads to a fixed-point equilibrium; for example, market prices always fluctuate around the equilibrium level, high prices lead to fewer consumers, and fewer consumers lead to lower prices; positive feedback leads to instability, such as avalanches and stock market crashes.

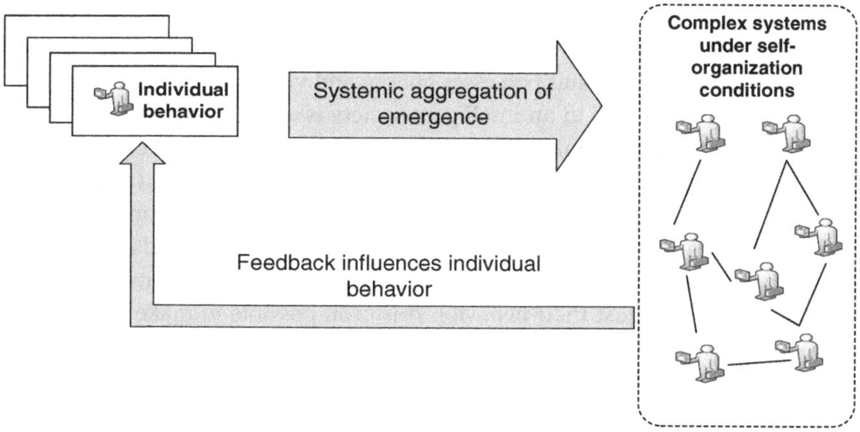

Figure 1.3 Schematic diagram of second-order emergence.

System complexity is the dialectical unity of diversity and consistency. Diversity refers to the heterogeneity of system composition and is also the basis for the system's vitality, creativity, and evolutionary ability, while consistency is the basic condition for the system to form a certain order and function. Studying social systems from the perspective of system complexity focuses more on the diverse behaviors of heterogeneous subjects and the dynamic equilibrium process of interconnection and interaction between heterogeneous subjects. This perspective of system complexity, which emphasizes the dialectical unity of diversity and consistency, helps to deeply understand and grasp the evolutionary and regulatory laws of the unity of "order and vitality" of the system and to overcome the limitations of traditional research methods, such as localization, linearity, and static analysis. Although we cannot accurately predict the behavior of each individual or even society, we can still make correct decisions.

1.1.2 Complex Social System: Cyber, Physical, and Social System

Nowadays, intelligent technologies are developing rapidly. Related technologies include information and communication technologies such as the Internet of Things (IoT), big data, cloud computing, virtual reality, blockchain, and artificial intelligence (AI). Moreover, new technologies or products are constantly emerging. Human society is like a container, within which these intelligent technology clusters are constantly undergoing various combinations and integrations to form various services that meet complex application scenarios [3]. "Data + computing power + AI algorithms" is forming a new type of infrastructure for an intelligent society. Based on this, the entire society is gradually evolving into a complex functional entity that matches demands with services. It is jointly created and operated by all members of society. Driven by data logic, it achieves iterative evolution and self-growth and possesses energy and vitality beyond imagination [4]. The term "intelligent" in an intelligent society is a personified metaphorical usage, meaning that the elements in society are no longer merely human intelligent entities but include an increasing number of machine intelligent entities and elements that integrate the two [5]. When various problems and changes arise in society, machine intelligence entities can collect all kinds of data and information through technical means, respond based on in-depth data mining, and constantly receive feedback to adjust their behavior. It is even possible to make predictions and take preventive measures before problems occur. This represents a leap from quantitative change to qualitative change compared with the sluggish response in traditional society.

The essence of the intelligent society is a complex system with increasingly strong coupling of "cyber, physical, and social system" (CPSS), that is, the integration of the physical world with human society, the integration of the real world with the virtual world, and the continuous improvement of the level of human–machine fusion intelligence [6]. As shown in Figure 1.4, the operating logic of CPSS includes three parts. (1) Physical space: various online or offline resources (applications, platforms, data, algorithms, and facilities) in the physical space are virtualized and released in the form of services. (2) Cyberspace: supporting intelligent data processing for informed decision-making, which can not only create "virtual organizations" in various fields (e.g., families, enterprises, and governments) but also define the workflow of various public affairs (e.g., smart government affairs, smart transportation, and smart security). (3) Social space: involving the coordination of interests among the government, enterprises, and citizens, to achieve value creation and circulation between demanders, suppliers, and service operators. Table 1.1 summarizes the current status of the intelligent society according to the three development stages, including both the advantages of intelligent technology in dealing with various social governance issues and the potential risks of intelligent technology when confronted with complexity.

In order to achieve this goal, intelligent social governance requires consideration from two aspects: on the one hand, the emergence of intelligent society is a bottom-up, self-organizing, and self-growing emergence process, and the overall state of the system depends on the spontaneous interaction between different intelligent subjects; on the other hand, intelligent society is also a product of top-down planning, and the laws obtained by big data analysis can be used to change the design and evolution path of the social ecology. Although a lot of progress has been made in the research of intelligent social governance, it still faces a series of problems and challenges caused by complexity [14]. It is necessary to complete the transformation from "comprehensive coordination" to "game evolution," from

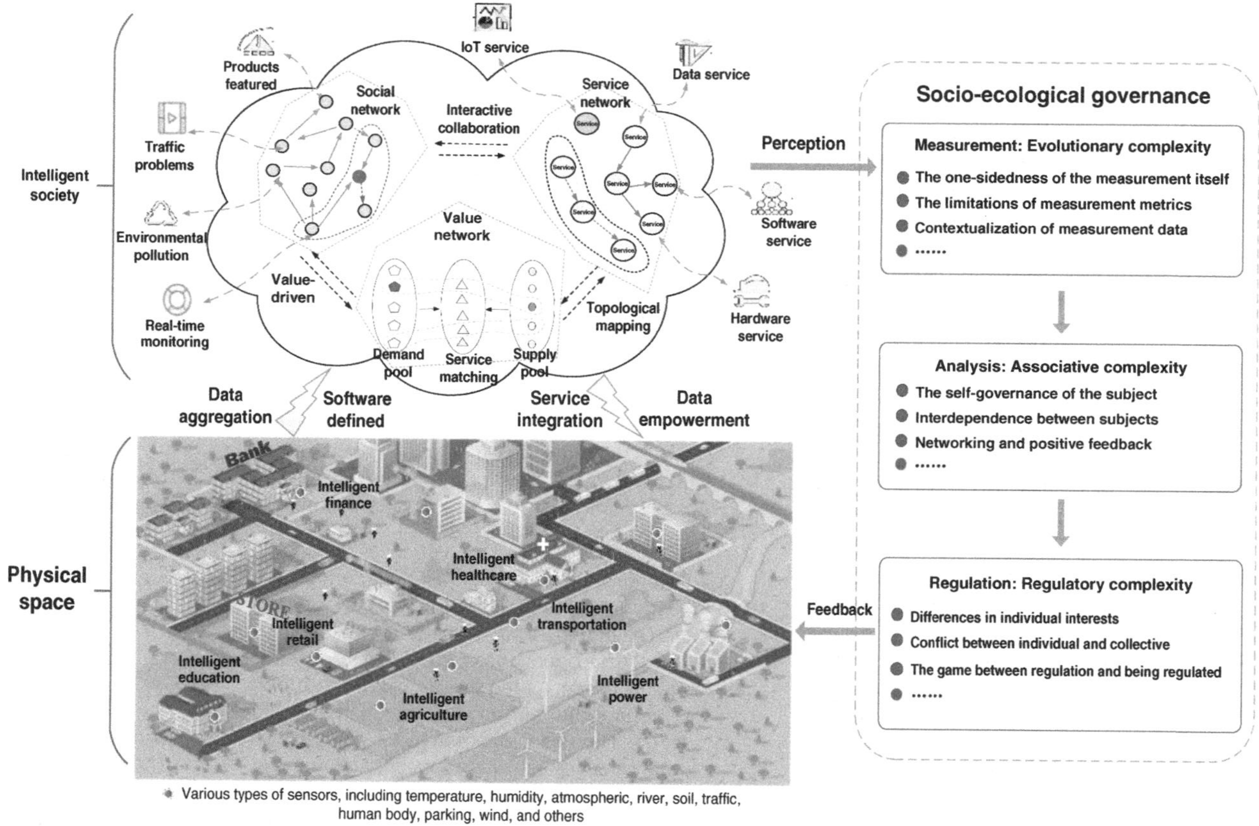

Figure 1.4 Intelligent society with CPSS characteristics.

Table 1.1 Successful Application and Potential Threats in the Intelligent Society.

Application Scenario	Successful Applications	Potential Hidden Dangers
Phase 1 (1998–2012) Internet Intelligence	Machine intelligence mainly stays in the Internet world. In response to the information overload problem in the Internet era, an information intermediary platform is built between suppliers and demanders to provide accurate recommendations for users. Representatives in the e-commerce field include Amazon, Taobao, etc.; representatives in the information field include Google, Netflix, Toutiao, etc.	Technology: Users may suffer from information cocoons [7], algorithmic discrimination [8], etc.; algorithmic drift may also lead to inaccurate predictions [9, 10]. Society: The centralization of data leads to excessive power in a small number of platform companies, which poses a risk of manipulating users and exacerbates the wealth gap and conflicts between third-party suppliers and platforms [11].
Phase 2 (2011 to Present) The Physical World is Intelligent	Machine intelligence has begun to penetrate into the physical world. In the physical world, there is a common contradiction between limited service capabilities and high-speed growth demand. Traffic congestion, retail, energy shortages, and environmental pollution all reflect the lack of dynamic service capabilities and efficiency of infrastructure. Problems with public facilities such as finance, education, and medical care reflect the lack of public service layout, supply, and service quality. The goal of intelligence is to change the current supply–demand matching situation, or to eliminate the information asymmetry between supply and demand, or to significantly improve supply capacity. Typical representatives include Uber, Meituan, and Hema Fresh in O2O life services; Alipay and PayPal in financial services; Alphabet's "Sidewalk Toronto" project in smart cities, and Ali's city brain.	Technology: Machine intelligence mostly makes decisions based on historical circumstances and is powerless to respond to emergencies, such as the impact of a financial crisis on personal credit or the impact of a sudden outbreak on the scheduling of medical resources. Society: In many cases, smart cities hope to achieve optimal scheduling of supply and demand matching without increasing supply. However, local optimization in complex networks may not necessarily lead to global optimization. Whether the smart service system can really solve the problem remains in doubt. For example, smart navigation services may fail during peak hours due to big data traps because it will lead to a competition between a large number of users for the same road resources [12].

Phase 3 (2015 to Present) Autonomous Intelligence	Machine intelligence is gradually achieving autonomous decision-making and continuously driving the accelerated evolution of human society. Its characteristics are as follows. (1) Enhanced integration: All human, material, and environmental factors are incorporated into the intelligent network and are fully perceived, recognized, and calculated. (2) Comprehensive intelligence: Machine intelligence will be widely distributed in society and seamlessly integrated with human intelligence. (3) Detailed planning: While allowing for unknown risks, machine intelligence can effectively provide public service products.	Technology: Socioeconomic systems are all super-complex systems, with many internal layers, complex influencing factors, and intricate relationship networks. The larger the scope of the problem, the greater the complexity of the data associations, and the possible scenarios increase exponentially. Simple, unified, and average centralized interventions lack adaptability and are difficult to deal with uncertain and global problems. The failure to predict financial crises and stock market fluctuations is a classic case, the reason for which is the endless emergence of various black swan events [13]. Society: There is a contradiction between the macro-control of the social and economic system and the autonomous behavior of individuals. Both sides are always in an evolutionary game, and it is difficult to have effective measures from beginning to end, such as urban planning and social ethical conflicts.

"macro description" to "causal analysis," and from "single indicator" to "value coordination," as specifically manifested in the following three aspects.

1.1.2.1 Evolutionary Complexity (Difficult to Measure)

The subjects in the social ecology not only are people with autonomous consciousness and strategic behavior but also gradually include various intelligent machines, human–machine fusion systems, social organizations, and digital platform companies. The AI system with increasing capabilities and the people and organizations empowered by AI are both autonomous and passive, both conflicting and cooperative, forming an evolutionary system of multi-agent game. As the number of collaborating subjects continues to increase, the dynamic interactions between the subjects become increasingly complex, and the probability of random deviations will increase. In addition, the transparency of the service process itself is very low, and the visibility of risks is easily obscured by various links and will eventually accumulate until they erupt. Without an appropriate performance measurement system (the comprehensive ability to do the right thing in the right way) to explicitly identify the risks in the service process, it is impossible to continuously improve and govern the social system.

1.1.2.2 Association Complexity (Difficult to Analyze)

The evolutionary complexity of social–ecological systems is essentially the organization and evolution from multi-subject behaviors and their associated service networks toward overall functions. With the accessibility and ubiquity of information technology, the associations and autonomous evolutionary behaviors among various subjects are constantly increasing, which often leads to unexpected emergent phenomena. Any governance strategy, through networked connections and positive feedback amplification, may quickly be associated with other events to produce a cascading "$1+1>2$" emergence, and even trigger "polarization" or "butterfly effect," which spills over into systemic risks. However, most of the current analytical methods focus on "description" rather than "explanation." They can present the evolutionary process, but it is difficult to provide insightful causal analysis to understand how the system operates.

1.1.2.3 Regulatory Complexity (Difficulty in Regulation)

As far as the social–ecological system is concerned, different individuals have different interests and needs. Sometimes, individual interests are inconsistent with the overall interests of the system, and there may even be conflicts. In addition, its internal subjects are diverse, including not only people and institutions but also AI algorithm-based platforms. They all have autonomous consciousness and learning ability and will change as the external environment changes. There is a two-way feedback mechanism between the social system and the individual behavior, which will cause the system to produce complex second-order emergence. Therefore, there will often be "a situation where subordinates find ways to circumvent superior policies," which will greatly reduce the effectiveness of strategy implementation. This requires that the governance strategy should not only consider individuals' pursuit of interests to ensure the motivation and vitality of system development but also resolutely prevent the interest imbalances and avoid threatening the order and stability of the system. These have posed challenges to the existing social–ecological governance capabilities.

1.1.3 The Rise of Computational Social Science

In order to model, analyze, design, and manage complex social systems, we need to shift through the myriad and complex phenomena to find unique perspectives for observing and understanding the intelligent society. Against this background, in February 2009, 15 top scholars from world-class universities such as Harvard University and MIT jointly published a paper "The Advent of the Era of Computational Social Sciences" in *Science* [15]. In 2012, 14 famous European and American scholars jointly issued a "Declaration on Computational Social Sciences," and the voice of *Computational Social Sciences* overshadowed other terms [16]. In 2020, these scholars once again published a joint paper in the journal *Science*, highlighting the problems and challenges faced by the development of computational social sciences [17]. In the development of computational social science, many subdisciplines and subfields, such as computational economics [18], computational finance [19], computational histology [20], and computational epidemiology [21], have evolved.

Researchers in computational social science come from both the social sciences and computational sciences. Traditionally, social scientists have prioritized providing satisfactory explanations for the behavior of social systems, often relying on causal mechanisms in substantive theories. In contrast, computer scientists have traditionally focused on developing accurate predictive models, regardless of whether these models fit causal mechanisms or even whether they can provide explanations. Therefore, the standard practice for social scientists is to fit their models entirely "in sample" because they try to explain social processes rather than predict outcomes, while for computer scientists, evaluation using "out-of-sample" data is necessary. In addition, computer scientists typically allow the complexity of the model to increase as long as it continues to improve predictive performance, while for social scientists, the model should be based on substantive theory and is, therefore, constrained by substantive theory [22].

Both approaches are valid under their respective conditions, and both have produced a large number of fruitful scientific results. However, both approaches have also been criticized: on the one hand, theory-driven empirical social sciences have

Whether intervene / Whether predict	No intervention or allocation changes	Under intervention or allocation changes
Focus on specific features or effects	Quadrant 1: Descriptive modeling describes past or current situations (Basic activities, such as statistical surveys network community detection)	Quadrant 2: Explanatory modeling estimates the impact of changes (Social sciences, such as field experiments)
Focus on predicting outcomes	Quadrant 3: Predictive modeling predicts outcomes for similar future situations (Intelligent sciences, such as recommendation algorithms)	Quadrant 4: Integrative modeling predicts outcomes and estimates the impact of unseen situations (Intelligent sciences + Social sciences) ?

Figure 1.5 Classification of research methods in computational social science.

been criticized for being unable to predict results of interest and unable to provide solutions to real-world problems; on the other hand, the predictive models given by computational science have been criticized for being nongeneralizable, inexplicable, and biased. In order to promote the development of computational social science, we need to incorporate the thinking and methodologies of these two classical traditions, while also reconciling the different cognitive values of social science and computational science to produce new methods that both sides can agree on. Figure 1.5 shows four quadrants, each describing the main research methods of computational social science.

Descriptive modeling (quadrant 1) involves thinking about, defining, measuring, collecting, and describing relationships between quantities of interest. Activities in this quadrant include traditional statistical and survey research, as well as computational methods such as topic modeling and community detection in networks. Much of what we know about public opinion, economic conditions, and everyday life experiences comes from survey research. For example, recent studies have documented significant differences in mortality rates, wealth gaps, and intergenerational economic mobility among racial and ethnic groups. Qualitative and comparative methods popular in sociology, communication, and anthropology also belong to this quadrant. To date, most of the progress in computational social science has been made in using digital signals and Internet platforms to study concepts that were previously unmeasurable. In other words, descriptive work, both qualitative and quantitative, is useful and interesting in its own right and is also fundamental to activities in the other three quadrants.

Explanatory modeling (quadrant 2) refers to activities aimed at identifying and estimating causal relationships but are not directly concerned with predicting outcomes. Most traditional empirical sociology, political science, economics, and psychology fall into this quadrant, including statistical modeling of observational data, laboratory experiments, field experiments, and qualitative methods. Some methods (e.g., randomized or natural experiments) isolate causal effects by design, while other methods (e.g., regression modeling, qualitative methods) invoke causal explanations based on theory. Methods in this quadrant tend to prioritize simplicity, considering only one or a few characteristics that may affect the outcome of interest. These methods are very useful for understanding individual causal relationships, forming theoretical models, and even guiding policy. For example, social experiments show that job applicants with "black" names are less likely to get interviews than those with "white" names. This reveals the existence of structural racism and can inspire specific policy interventions.

Predictive modeling (quadrant 3) refers to activities that attempt to directly predict outcomes of interest but do not explicitly focus on identifying causal relationships. "Predictions" in this quadrant may or may not be about actual future events; however, in contrast to quadrants 1 and 2, they refer specifically to "out-of-sample" predictions, meaning that the data on which the model is evaluated (the test data) is different from the data on which the model is estimated (the training data). Activities in this quadrant include time series modeling, forecasting competitions, and many supervised machine learning activities, ranging from simple linear regression to complex deep neural networks. From a policy perspective, it is helpful to have high-quality predictions about future events, even if these predictions are not causal in nature. For example, applications of machine learning abound in Internet advertising and recommender systems. Although these algorithms cannot determine what caused people to click or content to go viral, they can still provide useful inputs to policymakers. That said, there is often an implicit assumption that the data used to train and test the model comes from the same data-generating process, akin to making predictions in a static (albeit potentially noisy) world. Thus, while these methods are generally applicable to fixed data distributions, they may not be applicable to settings where features or inputs are actively manipulated (such as in controlled experiments or policy changes).

Integrative modeling (quadrant 4) refers to activities that attempt to predict as-yet unseen outcomes based on causal relationships. More specifically, quadrant 3 focuses on data that are out-of-sample but still come from the same (statistical) distribution, and the focus here is on generalizing "out-of-distribution" to scenarios that may vary naturally. This category includes changes in the distribution that we have observed before as well as completely new and more extreme situations (due to some uncontrollable factors, or due to some deliberate intervention, such as an experiment or policy change). Integrative modeling, therefore, requires focusing on estimating causal effects (quadrant 2) rather than simply associative effects and considering the impact of all these effects simultaneously to predict outcomes as accurately as possible (quadrant 3). Ideally, work in this quadrant will give high-quality predictions of future outcomes in a (potentially) changing world. Research in this quadrant is relatively rare and more challenging.

1.2 Methodological Framework of Computational Experiments

1.2.1 Computational Experiment = Agent Based Modeling (ABM) + Experiment

In response to the challenges faced by ABM in the field of causal inference, Professor Fei-Yue Wang systematically proposed the basic ideas, concepts, and methods of computational experiments in 2004, emphasizing the circular feedback relationship between artificial systems and actual systems [23, 24]. Computational experiment method is an improved and extended version of ABM. ABM is not simply used as a simulation tool but as an "artificial social laboratory" that "spontaneously grows and cultivates" an alternative version of the actual system and can conduct various "counterfactual experiments" on system behavior and decision analysis. On the one hand, computational experiments use ABM to emphasize vertical causality, pay attention to the causes and formation process of phenomena, and give causal significance to the cross-level correlation information generated by ABM (the connection between micro-mechanisms and macro-emergence). On the other hand, computational experiments "use counterfactual algorithms to construct multiple worlds" to emphasize horizontal causality, establish a probabilistic relationship between model parameters and macro-emergence through experimental mechanisms, reveal the patterns and laws within the system, and provide "description," "prediction," and "prescription" for the possible operation of the actual system. Because the virtual foundation on which the experimental mechanism is based, that is, the attributes, behaviors, and local environment of the agent, is a replica of its real-world counterpart. Therefore, the probabilistic relationships established between different levels of analysis can be viewed as counterfactual connections across levels of analysis in the real world.

Computational experiments belong to the research path of "generative explanation," that is, to trace the causes by exploring the mechanisms that can show how social phenomena arise [25]. Traditional inductive interpretation uses statistical methods to derive general rules from data, and these rules are used to process data and are independent of data. However, generative explanations are based on the use of micro-interaction rules to derive macroscopic emergent structures, that is, general laws necessarily depend on the basis on which they are generated. Thus, generative explanations methodologically offer the idea that "how we know" is the basis of "whether we know," a connection that is precisely what inductive and deductive methods lack. Generative explanations are different from traditional deductive or inductive paths and place individual differences, local interactions, and the evolution of rules, which are ignored by traditional equilibrium analysis, at the core of the discussion. As shown in Figure 1.6, the computational experiment method is an "experiment + deduction" mechanism to explore the dynamic patterns, which consists of the following four core steps.

1.2.1.1 Preparation

The first step in a computational experiment is preparation, and the artificial society model can only be used for the computational experiment if it is proven to be consistent with the actual scenario. In the process of research, the artificial society model needs to be continuously adjusted, and occurred situations can be used as a benchmark for model adjustment, similar to the standard sample in chemical experiments, until the simulated situation and the real situation match in terms of statistical distribution characteristics. In order to explore as many unknown scenarios as possible in the real world, it is necessary to analyze the conditions under which these unknown scenarios occur and the likelihood of these conditions, including (1) an appropriate number of agents, with neither all agents nor some key agents ignored; (2) clear boundaries, abstracting the random perturbation factors that affect the state and properties of the system into boundaries, and eliminating the rest of the irrelevant parts; (3) agent–agent, agent–environment, and environment–environment interactions are based on knowledge rather than data, i.e., those linear/nonlinear, deterministic/random, static/dynamic function descriptions are based on the abstract relationship of facts; and (4) eigenvectors are set up to represent the state of the system based on the research objectives, and these variables will be combined according to a certain relationship to achieve a measurement of the state of the system.

1.2.1.2 Design of Experiments

Computational experiments take the "multi-world" view of complex systems research, where the results of the experiment are treated as a possible "reality" that behaves "differently" but "equivalently" from reality. Experimental design follows an experimental methodology, including the setting of initial rules and the testing and selection of models, either to reproduce some real-world settings or phenomena or to construct and observe possible worlds. Generative experiments deal with multivariate realizable cases: what if multiple microscopic mechanisms generate macroscopic patterns P (or rules) to be explained? The role of "experiments" is to assess the sufficiency of the generation of microscopic mechanisms: if multiple microscopic mechanisms can derive P, they can only be used as candidate explanations, and the most reliable explanation needs to be determined at the microscopic level. Therefore, it is necessary to continuously

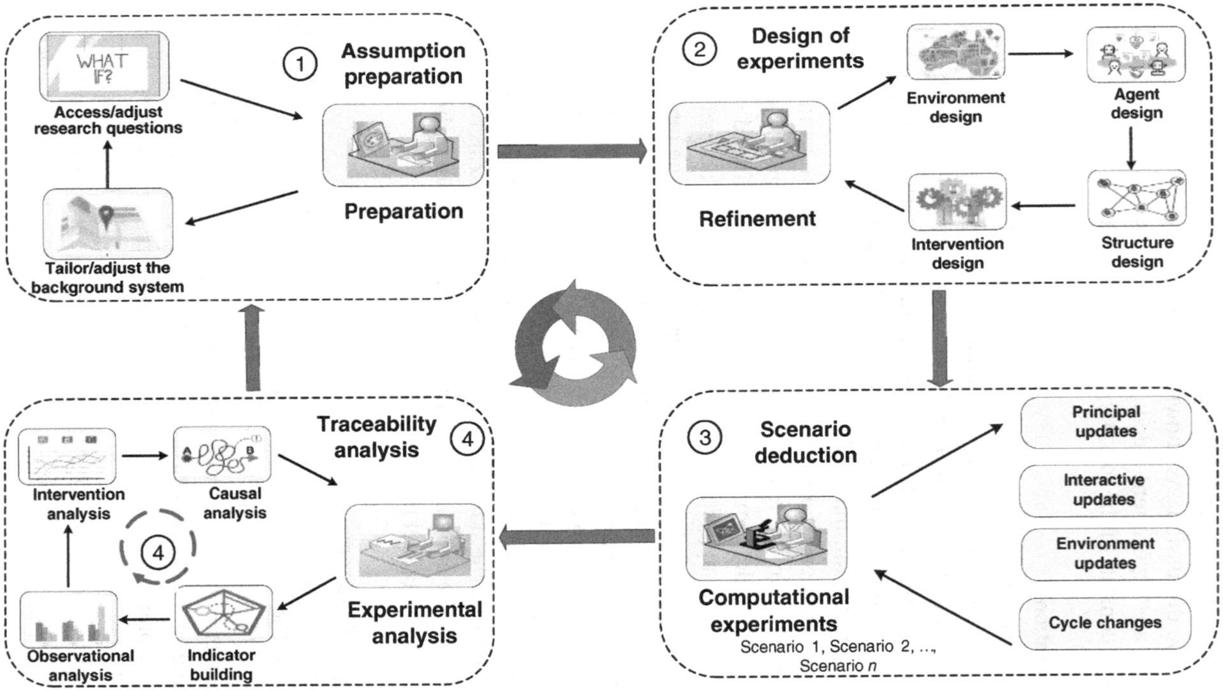

Figure 1.6 The process model of the computational experiment method.

deduce according to the experimental design scheme, and the whole tracing process is a process of continuous experimentation. Further, the same initial setup may evolve into different possible outcomes. By designing counterfactual scenarios that have never happened in reality, it is possible to identify potential adverse effects that may occur and provide "references," "predictions," and "guidance" for possible scenarios.

1.2.1.3 Scenario Deduction

Computational experiment scheme deduction is deductive, which refers to the computational process of deriving macro rules from the initial configuration of the agents. ABM is particularly suitable for generative inference, which is to convert the generation process into an agent-based computational model and reveal the micro-level generation mechanism of macroscopic rules through computational simulation. In this process, the ability to generate the macroscopic results of interest is key. Because each agent model is a computational program, its running process can be expressed by a loop function. Assuming the n-th state of the system, then the $(n+1)$ state is deduced in a strictly deductive and computational manner. From the initial setup of the agents to every subsequent step, the computer implementation can be equated with the deduction of theorems. If x is considered the initial state of a set of agents, then it can be updated repeatedly according to the local interaction rules: if x, then y.

1.2.1.4 Traceability Analysis

From the perspective of mechanism, the correlation and continuous covariation between social phenomena and variables cannot explain social phenomena but need to be explained by a series of agents and their behaviors that cause phenomenal changes. Mechanisms can help us understand why a particular cause is of explanatory relevance. Something like: "Surprising fact C has been observed; if A is true, C is certainly true as well; therefore, there is reason to guess that A is true." Here, hypothesis A is an attempted guess, and whether or not it can be recognized depends on A being able to explain phenomenon C. So, the slogan of generative social science is: if you can't generate it, you can't explain its emergence. In other words, the emergence of macroscopic outcomes depends on the micro-mechanism of local interactions, and this sufficiency of generation is a necessary condition for explaining how macroscopic outcomes are generated.

1.2.2 Computational Experiments + Emerging Technologies

In the field of social simulation, customized methods and solutions are the norm, and the lack of a unified methodological framework limits the use of computational experiment methods by many interdisciplinary researchers. The individual steps of computational experiments are the vehicles through which emerging technologies come into play [26]. In recent years, various intelligent technologies, such as reinforcement learning, digital twins, large language models, and causal inference, have had a considerable impact on computational experiment methods. The following are some of the interdisciplinary research studies carried out by computational experiments and various intelligent technologies:

1.2.2.1 Computational Experiments + Machine Learning

When dealing with a large number of individual agents with different behaviors, it is not easy to design the decision rules for each agent. Generally speaking, we assign some simple deductive rules to each agent to govern their decision-making process, and the resulting macroscopic phenomenon is called emergence. However, it is generally accepted that such models are often not fully representative of the true way individuals make decisions on behavior. There are a variety of factors that affect decision-making, including the complex internal state of the agent, the agent's understanding of the environment in which it operates, and the corresponding dynamic rewards and risks. For human society, there may also be complex social factors, different knowledge bases, and agent learning curves [27]. Therefore, the necessity of combining learning logic, uncertainty, and induction has received increasing attention in the field of intelligence, and related technologies include knowledge discovery and representation, uncertainty reasoning, and machine learning [28]. ABM's developers have been facing difficulties in designing behavioral rules for agents. By employing machine learning techniques, traditional agent behavior rules can be transformed from hard-coded rules into adaptive models [29]. Although there is a lot of research on applying machine learning to ABM, the universally applicable scenarios, frameworks, and implementation processes have not been well addressed. Among them, representative works include Ref. [30], which introduces the application of machine learning techniques based on classification and regression in ABM, and Ref. [31], which focuses on the application of reinforcement learning to ABM.

1.2.2.2 Computational Experiments + Large Language Models

In an artificial society, the ideal model for an agent is an autonomous agent, capable of running in a loop in its simplest form, with each iteration generating self-directed instructions and actions. They do not rely on humans to direct conversations and are highly scalable. In the past, academia focused on how reinforcement learning could be used to improve AI agents. The emergence of large language models has undoubtedly brought new possibilities to the development of AI agents. In 2023, AI agents ushered in a big explosion, and multiple AI agents such as BabyAGI and AutoGPT sprang up like mushrooms after a rain. One of the most intriguing projects is the "virtual town" built by Stanford researchers, where 25 AI agents survive and can not only engage in complex behaviors (such as throwing a Valentine's Day party) but also demonstrating human-like social behaviors [32]. In social simulation, the key to using large language model agents is their excellent role-playing ability, which is essential for realistically portraying various characters and perspectives in reality. Research in this field is growing rapidly, encompassing a variety of fields such as social sciences, games, psychology, economics, and policymaking [33–36]. In the future, AI agents may not be individual entities, but many AI agent organizations, and even an AI agent civilization, will emerge. How to combine large language models with computational experiments is still in the stage of experimentation and exploration, and various technical challenges need to be faced [37].

1.2.2.3 Computational Experiments + Social Networks

The evolution of complex social systems is determined not by individual behavior but also by the behavior of complex networks embedded in individual relationships, communities, and markets. In order to understand these complex social processes, researchers must focus on the dynamic interaction of individual behaviors and social network structures. Social network analysis (SNA) rigorously quantifies the patterns of relationships between social entities through a formally defined graph theory approach, which can be used to understand various social phenomena [38]. ABM has also proven to be a valuable method for solving complex tasks in analyzing interactions between individuals or groups. Since the interaction between agents can be mapped as nodes and links in network science, the combination of the two approaches is easy to achieve. Embedding the network into the ABM makes the interaction between agents no longer limited to spatial relationships but also through social networks, which provides new opportunities to gain insight into the behavior of social systems. The integration of social networks into computational experiments can help understand these processes [39]. In addition, ABM can supplement the sampling bias of network structure that is common in empirical methods in social science. Since there is little complete network data available, it is often not possible to get a complete picture of all node relationships. Computational experiments can be used as "virtual laboratories" to explore networks in space and time and to test hypotheses about causality [40]. Therefore, the combination of computational experiment method and network analysis methods can help bridge the gap between the two types of methods (theoretical and empirical), opening up many research possibilities. Many interesting examples demonstrate this potential, such as epidemiology, marketing, or social dynamics.

1.2.2.4 Computational Experiment + Experimental Design

For social simulation, each scenario is multi-node and nonlinear, and each AI agent has N choices at any node, resulting in countless storylines. As the complexity of the system increases, the number of influencing factors increases exponentially. Moreover, there are also human and social factors, not only randomness but also human initiative, which leads to greater complexity. Each experiment will have different influencing factors and different research objects, so the core of the whole experiment lies in the experimental design, and it is necessary to determine the experimental objects, experimental objectives, influencing factors, and intervention methods. For example, whether the experiment focuses on variance or treatment effects, whether the experimental population is homogeneous or heterogeneous, or whether the factors of the experiment are single or multifactorial. These differences mean that traditional experimental design (design of experiments [DOE]) approaches may not adequately address the problems faced by computational experiments [41]. If we cannot identify appropriate response variables and influencing factors, we will not be able to explicate these emergence processes, let alone continuously improve and govern complex social systems. Compared with traditional simulation design, the experimental elements emphasized in computational experiments are not limited to the value of a certain variable, but may involve different data structures, and are difficult to quantify, which is a challenging problem at present.

1.2.2.5 Computational Experiment + Causal Inference

An interesting feature of computational experiments is that "the patterns, structures, and behaviors that emerge from the model are not the result of explicit programming but are generated through the interaction of agents." Sensitivity analysis techniques are commonly used to determine the effect of changes in input parameters on the output variables of artificial society models [42]. However, sensitivity analysis does not elucidate causal relationships between different elements in the model. These causal relationships are essential for a proper understanding of the emergent properties of the modeled system. In order to analyze causality, the field of causal inference has been pioneered by Rubin [43] and Pearl [44]. The focus in this area is on the construction and analysis of so-called cause–effect diagrams, such as directed acyclic graphs (DAGs). By encoding the causal structure between different variables, these diagrams can be used to analyze the (causal) independence of variables, as well as the causal paths between variables. In computational experiments, there are two advantages to using causal findings to analyze systemic emergences [45]. First, it provides researchers with an alternative way to understand the emergent properties of artificial society models and the real-world systems they mimic; it also provides insight into what factors lead to the emergent behavior of the model. This can lead to potentially unforeseen explanations for emergent attributes that cannot be revealed by sensitivity analysis or machine learning techniques alone. Second, the causal diagram provides a clear representation of the underlying mechanisms in the artificial society model. This allows the modeler to effectively communicate the results externally to nontechnical audiences. However, research in this area is still in its infancy.

1.2.3 Computational Experiments + Applications

With the help of computational experiments, researchers can expand their research scope to the intersection of social science and intelligent science from their own disciplinary standpoint and can start from a "problem-oriented" approach, cooperate with experts in other fields to study many interesting and novel topics, and deal with highly characteristic social big data. At present, computational experiment methods have shown great application prospects in many social fields, especially in many important fields related to the management and regulation of complex systems, such as social governance, public health, and financial markets. The following are some of the interdisciplinary research efforts carried out by computational experiments with various social scenarios:

1.2.3.1 Computational Experiments + Public Health

Computational epidemiology is a multidisciplinary field that aims to use computers to understand and control the spatial spread of diseases (e.g., SARS, H1N1, and COVID-19) in populations. Traditional forecasting models rely on a large number of variable choices and differences in knowledge, attempting to make the best estimates with uncertain information. However, the epidemic spread is a dynamic and uncertain process, which has not only many influencing factors but also many unknown areas and emergencies, which makes it extremely difficult to predict and research. At present, computational experiments have become an important method to study the large-scale spread of infectious diseases and are mainly used to predict the spread trend, pre-assess the impact, and optimize intervention strategies [46]. However, there is no computational model that is completely consistent with the real physical propagation process, and all models are only approximate abstractions of the actual propagation process. It is necessary to continuously reduce and avoid the influence of uncertainty from the aspects of data, model, computation, and result analysis, so as to obtain simulation results that are closer and closer to the real propagation process.

1.2.3.2 Computational Experiments + Economics

Computational economics is a branch of economics that uses computers as tools to study human and socioeconomic behaviors. Computational economics mainly uses computational experiment methods to find analytical and statistical solutions to complex economic problems. In recent years, complex economics has become a popular research direction, specializing in the study of the overall economic process as a dynamic system of interaction between agents, so it is a way of studying economic adaptation from the perspective of complex adaptive systems. Tesfatsion and Judd provide a comprehensive introduction to agent-based computational economics methods and their applications in various economic fields [18]. In 2021, Salesforce presented a new study called "The AI Economist" that proposes a two-layer reinforcement learning method based on economic simulation to discover tax strategies that can efficiently find a balance between economic equality and productivity [47]. Tilbury reviewed the historical obstacles faced by classical ABM in economic modeling [48].

1.2.3.3 Computational Experiments + Sociology

Digital technology is being fully integrated into human society with new ideas, new business forms, and models, exerting an extensive and profound impact on our production and life. On the one hand, social forces and market entities have participated in social governance more extensively with the help of digital networks and platforms, and governance entities have become more diversified. On the other hand, the transformation of social behavior in the digital era has had an impact on the traditional hierarchical regulation, and the governance model and means urgently need to be modernized. At the same time, civic awareness is constantly rising, the demands of social subjects are becoming more complex, and the governance goals are diversified. This poses new challenges to the unified regulation of diversity and consistency in the digital social system and brings higher requirements for the fairness and efficiency of social governance [49]. Computational experiments play a crucial role in the study of intelligent social governance, relying on the advantages of precision, controllability, ease of operation, and repeatability, which can simulate how the interaction between individual behavior and algorithmic behavior produces the overall social macro law and help us understand the role of information in the operation of the social system, such as social phenomena like information cocoons, platform monopolies, and algorithmic discrimination [50].

1.2.3.4 Computational Experiments + Political Science

Political theory, electoral mechanisms, geopolitics, and international relations have always been hot research directions in the field of political science. With the rise of technologies such as social media, recommendation algorithms, and intelligent assistants, people's decision-making may be invisibly manipulated and directed toward interest-concentrated areas through various forms of algorithmic incentives. According to a recent study [51], when the proportion of people in a society exceeds a threshold, group selection is not necessarily optimal. The so-called "autonomous" behavior, which is constantly induced, is challenging the democratic foundations of social choice. Internationally, strategic game deduction is a very complex problem, involving many roles, and the influencing factors of each role's decision-making are also very complex. They will evolve over time, with challenges such as revenue ambiguity, diversity of strategies, and uncertainty of behavior. Computational experiments can deduce, analyze, and predict various political phenomena under various hypothetical pressures or extreme scenarios, such as the spread of political tendencies in social media, opinion conflict games and election predictions, and international strategic game deduction [52].

1.2.3.5 Computational Experiments + Finance

The complexity and inherent speculative nature of financial markets determine that they are often in a state of unbalanced fluctuations, and the implementation effect of financial regulatory policies is, therefore, facing great uncertainty. The financial market is a complex adaptive system, which is linked by a large number of parallel and interactive adaptive agents through economic behavior. Each agent can only obtain partial information of the whole system, but the agent has "adaptability" and can continuously learn and change its investment strategy through adapting to the environment. In 1987, Arthur et al. of the SFI research group constructed an artificial stock market that used a multi-agent model to explore financial market issues such as asset pricing, investor behavior, and market volatility. This model is a breakthrough achievement in the research process of agent-based financial market modeling, which pioneered agent-based computational finance (ACF) and became the real beginning of agent-based financial market modeling research [53]. Prof. Zhang Wei from Tianjin University has used computational experiments to study the possible financial market reactions of various regulatory measures and policies in depth, so as to evaluate the effect of policies and the feasibility of institutional construction [54].

1.3 Organizational Structure of the Book

As a research method that integrates social science and intelligence science, computational experiment methods integrate multiple disciplines such as economic science, computational science, complexity science, and intelligence science. Despite the efforts of researchers, computational experiment methods are not yet a mature set of tools, nor are they a complete set of theoretical systems, and there is still a huge gap between their prospects and practical applications. Therefore, in order to promote the research development of this field, this book will address the following three questions:

1) **How is it modeled?** The agent model is the core of computational experiments, and it also serves as the container for different intelligent knowledge to come into play. At present, most researchers manage the decision-making

process of each agent by assigning some simple deductive rules to them, resulting in macroscopic emergence. However, researchers generally agree that this model is often not fully representative of the real way in which social agents make decisions. A good agent model should have human-like characteristics, and there are various factors that affect its decision-making, including internal state, environment, and corresponding dynamic rewards and risks, and may even involve complex social factors, different cultural backgrounds, and the agent's learning curve. Therefore, how to use various AI technologies (such as large language models, reinforcement learning, and swarm intelligence) to improve agent model design (including heterogeneity, bounded rationality, learning evolution, and interaction) has become a key challenge for the sustainable development of this field.

2) **How to experiment?** In the computational experiment system, each scene is multi-node and nonlinear, and each AI agent has N choices on any node, resulting in countless story lines. As more agents are involved, the more diverse the agents are, the more complex the interactions will be, the number of influencing factors will increase exponentially, the probability of random deviation will also be greater, and the visibility of emergent phenomena will be easily overshadowed by various details. Researchers mostly focus on "description" rather than "experiment." They can deduce the evolution process of the system, but it is difficult for them to provide insightful causal analysis of emergent phenomena to understand the operation of the system and cannot give the laws behind the phenomena. Therefore, this book will focus on how to use computational experimental design to realize causal relationship mining and causal mechanism inference, so as to provide a decision-making basis for subsequent regulatory strategy optimization.
3) **How to develop?** As a multidisciplinary field, computational experiments involve a lot of domain knowledge. It is often difficult for researchers to clarify the relationship between AI technology and sand table models, between simulation and computational experiments, and between general settings and specific scenarios, and this dilemma is exacerbated by the emergence of various AI technologies. At present, most researchers focus on "explanation" rather than "prediction," and sand table deduction is relatively simple, which can often only be used to explain phenomena, but it is difficult to play the role of intelligent decision-making. In order to solve this problem, this book will focus on how to use a large language model technology to support the development of computational experiment methods, hoping to construct a suitable platform support for the implementation of computational experimental theories.

The organizational structure of the book is shown in Figure 1.7, which is mainly divided into four parts: basic theory, method framework, key technologies, and

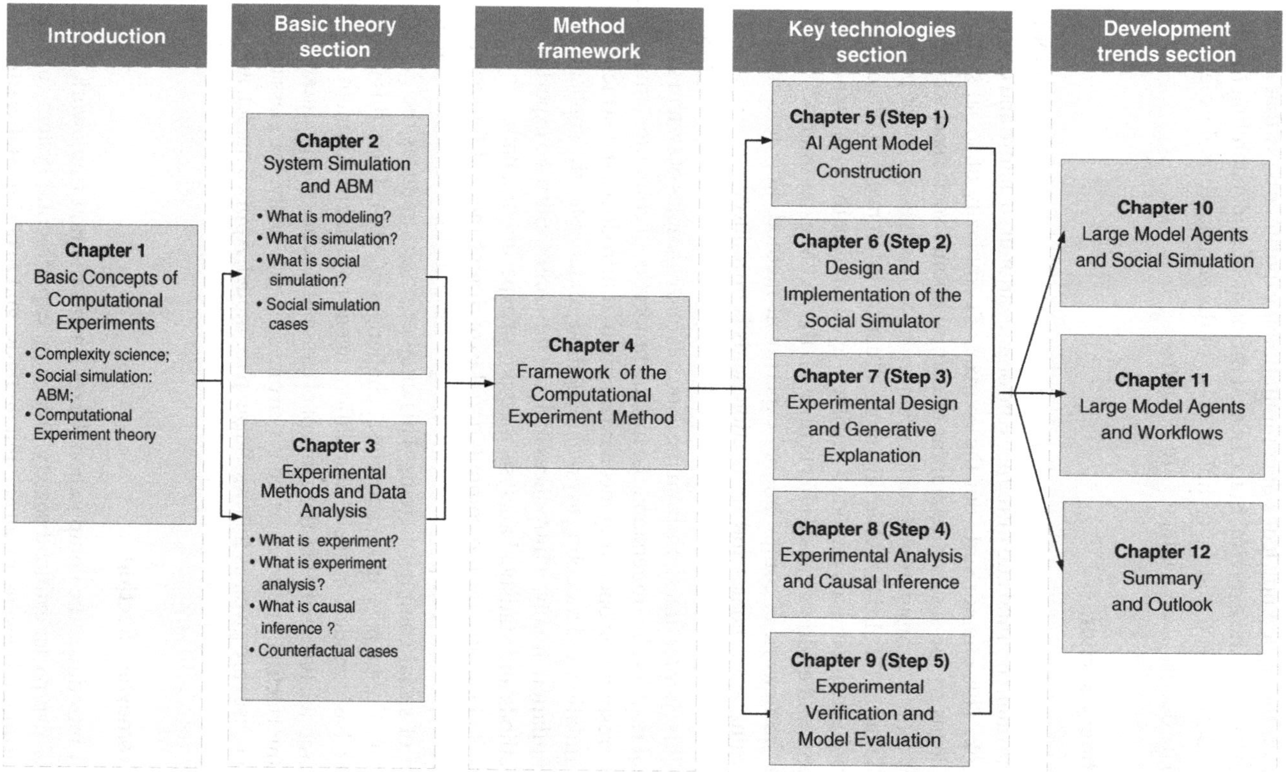

Figure 1.7 The structure of the content of the book.

development trends. The book adopts the research idea of "theory + practice," starting from the basic knowledge of computational experiments, so that readers can further analyze the experimental cases on the basis of understanding the theoretical models, so as to have a complete grasp of the computational experiment method. The details of each chapter are as follows.

1.3.1 Theory Section

This chapter mainly introduces the origin and ideological essence of computational experiment method and introduces the organizational structure and content of this book. This chapter sorts out the sources of complexity of social systems and thus leads to the research quadrant of computational social science. As an important tool in computational social science, ABM has made a lot of progress in many fields, but it also faces the dilemma of "causal inference." In this context, computational experiments provide a feasible way to analyze the behavior of complex systems and evaluate causal effects through the "ABM experiments." Finally, under the framework of computational experiment method, how to combine it with emerging technologies and application fields is introduced.

Chapter 2 mainly introduces the computational experiment method for physical systems. This chapter first starts from the concepts and examples of system simulation so that readers can understand the unique properties of computer simulation and the corresponding operating mechanism; then, it introduces the principles of system simulation and modeling to provide method guidance for the application of system simulation; finally, it introduces in detail the commonly used experimental design methods and statistical analysis methods of experimental data.

Chapter 3 elaborates on the computational experiment method for social systems in detail, which can provide some unique insights for the simulation and deduction of social complex systems. This chapter first introduces the development history, design principles, and causal challenges of ABM. Then, this chapter gives three basic models in social simulation: agent model, environment model, and rules model. Based on this, the concept of causal inference is introduced, especially counterfactual inference based on virtual scenes. Finally, the classic social simulation cases are given as follows: Sugarspace model, AppEco model, artificial stock market model, service bridge model, etc.

1.3.2 Framework Section

Chapter 4 introduces the computational experiment method framework, which consists of five parts: artificial society modeling, experimental system construction, experimental design, experimental analysis, and experimental verification. Then, the implementation platform of the method is given, which mainly includes four core parts. (1) Sand table design module: builds AI agent and AI society. (2) Computational experiment module: provides tools for governance experiment design and experimental operation analysis. (3) Simulation engine module: decouples the design of sand table scene construction from multi-agent operation management. (4) Scene docking module: is a hub that connects practical applications and sand table scenes and can use 2D scenes, GIS modeling, 3D scenes, etc. according to needs.

1.3.3 Technical Section

Chapter 5 focuses on the first step of the computational experimental method: artificial society modeling. The difficulty of social system modeling lies in the modeling of human characteristics, including heterogeneity, bounded rationality, learning evolution, and interaction. In view of these difficulties, this chapter introduces in detail the four levels of artificial society modeling framework: autonomous feature modeling of AI agent, evolutionary feature modeling of AI agent, interaction feature modeling of AI society, and emergent feature modeling of AI society. The model framework is customizable and can use a variety of popular intelligent technologies to provide support for the construction of computational experiment systems.

Chapter 6 focuses on the second step of computational experiment method: experimental system construction. This stage mainly solves two problems: (1) how to map a specific domain model to a general artificial society model, including the agent model, the environment model, and the rule model, and provides the corresponding model docking cases and (2) how to use the control flow to integrate many model components of the artificial society, mainly the integration of the agent model and the environment model, and make the whole artificial society operate dynamically through the system engine, so as to lay the foundation for the subsequent simulation and deduction.

Chapter 7 focuses on the third step of computational experiment method: computational experimental design. Traditional experimental design methods cannot solve the problems of randomness and initiative in social simulation. In order to address this challenge, this chapter proposes a framework for computational experimental design: (1) to determine the influencing factors and response variables of the system through descriptive modeling of the artificial society, (2) to describe deterministic and nondeterministic factorial experiment method through interpretive modeling of experimental design, and (3) to obtain a predictive meta-model that understands the dynamics of the system based on the data between the inputs and outputs of the artificial society. Finally, an automated scenario generation algorithm is proposed, and the effectiveness of the proposed method is illustrated by the case of API service market.

Chapter 8 focuses on the fourth step of the computational experiment method: computational experiment analysis. In view of the limitations of traditional analysis methods in explaining the behavior of social systems, this chapter proposes a causal inference framework for computational experiments, which mainly includes three levels. (1) Observational analysis: focusing on the possibility space modeling of the evolutionary path of social systems, identifying all the states that social systems may experience and the probability distributions of their occurrences. (2) Intervention analysis: Randomized controlled experiments are used as a means to observe how the change of the cause variable caused the change of the outcome variable, so as to mine the robust structural causal model between the variables. (3) Mechanism analysis: Using the evolution of the world model within the agent as a tool, we can understand how individuals at the micro level lead to the emergence of phenomena at the macro level by analyzing the agent's behavior trajectory and thinking chain.

Chapter 9 focuses on the fifth step of the computational experimental method: computational experimental validation. How to evaluate the accuracy and validity of computational models is a difficult problem, which hinders computational experiments from becoming a more powerful tool for studying the operation of complex systems. In order to address this challenge, this chapter introduces a comprehensive capability maturity assessment framework for computational experiments, which provides a comprehensive assessment in terms of expectations, theoretical basis, data sources, intermediate stages, and result realization and evaluates whether the model meets expectations by comparing expectations and implementations. Finally, this evaluation framework is used to analyze and evaluate several representative models of COVID-19 to help decision-makers identify more mature and valuable models and point out to modelers the direction in which the models can be improved in the future.

1.3.4 Trending Section

Chapter 10 introduces the convergence trend of large language model agents and social simulations. The biggest advantage of large language model agents is that simulation agents with human-like characteristics (context learning, continuous learning, inference, etc.) can be developed so that the constructed artificial society also exhibits similar characteristics to real social systems (organized cooperation, reasonable competition, information dissemination, and group emergence). Starting from "Stanford Smallville," this chapter first explains how to use a large language model technology to achieve a sandbox environment similar to "Sims," in which each agent has a different portrait and can communicate with each other to generate rich dialogue and behavior data. Then, under the guidance of the hierarchical modeling framework of artificial society (Chapter 5), the key feature implementation techniques of generative agent are summarized. Finally, a typical case of a large language model agent used in social simulation is given.

Chapter 11 describes the convergence trend of large language model agents and workflows. The autonomous agent uses large language models (LLMs) as a "powerful general-purpose problem solver" for more intelligent and automated problem-solving. Compared with a single agent, the multi-agent system can effectively simulate the cooperation mode of human group problem-solving, emphasizing diverse agent configurations and collaborative workflows between agents, so as to solve more dynamic and complex tasks. However, the division of labor and collaboration also brings communication costs and accumulated errors. This chapter focuses on the workflows, collaboration models, and technical challenges of autonomous agents to lay the groundwork for automating computational experiments.

Chapter 12 summarizes the content of the book and gives the focus of further research.

1.4 Chapter Summary

This chapter begins with the concept of complex systems, i.e., the reasons for the emergence of computational experiment method. Then, the ideas and application fields of computational experiment methods are introduced, and it is clearly pointed out that computational experiments are a higher level of computational simulation with the further development of computing technology and analysis methods, a feasible way to analyze the behavior of complex systems and evaluate the effects of various decisions, and an effective means to make up for the difficulty or even inability of traditional methods to experiment on complex systems. In addition, this chapter explains the organizational structure of the book and provides a guide for subsequent study of the book.

References

1 Doerner, D. (1990). The logic of failure. *Philosophical Transactions of the Royal Society of London. B, Biological Sciences* 327 (1241): 463–473.

2 Mitchell Waldrop, M. (1992). *Complexity: The Emerging Science at the Edge of Order and Chaos*. New York: Touchstone.

3 Papazoglou, M.P., Traverso, P., Dustdar, S. et al. (2008). Service-oriented computing: a research roadmap. *International Journal of Cooperative Information Systems* 17 (02): 223–255.

4 Xue, X., Li, G., Zhou, D. et al. (2022). The research roadmap of service ecosystem: a crowd intelligence perspective. *International Journal of Crowd Science* 6 (4): 195–222.

5 Rahwan, I., Cebrian, M., Obradovich, N. et al. (2019). Machine behaviour. *Nature* 568: 477–486.

6 Wang, F.Y. (2010). The emergence of intelligent enterprises: from CPS to CPSS. *IEEE Intelligent Systems* 25 (4): 85–88.

7 Flaxman, S., Goel, S., and Rao, J.M. (2016). Filter bubbles, echo chambers, and online news consumption. *Public Opinion Quarterly* 80 (S1): 298–320.

8 Hajian, S., Bonchi, F., and Castillo, C. (2016). Algorithmic bias: from discrimination discovery to fairness-aware data mining. *Proceedings of the 22nd ACM SIGKDD International Conference on Knowledge Discovery and Data Mining*, pp. 2125–2126.

9 Ginsberg, J., Mohebbi, M., Patel, R. et al. (2009). Detecting influenza epidemics using search engine query data. *Nature* 457: 1012–1014.

10 Lazer, D., Kennedy, R., King, G., and Vespignani, A. (2014). The parable of Google flu: traps in big data analysis. *Science* 343 (6176): 1203–1205.

11 Ford, M. (2015). *Rise of the Robots: Technology and the Threat of a Jobless Future*. Basic Books.

12 Xue, X., Huang, F., Wang, S., and Feng, Z. (2021). Research on escaping the big-data traps in O2O service recommendation strategy. *IEEE Transactions on Big Data* 7 (1): 199–213.

13 Taleb, N.N. (2007). Black swans and the domains of statistics. *The American Statistician* 61 (3): 198–200.

14 Fang, W. and Lei, G. (2022). Research on systemic complexity in the digital society. *Management World* 38 (9): 208–221.

15 Lazer, D., Pentland, A., Adamic, L. et al. (2009). Computational social science. *Science* 323 (1): 721–723.

16 Conte, R., Gilbert, N., Bonelli, G. et al. (2012). Manifesto of computational social science. *The European Physical Journal Special Topics* 214 (1): 325–346.

17 Lazer, D.M.J., Pentland, A., Watts, D.J. et al. (2020). Computational social science: obstacles and opportunities. *Science* 369 (6507): 1060–1062.

18 Tesfatsion, L. (2006). Agent-based computational economics: a constructive approach to economic theory. *Handbook of Computational Economics* 2: 831–880.

19 LeBaron, B. (2006). Agent-based computational finance. *Handbook of Computational Economics* 2: 1187–1233.

20 Carley, K.M. and Gasser, L. (1999). Computational organization theory. In: *Multiagent Systems: A Modern Approach to Distributed Artificial Intelligence* (ed. G. Weiss), 299–330. MIT Press.

21 Marathe, M. and Vullikanti, A.K.S. (2013). Computational epidemiology. *Communications of the ACM* 56 (7): 88–96.

22 Hofman, J.M., Watts, D.J., Athey, S. et al. (2021). Integrating explanation and prediction in computational social science. *Nature* 595 (7866): 181–188.

23 Wang, F.-Y. (2004). The parallel system method and the management and control of complex systems. *Control and Decision* 19 (5): 485–489.

24 Wang, F.-Y. (2004). Computational experiment methods and the analysis of complex system behavior and decision evaluation. *Journal of System Simulation* 05: 893–897.

25 Yuan, J. (2020). The problem of generative explanation in computational social science. *Studies in Dialectics of Nature* 36 (4): 103–107.

26 Xue, X., Yu, X., Zhou, D. et al. (2023). The origin, current status, and prospects of computational experiment methods. *Acta Automatica Sinica* 49 (2): 246–271.

27 DeAngelis, D.L. and Diaz, S.G. (2019). Decision-making in agent-based modeling: a current review and future prospectus. *Frontiers in Ecology and Evolution* 6: 237.

28 Conte, R. and Paolucci, M. (2014). On agent-based modeling and computational social science. *Frontiers in Psychology* 5: 668.

29 Zhang, W., Valencia, A., and Chang, N.B. (2021). Synergistic integration between machine learning and agent-based modeling: a multidisciplinary review. *IEEE Transactions on Neural Networks and Learning Systems* 34 (5): 2170–2190.

30 Pereda, M., Santos, J.I., and Galán, J.M. (2017). A brief introduction to the use of machine learning techniques in the analysis of agent-based models. In: *Advances in Management Engineering* (ed. C. Hernández), 179–186. Cham: Springer.

31 Prasanna, A., Holzhauer, S., and Krebs, F. (2019). Overview of machine learning and data-driven methods in agent-based modeling of energy markets. *Proceedings Informatik*, Kassel, Germany, pp. 571–584.

32 Park, J.S., O'Brien, J., Cai, C.J. et al. (2023). Generative agents: interactive simulacra of human behavior. *Proceedings of the 36th Annual ACM Symposium on User Interface Software and Technology*, pp. 1–22.

33 Guo, T., Chen, X., Wang, Y. et al. (2024). Large language model based multi-agents: a survey of progress and challenges. arXiv preprint arXiv:2402.01680.

34 Wang, L., Ma, C., Feng, X. et al. (2024). A survey on large language model based autonomous agents. *Frontiers of Computer Science* 18 (6): 186345.

35 Ma, Q., Xue, X., Zhou, D. et al. (2024). Computational experiments meet large language model based agents: a survey and perspective. arXiv preprint arXiv:2402.00262.

36 Xi, Z., Chen, W., Guo, X. et al. (2025). The rise and potential of large language model based agents: a survey. arXiv 2023. arXiv preprint arXiv:2309.07864, 10.

37 Xue, X., Xiangning, Y., and Wang, F.-Y. (2023). ChatGPT chats on computational experiments: from interactive intelligence to imaginative intelligence for design of artificial societies and optimization of foundational models. *IEEE/CAA Journal of Automatica Sinica* 10 (6): 1357–1360.

38 Scott, J. (2011). Social network analysis: developments, advances, and prospects. *Social Network Analysis and Mining* 1: 21–26.

39 Squazzoni, F., Jager, W., and Edmonds, B. (2014). Social simulation in the social sciences: a brief overview. *Social Science Computer Review* 32 (3): 279–294.

40 Carley, K.M. (2009). Computational modeling for reasoning about the social behavior of humans. *Computational and Mathematical Organization Theory* 15: 47–59.

41 Xue, X., Zhou, D., Xiangning, Y. et al. (2024). Computational experiments for complex social systems: experiment design and generative explanation. *IEEE/CAA Journal of Automatica Sinica* 11 (4): 1022–1038.

42 Thiele, J.C., Kurth, W., and Grimm, V. (2014). Facilitating parameter estimation and sensitivity analysis of agent-based models: a cookbook using NetLogo and R. *Journal of Artificial Societies and Social Simulation* 17 (3): 11.

43 Rubin, D.B. (1974). Estimating causal effects of treatments in randomized and nonrandomized studies. *Journal of Educational Psychology* 66 (5): 688–701.

44 Pearl, J. (2009). *Causality*. Cambridge University Press.

45 Sharpanskykh, A., Curran, R. et al. (2019). Using causal discovery to analyze emergence in agent-based models. *Simulation Modelling Practice and Theory* 96: 101940.

46 Bin, C., Mei, Y., Chuan, A. et al. (2020). Prediction of epidemic transmission and evaluation of prevention and control measures based on artificial society. *Journal of System Simulation* 32 (12): 2507.

47 Zheng, S., Trott, A., Srinivasa, S. et al. (2022). The AI economist: taxation policy design via two-level deep multiagent reinforcement learning. *Science Advances* 8 (18): eabk2607.

48 Tilbury, C.R. (2022). Reinforcement learning for economic policy: a new frontier?

49 Yang, X., Xu, Z., and Guo, L. (2022). Computational governance: an emerging interdisciplinary field worthy of attention. *China Science Foundation* 36 (Issue 1): 94–99.

50 Salganik, M.J. (2017). *Bit by Bit: Social Research in the Digital Age*. Princeton University Press.

51 Carr, N. (2015). *The Glass Cage: Where Automation is Taking Us*, 67. London: Random House.

52 Cioffi-Revilla, C. and Rouleau, M. (2010). MASON RebeLand: an agent-based model of politics, environment, and insurgency. *International Studies Review* 12 (1): 31–52.

53 Arthur, W.B., Holland, J.H., LeBaron, B. et al. (1997). Asset pricing under endogenous expectations in an artificial stock market. In: *The Economy as an Evolving Complex System II* (ed. W.B. Arthur, D. Lane, and S.N. Durlauf), 15–44. Addison-Wesley.

54 Zhang, W. (2019). *Computational Experimental Research in Finance*. Science Press.

2

Computational Experiments for Physical Systems

In the face of complex systems in natural sciences, traditional research methods are difficult to effectively tackle their nonlinear and emergent characteristics. System simulation technology has become a key tool for understanding and regulating complex systems. This chapter introduces core concepts of computer modeling and simulation, laying the groundwork for social modeling and computational experiments. First, it explains the principle of similarity in system modeling and the specific steps involved in modeling and discusses the three core elements of computer simulation: system, model, and simulation. Then, the concepts and steps of system simulation are introduced, and some application examples of computer simulation technology in various fields are provided. Finally, the chapter provides a systematic overview of experimental method and data analysis frameworks, helping readers distinguish between deduction and experimentation.

2.1 The Principle of Similarity and System Modeling

2.1.1 The Principle of Similarity

Similarity is a simple and highly universal concept, and it is also a widely prevalent phenomenon in nature. The principle of similarity [1] is one of the most fundamental theories in the discipline of system simulation. It states that for any system in nature, there exists another corresponding system that can, in a certain sense, possess similar mathematical descriptions or physical properties. This principle mainly includes three theorems. (1) First theorem: For similar phenomena, there are similarities among the physical variables of their single-valued conditions, and the values of similar characteristic numbers (i.e., similarity constants) are the same. (2) Second theorem: The functional relationships among the physical variables of a certain phenomenon can be expressed as functional relationships among similar characteristic numbers. (3) Third theorem: Phenomena are similar when their single-valued conditions are similar, and the number of similar features composed of single-valued conditions is equal. The similarity principle enables studying complex phenomena via simplified model experiments.

Many scientists have applied the principle of similarity in their research. For example, in 1638, Galileo applied the principle of geometric similarity in the field of shipbuilding, demonstrating its practical application; in 1686, Newton proposed similarity laws for the motion of solid bodies, expanding the scope of the similarity principle; in 1848, Cauchy derived acoustic phenomena and laws in geometrically similar elastic bodies from the motion equations of elastic objects, proving the applicability of the similarity principle across different fields; and in 1920, Kirchhoff's elastic similarity theorem was formally finalized by academia, further deepening the understanding of the similarity principle. Therefore, the principle of similarity serves as a bridge between scientific innovation and technological application.

There are many types of similarity, including geometric similarity, environmental similarity, and performance similarity. As shown in Figure 2.1(a), geometric similarity refers to the proportional scaling of geometric dimensions, where a model obtained by proportionally reducing structural dimensions is called a scaled-down model. As shown in Figure 2.1(b), environmental similarity refers to

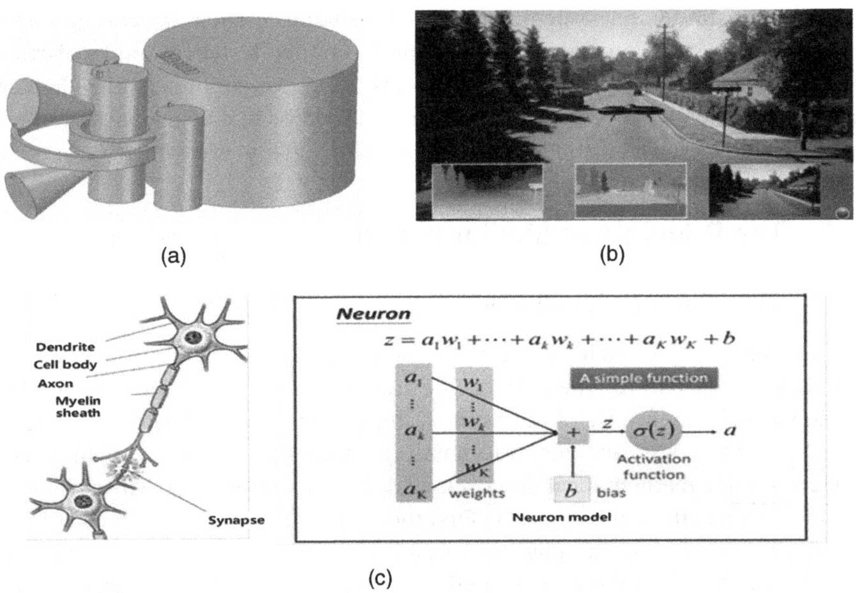

Figure 2.1 Simulation example diagram. (a) Geometric similarity simulation diagram. (b) Environmental similarity simulation diagram. *Source:* [4] / with permission of Springer Nature. (c) Neuron model modeling and simulation.

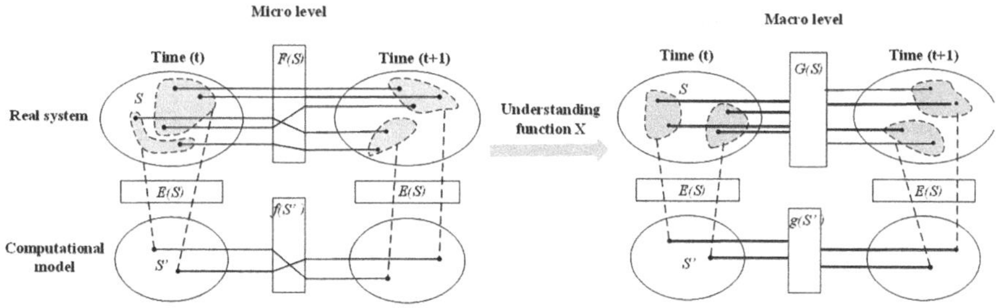

Figure 2.2 Concept of modeling in complex systems.

the pursuit of realism in sensory experiences, such as hearing, sight, and smell, within simulation systems involving human participation (e.g., virtual reality). Artificial neural networks are a typical example of logical similarity. Based on the way the human brain processes information from the external world, researchers have simulated various parts of biological neurons (such as synapses, axons, nerve endings, and intercellular connections) to construct artificial neural network models, which are used to study and simulate the thinking process of the human brain. A diagram of the neuron model's modeling and simulation is shown in Figure 2.1(c).

2.1.2 System Modeling

Modeling simplifies systems by identifying key abstraction conditions. It is a process of deciding which simplification conditions to impose on the entities of the system and then using these abstractions to reveal their underlying characteristics. Figure 2.2 illustrates the modeling concept of complex systems: the upper part represents the real system being modeled, while the lower part shows the constructed model [2]. We assume that the real system contains a set of states S and a state transition function $F(S)$, which maps a given state at time t to a new state at time $t+1$. The function $F(S)$ is unknown to the modeler, and due to the dimensionality of the state space and the potential complexity of $F(S)$, it is usually impossible for the modeler to solve it directly. As an alternative, the modeler can reduce the size of the state space and seek a simpler transition function. One way to reduce the state space is by generating equivalence classes: mapping a subset of the real system's states S to a model state S'. This mapping is denoted as $E(S)$. Based on this new model state, the modeler then seeks a useful transition function $f(S')$.

In the modeling process, special attention must be paid to how well the model captures the behavior of the real system. As shown in Figure 2.2, suppose we

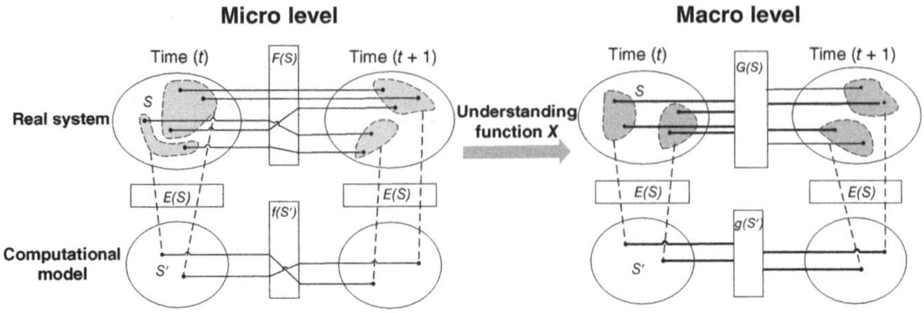

Figure 2.3 Concept of modeling in complex systems.

start from a state of the real system. The model maps this state to a simplified one, then predicts the next state using $f(S')$. In the real system, the state evolves into $F(S)$ over time. Therefore, if the condition $f(E(S)) = E(F(S))$ is satisfied, the model aligns with the real system. If the mapping between the model and the real system possesses this commutative property, it is called a homomorphism. From this perspective, the goal of modeling is to find a set of equivalence classes and a transformation function that result in a useful homomorphism. For any given real-world system, there may exist multiple equivalence mappings (and associated transformation functions) that produce homomorphisms. The art of modeling lies in selecting the appropriate equivalence classes and transformation functions. Choosing overly broad equivalence classes can make it easier to find a suitable transformation function, but this may reduce the explanatory power of the model.

For complex systems, reductionism fails: the whole is not merely greater than the sum of its parts but exhibits fundamentally emergent properties. As shown in Figure 2.3, the left side represents the microscopic level of a system, while the right side represents the macroscopic level. The macroscopic behavior of the system emerges from the interactions among entities at the microscopic level, and this macroscopic level is governed by a transformation function $G()$. It is important to note that knowledge about the microscopic level of the system (such as understanding $F()$ or the details of the microscopic model) does not directly help in understanding the system at the macroscopic modeling level. In this context, understanding the function $X()$—that is, "how the macroscopic level emerges from the microscopic level"—becomes particularly crucial. Researchers aim to develop a theory that helps explain how a system composed of micro-level entities and interaction rules transforms into a macro-level system state.

Elevating real-world phenomena and problems to the theoretical level of mathematical models forms the foundation of modern scientific discovery and technological innovation. For example, the Pythagorean theorem in ancient China revealed a specific relationship among the sides of right-angled triangles, while Western scholars further abstracted such mathematical principles to a universal theoretical level. These mathematical models have been widely applied in fields such as engineering mechanics and electromagnetism, exerting a profound influence on the development of science and technology. The importance of modeling is also evident in the discovery of electromagnetic waves. Faraday's "law of electromagnetic induction" laid the groundwork for the development of electromagnetism. Building on this, Maxwell formulated Maxwell's equations through inheritance, induction, and deduction, thereby elevating electromagnetism to a theoretical level of "mathematical abstraction/mathematical modeling." As a result, he is honored as the "father of electromagnetism." The discovery of electromagnetic waves was derived through deductions based on mathematical models, highlighting the crucial role of mathematical modeling in scientific and technological research.

As shown in Figure 2.4, experiment, induction, and deduction are key approaches and methods for establishing mathematical models of systems. A mathematical model represents an abstract understanding of the natural world and is characterized by behavioral similarity to natural phenomena or problems. By using mathematical models, people can investigate and analyze issues and phenomena in the natural world, thereby achieving the goals of understanding and transforming the world. System modeling is a complex and meticulous task that requires careful attention to every step of the process. Purpose, method, and validation are the three essential elements of modeling work.

2.1.2.1 The Purpose Must Be Clearly Defined

In the modeling process, the first step is to clearly define the purpose of modeling. For the same system, different research objectives will lead to different modeling mechanisms, methods, and resulting models. We need to engage in auxiliary thought processes to solve problems and choose hypotheses. This thinking process may involve analyzing the problem, inferring hypotheses, and evaluating possible solutions. Such a process helps to better understand the essence of the problem and formulate more effective solutions. Once a mathematical model is integrated into a set of axioms and theorems, it can serve as a powerful tool for understanding the objective world. This enables the derivation of conclusions about the real world from the model—conclusions that are both reliable and valid. By constructing and validating mathematical models, we can gain a deeper understanding of the laws governing the operation of the objective world, thereby enhancing our cognitive capabilities.

2.1.2.2 The Method Must Be Appropriate

Figure 2.5 illustrates the process of system modeling. Mechanism-based modeling is suitable for systems with clearly understood internal structures or properties

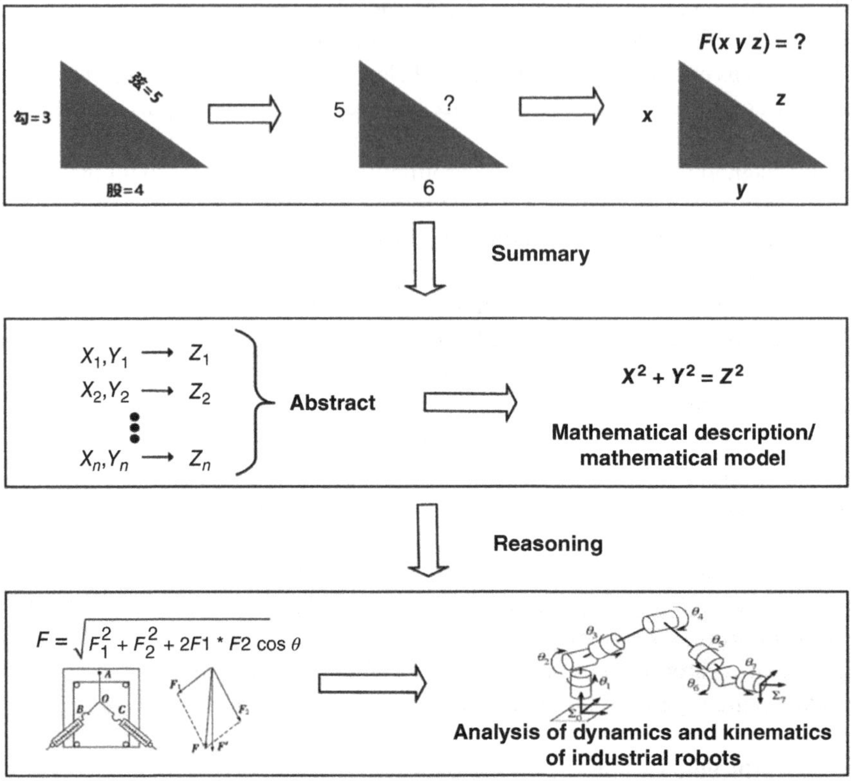

Figure 2.4 System simulation modeling diagram.

(known as white-box problems). In such cases, system models can be constructed through reasoning processes such as analysis, deduction, and transplantation, based on known fundamental laws and theorems. Experimental modeling applies to systems whose internal structures and properties are unknown or not well understood (known as black-box problems). These models are typically developed using empirical knowledge or experimental data, employing reasoning methods such as analysis, induction, and analogy. In practice, it is often difficult to achieve satisfactory results using a single approach. Therefore, multiple methods are typically combined—referred to as integrated modeling.

Induction is the process of moving from specific, concrete observations to general, universal knowledge, based on "observation, experience, and experimentation" as well as large amounts of basic data derived from them. Kepler's third law of planetary motion is a classic example of inductive modeling. Deduction refers to deriving specific propositions from general ones, relying on known mechanisms

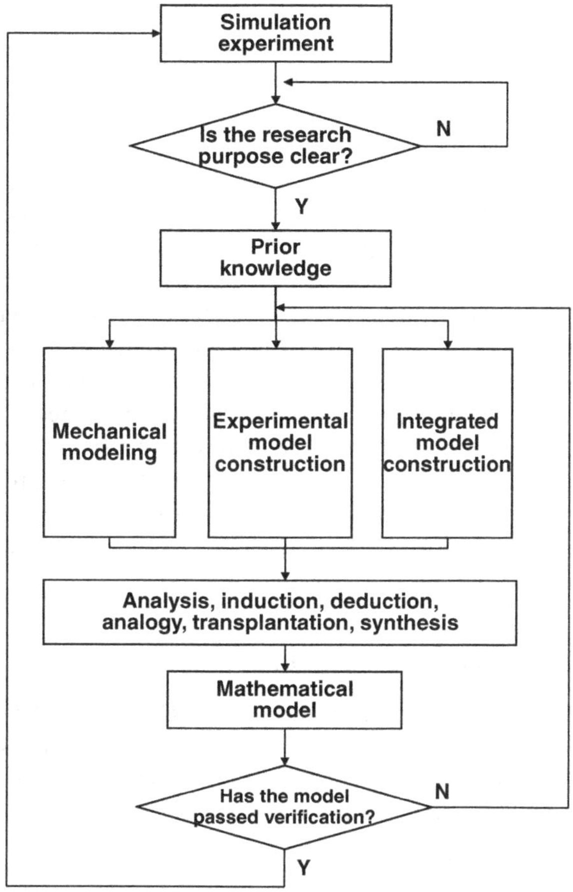

Figure 2.5 System modeling process.

or theorems to deduce particular problems. Newton's law of universal gravitation is a result of deduction—it was derived using calculus as a tool, based on Kepler's third law of planetary motion and Newton's second law of motion. Analogy makes use of the behavioral similarities between different types of objects. It infers or characterizes certain features of a relatively "complex object" by analyzing a relatively "simple object." Transplantation refers to applying theories and methods from one discipline or domain to another. It reflects the interdisciplinary, interactive, and interconnected nature of scientific development.

2.1.2.3 The Results Must Be Validated

In simulation experiments, the validity of the results depends on the reliability of the system model. To verify whether a system model can accurately describe the performance and behavior of the actual system, the following model validation methods are commonly used. (1) Theoretical validation method: This method evaluates the credibility and effectiveness of the system model by examining how well the simulation results conform to necessary conditions (such as various characteristics and governing laws). (2) Mathematical statistics method based on experimental modeling: This approach assesses the model's reliability and accuracy by analyzing the statistical similarity between the outputs of the system model and those of the real system. (3) Physical model validation method: This involves applying physical (or semi-physical) simulation techniques to achieve the highest possible model validation accuracy under feasible conditions.

Model validation should pay attention to the following points. Model validation is a process: The modeler's understanding of the problem needs to evolve from a perceptual to a rational level, often requiring multiple iterations to complete. Model validation involves uncertainty: The degree of similarity or approximation between the system model and the actual system depends on the modeler's understanding and knowledge of the real system. Comprehensive model validation is often difficult to achieve: For complex system models and simulation problems (such as social or ecological systems), validation typically requires a large amount of statistical analysis data.

2.1.3 Model Formalization

In mathematics, the concept of a set is built upon abstraction. Set operations allow us to handle relationships at an abstract level without specifying all the details. Therefore, set theory is highly useful in modeling. A mathematical model is generally defined as a seven-tuple set structure [3]:

$$S = \{T, X, \Omega, Q, \delta, \lambda, Y\} \tag{2.1}$$

where T denotes the time set, representing the time coordinate of system evolution; integers correspond to discrete-time systems, and real numbers correspond to continuous-time systems. X denotes the input set, representing the influence of the external environment on the system. Ω denotes the input segment set, which describes the input over a certain time interval and is a subset of (X, T). Q denotes the internal state set, which is the core of modeling the internal structure of the system. δ denotes the state transition function, which defines how the internal state of the system evolves. λ denotes the output function, i.e., a mapping $Q \times T \times X \rightarrow Y$. Y denotes the output set, through which the system interacts with the environment.

As shown in Figure 2.6(a), a behavior-level description of a model treats the system as a black box, where an input signal is applied to the system, and its output signal is measured and recorded. The system description requires at least a time set, which is typically either an interval of real numbers (for continuous time) or an interval of integers (for discrete time). The basic unit of system description is

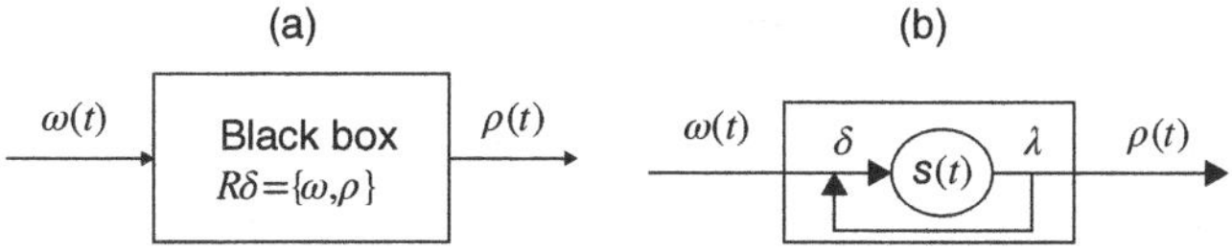

Figure 2.6 System behavior black-box and state-structure-level diagram. (a) System behavior black box diagram. (b) State-structure level diagram.

the trajectory, which represents a mapping from a time interval to a set of possible observable outcomes.

As shown in Figure 2.6(b), a state-structure-level description of a model requires a clear understanding of the internal functioning of the real system, including its internal states and overall structure. Over time, this type of description enables the model to automatically generate a behavioral trajectory. The generation of this trajectory is based on the existence of a state set and a state transition function in the system description. The state set represents all possible outcomes at any given time, while the state transition function provides the rules for computing future states from the current state. To produce an output set, an output function must also be defined. Compared to the behavior-level description, a state-structure-level model is more representative. The presence of a state set is sufficient to compute the system's behavior. As illustrated in Figure 2.7, a complex system can be viewed as an integrated whole formed by interconnecting many basic black boxes, each referred to as a component or subsystem. This modeling approach allows the modeler to understand not only the operational relationships within the system but also the internal decomposition structure of the actual system.

Each system has its own domain-specific characteristics, and the complexity of the problems being studied can vary greatly. Some models aim merely to reveal the underlying patterns of phenomena in the real world, while others are intended to provide concrete, actionable recommendations for intervening in reality. Due to the diversity and uncertainty of research problems, along with the flexibility of computational models, it is critically important to ensure consistency between

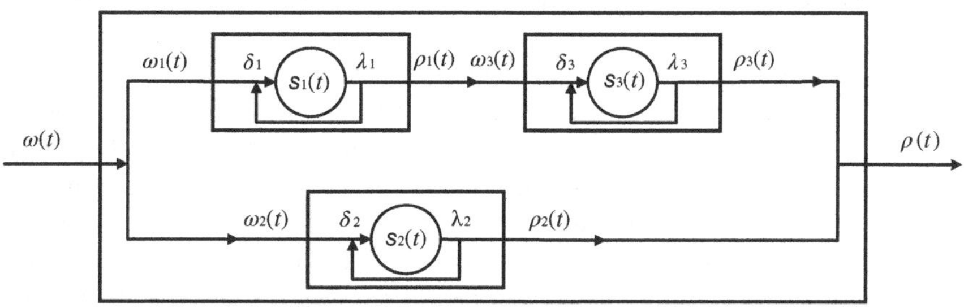

Figure 2.7 Decomposition structure-level diagram.

computational models and their corresponding reference systems. In general, computational models are relatively flexible in form, while reference systems are often complex in operation. The key differences between them typically manifest in two aspects. (1) If the computational model is too abstract, it may fail to capture the operational laws of the reference system. (2) If the computational model is not abstract enough, the modeling process can become excessively complex, encountering challenges such as data scarcity, limited resources and time, and an incomplete knowledge framework. Therefore, achieving a precise mapping between the reference system and the computational model becomes a central concern in the formalization of models.

2.2 Simulation Examples and Concepts

2.2.1 Concept of Simulation

In 1961, G.W. Morgenthaler first defined simulation as "the realization of the essence of a system or activity in the absence of the actual system" [4]. In 1978, Körn described continuous system simulation as "conducting experiments with a model that can represent the system under study" [5]. In 1982, Spriet further expanded the concept of simulation, proposing that all activities supporting model construction and model analysis fall under the scope of simulation [6]. In 1984, Orën introduced the basic conceptual framework of simulation—modeling–experimentation–analysis—and emphasized that "simulation is a model-based activity" [7]. Computer simulation consists of three core elements: system, model, and simulation. The system is the object of the simulation, the model is an abstraction of the system, and the simulation is the experimentation conducted on the model. The details of these elements are described below (Figure 2.8):

2.2.1.1 System

A system is a collection of entities with specific functions that are combined in accordance with certain rules and interact and depend on each other. The "three elements" of a system include entities (determining the composition and boundaries of the system), attributes (describing the characteristics of entities), and activities (the interaction and change processes among entities within the system). The system has integrity (cannot be divided at will), correlation (entities are connected according to certain rules), and subordination (defining boundaries between internal factors and the environment). According to their temporal characteristics, systems can be classified into continuous systems, discrete-time systems, and discrete event systems. Table 2.1 provides specific details.

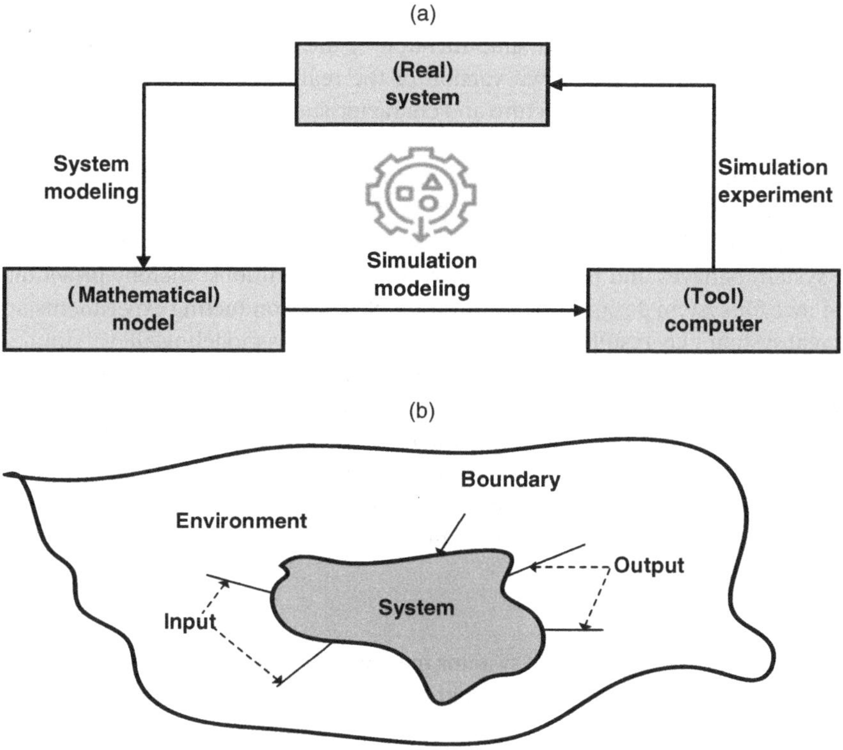

Figure 2.8 Schematic diagram of a simulation system. (a) System modeling diagram. (b) System affiliation diagram.

Table 2.1 System Model Description Table.

Name	Definition	Model Description
Continuous System	A system in which the system state variables change continuously over time, such as a boiler temperature regulation system.	Ordinary differential equations, partial differential equations
Discrete-Time System	The changes in system state variables only occur in a set of discrete moments in a system, such as a sampling control system.	Difference equation, finite state machine Markov chain
Discrete Event System	Event-driven systems in which the system state undergoes discrete changes at certain random time points, such as assembly production systems.	Probability distribution queuing theory

2.2.1.2 Model

A model is an abstraction and simplification of the essence of a real system. A physical model is a scaled-down version of the real system, such as a sandbox model, used to display the structure and characteristics of the system according to a certain scale. A mathematical model, on the other hand, describes the internal laws and behavior of the system using mathematical expressions. System modeling is crucial for system design and performance prediction. First, before the actual system is built, modeling enables a more accurate understanding of how the system behaves and performs under different conditions, thereby providing guidance for system design and optimization. Second, conducting experiments on the real system may result in damage or failure, whereas modeling allows simulation of the system's operation in a virtual environment, avoiding potential risks and hazards. In addition, system modeling helps ensure system reliability. When multiple experiments are required, it is difficult to guarantee that the conditions are exactly the same each time, which may lead to uncertainty in the results. Through modeling, multiple simulation experiments can be conducted in a controlled environment, enabling more accurate assessment of system performance and strengths and weaknesses.

2.2.1.3 Simulation

Conducting experiments on real systems may require a significant amount of time and cost, while through modeling, rapid and low-cost experiments can be carried out in a virtual environment, saving time and resource costs. According to the attributes of the model, simulation can be classified into three categories: physical simulation, mathematical simulation, and semi-physical simulation [5]. According to the proportional relationship between the simulated clock and the actual clock, simulation can be further classified into real-time simulation (where the speed of the simulated clock is the same as that of the actual clock), sub-real-time simulation (where the simulated clock is slower than the actual clock), and ultra-real-time simulation (where the simulated clock is faster than the actual clock). According to the characteristics of the system model, system simulation can be divided into continuous system simulation and discrete event system simulation. Table 2.2 presents the differences between computer simulation and agent-based modeling (ABM) as well as computational experiments.

2.2.2 Simulation Steps

Through the execution of simulation activities, from modeling to experimentation and then to result analysis, it is helpful to comprehensively grasp the operation status and performance of the system. Such a systematic simulation process can help optimize system design, enhance system performance, and provide important references for system optimization and improvement.

Table 2.2 Comparison of Characteristics of Computer Simulation and Computational Experiments.

	Computer Simulation	Agent-Based Modeling (ABM)	Computational Experiment Method
Research Objective	Establish a simulation model of the physical system to pursue the reproduction of system behavior, thereby supporting the analysis, testing, and training of the system.	Establish simulation models of social systems and pursue explanations for various complex social phenomena, including those in the political, economic, financial, and social fields.	By relying on social simulation and experiment method, causal inference can be achieved, and the operational rules of the system can be explored, which can provide intelligent decision support.
Theoretical Basis	According to the principle of similarity, the output of the simulation system can represent the output of the actual system. The results can be repeated. As long as there are the same inputs, the same output results will be obtained.	ABM can reproduce individuals in the real world in an artificial environment, observe the emergent consequences of their group behaviors, and explain complex phenomena in a bottom-up manner. ABM establishes a "connection" between individual interaction at the micro level and the emergence of organizations at the macro level.	Social simulation deduction is adopted to emphasize vertical causality, focusing on the causes and formation processes of phenomena. By using "counterfactual algorithms to construct multiple worlds," horizontal causality is emphasized through experimental mechanisms. Build a causal model of system evolution to reveal the patterns and laws within the system.
Implementation Technology	Discrete simulation, continuous simulation, parallel simulation, distributed simulation, etc.	ABM+ machine learning, ABM+ social networks, ABM+ large language models, etc.	Social simulation + experimental design, social simulation + causal inference, social simulation + regulatory optimization, etc.

(Continued)

Table 2.2 (Continued)

	Computer Simulation	Agent-Based Modeling (ABM)	Computational Experiment Method
Application Field	Multiple fields such as transportation, aerospace, industrial manufacturing, weather forecasting, and electronic information industry.	ABM can assist people in intuitive reasoning and is particularly suitable for the research of social issues where empirical data is difficult to obtain, such as racial segregation, wargame simulation, and the spread of epidemics.	It is playing an increasingly important role in the measurement, analysis and regulation of CPSS in smart cities, smart factories, driverless vehicles, social governance, etc.
Defect	Due to the lack of sufficient theoretical and prior knowledge, it is difficult to accurately describe and analyze complex systems using top-down modeling. Due to the lack of consideration of "human factors," it is unable to explain and analyze complex social phenomena.	When there is a lack of a unified and standardized modeling method, users often rely on their own experience to build the system. There is no consensus yet on how to prove the validity and equivalence of the computational model. This will greatly increase the difficulty of application and reduce the credibility of the results.	Although a unified framework has been formed, it is still necessary to enhance the integration of artificial social modeling and experimental design analysis. In addition, further research is needed on how to integrate it with various new technologies and new fields.

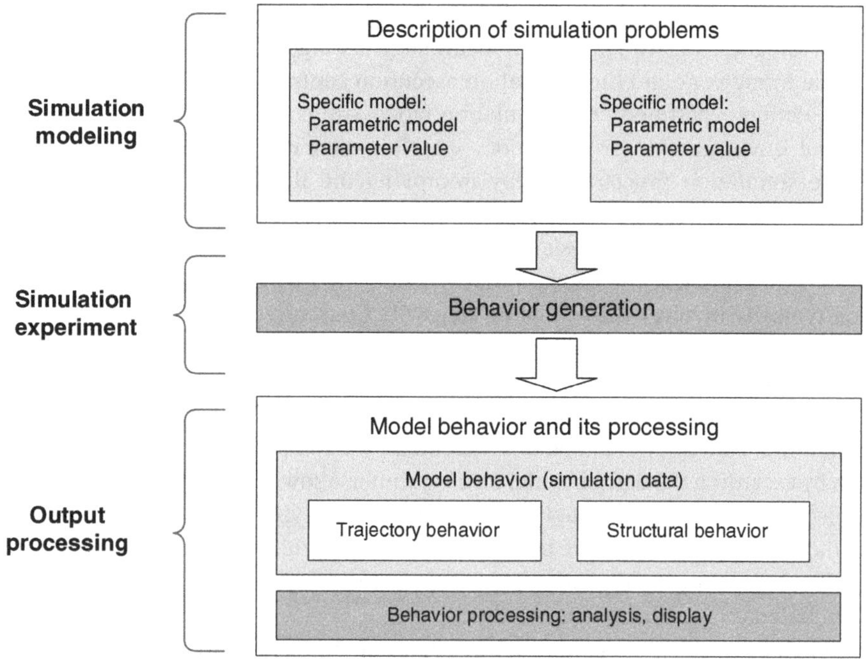

Figure 2.9 Simulation procedure diagram.

In the system modeling phase, system identification techniques are used to describe the characteristics and behavior of the real system. The transfer function (behavior level) treats the system as a "black box," where only the output of the system is measured under the influence of input signals to establish a transfer function model. The structure diagram (decomposition structure level) decomposes the system into several black boxes and defines the input and output of each box, as well as their interconnections, in order to establish a structural diagram model. The state equation (state structure level) defines not only the inputs and outputs of the system but also the internal state set and the state transition function, thereby forming a state equation model (Figure 2.9).

Simulation experiments observe the behavior and performance of a system through both physical experimentation and model-based simulation. In an experiment, the system is placed under controlled or specific conditions for observation and measurement, whereas in a simulation experiment, the established model is operated and observed under the same conditions. Both the physical experiments and the simulation experiments must account for the impact of errors on the results. Therefore, it is necessary to select appropriate equipment and methods to ensure the accuracy of the outcomes. Simulation experiments allow for the evaluation of system performance and provide references for optimization and improvement. Modern simulation technology distinguishes between the experimental framework and the simulation execution control. This means that during a simulation experiment, the simulation model can be placed in an independent control environment for operation, enabling better control and adjustment of the simulation process, thereby improving the flexibility and efficiency of simulation.

The core principle of physical simulation is to represent real-world behavior using digital models and execute these models on a computer. Physical simulation typically involves the following steps. (1) Construct a physical model based on the physical laws (knowledge) of the real world, in order to create a mathematical representation of real-world phenomena. (2) Convert the physical model into a digital model, meaning that the mathematical description is transformed into a format that can be understood and executed by a computer. (3) Run the simulation by executing the digital model on a computer, allowing researchers to observe behaviors within the simulated environment. (4) Analyze the experimental results to understand and predict the behavior of the real world. In essence, physical simulation is the process of generating simulation results that comply with physical laws, based on established knowledge.

To understand the concept of system simulation, let us start with an example of analyzing a mechanical displacement system. Consider a mechanical displacement system composed of a spring, mass, and damper, where the mass of the object is $m = 1$ and the spring stiffness coefficient is $k = 4$. The system is subjected to an external force $r(t) = u(t)$, i.e., a unit step input. The objective is to determine the range of the damping coefficient f (where $0 \leq f \leq 10$) such that the unit step response of the system does not exhibit oscillation (Figure 2.10).

1) **Establish the mathematical model of the system**

   In the given system, the input is the external force r, and the output is the displacement y, which is a function of time t. Let V represent the velocity of motion. According to Newton's law, we have

$$ma = \sum F = r - F_v - F_k \tag{2.2}$$

where $a = \frac{d^2y}{d^2t}$, $F_v = fv = f\frac{dy}{dt}$, $F_k = ky$. Substituting these into the original equation gives the intermediate variable:

$$m\frac{d^2y}{d^2t} = r - f\frac{dy}{dt} - ky \tag{2.3}$$

$$m\frac{d^2y}{d^2t} + f\frac{dy}{dt} + ky = r \tag{2.4}$$

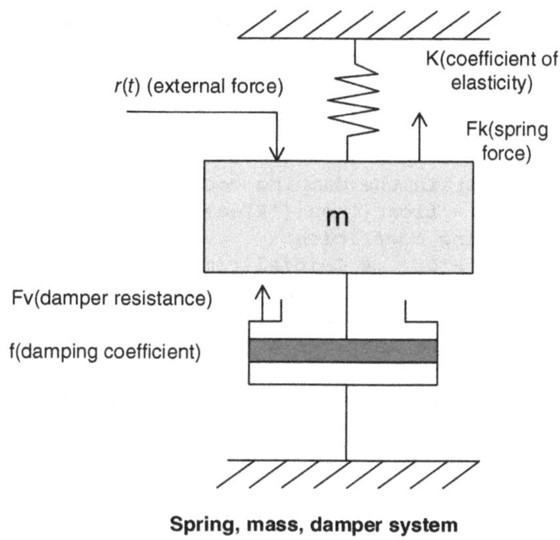

Figure 2.10 Mechanical displacement system diagram.

2) **Establish the simulation model**

For the needs of the subsequent steps, convert it into the state equation and output equation:

$$\begin{bmatrix} x_1' \\ x_2' \end{bmatrix} = \begin{bmatrix} 0 & 1 \\ -\frac{k}{m} & -\frac{f}{m} \end{bmatrix} \begin{bmatrix} x_1 \\ x_2 \end{bmatrix} + \begin{bmatrix} 0 \\ \frac{1}{m} \end{bmatrix} r \tag{2.5}$$

$$y = [1 \quad 0] \begin{bmatrix} x_1 \\ x_2 \end{bmatrix} \tag{2.6}$$

Convert them into a model that can be executed on a computer—a simulation model. Continuous systems are typically represented using difference equations (sets), with T being the step size. By directly applying the Euler formula, the discrete state equation and output equation can be obtained as follows:

$$\begin{bmatrix} x_1((n+1)T) \\ x_2((n+1)T) \end{bmatrix} = \begin{bmatrix} x_1(nT) \\ x_2(nT) \end{bmatrix} + \left\{ \begin{bmatrix} 0 & 1 \\ -\frac{k}{m} & -\frac{f}{m} \end{bmatrix} \begin{bmatrix} x_1(nT) \\ x_2(nT) \end{bmatrix} + \begin{bmatrix} 0 \\ \frac{1}{m} \end{bmatrix} r(nT) \right\} T \tag{2.7}$$

$$y[(n+1)T] = \begin{bmatrix} 1 & 0 \end{bmatrix} \begin{bmatrix} x_1[(n+1)T] \\ x_2[(n+1)T] \end{bmatrix} \tag{2.8}$$

3) **Programming and debugging**

To enable the simulation model to run on a computer, it must be described using an algorithmic language—that is, by writing a computer program and performing debugging. Python is used for programming. The pseudocode is as follows:

Module 1: Pseudocode for Running the Simulation Mode

```
import numpy as np
import matplotlib.pyplot as plt
# Initialize parameters
  m = 1    # Quality
  k = 4    # Spring coefficient
# Obtain the damping coefficient input by the user
  f = float(input('Please enter the damping coefficient f:'))
# Damping coefficient
  t = 0    # Initial time
  T = 0.01  # Time step
  Tmax = 10   # Maximum time
# Define matrices A and B
  A = np.array([[0, 1], [-k/m, -f/m]])
  B = np.array([0], [1/m]).reshape(2, 1)
# Initialize the state vector x
  x = np.array([0], [0]).reshape(2, 1)  # Initial position
and speed
# Initialize the list for storing the results
  Y = [0]  # Location record
  H = [0]  # Time record
# Iterative solution using the Euler method
  while t < Tmax:
    xs = x + (np.dot(A, x) + B) * T     # Calculate the state
of the next moment
    y = xs[0, 0]     # Current position
    t = t + T    # Update time
    Y.append(y)   # Record location
    H.append(t)   # Record the time
     x = xs   # Update the current status
# Draw graphics
  plt.plot(H, Y, 'k')
  plt.grid(True)          # Add grid lines
  plt.xlabel('Time t')    # Set the coordinate axis labels
and titles
  plt.ylabel('Response Y')
  plt.title('System time response curve')
  plt.show()          # Display graphics
```

This code first sets some parameters, such as mass, damping coefficient, spring coefficient, initial position, and initial speed. Then, a time step of 0.01 seconds is set to perform the simulation of the spring–mass–damper system. Finally, the relevant functions are called to perform the simulation and draw the position curve of the spring–mass–damper system.

4) **Verification and validation of the simulation model**
To make the simulation research more effective, it is necessary to compare the data from the simulation program operation with the data observed from

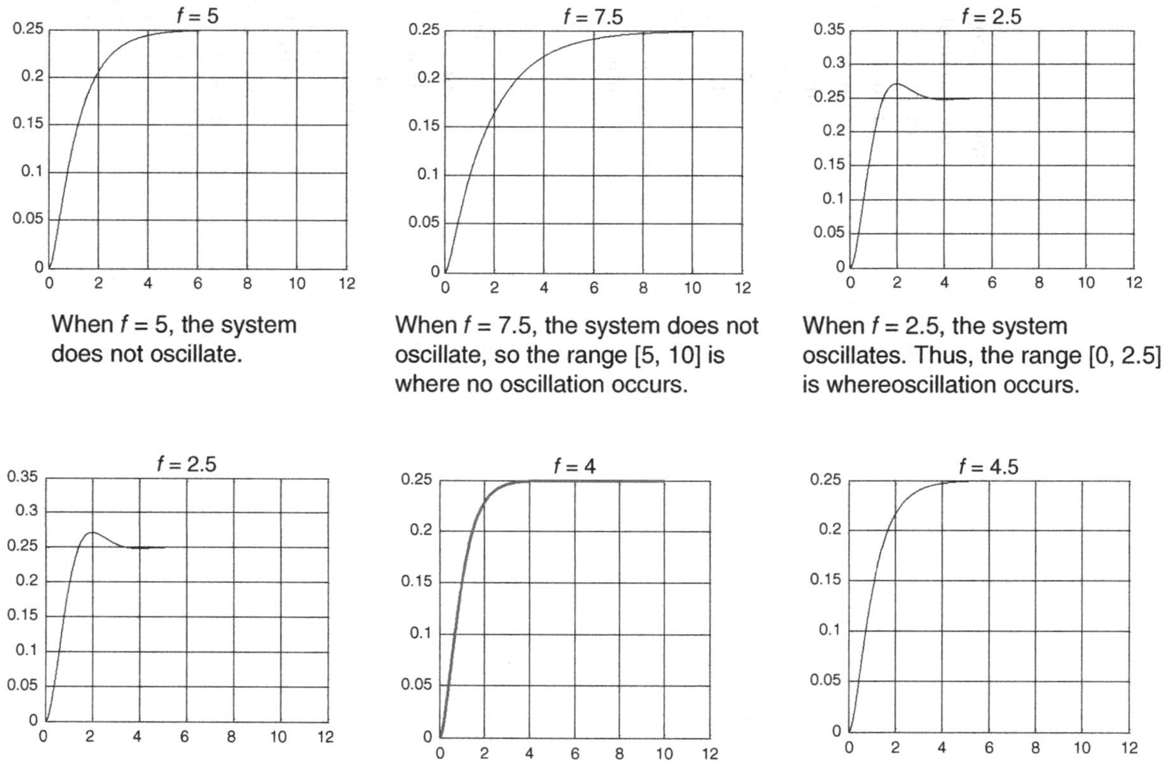

Figure 2.11 Simulation diagram of the critical point of the mechanical displacement system.

the actual system operation to confirm the correctness of the mathematical model.

5) **Simulation experiments are conducted on the computer and the simulation results are analyzed** (Figure 2.11)

2.2.3 Simulation Case

2.2.3.1 Manufacturing Simulation

Manufacturing simulation is a technology that utilizes computer technology, information technology, and other technologies to conduct high-precision digital modeling and simulation of industrial products, their production processes, and manufacturing systems and to predict, analyze, and optimize actual manufacturing processes in a virtual environment. Production line simulation is a key area of manufacturing simulation. It mainly uses digital technology to build a mathematical model of the physical production line and simulates the activities and states of the actual production line through the corresponding simulation algorithms, providing a basis for the layout design and scheduling decisions of the production line. It covers many functions such as planning the production system, optimizing the layout, and determining the production capacity. Figure 2.12 presents the corresponding logistics simulation system.

2.2.3.2 Scientific Exploration

Scientific exploratory simulation refers to the simulation of various phenomena, processes, and systems in fields such as natural sciences, engineering, and social sciences. It helps researchers understand, analyze, and predict the behavior and characteristics of research objects, thereby promoting scientific innovation. From a practical perspective, it first requires abstraction and simplification of the research object, extracting key features and parameters, and establishing mathematical or logical models that reflect its essential laws. Then, with the powerful computational capabilities of computers, the model is solved and calculated according to predefined rules and algorithms, simulating the operating states and dynamic processes of the research object under different conditions. Finally, visualization techniques are used to present the simulation results in intuitive forms such as graphs, images, and animations, making it easier for researchers to observe, analyze, and evaluate. For example, in astronomy, when simulating the evolution of galaxies, researchers establish equations of motion for stars, gases, and other materials in the galaxy based on celestial mechanics principles. Through long-term computer simulations, the formation, development, and changes of galaxies over billions of years can be demonstrated (Figure 2.13).

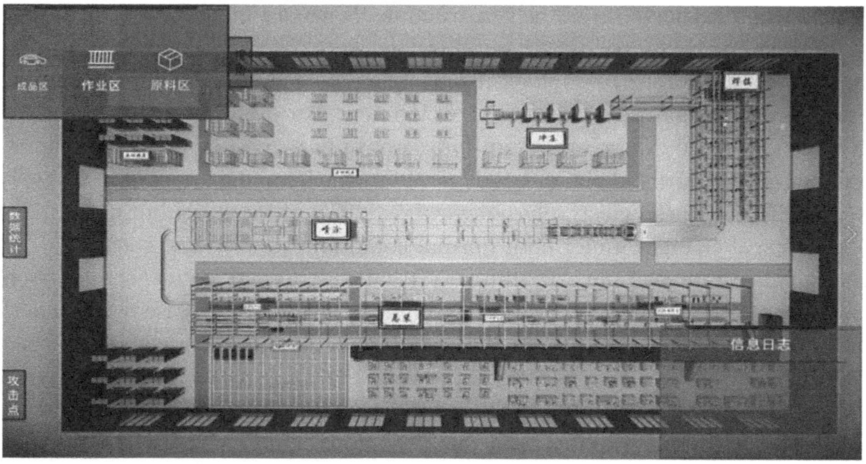

Figure 2.12 Simulation of automobile production line.

Figure 2.13 Gregory Rudnick of the University of Kansas led an international collaborative team that utilized advanced computer simulations and a vast amount of observational data to study the formation and evolutionary mechanisms of galaxies in the universe. This work uses images of approximately 14,000 galaxies from the DESI Heritage Survey, the Wide-Field Infrared Survey Explorer (WISE), and NASA's Galaxy Evolution Explorer (GALEX). *Source:* [8]/ National Science Foundation / Public domain.

2.2.3.3 Virtual Patient

Early virtual patient research emerged in the 1970s, with systematic development beginning in 1989 in the United States, and is capable of simulating the physiological responses of real patients under different health conditions, the progression of diseases, and their responses to various treatment methods [9].

From a technical perspective, the construction of virtual patient models integrates multidisciplinary knowledge, including precise digital modeling of human anatomical structures, physiological principles of organ system functions, and pathological mechanisms of disease development. For example, a virtual patient model of the cardiovascular system must not only accurately present the morphology of the heart and blood vessels, simulate the hemodynamics of blood flow within them, and the electrophysiological activities of the heart but also reflect physiological changes under both normal conditions and disease states such as coronary heart disease and arrhythmia. Virtual patients can be categorized based on the application needs and the level of modeling detail. By disease type, models may focus on a single disease (such as a virtual patient for diabetes) or multiple coexisting diseases; by application scenario, there are basic teaching virtual patients for medical education, which emphasize the demonstration of normal physiological processes and typical manifestations of common diseases to help medical students build foundational understanding, as well as complex virtual patients for clinical research, which simulate personalized disease characteristics and individual patient differences to support advanced research needs such as drug development and treatment strategy evaluation.

2.3 Experimental Theory

2.3.1 What's Experiment?

An experiment is "a process of introducing artificial interventions to verify a certain theory or hypothesis." Therefore, the direct goal of an experiment is to test hypotheses. The experimental analysis process includes the following steps: first, form an experimental hypothesis based on the problem, which should have a logical basis and be quantifiable; second, design the experiment to clarify the experimental subjects, objectives, and intervention methods; third, implement the experimental plan and conduct continuous observation; finally, conduct in-depth analysis of the experimental results and make final decisions accordingly. Since the influencing factors and research subjects of each experiment are different, the experimental design is the core of the entire process. In the design stage, it is necessary to clarify key elements such as experimental subjects, research goals, and intervention methods. For example, whether the experiment focuses on variance or treatment effects, whether the experimental population is homogeneous or heterogeneous, and whether the experimental factors are single or multifactorial.

Different combinations of the abovementioned elements constitute various experimental types, such as nonrandomized and randomized experiments. Nonrandomized experiments refer to those lacking experimental and control groups or with nonrandom group assignments, usually used for preliminary testing of

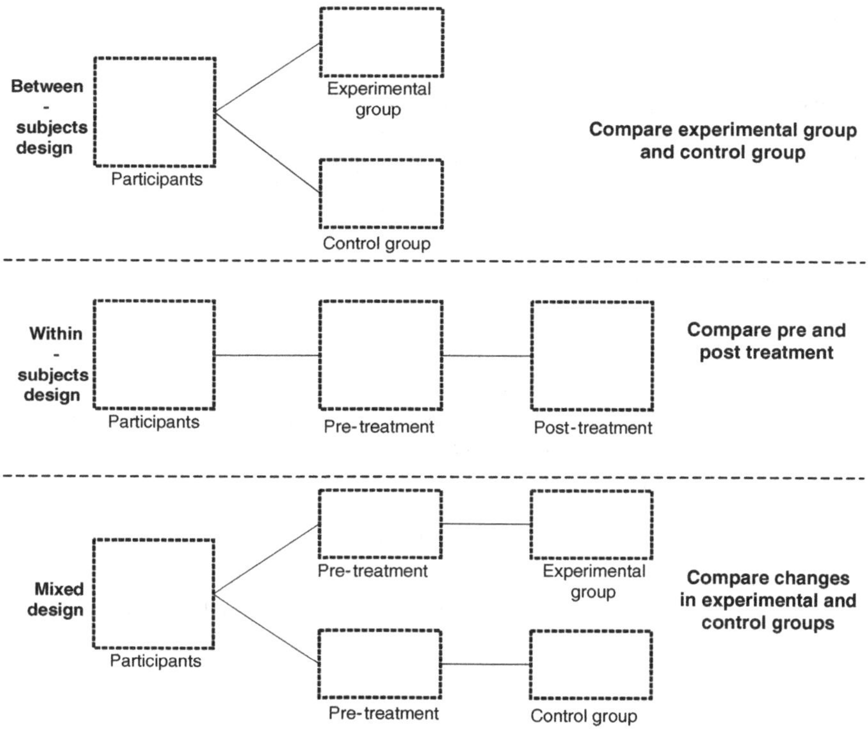

Figure 2.14 Common types of experimental designs.

variable reliability, intervariable relationships, and potential external interference factors. Randomized experiments refer to experiments in which subjects are assigned randomly, meaning every subject has an equal chance to enter a certain group. The goal is to examine interaction effects between variables. If the purpose is to test the effect of a single independent variable, a single-factor design should be used; if it involves two or more independent variables, a multifactorial design should be used to evaluate both the independent effects of each factor and their interaction effects on the dependent variable.

The commonly used symbols in experimental design include R for random sampling or random assignment, X for treatment or manipulation of the independent variable, O for observation or measurement of the dependent variable (with the number of observations indicated), and M for matched-group design. Figure 2.14 presents the common types of experimental designs, and Table 2.3 provides the corresponding detailed explanations.

After completing the experiment, analyzing the experimental data can reveal the relationships between variables, which mainly fall into three types. (1) Correlation refers to the tendency of two variables, X and Y, to change together, which can

Table 2.3 Common Experimental Designs and Their Descriptions.

Experimental Design	Symbolic Representation	Description
Posttest only Control Group Design	Random $X \rightarrow O_1$ $X \rightarrow O_2$	Random assignment with a control group allows group comparisons. Randomly assigning participants ensures comparability between groups.
Pretest–Posttest Design	Nonrandom $O_1 \rightarrow X \rightarrow O_2$	Examines the effect of variable X on the same group (e.g., purchase intention → ad → purchase intention). Two main flaws are (1) pretest may sensitize posttest and (2) uncontrolled external factors between O_1 and O_2.
Pretest–Posttest Randomized Control Group Design	Random $R: O_1 \rightarrow X \rightarrow O_2$ $R: O_3 \rightarrow \quad \rightarrow O_4$	Enables estimation of X's effect by removing pretest and external effects: $O_2 - O_1 = X +$ external factors + pretest + interaction; $O_4 - O_3 =$ external + pretest + interaction; Thus, $(O_2 - O_1) - (O_4 - O_3) = X +$ interaction. Interaction refers to the combined effect of two or more factors on the variable.

be either positive or negative. (2) Spurious correlation indicates that the observed relationship between X and Y may be caused by the influence of a third variable Z. In this case, the covariation between X and Y may be genuine or false, with the latter being spurious. (3) Causality means that a cause X "affects" or "leads to" an effect Y. Three necessary conditions must be met to establish causality, namely, logical covariation, temporal precedence, and the uniqueness of the cause.

The quality of an experiment is primarily evaluated through three indicators. (1) Heterogeneity of treatment effects: Although experiments typically measure the average treatment effect, the effect may vary significantly across individuals. Focusing solely on the average may obscure important information about how the treatment works, how to improve it, and how to target those who would benefit most. (2) Statistical conclusion validity: This concerns whether the statistical analysis of the experiment is conducted correctly. A key aspect is whether the p-value is accurately calculated—that is, the probability of obtaining the observed sample result, or a more extreme one, assuming the null hypothesis is true. Internal validity refers to whether the experimental steps are properly executed. External validity concerns whether the results can be generalized to other contexts. For most carefully designed and well-executed experiments, concerns about external validity remain the most challenging to address. (3) Mechanism analysis: While

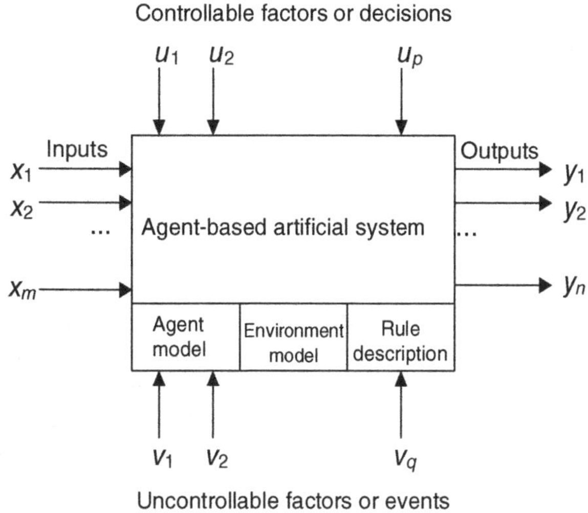

Figure 2.15 Schematic diagram of computational experiment design.

experiments reveal what happens, mechanism analysis explains why and how it happens. This type of analysis is sometimes referred to as identifying mediating variables. In many social contexts, treatments are likely to exert influence through multiple, interrelated pathways. Therefore, uncovering the underlying mechanisms can be very challenging. Common approaches include collecting process data and using factorial experiments.

2.3.2 What Is Experimental Design?

Experiments are generally used to study the performance of various processes and systems. As shown in Figure 2.15, a process can be represented as a combination of multiple operations, machines, methods, personnel, and other resources used to transform multiple inputs (typically a combination) into one or more observable response variables as output. Among them, $x_1, x_2 \ldots x_m$ are the system inputs, $y_1, y_2 \ldots y_n$ are the system outputs, $u_1, u_2 \ldots u_p$ are controllable factors or decisions, and $v_1, v_2 \ldots v_q$ are uncontrollable factors or events.

The main objectives of experimental design include (1) identifying the variables that most significantly affect the system output y, (2) determining the optimal settings for controllable variables u so that the output y approaches the ideal target, (3) minimizing the variability of output y, thereby improving the stability of the system, and (4) reducing the impact of uncontrollable factors or events v_i on the system, thus enhancing the robustness of the system.

Experiments typically involve multiple factors. In order to determine how these factors influence the system's output response, a widely used experimental strategy is the one-factor-at-a-time (OFAT) method. This approach selects a starting point or baseline level for each factor and then varies the values of each factor individually while holding all others constant. In the end, this method yields the effect of each factor on the response variable. However, a major limitation of the OFAT strategy is that it cannot account for potential interactions between factors, which are often common. Furthermore, in terms of experimental design efficiency, this strategy is inferior to alternative methods.

Factorial design is a strategy that properly handles multiple factors, where each factor has two or more levels, and it allows for the investigation of interactions among these factors. This design decomposes the variation in the experiment into contributions from different sources, including main effects (the effect of each individual factor) and interaction effects (the combined influence of factors). The fundamental assumption of factorial design is that the effects of different factors are independent, meaning each factor's influence is not confounded by others. While main and interaction effects are the primary focus in factorial designs, there may also be confounding factors that introduce additional error. When experimental data are influenced by such errors, the use of statistical methods for experimental design becomes the only objective approach. Therefore, factorial designs adhere to three basic principles: randomization, replication, and blocking [10].

1) Randomization refers to the random assignment of experimental materials and the random order of trials. Statistical methods typically assume that observations (or errors) are independently distributed random variables. Randomization supports this assumption and helps "average out" potential confounding effects, ensuring the credibility of experimental results.
2) Replication refers to the independent repetition of each factor-level combination. It allows experimenters to estimate experimental error and determine whether observed differences are statistically significant. When estimating the true value of a factor level using sample means, conclusions are only reliable if the number of replications is large enough and the error is sufficiently small.
3) Blocking is a design technique used to increase experimental precision and is often used to reduce or eliminate variability caused by nuisance factors. Nuisance factors refer to factors that may affect the experimental response but are not of direct interest to us. Generally, a block is a group of relatively similar experimental conditions. For instance, in agent-based simulations, the initial position of agents may affect results; placing data generated from different positions within the same block can significantly reduce its influence on outcomes.

Figure 2.16 illustrates a two-factor factorial experiment: the joint effect on golf scores. This experiment involves two factors (club size and golf ball type), each at two levels, and uses all possible combinations of the factor levels—resulting in a 2×2 factorial design (two factors, each with two levels). This design allows researchers to assess the individual (main) effects of each factor and whether

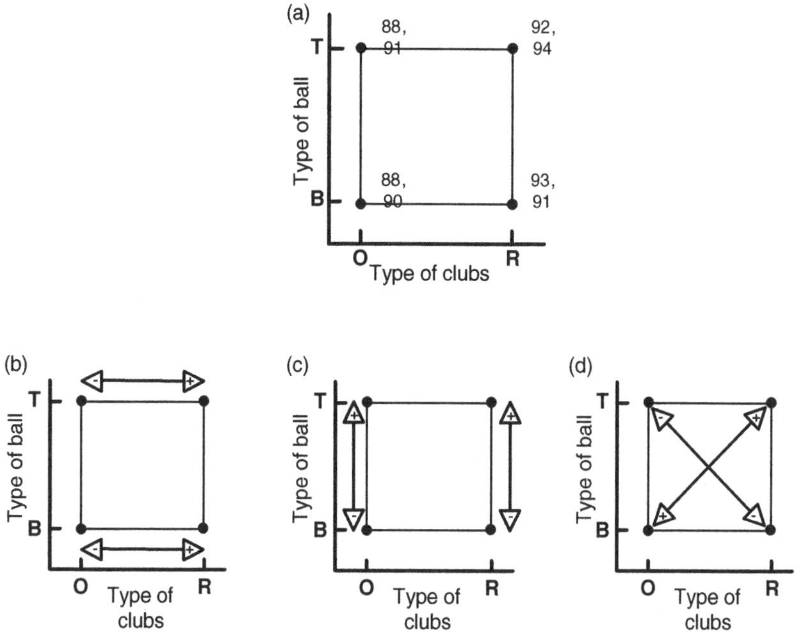

Figure 2.16 Factor effect calculation in the golf experiment. (a) Scores from the golf experiment. (b) Comparison of scores leading to the club effect. (c) Comparison of scores leading to the ball effect. (d) Comparison of scores leading to the ball–club interaction effect.

interactions exist between them. The experimenter played two rounds of golf under each condition shown in Figure 2.16(a), meaning each treatment was replicated twice.

Figure 2.16(b)–(d) present the results of the factorial experiment shown in Figure 2.16(a). The scores for each treatment combination are displayed at the corners of the squares. By comparing the average score differences on the left and right sides of the square in Figure 2.16(b), one can estimate the effect of switching from an oversized club to a regular club—i.e., an average improvement of 3.25 strokes per round.

$$\text{Club effect} = \frac{92+94+93+91}{4} - \frac{88+91+99+90}{4} = 3.25 \tag{2.9}$$

Similarly, by comparing the average difference between the four scores at the top and bottom of the square (Figure 2.16(c)), we can estimate the effect of the type of golf ball used.

$$\text{Ball effect} = \frac{88+91+92+94}{4} - \frac{88+90+93+91}{4} = 0.75 \tag{2.10}$$

Finally, the interaction effect between the type of golf ball and the type of club is measured as: the average score along the "right-to-left" diagonal of the square minus the average score along the "left-to-right" diagonal (see Figure 2.16(d)), that is:

$$\text{Ball} - \text{driver interaction effect} = \frac{92 + 94 + 88 + 90}{4} - \frac{88 + 91 + 93 + 91}{4} = 0.25 \tag{2.11}$$

The results of this factorial experiment indicate that the main effect of club type on scores is significantly greater than the effect of golf ball type, and there is also an interaction effect between the two. To determine whether these effects are statistically significant, a hypothesis test can be conducted. In fact, some statistical evidence suggests that the effect of the bat is significant, but the effect of the golf ball or their interaction is not significant. Therefore, the experimenter should consistently use oversized clubs when hitting the ball. This example clearly demonstrates a key feature of factorial experiments—namely, their ability to most efficiently utilize experimental data.

2.3.3 Mainstream Experimental Design Methods

Classical experimental design methods were proposed in the 1920s by Fisher to address experimental design issues in agriculture. Subsequently, to meet various experimental goals and application needs, researchers have successively introduced design methods such as Monte Carlo sampling, response surface method (RSM), importance sampling, and supersaturated designs. The development process is illustrated in Figure 2.17 [11]. As shown in the figure, experimental design methods evolved from solving real experimental design problems in agriculture and industry based on theoretical derivations, to addressing simulation-based experimental design problems more efficiently, and eventually to handling high-dimensional and complex simulation problems. Since their inception to the present, experimental design methods have primarily gone through four major stages:

1) **1920s**
   Fisher, through statistical analysis of data from agricultural experiment stations, summarized the three basic principles of experimental design: randomization, replication, and blocking. Consequently, Fisher systematically introduced statistical thinking and principles into experimental design research, including factorial design and analysis of variance, and established the dominant experimental design methodology of the time. For factorial design, one can categorize it into full factorial and fractional factorial designs based on whether all factor-level combinations are considered.

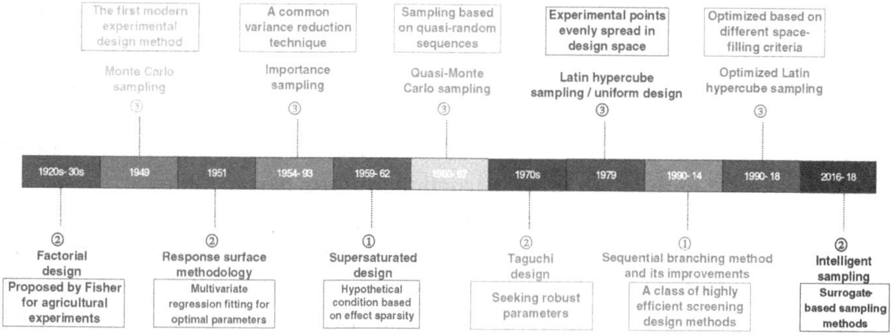

Figure 2.17 Development trajectory of experimental design.

2) **1950s**
   In 1951, Box and Wilson, while addressing industrial experimentation issues, proposed the RSM. This method aims to build regression models between factors and responses and uses the model to find the optimal factor levels [12]. Unlike Fisher's work, which focused on agricultural experiments, industrial experiments are characterized by timeliness and sequentiality, and RSM was proposed based on these characteristics. Over the following 30 years, RSM was widely applied in various fields, such as chemical engineering and processing industries.
3) **Late 1970s**
   Japanese statistician Genichi Taguchi proposed the Taguchi method, aiming to address quality fluctuations in product manufacturing processes and minimize variation around the target value [13]. Typically, this method designs experiments using orthogonal arrays and then analyzes the results with analysis of variance. Although the method performs well under certain quality control scenarios, studies have shown that in complex systems or high-dimensional parameter spaces, its performance is often suboptimal, or even ineffective in some situations. Therefore, subsequent researchers sought new solutions for robust parameter design.
4) **Post-1980s**
   With the advancement of system engineering and computational technologies, the demand for computer-based simulation experiments surged, requiring more efficient experimental design methods. In 1979, McKay et al. proposed the Latin Hypercube Sampling (LHS) method [14], which was soon widely adopted. Simultaneously, Fang Kaitai and Wang Yuan introduced the Uniform Design (UD) method [15], marking the extension of experimental design from physical experiments to computational experiments. Compared to orthogonal design, which emphasizes orthogonality and comparability, both LHS and UD

focus on even dispersion, allowing for a significant reduction in the number of required experimental runs. These methods laid the foundation for systematic research and development in the field of sampling-based experiment design. While LHS emphasizes axial uniformity to ensure better coverage in sample space, UD focuses more on theoretical uniformity, with a well-defined mathematical foundation for constructing uniform design tables.

In experimental design, causal relationships are inferred by observing the influence of factors on outcomes. If, under experimental conditions, the change in a certain factor consistently leads to changes in the results across different conditions, one may temporarily infer that this factor is possibly one of the causes of the outcome. However, to more accurately confirm the causal relationship, further research is needed to rule out other possible explanations and confounding factors. Table 2.4 presents the factorial design methods commonly used in statistics.

2.4 Statistical Methods

2.4.1 Descriptive Statistics and Probability Distribution

As an important academic discipline, statistics is widely applied across various fields. Descriptive statistics is the first step in statistics, involving the summarization, organization, and visualization of data to better understand its characteristics. It mainly provides fundamental data concepts, including central tendency, variability, and distribution (see **Table A.2.1**). In business, descriptive statistics can be used to analyze sales data, market research, and customer satisfaction surveys to support decision-making. In the medical field, it can help doctors understand patients' physiological indicators such as weight, blood pressure, and body temperature. In education, it can be used to assess students' exam performance and learning progress.

Probability is used to describe the likelihood or uncertainty of an event occurring. It can quantify the extent to which an event may happen or predict the relative frequency of an event among a series of possible outcomes. Probability is used to depict the likelihood of an event occurring under uncertain conditions and is usually represented by a number between 0 and 1. Probability distribution describes the probability that a random variable will take on different values. For discrete random variables, they can take on a series of distinct and countable values, such as the outcome of a die roll (1, 2, 3, 4, 5, 6). Discrete probability distributions are described by a probability mass function (PMF), which gives the probability of each possible value. For continuous random variables, they can take on any value within a continuous range, such as human height, temperature, or stock prices.

Table 2.4 Common Experimental Design Methods in Statistics.

Name	Explanation	Example
Latin Square Design	In the simplest case, Latin square design is used to control two confounding factors, often involving spatial location and time effects. It is an $n \times n$ Latin square where each treatment appears exactly once in each row and each column. This method helps researchers control potential confounding variables and more accurately estimate the impact of the studied factors on results.	Suppose we want to test the effects of three different fertilizers (A, B, C) on crop growth, using Latin square design to account for variations in soil and weather. We choose three different test sites and plant three types of crops, ensuring that each fertilizer is tested once at each site. The Latin square is constructed as follows: | 1 | 2 | 3 | --+---+---+---+ A | C | B | A | B | A | C | B | C | B | A | C |
Crossover Design	Typically used to compare the effects of different treatments on the same subject, aiming to eliminate individual differences. In a crossover design, each subject receives different treatments in a random order to reduce the impact of confounding variables and increase experiment reliability. This method is common in medicine and biology, especially drug trials, where it's important to evaluate the effectiveness of drugs or therapies. However, crossover designs may not be suitable when treatments have lasting or cumulative effects.	Suppose we are studying the effects of two different drugs on headaches. A crossover design is used to compare the two drugs on the same group of participants. The scatter plot helps us better observe changes in headache scores for each subject under different conditions. 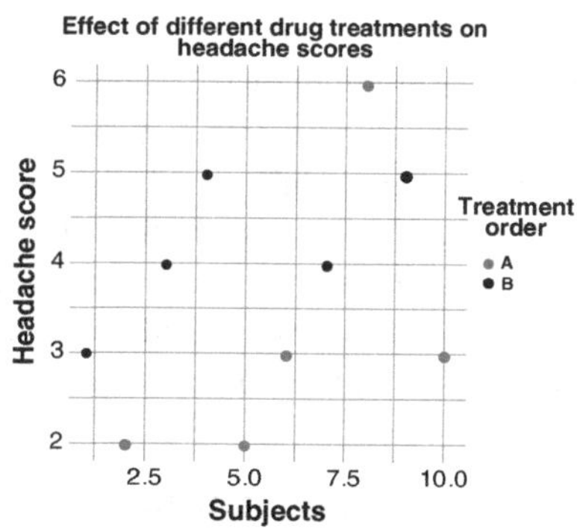

(Continued)

Table 2.4 (Continued)

Name	Explanation	Example
Randomized Complete Block Design (RCBD)	Experimental units are divided into k blocks. Within each block, treatments are randomly assigned to control for variation under different conditions and to reduce experimental error. This method assumes there may be confounding variables that affect outcomes and uses variance analysis to determine whether differences in treatment levels are significant. It's suitable when the impact of confounding variables varies across blocks and helps to assess whether treatments are effective.	Suppose the experimental field is divided into k blocks (e.g., different plots of land), and each block is randomly assigned three types of fertilizer. The plant height in each block is then recorded. RCBD is used to analyze whether different fertilizers significantly affect plant height. 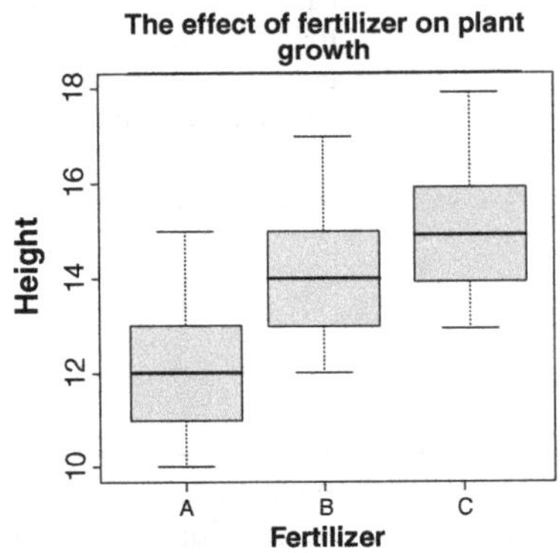

Continuous probability distributions are described by a probability density function (PDF), where the PDF indicates the probability density over a specific range of values. **Table A.2.2** gives the definition of the related core concepts.

The PMF describes the probability that a random variable takes on specific values under different distributions. Depending on the specific context and problem, selecting an appropriate distribution and using the corresponding PMF can help us conduct probability analysis and statistical inference. The PDF describes the probability density distribution of a random variable under different distributions. Normal distribution, exponential distribution, and uniform distribution are widely applied in various fields and can be used to model the probability distributions of various phenomena and random events. **Table A.2.3** gives the typical PDFs.

2.4.2 Inferential Statistics and Sampling Methods

Inferential statistics represent a further development in the field of statistics, involving drawing general characteristics or conclusions from sample data. It is used to analyze and interpret data, making inferences and predictions about unknown phenomena or populations. Inferential statistics have broad applications in practice, and their core value lies in its wide applicability and strong explanatory power. First, inferential statistics can be used to describe and explain data. For example, researchers often use correlation analysis and regression modeling to explore the relationships and patterns among variables. Second, inferential statistics can also be used to predict and control future events. For instance, in the financial domain, we can use historical stock data to analyze and forecast future stock market trends. Furthermore, inferential statistics can be applied in decision-making and problem-solving. For example, in industrial production, we can analyze data from the production process to identify factors influencing product quality, thereby improving processes and enhancing product quality. As shown in **Table A.2.4**, the main tools of inferential statistics include sampling, confidence intervals, and hypothesis testing.

In statistics, sample data is often used to infer the characteristics of a population because it is usually difficult or impractical to obtain data from the entire population directly. By calculating statistics from the sample, we can estimate parameters of the population and make inferences about its characteristics. In order to ensure that the sample adequately represents the population, it is necessary to use scientifically sound and reasonable sampling methods. Sampling determines the quality and representativeness of the data obtained from the population. Randomness plays a key role in sampling. By using random sampling, we ensure that each member of the population has an equal chance of being selected, thereby reducing sampling bias and improving the representativeness of the sample. Randomness also plays an important role in the sampling distributions and confidence intervals in statistical inference. Statisticians study and develop various sampling methods, including simple random sampling, stratified sampling, and cluster sampling, to meet the needs of different research questions and resource constraints. The core concepts related to sampling are illustrated in **Table A.2.5**.

In practical scenarios, if historical data over a certain period is already available, it can serve as the basis for generating datasets for experiments. However, such data is often limited, and overfitting is likely to occur. As shown in Figure 2.18, to address this situation, data generation techniques are widely applied, mainly including resampling and Monte Carlo methods [16]. Resampling involves repeatedly drawing samples from an existing dataset using random sampling and then analyzing these samples to obtain information about the population. Resampling includes two main methods: sampling with replacement and sampling without replacement. The key difference lies in whether repeated selection of the same elements is allowed during the sampling process. Sampling with replacement means that the same element may appear multiple times in different sample sets. Typical methods include bootstrapping, Monte Carlo simulation, etc., which can generate multiple bootstrap samples with similar properties and are used to evaluate the

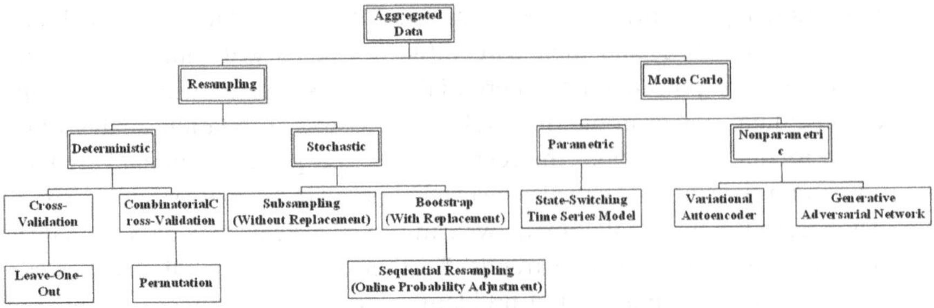

Figure 2.18 Methods for generating datasets in computational experiments.

uncertainty and distribution of statistics. Sampling without replacement ensures that each element appears only once in a sample set. Examples include simple random sampling, systematic sampling, and stratified sampling, which are commonly used in surveys and experimental research.

For variables that follow a probability distribution or PDF, the Monte Carlo method can be used to construct a series of sample sets that satisfy various probabilistic conditions. Monte Carlo experiments can be either parametric or nonparametric. Parametric methods can replicate the statistical properties of observed datasets into unobserved ones [17]. Markov Chain Monte Carlo (MCMC) performs approximate sampling from a probability distribution by conducting a random walk in the parameter space based on the properties of Markov chains. MCMC methods are commonly used for posterior distribution sampling in Bayesian statistics, where the posterior distribution of parameters is usually derived from the prior distribution and observed data. However, a drawback of parametric Monte Carlo is that the data generation process may be more complex than the replication of a finite set of algebraic functions. In such cases, nonparametric Monte Carlo experiments may be helpful, such as variational autoencoders, self-organizing maps, or generative adversarial networks [18]. The common resampling methods used in statistics are illustrated in **Table A.2.6.**

2.4.3 Parameter Estimation and Hypothesis Testing

In statistics, parameter estimation and hypothesis testing are two core concepts that play a vital role in fields such as research, business, and medicine. Parameter estimation is the foundation of statistical inference, aiming to use sample information to infer unknown parameter values of a population. Point estimation is the most basic method of parameter estimation, which uses a single value to estimate a population parameter. However, point estimation does not indicate the uncertainty of the estimate. To gain a more comprehensive understanding of parameter estimation, interval estimation needs to be introduced. Interval estimation provides a range of values—called a confidence interval—within which the population parameter is likely to fall. Confidence intervals are typically expressed with a confidence level (such as 95%) and are calculated based on sample statistics and their distributional properties.

There are various methods for parameter estimation, and the choice among them typically depends on the nature of the problem, the type of data, and the researcher's preference. **Table A.2.8** summarizes some commonly used parameter estimation methods.

Hypothesis testing is a statistical method used to test whether a population parameter conforms to a specific hypothesis. It typically involves two hypotheses: the null hypothesis (H0) and the alternative hypothesis (H1). These two hypotheses are mutually exclusive. The goal of hypothesis testing is to determine, based on the data, whether there is sufficient evidence to reject the null hypothesis. Hypothesis testing generally includes the following steps. (1) Formulate hypotheses: clearly define the null and alternative hypotheses; (2) Collect data: gather appropriate sample data; (3) Calculate test statistic: compute a statistic based on the data to measure the difference between the observed effect and the expected effect under the null hypothesis; (4) Determine significance level (α): choose an appropriate significance level, typically 0.05; (5) Perform the hypothesis test: use the significance level and the test statistic's distribution to determine whether to reject the null hypothesis; and (6) Make a decision: based on the result of the hypothesis test, decide whether to accept or reject the null hypothesis. **Table A.2.9** gives some key concepts in statistical hypothesis testing.

These concepts collectively form the foundation of hypothesis testing. Researchers use them to evaluate data, make decisions, and understand whether observed effects are practically meaningful. Hypothesis testing helps determine whether there is sufficient evidence to support a hypothesis and provides a statistical framework for inference in scientific research and decision-making. In hypothesis testing, we typically aim to control the probability of Type I error (α), while striving to reduce the probability of Type II error (β). It is important to note that statistical significance does not imply certainty. Even if a result is statistically significant, there is still some uncertainty. It provides a probabilistic explanation of the observed data under the null hypothesis. Properly understanding statistical significance requires treating it as a tool to assist researchers in evaluating whether the observed data is unlikely under the null hypothesis. However, to make a comprehensive decision, researchers must also consider effect size, real-world context, and control of Type I error. Statistical significance is useful in scientific research, but it should not be used as the sole criterion; it must be considered together with other information. **Table A.2.10** gives some common hypothesis testing methods in statistics.

2.5 Chapter Summary

This chapter aims to equip readers with foundational knowledge of computer simulation technology, including the principle of similarity and system modeling, as well as the concepts, procedures, and examples of simulation. It further transitions from physical system simulation to social system simulation, introducing the fundamental models used in social simulation, including agent models, environment models, and rule-based models. Finally, several classic cases of social simulation are presented, covering three types: thought experiments, mechanism exploration, and parallel optimization.

References

1 Law, A.M., Kelton, W.D., and Kelton, W.D. (2007). *Simulation Modeling and Analysis*. New York: Mcgraw-hill.

2 Xue, X., Chen, F., Zhou, D. et al. (2022). Computational Experiments for Complex Social Systems—Part l: the customization of computational model. *IEEE Transactions on Computational Social Systems* 9 (5): 1330–1344.

3 Meerschaert, M. (2013). *Mathematical Modeling*. Academic press.

4 Morgenthaler, G.W. (1961). The theory and application of simulation in operations research. In: *Progress in Operations Research* (ed. R.L. Ackoff), 1–363. New York: Wiley.

5 Korn, G.A. and Wait, J.V. (1978). *Digital Continuous-system Simulation*. Prentice Hall.

6 Spriet, J.A. and Vansteenkiste, G.C. (1982). *Computer-Aided Modeling and Simulation*. Cambridge, MA: Academic Press Inc.

7 Ören, T.I. (1984). GEST—a modelling and simulation language based on system theoretic concepts. In: *Simulation and Model-based Methodologies: An Integrative View* (ed. T. Ören I, B.P. Zeigler, and M.S. Elzas), 281–335. Berlin, Heidelberg: Springer Berlin Heidelberg.

8 Rudnick, G.H. (2012). Characterizing the molecular gas contents of high redshift galaxies J. NSF Award Number 1211358. *Directorate for Mathematical and Physical Sciences* 12 (1211358): 11358.

9 Cook, D.A. and Triola, M.M. (2009). Virtual patients: a critical literature review and proposed next steps. *Medical Education* 43 (4): 303311.

10 Montgomery, D.C. (2017). *Design and Analysis of Experiments*. John Wiley & Sons.

11 Lingyun, L. (2018). *Research on Efficient Experimental Design Methods for Complex Simulation Systems*. Harbin Institute of Technology.

12 Draper, N.R. (1992). Introduction to Box and Wilson (1951) on the experimental attainment of optimum conditions. In: *Breakthroughs in Statistics: Methodology and Distribution* (ed. S. Kotz and N.L. Johnson), 267–269. New York, NY: Springer New York.

13 Tsui, K.L. (1992). An overview of Taguchi method and newly developed statistical methods for robust design. *IIE Transactions* 24 (5): 44–57.

14 McKay, M.D., Beckman, R.J., and Conover, W.J. (1979). A comparison of three methods for selecting values of input variables in the analysis of output from a computer code. *Technometrics* 21 (2): 239–245.

15 Wang, Y. and Fang, K.T. (1981). A note on uniform distribution and experimental design. *Ke Xue Tong Bao* 26 (6): 485–489.

16 De Prado, M.M.L. *Machine Learning for Asset Managers*. Cambridge: Cambridge University Press.

17 Epstein, J.M. (2009). Modelling to contain pandemics. *Nature* 460 (7256): 687–687.

18 Goodfellow, I.J., Pouget-Abadie, J., Mirza, M. et al. (2014). Generative adversarial networks. arXiv preprint arXiv: 1406.2661.

3

Computational Experiments for Social Systems

This chapter introduces the social simulation foundation of computational experiments, offering insights into modeling complex social systems to support interdisciplinary research. It begins by discussing the development history and design principles of agent-based modeling (ABM), helping readers understand the significant challenges that traditional physical simulation methods face in the social domain. Then, it provides a detailed introduction to foundational models for simulating social systems, including individual models, environmental models, and rule-based models. Furthermore, causal inference and algorithmic counterfactuals are introduced to help readers distinguish between ABM and computational experiments. Finally, it presents classic cases of social simulation, covering three types: thought experiments, mechanism exploration, and parallel optimization.

3.1 Social Simulation Method

3.1.1 Development History of ABM

In the past decade, Internet technology has effectively digitized the social, economic, political, and cultural activities of billions of people, generating a massive amount of data. Using these data, researchers model social systems as interaction networks and apply complex network analysis to uncover system characteristics, yielding landmark results. In 1998, D.J. Watts, a doctoral student at Cornell University, and his mentor published a paper titled "Collective dynamics of 'small-world' networks" in *Nature* [1], and in 1999, A.L. Barabási, a professor at the University of Notre Dame, and his doctoral student R. Albert published a paper titled "Emergence of Scaling in random networks" in *Science* [2]. The two papers revealed the small-world characteristics and scale-free properties of complex networks, respectively, marking the beginning of a new era in complex network research.

However, such data are rarely collected for specific research questions, unlike traditional survey or lab data, which are designed systematically. Consequently, simple data analysis faces methodological dilemmas. (1) Data representativeness: Big data is also a form of sampling, and its representativeness is difficult to evaluate. People who use the Internet and various social platforms are not randomly distributed in the total population. For example, the elderly use the Internet less. When representativeness cannot be guaranteed, the huge sample size of big data cannot increase the reliability of research conclusions. (2) Data authenticity: The individual or group behavior patterns obtained from big data may not be consistent with those in real life. For example, some people are very active on the Internet, but they are taciturn and unsociable in real life. In addition, big data only records some aspects of people's lives, not all aspects. (3) Data reliability: In many cases, data are records left when we use a certain product (such as search engines, Toutiao). It should be noted that when using these products, our behavior patterns may be guided by recommendation algorithms and undergo some changes, which will cause the corresponding data to be "distorted."

Field experiments could address these issues but face significant practical challenges. The reasons can be attributed to the following aspects: (1) economically, due to the scale and cost factors of social systems, the cost of experiments is too high to be affordable; (2) legally, issues such as national defense, military preparedness, and social security are protected by legislation, so it is impossible to experiment on the systems under research or rebuild these systems for experiments; and (3) morally, social systems often involve a large number of people. Experimenting on these systems may have an impact on people's normal lives and even endanger people's lives and property, so such experiments are morally unacceptable. In this context, social modeling and simulation methods have gradually developed into another mainstream method in computational social science research. As shown in Figure 3.1, social simulation uses computers as "artificial laboratories" to "cultivate" macro phenomena that may appear in actual systems, explore the laws behind them, and provide a feasible way to analyze the behavior of complex systems and evaluate the effects of interventions. There are two types of social modeling and simulation methods:

Top-down modeling method: The typical representative is the system dynamics method, which uses mathematical language to express the characteristics, states, relationship, and process of a specific object or problem; forms

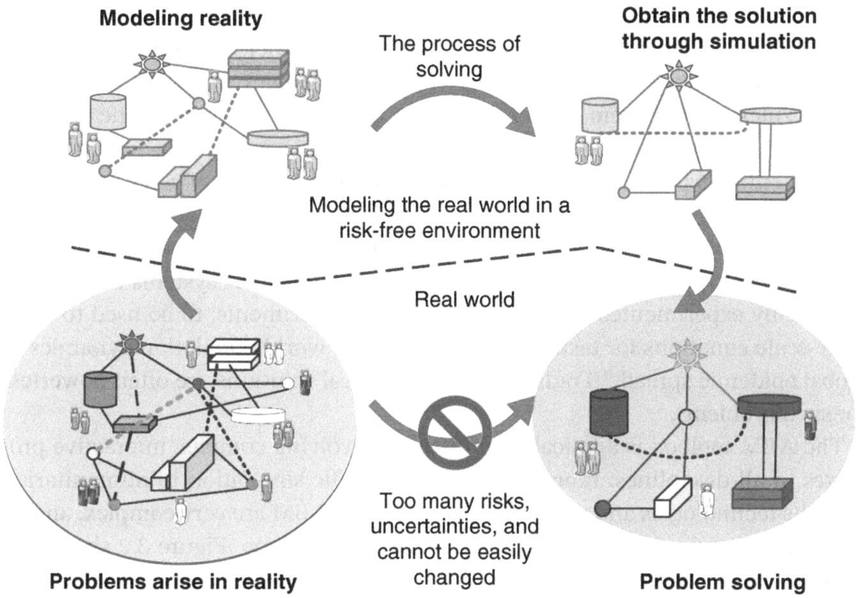

Figure 3.1 The core idea of social simulation.

explanations; and makes judgments and predictions through deduction, calculation, and analysis. This type of method is highly abstract and logically rigorous, but when faced with nonlinear systems, it may be the case that the system can be represented, but not solved. At the same time, the model results are very sensitive to the model assumptions, resulting in the risk of "a slight error leading to a huge mistake."

Bottom-up modeling method: The typical representative is ABM, which abstracts and simplifies the behavior of the subjects in the system into agents and produces interesting macro results through simple interactions between them. It is considered to be the third scientific method, besides classical deductive reasoning and inductive reasoning [3]. Compared with the top-down modeling method, the advantage of ABM is mainly reflected in the inclusion of the framework of emergence. The basic elements of the model need to be flexible enough, including the heterogeneity and adaptability of the subjects, scalability of quantity, etc., so that new and unexpected features will naturally appear in the model.

In the 1990s, the Santa Fe Institute (SFI) first began to use computer simulation to address economic system problems, pioneering social simulation research. For the first time, Epstein and Axtell's "Growing Artificial Society" systematically used ABM as a modeling tool for social sciences, to build an artificial social laboratory through the KISS principle (Keep It Simple and Stupid) [4]. ABM is related to the concept of multiagent systems but differs from it. Its goal is to provide explanatory insights into the collective behavior of agents that follow simple rules in the system, rather than to design agents or solve specific practical or engineering problems. ABM is mainly used to achieve the following purposes: (1) existential experiments: to observe whether a certain set of rules can lead to the emergence of a certain complex behavior; (2) understanding experiments: to try to simulate the system from the bottom up to understand the operation process of complex systems, especially those social complex systems that cannot be actually experimented on; and (3) deductive experiments: to be used to solve large-scale equations for tasks such as "predicting world population dynamics or global epidemic spread." Traditional mathematical solutions are often powerless for such problems.

The ABM toolbox is applicable to research involving complex interactive processes in all disciplines, ranging from urban traffic simulation to humanitarian aid. The technologies and applications involved in ABM are very complex, and its connotation and scope are constantly changing over time. Figure 3.2 shows the important topics in ABM research, which can be divided into three periods based on the timeline of events: the first stage (1970–1995) is the formation stage of the basic concept of social simulation, including "artificial life" and "artificial society"; the second stage (1995–2004) sees the emergence of artificial society models in various fields (such as sociology, business and marketing, land use, and finance), and the problems existing in ABM itself have also begun to attract attention; and the key event of the third stage (2004 to present) is the emergence of the ACP method, which marks the beginning of the shift of ABM from "toy models for verifying hypotheses" to "how to empower reality." With the emergence of large language model technology, the field of agent-based simulation is entering a stage of rapid development. Table 3.1 shows the landmark events in this process.

3.1.2 Design Principles of ABM

All social simulations are designed to solve one or more research problems defined based on a reference system. Research problems are the main engine of scientific exploration and largely determine the degree of consistency between social simulations and the real world. Generally speaking, the representation of social simulations is relatively simple, while the operation of actual systems is relatively complex. The contradiction between these two is mainly manifested in two aspects: (1) if the level of abstraction of the simulation system is too high, it will be difficult to reflect the operational laws of the real world and (2) if the level of abstraction of the simulation system is too low, the complexity of model building will be very high, and a series of problems such as lack of data, insufficient

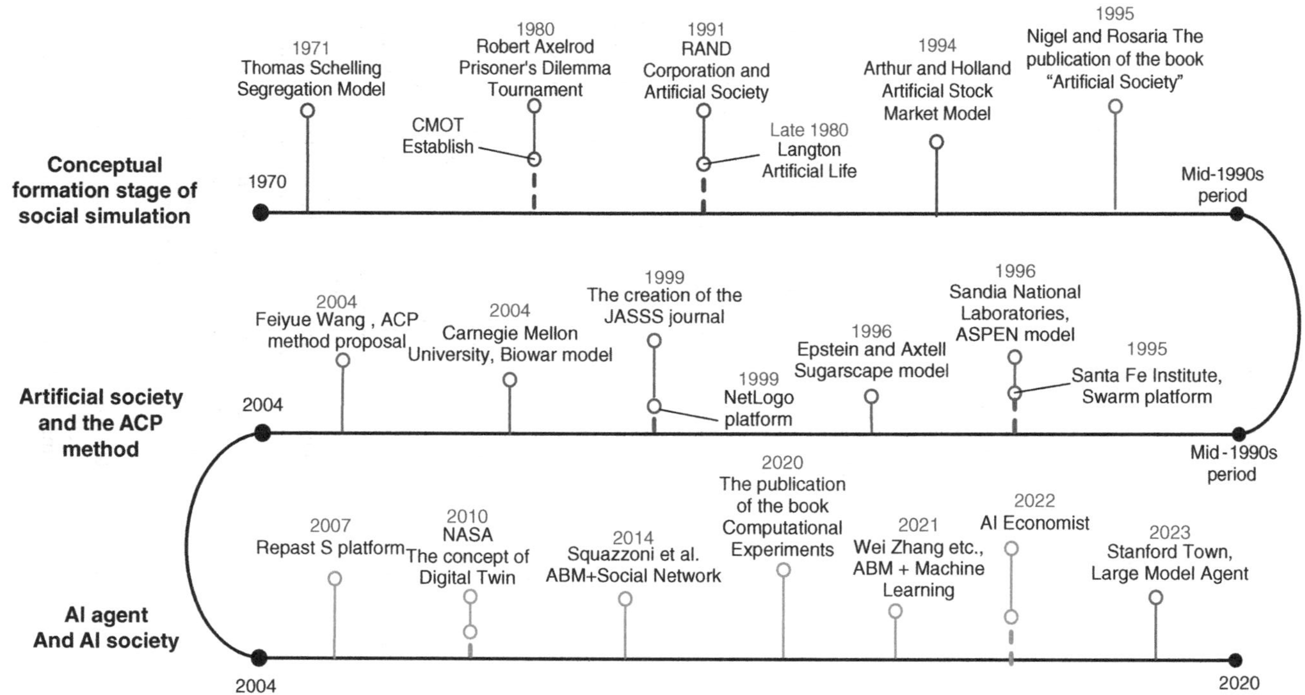

Figure 3.2 The development of ABM.

Table 3.1 ABM Development History.

Years	Figure	Representative Work
1970	Conway	Wrote the "Game of Life" program, opening the prelude to artificial life research [5].
1971	Thomas Schelling	The classic Segregation model [6] was proposed, which embodies the basic concept of the agent model, that is, autonomous agents interacting in a shared environment can eventually observe aggregation and emergence phenomena.
Early 1980	Robert Axelrod	Focused on modeling political systems, covering phenomena such as racism and cultural communication. Hosted the Prisoner's Dilemma strategy competition [3].
Mid-1980	CMOT	CMOT (Computational and Mathematical Organization Theory) was founded, and the field grew as a special interest group of the Management Science Institute and its sister society, the Operations Research Society.
Late 1980	Christopher Langton	Proposed the theory of artificial life and, together with other researchers, used computers to construct model systems with behavioral characteristics of natural life systems [7], such as self-reproducing cellular automata, Boids model, ant colony models, Tierra models, and "Amoeba World."
1991	RAND Corporation, USA	For the first time, the concept of "artificial society" was formally proposed, which is to use agent technology to build a social laboratory (subjects, environments, and rules) in a computer to conduct experimental evaluations on different policies to ensure the effectiveness of the policies [8].
1994	Arthur and Holland	Based on the theory of complex adaptive systems, a research framework for artificial stock market was proposed [9], emphasizing that financial markets are complex dynamic systems formed by the interaction of heterogeneous entities with learning capabilities.
1995	Nigel Gilbert and Rosaria Conte	The book *Artificial Societies: The Computer Simulation of Social Life* was published [10], and the concept of artificial society was formally proposed and became a relatively independent field of social science.
Mid-1990s	Santa Fe, USA	Released the Swarm platform, which was initially developed by Chris Langton, Roger Burkhart, Nelson Minar, and others, and subsequently many people around the world participated in its open source follow-up development.

Table 3.1 (Continued)

Years	Figure	Representative Work
1996	Sandia National Laboratories, USA	The ASPEN model [11] is an agent-based microeconomic simulation model designed to simulate and study the US economy in detail and is widely used in the field of economic analysis.
1996	Epstein and Axtell, University of Chicago	Developed "Sugarscape" to simulate and explore social phenomena such as seasonal migration, sexual reproduction, genetic inheritance, and market formation [12].
Late 1990s	CASOS	CMOT was renamed as Computational Analysis of Social and Organizational Systems (CASOS). The North American Association for Computational Social and Organizational Sciences (NAACSOS), the European Social Simulation Association (ESSA), and the Asia-Pacific Social Simulation Association (PAAA) were also established.
1999	Nigel Gilbert	The book *Simulation for the Social Scientist* was published, and the journal *The Journal of Artificial Societies and Social Simulation* (*JASSS*) was founded, marking the gradual maturity of the field of sociological simulation.
Early 2000	Uri Wilensky, Northwestern University, USA	NetLogo was developed with the characteristics of easy operation and good visualization. It is very suitable for beginners to model and simulate multiagent systems and is widely used in education, scientific research, and other fields.
2004	Carnegie Mellon University (CMU), USA, Kathleen Carley	The Biowar model was developed to combine computational models such as social networks, transmission vectors, disease models, demographic models, spatial models, wind diffusion models, and diagnostic models into a single integrated system designed to simulate the impact of a bioterrorist attack on a city [13].
2004	Institute of automation, Chinese Academy of Sciences, Wang Fei-Yue	A set of ACP methods for studying complex systems, namely, "artificial system + computational experiment + parallel execution," was proposed, which uses computational experiments to provide "reference," "prediction," and "guidance" for possible situations of actual system operation [14, 15].
2007	Michael J. North et al.	The multiagent simulation tool Repast was developed [16]; Repast Simphony version 2.0 was released in 2012; and it is still being updated and the latest version was released in 2024.

(Continued)

Table 3.1 (Continued)

Years	Figure	Representative Work
2010	NASA	The concept of digital twins was proposed, emphasizing the establishment of a virtual-real two-way dynamic feedback mechanism, which can simulate, analyze, and optimize physical entities. This technology can greatly enhance the fidelity of environmental modeling.
2014	Squazzoni et al.	Emphasized the combination of ABM and network science [17], mapping the interactions between agents into nodes and links in the network can supplement the network structure sampling bias commonly found in social empirical methods.
2020	Xue Xiao	Published *Computational Experiment method for Complex Systems: Principles, Models and Cases* [18], the first book to systematically review computational experiment method.
2021	Wei Zhang et al.	Studied how to combine different machine learning (ML) techniques with ABM to model complex, large-scale systems from the perspective of improving model accuracy or robustness and providing better decision-making strategies [19].
2022	S Zheng et al.	Proposed a new study called "The AI Economist," which uses economic simulation methods to discover tax strategies that can effectively find a balance between economic equality and productivity [20].
2023	Stanford University, USA, Joon Sung Park, etc.	A generative agent architecture based on a large language model was proposed, and a virtual sandbox environment "Stanford Smallville" was designed [21]. Twenty-five agents with unique backgrounds run in it, and the complete experience of the agents can be stored through natural language, with features such as task decomposition, memory, and reflection.

resources, and inadequate knowledge system will be encountered. Therefore, it is very important to achieve an appropriate mapping between the real world and the simulation system. In the design of simulation models, the model design principles should be balanced with the actual situation, as shown in Figure 3.3.

Model design is a balance between reality and abstraction at a certain level, so it is not expected that the model will perform well at different levels. For example, a highly realistic model may have great policy value but little or no theoretical value; conversely, a highly abstract model may provide profound scientific insights

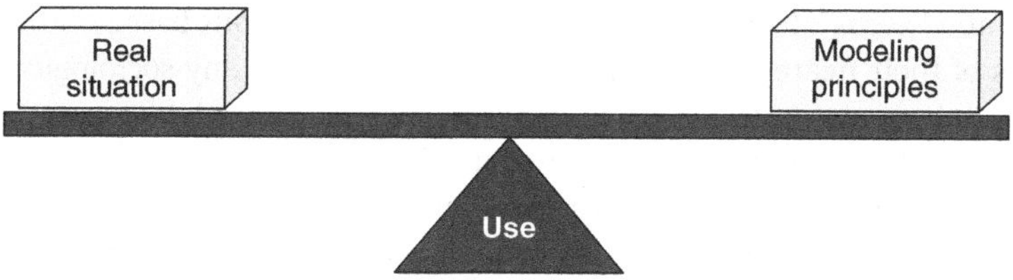

Figure 3.3 The balance between model design principles and actual conditions.

but fail to provide directly applicable results in terms of policy contribution. Two modeling principles are given, depending on the degree of closeness to a given reference system.

3.1.2.1 KISS Principle (Abstract)

The KISS principle is mainly applicable to theory building, where the model is small in scale and has few parameters, and the goal is to find the minimum counterintuitive conditions for a certain macro-emergent phenomenon [22]. The KISS-oriented ABM is regarded as a "thinking tool" or "intuitive engine," and the balance between the model design principles and the real situation is shown in Figure 3.4. This model is sometimes called a toy model, which is similar to the reference system only in a few qualitative aspects and does not attempt to reproduce any quantitative characteristics, but it can provide some applicable insights in the problem domain. Supporters of the KISS method agree with the following view: "If the goal of ABM is to enrich our understanding of the emergence process of a phenomenon…, then the simplicity of the assumptions is important, and the reproduction of the real details in a specific environment is not so important" [23].

Many early social simulations belong to this category, providing a unique way to understand the basic rules of human and social operations. The widely praised racial segregation model [6] assumes an ideal two-dimensional (2D) space in

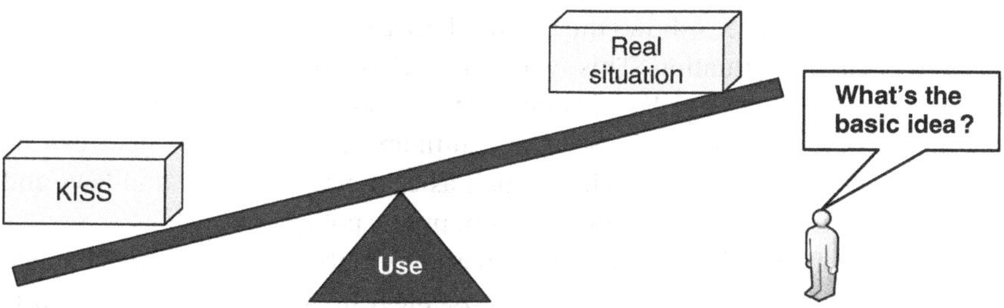

Figure 3.4 The balance between the design principles of the parsimony method model and the actual situation.

which "stars" and "zeros" decide whether to change their positions based on the characteristics of their nearest neighbors. Schelling did not use any sociological or psychological theory to prove his microscopic assumptions, nor did he use any empirical data to set the preferences of agents, neighborhood ranges, and group sizes. Instead, Schelling made (weak) structural analogies between the fictional mechanism and the segregation mechanism, in reality based on intuition and common sense. The results of the model are not specific to racial segregation data in a specific geographic area but can actually cover any phenomenon in which two groups have some tendency to separate. Here, the realism of the model is secondary, and the heuristic value of the model is what is really important.

Subsequently, the Sugarscape model designed by Epstein and Axtell can be seen as a typical case of the KISS principle. The assumptions about the agent rules in the model use some existing sociological/psychological theories, but they are weak; data are also not used to assess the extent to which the simulation results reproduce real-world regularity. For this reason, this type of ABM only has heuristic value, but it is not clear to what extent the model reflects the real world. Epstein and Axtell's book *Growing Artificial Society* has had a considerable impact on beginners to ABM [4]. Since then, many studies have been dominated by the Sugarscape style, including the evolution of cooperation [3], trust and reputation [24], and the emergence of norms [25]. For a long time, the field was dominated by abstract ABMs, far more than data-driven ABMs.

3.1.2.2 KIDS Principle (Realistic)

Although, historically, the KISS principle has had a clear impact on the field, the knowledge generated by such "unfounded" ABM, which lacks credibility, has faced a series of skepticism. Modelers, accustomed to the KISS principle, are beginning to seek ABMs that are more closely connected to the "real world." In this context, the KIDS (Keep It Descriptive and Stupid) modeling principle began to emerge. This type of model is oriented to solving application problems, as close to reality as possible. The model is large scale, has many parameters, requires testing of many modules, and its experimental output is most consistent with empirical data. Figure 3.5 shows the balance between the design principles of the model and the real situation. This type of model is widely used in commercial and government agencies. For example, public policy involves areas of high uncertainty (such as social networks and human behavior), and ABM can be used to analyze the effects of policies (such as economic stimulus and laws and regulations) to improve the scientific basis of public policymaking.

Proponents of KIDS believe that ABM has very high flexibility in mechanism design, and model simplification is only a possible result of analysis, not a starting point for analysis [26]. From the KIDS perspective, a model that is rich in scope should be first established to allow for the description of existing empirical

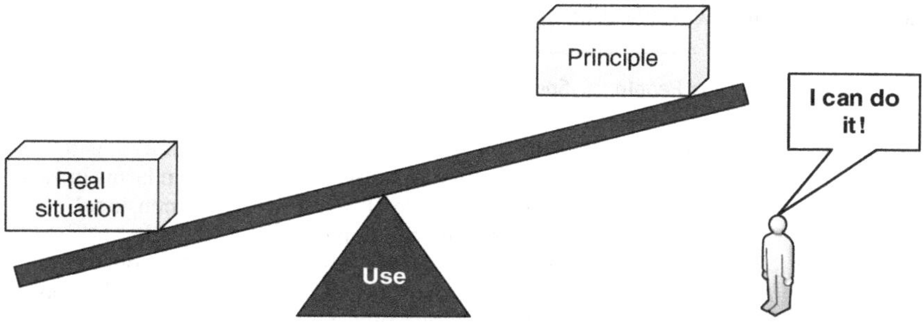

Figure 3.5 The balanced relationship between realistic method model design principles and real situations.

evidence and then simplified; the simplification process cannot be contrary to the empirical data of the phenomenon being studied and will not degrade the model's performance in interpreting the phenomenon. The goal is to design a "high fidelity" ABM [27]; that is, both the micro assumptions of the model and the macro objectives that are to be reproduced have "high-dimensional authenticity" characteristics [28].

The modeling method of KIDS has begun to deeply influence contemporary ABM research, based on real-world datasets, initializing the probability distribution of ABM core variables, formalizing the network structure between agents, and establishing agent behavior rules. Related cases include the differential fusion of innovation [29], reputation dynamics [30], and game theory of cooperative behavior [31]. In economics, ABM has also begun to combine with experimental research on economic decision-making [32] and empirical verification [33]. Therefore, while "calibration and verification of models" remains a research frontier in the field, ABM is increasingly moving from exploratory models to rigorous empirical verification (Table 3.2).

3.1.3 Causal Challenges for ABM

In general, the ABM model is very simple and (relatively) easy to fully examine and understand its internal functioning. Thus, the knowledge they produce can be explained from a generative perspective (vertical causality on the right side of Figure 3.6) in terms of causality, i.e., which dynamic chain of events relates the model's micro-hypotheses to a given set of macroscopic outcomes. In fact, "intelligibility" is thought to help enhance the "credibility" of abstract models [45], and Schelling's model of apartheid is one such model. If we look at it from the perspective of causal dependence (horizontal causality on the left side of Figure 3.6), this knowledge can also be explained causally. By manipulating ABM parameters (i.e., sensitivity analysis) and changing the model internals (i.e., robustness analysis), it

Table 3.2 Application Classification of ABM.

Case category	Year and People	Specific Case Content
Theoretical Support	1965, Torsten Hägerstrand	ABM was used to explain the agglomeration patterns that emerged in terms of innovation on farms in two regions of Sweden [34]. In the two-dimensional space, the distribution of farms in the relevant regions of Sweden is replicated, and the contact probability matrix is set up based on local telephone traffic and migration statistics.
	2003, Janssen and Jager	ABM was used to study the diffusion of products in the market [35]. Specific psychological theories of preference change are used to design the behavior rules of the agents. All the parameters and variables of ABM ultimately come from theoretical probability distributions, which have a certain degree of truth in terms of network composition.
	2007, Billari et al.	ABM was used to explain the similarity in the age distribution of first marriage across countries [36]. A cognitive heuristic agent was designed to simulate mate selection, and the relative explanatory power of each model variant was evaluated by comparing the simulation distribution with the empirical distribution.
	2013, Mäs and Flache	ABM was used to study polarization of opinion in the community [37]. The micro-hypothesis comes from the theory and empirical evidence of social psychology about attitudinal changes and memory processes; some model parameters were calibrated by experimental data. Finally, the simulated opinion trend is compared with the experimental results.
Parameter Calibration	2008, Dugundji and Gulyas	Individual commuting patterns in the city of Amsterdam were studied using ABM [38]. The agent's choice of transportation mode is based on the survey data, and a function of sociodemographic characteristics (such as gender, income, age, education, and place of residence) is constructed, which can predict the proportion of agents choosing different transportation modes. The agents' behavior does not refer to sociological or psychological theories.
	2011, DiMaggio and Garip	The adoption of ABM explains the inequality in the rate of adoption of new technologies among people with different educational backgrounds [39]. The size of the agent population, as well as individual attributes, such as income, education, ethnicity, and network size, are based on survey data from the United States.

Table 3.2 (Continued)

Case category	Year and People	Specific Case Content
	2014, Fountain and Stovel	ABM was used to study the role of network structure and referral in career instability [40]. The distribution of employee skills and company size is calibrated based on US empirical data.
	2011, Bruch and Mare	Employed ABM to explore levels of apartheid [41]. A function for estimating actors' racial preferences based on survey data from American films was used.
Verification of Results	1999, Lux and Marchesi	ABM was used to explain the recurring statistical characteristics of financial markets [42], such as the distribution of unconditional returns, the long tail, or the persistence of price fluctuations. Trading entities switch from one group to another based on the comparison of their respective profits, opinion indices, and price trends.
	1997, Arthur et al.	ABM was used to explain price fluctuations in the stock market [9]. Agents modify trading strategies by performing random mutations and cross-operations on trading rules. The ability of ABM to reproduce the statistical characteristics of real stock markets, as well as the presence of behavioral heterogeneity, is used to compensate for the lack of realism of micro-level assumptions and the lack of attention to input-level calibration.
Parameter Calibration + Verification of Results	2011, Poletti et al.	An epidemic on a national scale was simulated using ABM [43]. Based on various census data, the agents are assigned sociodemographic characteristics, geographical location, and probability of action. The geographical space at the city level as well as the physical location of the workplace are also reflected. This model is used to predict the evolution of similar diseases at multiple scales (i.e., national, regional, and urban) and in different populations.
	2011, Frias Martinez et al.	ABM was used to assess the impact of government epidemic prevention interventions [44]. The model was calibrated and tested based on data from the 2009 H1N1 influenza outbreak in Mexico. Used cell phone recordings to characterize the true network and flow of connections between agents, allowing for finer parameter calibration. These models often do not refer to specific sociological or psychological theories, leaving "theoretical realism" at the micro level close to zero.

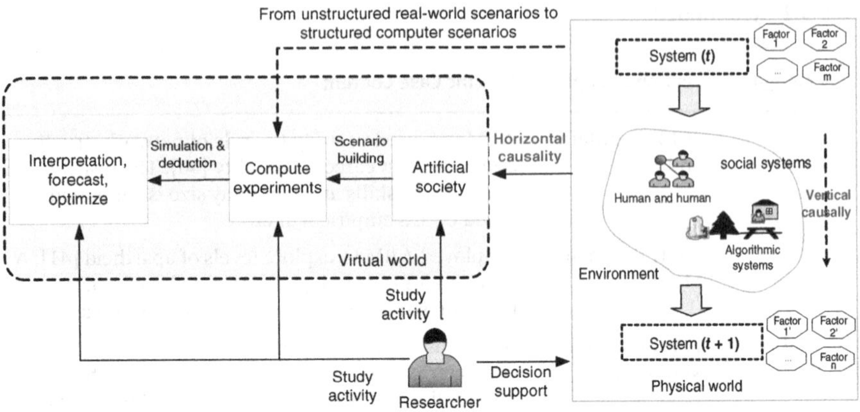

Figure 3.6 Schematic diagram of vertical and horizontal causality of ABM.

is indeed possible to determine that a certain parameter of the model is responsible for changes in the macro pattern.

However, the explainable causal knowledge generated by the abstract model is purely internal knowledge of the ABM instantiation system. In terms of inputs, it is not possible to argue that the micro-mechanisms of ABM mimic any aspect of social reality, as there are no specific theories or data to support hypotheses at the micro level. In terms of output, it is unclear what ABM is actually replicating because the model's macroscopic numerical results do not coincide with any clear empirical laws. Therefore, neither from the perspective of causal dependence nor from that of generation, do we have access to causal knowledge of any micro-level mechanism or the dynamic processes that arise from it. A highly abstract ABM obviously cannot be used directly for causal reasoning.

With the maturity of ABM, how to conduct causal inference based on simulation technology to provide decision support for real-world scenarios has become the focus of attention. Although many researchers (see [46–48]) all recognize ABM as a methodological tool based on mechanism explanations, there are doubts about the extent to which ABM can demonstrate that the hypothesized mechanisms can work in the real world. The literature [49] describes it this way:

> (…) There is a fundamental difference between causal inference based on observation (such as traditional epidemiology) and causal inference based on simulation modeling. Traditional epidemiological tools extract reasonable conclusions from incomplete real-world observations. (…) However, when we use ABM, we create a simulated world (based on prior knowledge or intuition), and then explore hypotheses about causes in this simulated

> world. These hypotheses are implemented through coding in the model. In ABM, we cannot directly determine whether X causes Y in the real world (because the virtual world is created by ourselves); In the model, we can only explore the possible impact of changing the X on the Y level. In the real world, we have facts and try to infer counterfactual conditions (what would we observe if handled differently). In the simulated world, everything is counterfactual because the world and all possible scenarios are artificially created by the researchers.

Diez Roux's statement proves a broad belief that observational and experiment methods are fundamentally different from ABM: The former deals with "facts," while the latter consists entirely of "hypotheses"; the former allows us to make causal connections in the real world, while the latter is considered by many to be impossible. Thus, the "observational approach" has an inherent superiority in causal inference, allowing for a unified counterfactual explanation within a latent causal framework [50, 51]. These methods rely on "partial and incomplete (often chaotic) observations," which are regarded as self-evident. However, it is important to note that observation methods are forced to rely on assumptions precisely because of data limitations. ABM, on the other hand, is seen as relying solely on "prior knowledge or intuition," which is considered "one of the most troubling problems" of the approach. How to systematically combine ABM with observational and experimental methods to maintain a balance between empirical data and theoretical hypotheses has become a problem for researchers.

For traditional observational experiments, the difficulty lies on how to construct counterfactual scenarios. But for ABM, the parameters of the simulation environment can be systematically controlled, and building counterfactual scenarios is no longer a challenge. Instead, how to construct a benchmark scenario that reflects reality becomes a new challenge, including whether the simulator truly reflects the data generation process and whether the final simulation results are reliable. In order to achieve this goal, ABM can also be anchored to data, and through parameter calibration and empirical verification, the theoretical realism and the realism of "input" and "output" can be achieved, respectively. This makes ABM a kind of "simulator" that allows the modeler to use ABM's knowledge across different levels to make inferences about the mechanics of the world outside the model. Due to data availability, ABM is often unable to perform a full empirical calibration, and reliability tools can only be used to quantify the degree to which ABM relies on modeling assumptions. Therefore, if an ABM achieves maximum empirical calibration and validation and is checked for system reliability, it is well placed to support the relevant causal inferences (i.e., connections between different levels in the real world). Table 3.3 shows the history of researchers in this field.

Table 3.3 ABM Works with Causal Inference.

Scholars	Main Points	Evolution of Perspectives
Grüne-Yanoff (2009)	ABM cannot provide a "complete" causal explanation because it is impossible to empirically verify all components of the model design mechanism; ABM cannot support "possible causal history" because there is a lack of an internal standard to select from all possible options [52].	Studied the Anasazi model, based on which they denied the usefulness of ABM for causal inference.
Elsenbroich (2012)	ABM can guide or complement causal inference methods that rely on empirical evidence, but it cannot independently determine the actual causal story, and complete understanding of micro-phenomena is impossible for causal explanations [53].	This is not a problem for ABM, but for social sciences as a whole.
Morgan and Winship (2015)	Simulation is at best a tool for "theory building," and its utility is limited due to the lack of transparency. Although ABM can be considered an analytical tool that bridges levels of analysis, it still lacks a solid methodology [51].	ABM lacks a reliable methodology.
Macy and Sato (2008)	Implicitly, a division of labor is proposed: ABM is a tool for theoretical exploration, while experimental and statistical methods are more suitable for observational data to support causal inference [54]. ABM can generate hypotheses for testing but cannot bear the burden of proof.	ABM and experimental statistics should be combined.
Bruch and Atwell (2015)	ABM can communicate with empirical data in various ways, allowing it to identify the realistic mechanisms behind the dependencies between variables [28].	ABM should absorb reality.
Casini (2014)	Whether a model is "credible" depends on the following three conditions [55]: "(1) the plausibility of the theoretical rationale, psychological assumptions, and functional analogies; (2) the robustness of the results to changes in initial conditions and parameter values; and (3) the robustness to changes in modeling assumptions." Robustness analysis is particularly important, demonstrating that the model in question captures the true features of the actual causal story, rather than accidental or artificial features.	The path to combining ABM with reality.

Table 3.3 (Continued)

Scholars	Main Points	Evolution of Perspectives
Ylikoski and Aydinonat (2014)	Systematic variations of initial model assumptions (i.e., robustness analysis) lead to "model clustering" that, on the one hand, helps to propose "core" feature hypotheses about actual causation and, on the other hand, helps to propose alternative explanations, positioning a given causal story within a set of possible alternative causal stories for empirical testing [56].	The specifics of ABM robustness analysis.
Törnberg (2019)	Because of the contingency of the social domain, it is impossible to know whether the simulated mechanisms are reflected in reality. The idea of carefully validating or calibrating the model by matching the regularities produced in the model with empirical regularities is "misleading" and "stems from confusing abstract categories with concrete phenomena" [57].	ABM should only be viewed as "an aid to enhanced intuition."
Manzo (2022)	Experimental and observational methods face the same demanding challenges as ABM when used to establish "horizontal" causal claims or counterfactual dependency claims. Experimental observational methods can only make credible causal inferences about whether changes in an independent variable produce differences in the dependent variable through a mixture of data and argument [58].	ABM and experimental observation methods complement each other.

3.2 Simulation of Social Systems

Social system simulation, similar to physical simulation, also involves stages such as the conceptual model, domain model, and computational model, with interdependent and constrained relationships among them. At the conceptual layer, agent models with human-like characteristics are constructed, including functional modules such as perception, cognition, decision-making, and behavior. The domain layer model is closely related to the conceptual layer and requires the development of a series of domain-specific models (such as geographic models, infrastructure, population models, intervention measures, and social relationships) to build an artificial society foundation consistent with the real world. At the numerical computation layer, agent strategy iteration equations based on neighborhood and memory information are constructed, reflecting a dynamic process of agent evolution and contextual updates. As shown in Figure 3.7, the digital thread serves as a bridge connecting different models, revealing the evolution of the system, historical configuration changes, and specific state transitions throughout the experimental cycle.

3.2.1 Agent Model

In social simulation, an agent refers to an individual with a certain degree of autonomy, corresponding to biological individuals or groups in real-world societies. The agent individual model serves as a carrier of knowledge from various domains and can be customized according to the application problem, including the structure of the agent, whether it possesses learning capabilities, the mechanisms of interaction among agents, and so on. Generally, all individuals in the system adopt the same agent structure. According to Russell and Norvig, agents can be classified into five types based on their perceptual capabilities and reasoning abilities [59] (Table 3.4).

As shown in Figure 3.8, the mainstream structure of an agent currently consists of four components: perception, decision, reaction, and optimization [60]. The information control flow within the agent structure connects these components into a unified whole. The formal expression of the agent structure, which is described by a set of attributes related to time t, is given as follows:

$$agent = < R, S_t, E_t, Y_t, V_t, N > \tag{3.1}$$

R represents the time-invariant characteristics of the agent, such as its identifier. S_t represents the time-varying characteristics of the agent, such as its role. E_t is the set of external events perceived by the agent that stimulate changes in its state and behavior. Y_t is the decision-making mechanism adopted by the agent in response to external event stimuli and during interactions with other agents. V_t is the set of behaviors of the agent, including both spontaneous actions and those triggered by external stimuli. Generally speaking, an agent follows the "benefit-increasing" principle of feedback learning: By continuously interacting with the environment, it gains experience in problem-solving and gradually optimizes its decision-making mechanism, thereby enabling it to take actions that are increasingly aligned with achieving its goals.

3.2.2 Environmental Model

In an artificial society, the environmental model is the mapping of the actual physical environment into the computer and serves as the activity space upon which agents depend. According to the modeling approach, environmental

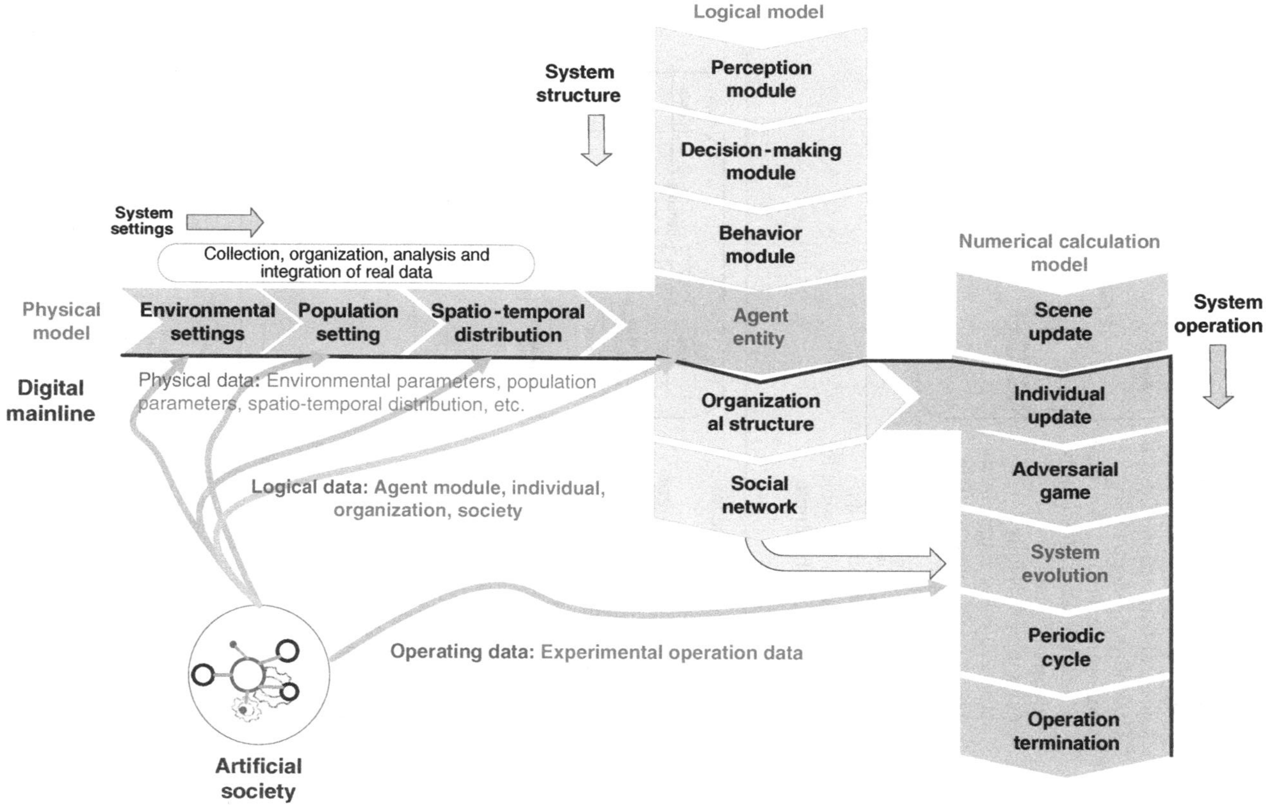

Figure 3.7 The operation process of the artificial society.

Table 3.4 Mainstream Agent Structures.

<table>
<tr><th>Name</th><th>Explanation</th><th>Illustration</th></tr>
<tr><td>Simple Reflex Agents</td><td>Agent functionality is based on condition–action rules: “If a condition is met, then perform an action.” This type of agent functionality is only successful when the environment is fully observable. For simple reflex agents operating in partially observable environments, infinite loops are often unavoidable. However, if the agent can randomize its actions, it may be possible to escape from such infinite loops.</td><td>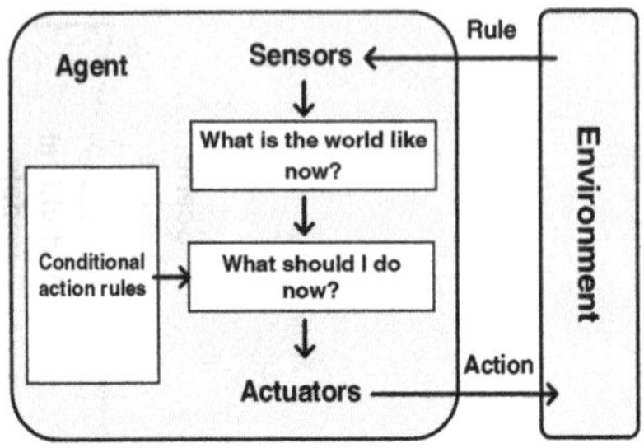
</td></tr>
<tr><td>Model-Based Reflex Agent</td><td>An agent can operate in partially observable environments. Its current state is stored internally in a certain structure that describes the parts of the “unseen world.” This knowledge about “how the world works” is referred to as a world model. The agent can use this internal model to determine how its perception history and actions have affected the environment. It then selects the next action in the same manner as a simple agent.</td><td>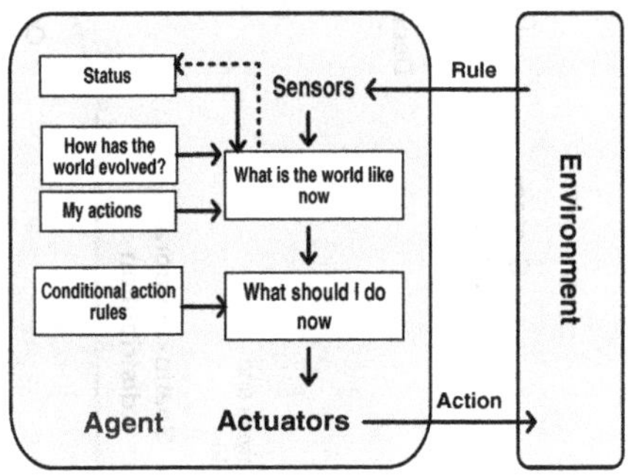
</td></tr>
<tr><td>Goal-Based Agents</td><td>An agent further extends its capabilities by using “goal” information. Goal information describes the desired state and provides the agent with a way to make choices among multiple possibilities—that is, to select the path most likely to reach the goal state. Search and planning are subfields of artificial intelligence that focus on finding action sequences that achieve the agent’s goals.</td><td>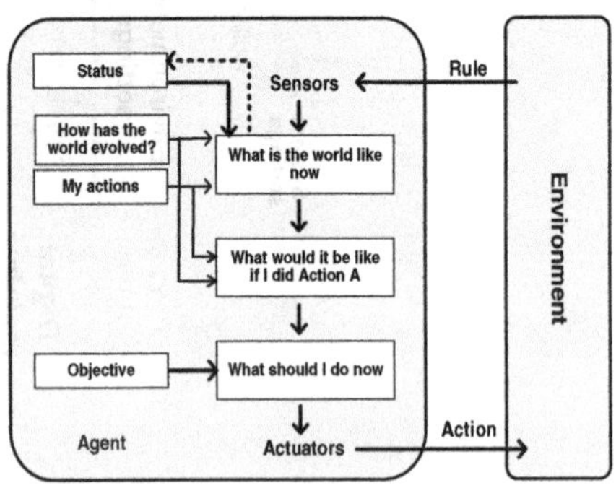
</td></tr>
</table>

Table 3.4 (Continued)

Name	Explanation	Illustration
Utility-Based Agents	An agent distinguishes between goal states and nongoal states. The utility function is used to measure the desirability of mapping the current state to a specific state. A more general performance measure should allow comparison of how well different environmental states satisfy the agent's goals. The term *utility* can be used to describe the agent's level of "happiness." A rational, utility-based agent will choose actions that maximize its expected utility. The agent must model and track its environment, which involves extensive research on perception, representation, reasoning, and learning.	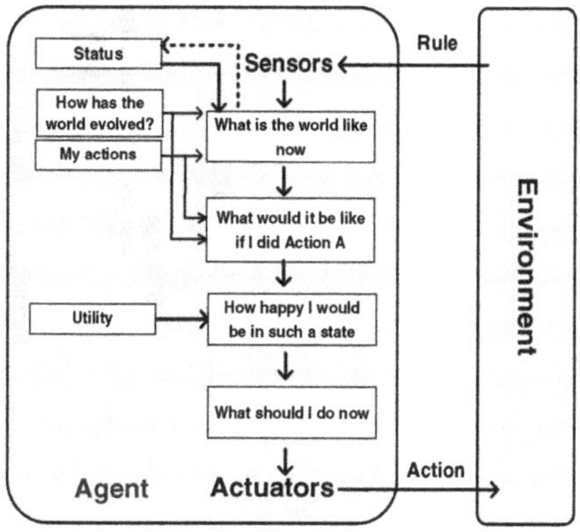
Learning Agents	The advantage of learning lies in enabling the agent to operate in unknown environments and acquire capabilities beyond its initial knowledge. The most important features of an agent are the learning element, which is responsible for improvement, and the performance element, which is responsible for selecting external actions. The learning element uses a critic to provide feedback on the agent's performance and determines how to modify the performance element or actor to perform better in the future. The problem generator is responsible for suggesting the next actions that will lead to new and informative experiences.	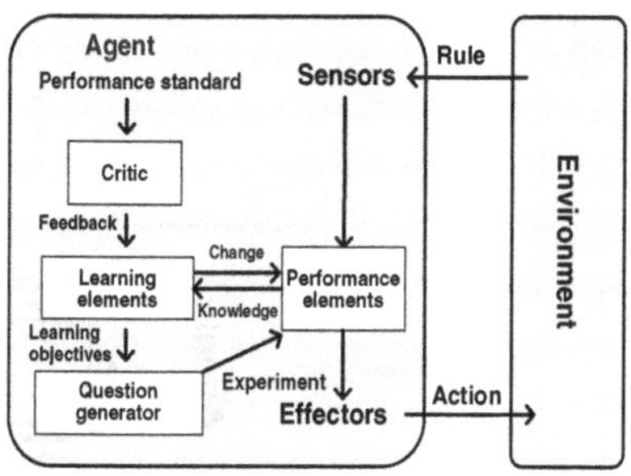

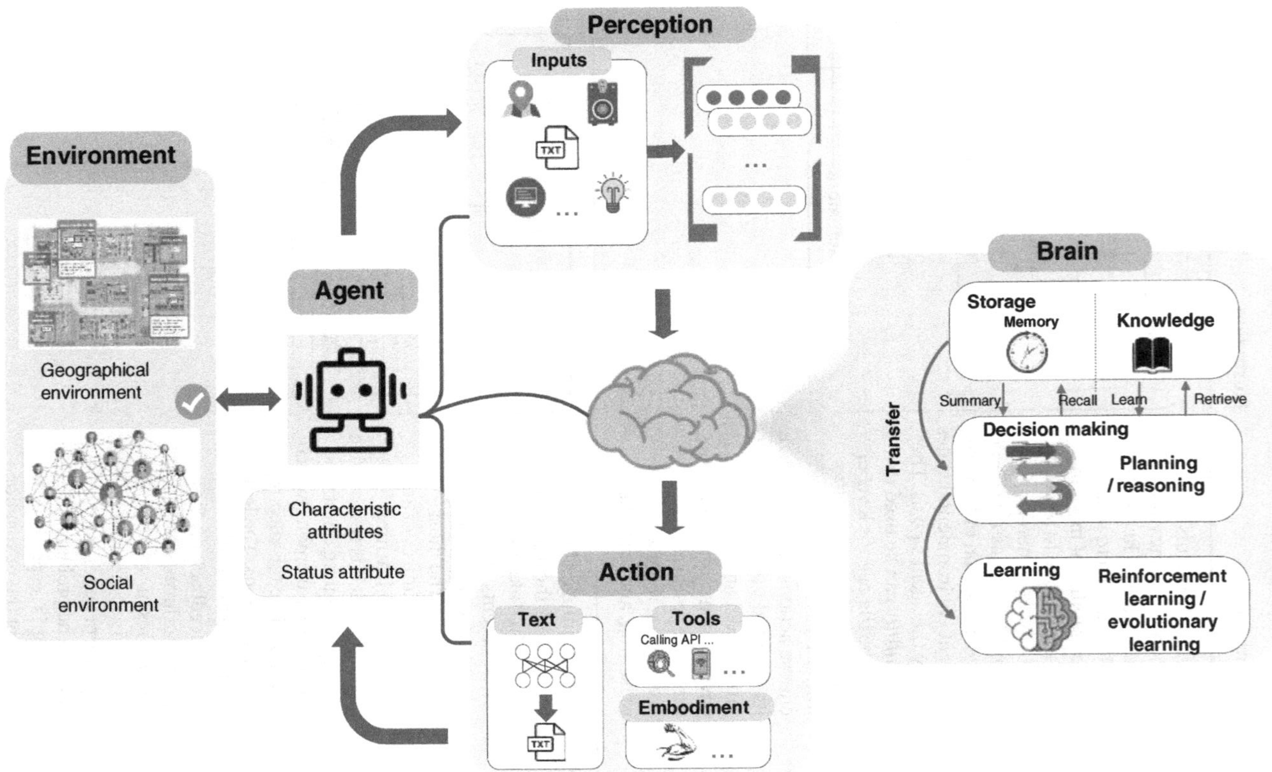

Figure 3.8 Structural model of individual agent.

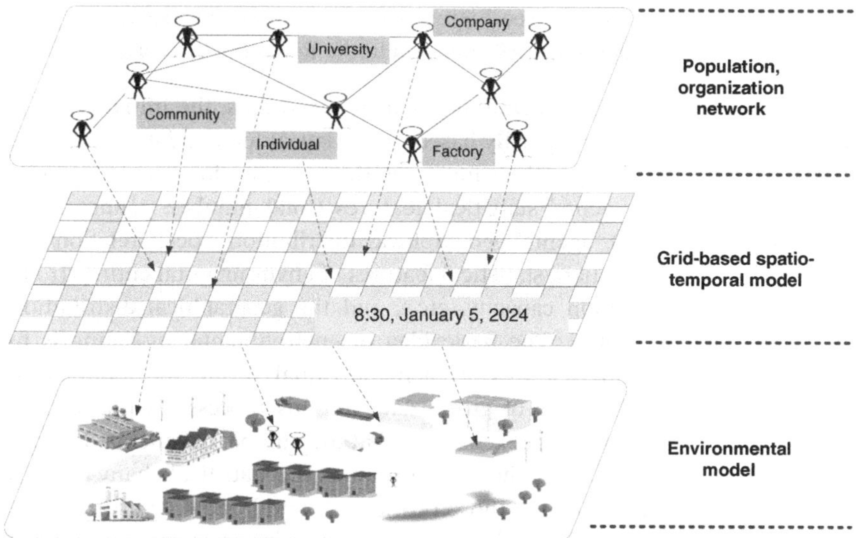

Figure 3.9 Abstract hierarchy of the environmental model.

models can be divided into entity-based modeling and grid-based modeling. Entity-based modeling refers to abstracting various environmental elements in the real world—such as buildings, road traffic, and climate conditions—into entity models. Many typical artificial society systems, such as EpiSimS [61], are implemented in this way. Grid-based modeling does not focus on specific environmental objects, but rather emphasizes modeling the spatial aspects of the environment. It uses discrete grids to describe spatial presence and environmental attributes, such as in the Sugarscape model [12].

Due to differences in the scale of artificial society scenarios, the granularity of environmental scene elements also varies. In large-scale artificial society scenarios—such as global spatial or city-level transmission analysis of diseases—the granularity of the environmental model is relatively coarse and generally focuses on transportation networks, such as airline networks. In small-scale artificial societies, the environmental model is relatively fine-grained and can include detailed models of natural environments, buildings, and road traffic. For large-scale artificial society scenarios, geographic information systems are typically used to build geospatial models. For small-scale artificial societies, 2D or 3D display technologies can be used to build visual scenes, while grid technologies establish the geospatial coordinate system to determine the relative positions in geographic space (see Figure 3.9). The interaction between agents and environmental models requires attention to the following three aspects:

Distribution characteristics: Due to practical constraints, the initial setup of environmental models is typically based on statistical feature data. Therefore, it is necessary to study algorithms for generating initialization data in artificial societies, including agents' statistical features (total number, gender ratio, age distribution, etc.), agents' geographical distribution, social relationship attributes of the population, statistical features of environmental entities (total number, type, population capacity, etc.), and the geographical distribution of environmental entities. The basic idea of environmental modeling is to reconstruct the specific features of each individual within the group from the statistical characteristics of group data. In this process, two aspects of consistency must be ensured: (1) ensuring that the number of generated agents is statistically consistent with the real world, i.e., the statistical characteristics of the generated agents match those of the real world and (2) ensuring that the internal logical structure of the generated agents is consistent with the real world, i.e., the organizational structure and relationships within the agents are in line with reality.

Spatial correlation: For individuals, although spatial mobility contains randomness, it more often follows patterns. Therefore, for each individual in the artificial society, it is necessary to establish association relationships with fixed activity locations. To establish matching associations between individuals and activity locations in a manner that conforms to human behavioral rules in the real world, two key tasks must be clarified: (1) determine possible related activity location types based on individual activity types, and establish a one-to-many mapping between activities and locations to ensure the agent's ability to randomly select activity venues and (2) search for locations that comply with travel patterns for each individual and match them accordingly, while optimizing the search algorithm to ensure that the computational time for matching large-scale populations and locations remains within an acceptable range. This endows individuals with behavior characteristics that reflect spatiotemporal correlations.

Spatiotemporal correlation: In artificial societies, agent behavior involves continuous spatial migration over time. To dynamically link agents with each other and with the environment, it is necessary to clearly define what type of activity an individual performs at what time, in what location, and what kind of interactions occur at that location. This ensures the temporal and spatial consistency and regularity of individual behavior. Three key issues need to be addressed here: (1) establish the mapping relationship between time and actions, i.e., define the behavioral temporal logic in the agent evolution process; (2) based on behavioral temporal logic, integrate the regularity and randomness of agent behavior, i.e., establish a stochastic model of generally performed and potentially performed actions during simulation evolution; and (3) under spatiotemporal constraints, establish a mechanism for individuals to select interaction partners when performing actions, i.e., clarify how to associate the social relationship network among individuals with behavioral models.

The development trend of environmental models is 3D digital twin visualization, which is closer to real-world cognition and more convenient for decision-makers to understand. In 1994 [62], Fei-Yue Wang proposed the concept of "Shadow systems," which is very similar to the idea of digital twins. In 2002, Michael Grieves of the University of Michigan proposed the "Twinning of systems" concept, aiming to construct digital twins equivalent to physical entities in virtual space, emphasizing the establishment of a bidirectional dynamic feedback mechanism to simulate, analyze, and optimize physical entities [63]. In 2004, Fei-Yue Wang proposed the concept of "Parallel systems" as a methodology for solving complex system problems [14]. In 2006, Grieves published *Product Lifecycle Management: Driving the Next Generation of Lean Thinking*, officially proposing the "Mirrored Space Model" [63]. In 2010, NASA introduced the concept of the digital twin in its Space Technology Roadmap (Area 11: Simulation-Based Systems Engineering): "A digital twin is an integrated, multi-physics, multi-scale simulation of a vehicle or system that uses the best available physical models, sensor updates, and operational history to mirror the life of its flying twin."

Didi Chuxing Company is a digital space application in smart travel, which helps illustrate the value of digital twins. As a basic digital twin infrastructure, it mirrors the real-time location of all vehicles, ride requests of passengers, and all drivers' language content, enabling unified, scenario-based application design. Before Didi, people hailed taxis from the street, resulting in random matching between supply and demand. Customers had to endure long waits, while drivers often worried about empty mileage. The emergence of Didi broke the constraints of line-of-sight range, greatly reducing matching time between supply and demand, and allowed users to easily rate services. It can be said that digital twins have significantly improved both the operational efficiency and quality of the real world. Table 3.5 illustrates the differences among digital twin companies in terms of product form, technical system, and value understanding.

3.2.3 Rule Model

Rule models describe the cyclical mechanisms of artificial societies, including the principles that govern how agents interact with each other, how environments interact, and how agents interact with the environment. These rules can be either mappings of real-world social rules or artificially assumed hypothetical rules. As shown in Figure 3.10, artificial societies adopt a feedback loop approach to establish a three-level evolutionary order:

Table 3.5 Technical Schools of Digital Twin.

Type	Product Form	Understanding of Digital Twin
Internet Digital Twin	Primarily PaaS platform + SDK. The PaaS platform mainly provides various case templates, reusable model resources, data panel components, and encapsulated animation APIs, data APIs, etc. Well-known domestic companies include Beijing Younuo Technology, 51World, Topology Software, etc.	Large screens, visualization, and low-code development, with a strong focus on geometric modeling, animation display, and data visualization of scenarios—that is, the twin of spatial form. The tech stack mainly revolves around WebGL or game engine technologies such as Three.js, Babylon.js, Unreal Engine, or Unity. Main application scenarios are typically in smart cities, with less involvement in industrial and equipment manufacturing sectors.
Industrial Software Digital Twin	Deeply rooted in the fields of automatic control and industrial simulation. Virtual scenes are built using simulation models, and real-time data is integrated to drive virtual simulation technology with real data to provide optimal processes and operational strategies for physical environments. Well-known companies include Siemens, Dassault, Schneider, etc.	Virtual products are constructed to validate product design, virtual production processes are created to validate manufacturing workflows, and parallel twins are used to verify and improve product performance. The tech stack mainly focuses on industrial simulation and IoT technologies, predominantly based on proprietary platforms. The main application areas are in industry and equipment manufacturing.

Individual evolution: At the bottom level is the individual evolution space, which simulates the phenomenon of individuals undergoing learning and evolution within a social system. In a social system, different individuals have different interests and needs, and individual interests may not align with the overall interests of the system or may even conflict. As a result, behaviors such as "countermeasures to top-down policies" often arise, which affect the implementation and effectiveness of policies. Cooperation among agents is not always smooth; conflicts may occur, including resource conflicts arising from multiple agents competing for the same limited resource; conflicts where one party's goal realization prevents the other from achieving theirs; and conflicts where different agents reach contradictory conclusions on the same issue.

Group emergence: If individual behavior is regarded as an input causal variable, then emergence appears as the output phenomenon. When individual behaviors change in response to system feedback, the system may evolve from

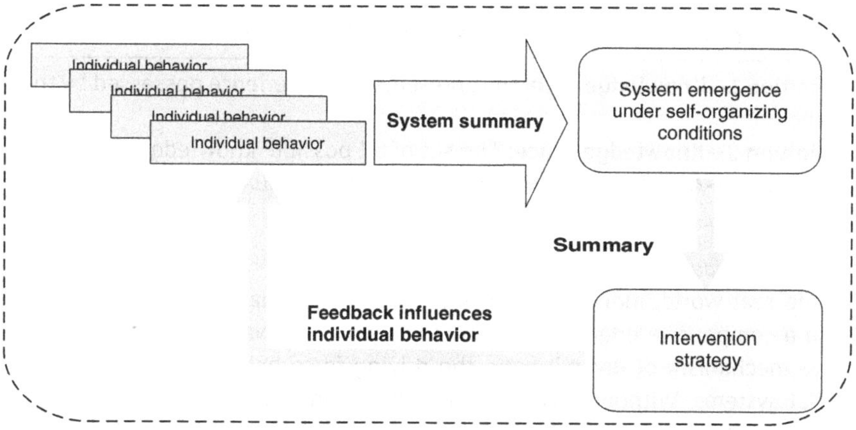

Figure 3.10 The operational logic framework of the rule model (replaced with game regulation).

first-order emergence to second-order emergence. In this way, a closed loop is formed between the overall system and individual behavior—namely, a bidirectional feedback mechanism (feed forward and feedback), which may lead to positive feedback loops and mutual reinforcement. Under such conditions, even a small deviation from equilibrium can lead to large and exaggerated effects.

Feedback regulation: The regulation mechanism refers to a mechanism by which macro-level phenomena exert adjustments on micro-level individual behavior. An effective regulation mechanism must consider the individual's pursuit of interests, to ensure the system's motivation and vitality for development; at the same time, it must firmly prevent interest imbalance to avoid threatening the system's order and stability. To ensure that groups composed of multiple agents can act in a coordinated manner, coordination becomes the core issue in multiagent systems. Coordination is the management of interdependencies among activities, including the appropriate arrangement of their goals, resources, and mental states. It is reflected in shared resource constraints, simultaneity constraints, producer/consumer relationships, and task/subtask relationships.

The mutual nesting and causality of the evolution mechanisms among various levels make this analytical framework exhibit the characteristics of complex system theory. These three levels and their interactions constitute a relatively complete and abstract circular analytical structure of the evolution process of artificial society. Each level focuses on different aspects of the socioeconomic system (such as individuals, organizations, clusters, and countries), and the relevant modeling techniques can be selected according to specific needs. The implementation logic of the rule model is shown in the following pseudocode:

The Implementation Logic Pseudocode of the Rule Model

Definition 1—Knowledge point: It represents the experience possessed by the agent.
Definition 2—Knowledge space: The set of all possible knowledge points.
Definition 3—Agent: An entity representing an individual within the artificial social system.

Step 1: Individual Evolution

In the real world, individuals must continuously enhance their abilities to gain a competitive edge in fierce competition. Thus, variation is the generative mechanism of diversity and the driving force behind the evolution of social systems. Without variation and innovation, evolution does not exist. In the modeling framework, the rules of variation and evolution for agent individuals can be defined using various machine learning (ML) or evolutionary algorithms. The simulation process emphasizes three characteristics: first, all variation processes are fundamentally uncertain, including both the random alteration of individual strategies and the generation of new strategies; second, variant individuals must be capable of acting quickly to reap benefits before other actors complete imitation; and third, variation is subject to numerous constraints at various levels, including various risk factors it may introduce.

Step 2: Organizational Evolution

The selection mechanism is the mechanism that reduces diversity. It evaluates the adaptability of individuals based on certain criteria, selects highly adaptive evolutionary units, and eliminates poorly adaptive ones. Without a selection mechanism, the social system would lose the ability to distinguish between good and bad, leading to a divergent, chaotic, and inefficient state. Therefore, a sound selection mechanism is a critical filtering process in the evolution of artificial society. Since defining fitness as a criterion for individuals to produce offspring (or undergo genetic replication) is essentially the same as the selection process, the evolution rules for organizations in the modeling framework can be defined according to evolutionary features of biological communities, such as Ant Colony Optimization (ACO) [64], Particle Swarm Optimization (PSO) [65], Artificial Bee Colony (ABC) algorithm, and others [66].

Step 3: Social Emergence

After intense competition, some elites stand out from the individual pool. Other individuals can enhance their survival capabilities in the ecosystem by imitating and learning from these elites—this marks the phase of social evolution. The social evolution process involves the diffusion of elite characteristics and the formation of culture. There are three typical diffusion models: the contagion model [67], the social threshold model [68], and the social learning model [69]. Factors influencing diffusion mechanisms also include the following. (1) The structure of the interaction network: In regular network structures, interaction clustering among individuals is high but diffusion speed is low. In random network structures, interaction clustering is low but diffusion speed is high. (2) The degree of heterogeneity among interacting agents (including differences in preferences, cognition, beliefs, knowledge stock, learning ability, and absorptive capacity): Generally, the greater the heterogeneity, the slower the diffusion speed. (3) Organizational rules: Different organizational rules shape different interaction patterns, thus influencing the speed of diffusion of variation.

Step 4: Feedback Regulation

The social space, in turn, acts upon the individual space—this is the mechanism by which macro-level phenomena affect micro-level individuals. In the real world, cooperation and competition among individuals shape the cultural atmosphere of the entire social ecosystem, which, in turn, influences individuals' decision-making mechanisms. To simulate the phenomenon that culture can accelerate individual evolution, feedback rules are designed in the modeling framework. These influences may be reflected in the following aspects: (1) an effective selection mechanism enables highly efficient individuals to acquire more resources, facilitating further technological accumulation and innovation; (2) the selection mechanism itself shapes the cognition, preferences, and internal constraints of variant individuals, thereby exerting systemic influence on variation; and (3) the selection mechanism of intervention strategies affects not only the variation level of specific individuals but also that of the entire system. Replicator dynamics is a typical deterministic and nonlinear evolutionary game model based on the selection mechanism. Adding random strategic variation to this model results in a composite evolutionary game model that includes both selection and variation mechanisms, commonly referred to as the replicator–mutator model.

Step 5: The Next Cycle

As time progresses, some elites may fall behind, while newly capable individuals may rise to become the new elites. Ultimately, the evolutionary equilibrium of the entire social system is broken, and a new cycle begins.

3.3 Causal Inference

3.3.1 Causal Inference

Causal inference is an important area in statistics. Its goal is to determine whether one variable (cause) has an effect on another variable (effect). Causal inference involves analyzing data and applying statistical methods to infer causal relationships. It is concerned not only with the association (correlation) between variables but also with the causal pathways among variables. The key to identifying which variables are causes and which are effects lies in how to control confounders to ensure the validity of causal inference. A direct causal relationship means that variable A directly causes changes in variable B, such as smoking (variable A) causing lung cancer (variable B). An indirect causal relationship means that variable A affects variable B through a third variable C; for example, education (variable A) affects health (variable B) through income (variable C).

Causal inference commonly involves the following concepts. (1) Confounders, also called confounding variables, are variables that affect both the independent and dependent variables, making it difficult to distinguish the causal relationship. For instance, high summer temperatures may lead to increased ice cream sales, which may also coincide with a rise in drowning incidents. (2) Counterfactuals: Assuming what would happen if a variable had taken a different value and are used to understand and infer causal relationships. (3) Causal graphs: Typically represented using directed acyclic graphs (DAGs), used to illustrate the causal relationships and structures between variables.

Correlation indicates that two variables are associated, but it does not necessarily mean that one variable causes the other. For example, correlation analysis might find a positive correlation between ice cream sales and drowning incidents, but this does not imply that ice cream sales cause drowning. The possible confounding variable might be temperature, which increases both ice cream purchases and swimming activity, leading to more drownings. Correlation is often measured using the Pearson's correlation coefficient or the Spearman's rank correlation coefficient. Establishing causal relationships is more complex than correlation and requires identifying and controlling for potential confounding variables. The thinking and methods of causal inference can be traced back to several centuries of philosophical exploration, but they have developed rapidly in modern statistics and epidemiology (Table 3.6).

The core of causal inference lies in "counterfactual thinking": simulating the outcome of "what would have happened without the treatment" through experimental design, statistical control, or causal models. In practical applications, it is necessary to consider data types, domain knowledge, and the

Table 3.6 Key Milestones in the Development of Causal Inference.

Stage	Description
Early Thoughts	• Ancient Greek philosophers such as Aristotle discussed the concept of causal relationships. • **Philosopher David Hume (18th century):** He proposed that causal relationships cannot be inferred directly from observed associations but must be established through experience and inductive reasoning.
Early 20th Century	• **Randomized controlled trials (RCTs):** Sir Ronald A. Fisher developed the concept of RCTs in agricultural experiments, emphasizing random assignment to control confounding variables. • **Karl Pearson and Francis Galton:** Developed correlation analysis and regression analysis, providing tools to understand relationships between variables, but mainly focused on correlation rather than causation.
Mid-20th Century	• **Donald Rubin and the Neyman–Rubin framework:** Introduced the counterfactual framework, formalizing causal inference and emphasizing randomization and counterfactual thinking. • **Judea Pearl:** In the late 20th century, Pearl proposed structural equation model (SEM) and causal diagrams and developed do-calculus, building a systematic methodology for representing and inferring causal relationships.
Modern Developments	• **Machine learning and causal inference:** In recent years, causal inference has been integrated with machine learning methods, leading to new techniques such as Causal Forests and deep learning-based causal models (e.g., TARNet and DragonNet). • **Applications in policy evaluation and economics:** Causal inference methods have been widely applied in policy evaluation, economics, medicine, and social sciences, advancing empirical research in various fields.

plausibility of assumptions in order to select the most robust method. For example, randomized controlled trials (RCTs) are the gold standard, but in observational data, difference-in-difference (DID) and instrumental variables are more commonly used; ML methods offer new tools for high-dimensional causal analysis. The main causal inference methods are shown in **Table A.3.1**.

Causal inference is a complex yet very important field that helps us understand the true relationships between variables. Through RCTs, observational studies, and structural equation modeling (SEM), we can infer causal relationships in different contexts while minimizing the influence of confounders and biases. In practical applications, the choice of method depends on the research question, data characteristics, and considerations such as ethics.

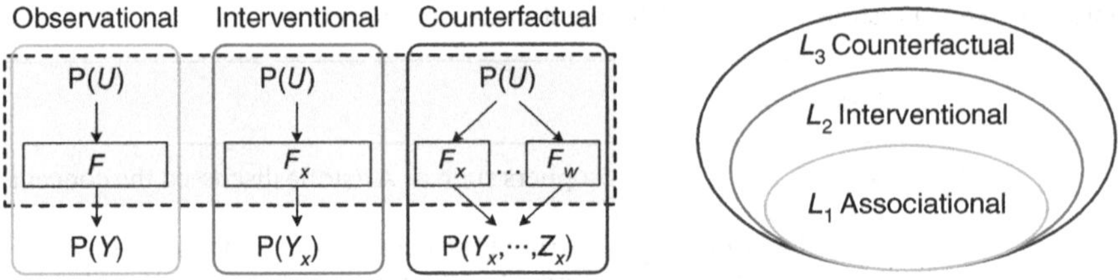

Figure 3.11 The three levels of the causal hierarchy are as follows. (1) Association: If we observe $X = x$, what can we tell about Y? (2) Intervention: If we perform the intervention $X{:} = x$, what can we tell about Y? (3) Counterfactuals: Given that a hypothetical event $E = e$ actually occurred, what would have happened to Y if $X{:} = x$ had been carried out? Left panel: $P(U)$ describes the natural states of the system in domain U; F represents the transition to the observational world (innermost ellipse), the interventional world F_X (middle ellipse) with interventions, and multiple counterfactual worlds $X, \ldots, W$ (outermost ellipse) with alternative actions. Right panel: The three levels of the causal hierarchy are theoretically separated in terms of identifiability: Data at a lower level is almost never sufficient to fully determine answers at a higher causal level.

3.3.2 Counterfactual Inference

Counterfactual inference is a specialized tool of causal inference used to evaluate hypothetical scenarios of "what if…." For example, it seeks to address questions such as: "What would the outcome Y have been if the event X had not occurred?" For instance, suppose we want to know whether a certain medical treatment is effective. We have two sets of data: one group that received the treatment and another that did not. Counterfactual inference tries to answer the question: What would the health outcomes of those who received treatment have been if they had not received it? The goal of counterfactual inference is to estimate what the outcome would have been under a hypothetical scenario (i.e., the counterfactual) and thereby estimate the causal effect of a treatment, policy, or action on the outcome (Figure 3.11).

Understanding the core of counterfactual inference lies in grasping the nature of "counterfactuals," namely, "if history were replayed and a certain event that has already occurred had not happened or had occurred differently, how would the outcome change?" This involves thinking about potential outcomes: We can only observe one outcome in the real world (the factual), but to make causal inferences, we must estimate what the outcome would have been under the unobserved scenario (the counterfactual). Since counterfactuals do not exist in reality, this leads to the "missing counterfactual" problem. The goal of counterfactual inference is to estimate the Individual Treatment Effect (ITE): ITE = Y(1) − Y(0). However, since only one of the outcomes can be observed, estimating this difference becomes difficult. In practice, researchers must adopt indirect approaches to construct counterfactual scenarios using observed data.

The digital revolution driven by Internet technologies has effectively digitized the social, economic, political, and cultural activities of billions of people, generating massive volumes of data. Against this backdrop, online experiments have increasingly been used for research purposes across various academic fields and are widely applied in areas such as user incentive design, recommendation strategy evaluation, and product functionality iteration [70]. This approach offers several advantages:

1) Scale can far exceed traditional lab or field experiments, while maintaining low cost. Most lab or field experiments may involve hundreds of participants, while digital experiments can expand the number of participants by hundreds of times without increasing costs.
2) Preprocessing can be used to address participant heterogeneity. Traditional experiments rarely obtain participants' background information, making group distinctions difficult; digital experiments typically incorporate such background data and perform preprocessing.
3) Extended duration is possible. Traditional experiments must be conducted and measured within a compressed time window; digital experiments can run continuously to collect outcome data.
4) Novel designs can be implemented to create highly precise and controllable parallel worlds, helping researchers understand the heterogeneity in treatment effects.

The "Artificial Music Market" is a typical case of an online experimental system [71]. In real life, popular songs, best-selling books, and blockbuster films tend to far exceed average popularity levels in their respective domains. However, it remains extremely difficult to predict which song, book, or film will succeed. This experiment focuses on the question: To what extent does the success of cultural products stem from quality, and to what extent from luck? In other words, if we could create parallel worlds that operate independently, would the same song still become popular in each of them? If not, what accounts for the differences? It is important to note that this question is not about predicting which song will become a hit.

The experiment created a website called MusicLab, where users could explore new music. A total of 14,341 participants took part. Using this site, a series of experiments were conducted. Participants in the experimental group were randomly assigned to one of eight parallel worlds. In each world, users could not see any information from other worlds and were not even aware that other worlds existed. Control group: participants selected and rated songs based solely on the given band names and song titles. Experimental group: followed the same procedure, but participants could see additional information (Figure 3.12).

Only experiments that construct parallel worlds can achieve this: The same song may be the most popular among 48 songs in one world, while ranking

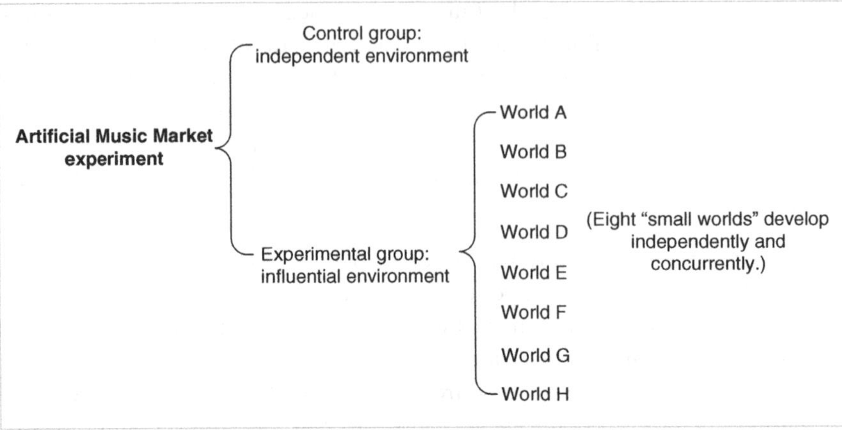

Figure 3.12 Parallel worlds of "Artificial Music Market."

40th in another. Surprisingly, for songs that are more appealing, their success is more susceptible to be influenced by chance factors. As shown in Table 3.7, the experimental results indicate that social influence significantly contributes to the inequality of song success and intensifies the unpredictability of outcomes. Moreover, under otherwise identical conditions, as the strength of the relevant influencing factor increases, both the inequality and unpredictability indices also increase. That is, when the influencing factor grows stronger (Posttest 2–Pretest 2) > (Posttest 1–Pretest 1), which indicates a rise in inequality and unpredictability.

Counterfactual inference has some similarities to the null hypothesis in hypothesis testing (both involve considering hypothetical scenarios), but they differ significantly in concept and application, primarily in the following aspects:

1) The goal of counterfactual inference is to estimate causal effects, i.e., the change in outcomes under different treatments; whereas hypothesis testing aims to determine whether there is sufficient statistical evidence to reject a hypothesis (the null hypothesis) about a population.
2) Counterfactual inference addresses potential and unobserved scenarios (counterfactual scenarios), while hypothesis testing concerns population-level assumptions based on observed data.
3) The core of counterfactual inference is to construct and estimate a hypothetical outcome (namely, the counterfactual) in order to understand causal relationships, whereas the core of hypothesis testing is to analyze data to determine whether to reject the null hypothesis.
4) Common methods for counterfactual inference include matching, regression, instrumental variables, and DID; hypothesis testing is primarily carried out by calculating test statistics (such as t-tests, z-tests, and chi-square tests).

Table 3.7 Parallel World Cases in the Artificial Music Market.

Type	Experimental Grouping	Experimental Subjects	Experimental Results
Experiment 1	Control Group 1	Participants could listen to songs, but there were no interfering factors (Pretest 1).	Inequality in song success: Posttest 1 > Pretest 1. Unpredictability in song success: Posttest 1 > Pretest 1.
	Experimental Group 1	Participants could listen to songs and also see the number of previous downloads (Posttest 1).	
Experiment 2	Control Group 2	Participants could listen to songs, but there were no interfering factors (Pretest 2).	Inequality in song success: Posttest 2 > Pretest 2. Unpredictability in song success: Posttest 2 > Pretest 2.
	Experimental Group 2	Participants could listen to songs, view previous download counts, and also observe ranking of popularity (Posttest 2).	

3.3.3 Algorithmic Counterfactuals

Why can scientists use computer simulations to predict weather, study disease spread, or even explore the mysteries of the universe? The underlying reason is simulation-based inference (SBI). In simple terms, SBI is a method that constructs simulation models of complex systems and then uses these models for reasoning and prediction. The core of this approach is that through large-scale computations and simulations, we can extract valuable information and thereby perform scientific inference. For example, weather forecasting is a typical application of SBI. Meteorologists use models that incorporate various meteorological variables to simulate atmospheric motion, and by repeatedly adjusting and computing, they can predict the weather for the coming days.

These simulators, as tools, construct models of system dynamics by integrating known (or assumed) mechanistic knowledge of the system. In terms of complexity, they range from extremely simple conceptual models (such as the Lotka–Volterra equations that describe predator–prey dynamics in ecosystems) to highly detailed and expensive simulations run on supercomputers (e.g., full-brain simulations).

Simulators provide virtual testing platforms for exploring complex real-world scenarios, and when equipped with causal ML tools (especially counterfactual inference), they can efficiently and economically support "what-if" analyses. That is to say, each run of the simulator constitutes a realization of a specific possible world. By altering independent parameters of the model and rerunning the simulation, we can study counterfactual "hypothetical" claims. A prevailing challenge is how to link these simulation models to empirical data. These numerical simulators often contain parameters whose precise values are not known a priori and must be inferred from data. Therefore, how to constrain numerical simulations using data becomes a key challenge in generating counterfactuals.

To address the challenge of connecting simulation models with data, statistical methods offer approaches to constrain model parameters using data, to select from different candidate models, and to evaluate whether hypotheses can be rejected based on empirical evidence. In particular, statistical inference aims to identify which parameters or parameter combinations are compatible with the empirical data and (possibly) prior assumptions. As shown in Figure 3.13, the general process comprises (1) sampling parameters from the prior and then using them to simulate synthetic data; (2) learning the (probabilistic) association between the data (or data features) and the latent parameters—i.e., establishing a probabilistic mapping between data features and parameters using simulated data. Various SBI methods (e.g., neural density estimation) can be used for this step; (3) applying

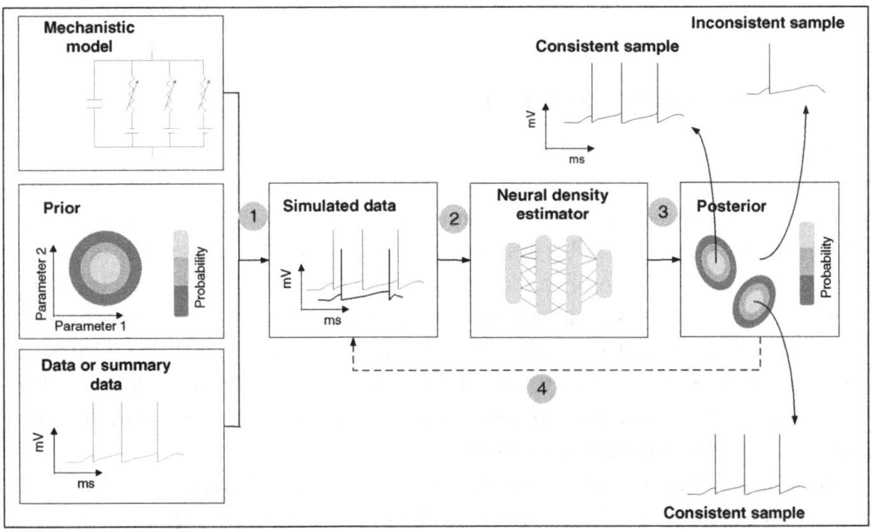

Figure 3.13 Flowchart of counterfactual generation. Counterfactual generation typically requires three inputs (on the left): candidate mechanistic models (e.g., biophysical neuron models), prior knowledge or constraints on model parameters, and observed data (or their summary statistics).

the learned model to empirical data to derive the full parameter space that is consistent with both the data and the prior, namely, the posterior distribution. The posterior may have a complex shape (e.g., multiple modes), and different parameter configurations may yield simulations consistent with the data. If needed, (4) the preliminary posterior estimate can be utilized to adaptively generate more informative simulations.

Abstractly, SBI can be regarded as lying between mechanistic models (e.g., described by differential equations) and statistical models (learned from data). In other words, mechanistic models describe the system dynamics and are manually encoded by human experts; statistical models (e.g., neural networks) learn mappings between variables from observed data and are only valid under fixed experimental conditions. SBI lies between the two, seeking to model "the effect of intervention-induced distributional change" by combining data-driven learning with assumptions not included in the statistical description of the system (Figure 3.14).

Counterfactual generative adversarial networks (CF-GANs) are a method for counterfactual generation in statistical causal inference, primarily employed to simulate individual potential outcomes under different treatments (or interventions). GANs consist of a Generator (G) and a Discriminator (D). The generator G attempts to generate realistic data, while the discriminator D tries to distinguish between real and generated data. This adversarial training allows G to gradually learn the true data distribution and generate plausible counterfactual outcomes, thereby providing support for estimating causal effects through simulated counterfactuals.

3.4 A Classic Case of Social Simulation

The application scope of social simulation includes three aspects: phenomenon interpretation, trend prediction, and strategy optimization. This section presents some classic cases in these three aspects to help readers deepen their understanding of social simulation.

3.4.1 Thought Experiments

Thought experiments do not present model-specific scenarios or real-world social systems, but rather describe the abstract logical relationships of general social systems. They aim to explore and quantitatively analyze the unpredictable consequences that certain assumptions may bring to human society through experiments. This type of research avoids the mapping issue from the actual society to the artificial society, typically offering metaphors, inspirations, and qualitative trends rather than precise solutions to complex issues.

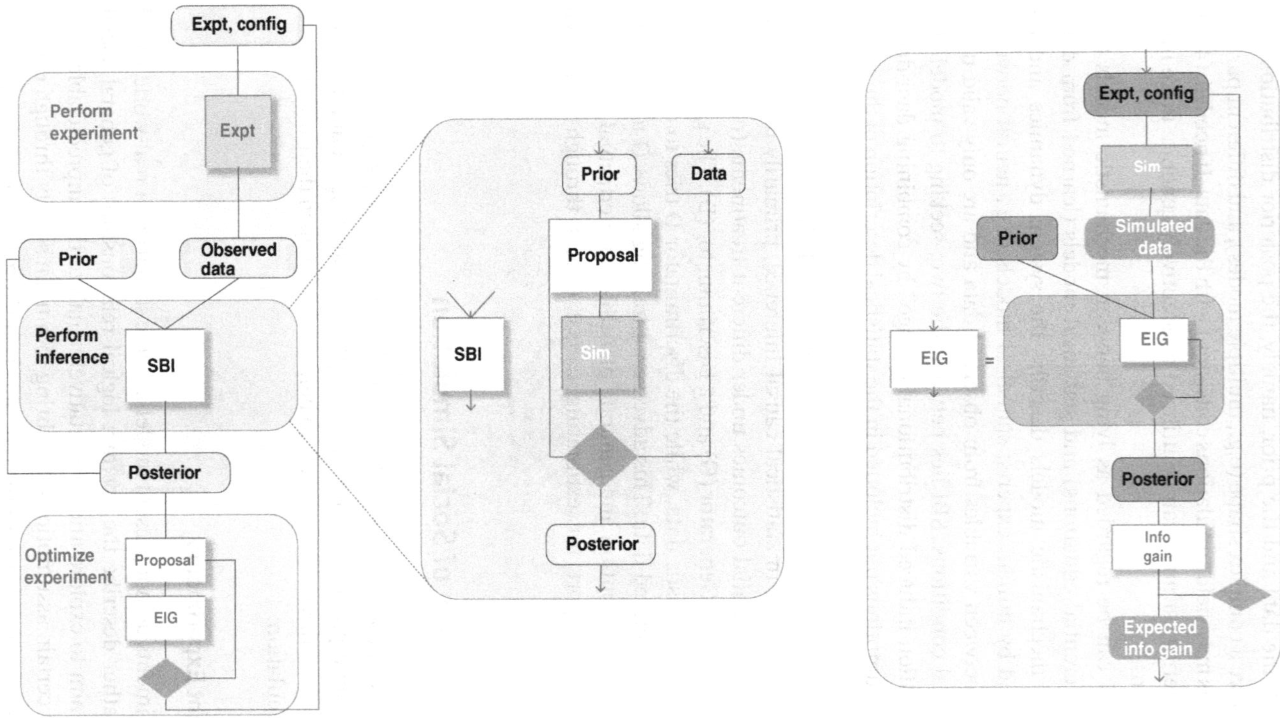

Figure 3.14 Simulation-based counterfactual generation method. The system input includes an external workflow (Square "Expt") that implements certain experimental protocols and a configuration for executing these experiments. The simulator depends on a set of theoretical models (Square "Sim" components). The right panel illustrates how experiments can be optimized through active learning and sequential design algorithms (i.e., Bayesian optimization, which balances exploration and exploitation to efficiently optimize expensive black-box objective functions). Here, the optimization objective is the expected information gain (EIG) regarding the hypothesized parameters. *Source:* [72] / Cornell University.

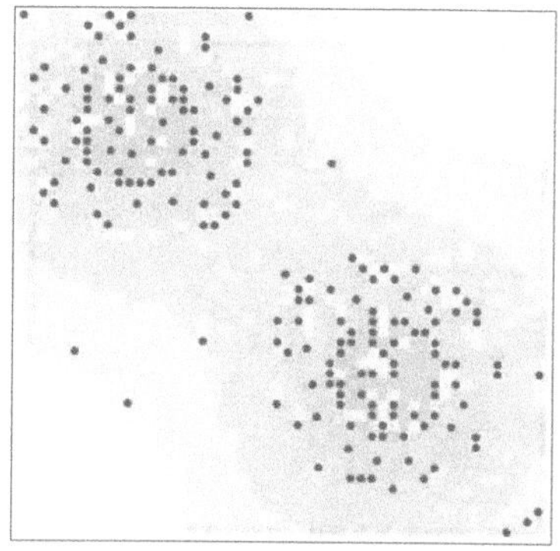

Figure 3.15 Distribution of sugar and agent in Sugarscape.

3.4.1.1 Sugarscape Model

In 1996, Joshua Epstein and Robert Axtell proposed an "artificial society" model, Sugarscape, which can conduct relevant experiments in economics and other social sciences [12]. As shown in Figure 3.15, the Sugarscape model is a closed world composed of squares: The dark grey dots represent the agent, which can only move around in this world; the yellow part represents social wealth—sugar; and the intensity of light grey indicates the concentration of sugar distribution. Each agent contains three attributes: visual range r, the resource metabolic rate v, and the amount of sugar s it possesses. Agents move around according to the following rules: (1) observe all cells within the visual range r and determine the cell with the highest sugar content; (2) if there is more than one cell with the maximum sugar content, choose the nearest one; (3) move to this square; and (4) Collect the sugar in this cell and update the corresponding variable.

As shown in Figure 3.15, as the experiment progresses, some agents—due to their superior individual abilities or advantages in local resource distribution—are able to collect more sugar; meanwhile, individuals lacking sugar gradually wither and die, resulting in most agents clustering in the two regions with higher sugar concentrations. Eventually, a small number of agents possess a large amount of sugar resources, while the majority own only a small amount, thereby confirming the well-known Matthew effect in social science. By adding multiple types of resources (e.g., spices) to the Sugarscape artificial society, researchers can study how individuals in real societies form markets through resource exchange. By altering the behavioral rules followed by agents, this model can be used to explore social phenomena such as environmental change, genetic inheritance, trading interactions, and market mechanisms. The Sugarscape model provides insights into wealth inequality, market formation, and other social phenomena by simulating agent interactions around resource collection.

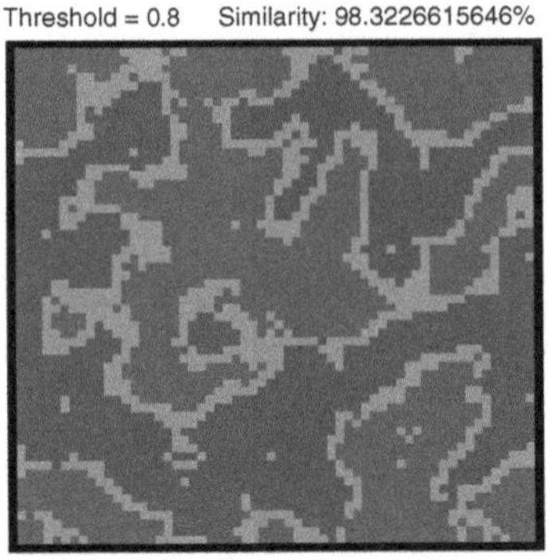

Figure 3.16 Racial segregation experiment based on the Schelling model.

3.4.1.2 Schelling Segregation Model

The Schelling segregation model, also known as the Schelling segregation model, was proposed by American economist Thomas Schelling [6]. This model illustrates the impact and role of homogeneity in spatial segregation, revealing the underlying mechanisms behind racial and income segregation. The model consists of three key elements: agents who generate behavior, the behavioral rules these agents follow, and the macro-level outcomes resulting from agent behavior. As shown in Figure 3.16, the experiment conceptualizes an entire city as a giant checkerboard, where each grid cell can be either occupied by an agent or remain vacant. There are two types of agents (medium gray and dark gray), with approximately equal numbers, and around 10% of the cells are left empty (light gray). Each agent has a minimum threshold—if the proportion of similar neighbors falls below this threshold, the agent will relocate to an unoccupied cell that meets its residential preference. The behavioral rules of the agents are as follows. (1) Count the number of same-type neighbors nearby. (2) If the number of similar neighbors is greater than or equal to the preference threshold, the agent is satisfied and stops moving; otherwise, it continues moving. (3) The agent will search for the nearest unoccupied cell that satisfies its threshold and move to that location.

Over time, the degree of segregation between different types of agents will become increasingly pronounced. By modifying the experimental settings—such as agent lifespan and the amount of empty space—it becomes evident that changing tolerance alone is not sufficient to prevent segregation, as nearly every agent prefers all neighbors to be of the same type. This phenomenon raises deeper questions about social issues: To what extent must racism persist before the entire societal structure shifts toward such a segregated pattern? How can racial segregation be reversed? Researchers have extended this type of model and applied it to studies of ethnic conflict in different regions, yielding encouraging results [73, 74].

3.4.1.3 Landscape Theory

Robert Axelrod of the University of Michigan proposed the landscape theory to study alliance formation in human societies and to predict the equilibrium states they are likely to reach [75]. This theory treats social units as entities that are drawn together by attractive forces and pushed apart by repulsive forces. This push-and-pull environment causes particles to cluster into alliances, and the final alliance configuration arranges these entities in the most stable manner. To describe this more effectively, Axelrod and his colleagues defined the concept of "total energy": For each possible configuration, the total of all attractive and repulsive forces between group members is calculated. The energy landscape is a graphical representation of all possible combinations and their corresponding energy levels—the more stable a combination, the lower its energy. Axelrod and his team used the landscape model to simulate potential alliance blocs on the eve of World War II, using six factors as the basis for determining interactions among countries: ethnic composition, religious belief, territorial disputes, ideology, economic conditions, and historical ties. Each factor was assigned a simple weight: +1 for convergent (cooperative) factors and −1 for conflictual or hostile factors.

Figure 3.17 sketches the scenarios of all the combinations. There are two basin structures in the figure, namely, the energy minima. The one that sinks deeper is very close to the actually formed alliance of the Central Powers and Axis Powers. Only Portugal and Poland have been classified into the "wrong" camp. Another basin foretold a very different scenario: The Soviet Union was in confrontation with all the other European countries. According to this energy landscape map, which "trough" Europe will eventually fall into is determined by where the starting point is located. The landscape theory tells us that for the special case of an alliance, it is possible to conduct actual calculations and obtain values with a certain degree of credibility, and the results are very similar. Through the calculation and experimental deduction of the landscape model, historical research becomes more concrete. When discussing the world situation, with a certain degree of quantitative language, the influencing factors in the historical development process can be identified, clarifying their scope and thus achieving the purpose of prediction.

3.4.2 Mechanism Exploration

Mechanism exploration focuses on modeling real-world social systems, emphasizing a high level of alignment between artificial social systems and real-world social systems, aiming to address existing or potentially emerging problems in society. Such applications often face the challenge of validating the accuracy of the mapping from the real society to the artificial one. Current researchers primarily aim to address this issue through rapidly evolving data acquisition and processing technologies, thereby enabling experimental results to address real-world questions.

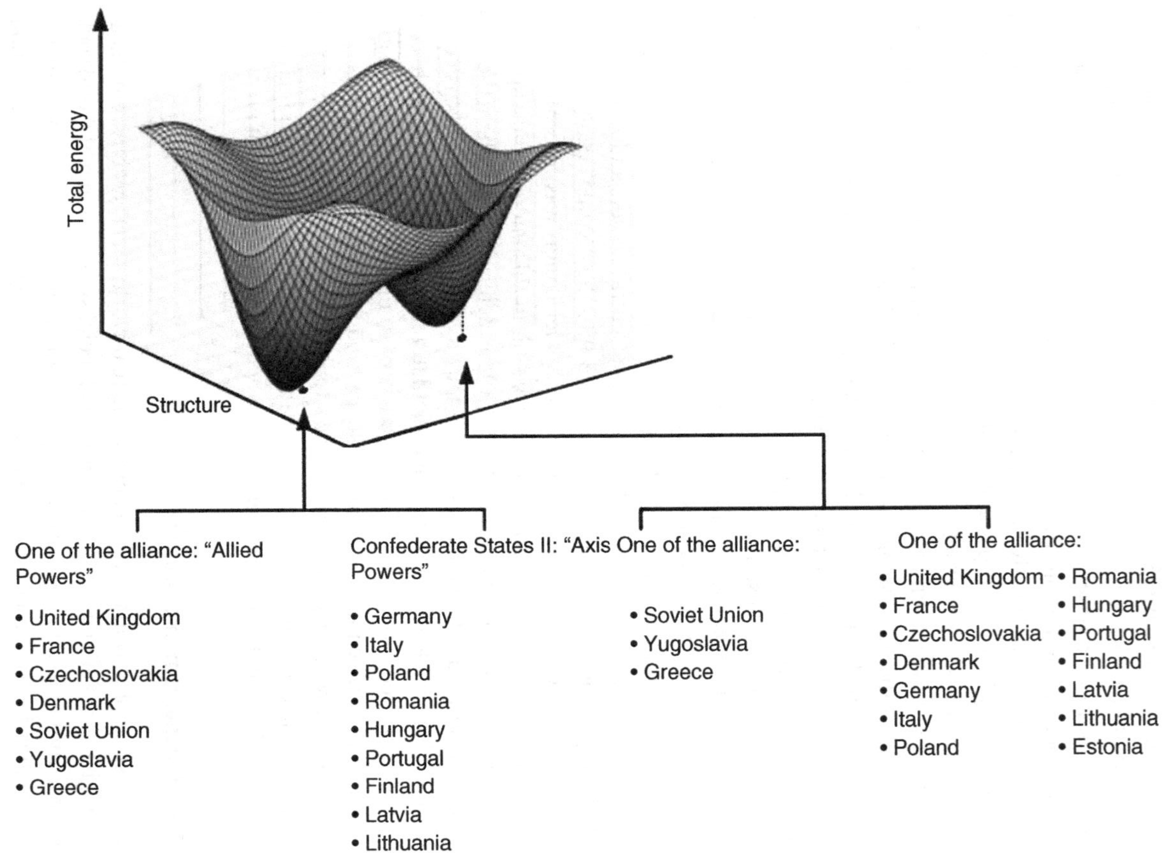

Figure 3.17 A picture of the Alliance camp on the eve of World War II.

3.4.2.1 Artificial Stock Market (ASM)

The stock market is clearly a complex system. Its multiple components and hierarchical structure make the stock market as a whole exhibit characteristics of complexity and unpredictability. To explore and understand how investors make portfolio choices, the SFI proposed the "artificial stock market" (ASM) model in 1987 [9]. It aims to understand the complexity of economic system from a novel perspective. This model abandons the assumption of a completely rational, omniscient, and omnipotent "economic man," and instead employs a bounded rational individual capable of learning, adapting to the environment and making decisions through induction. It regards the economic system as a complex system in which several interacting individuals exist. In this virtual market, several trading agents make predictions by observing the dynamic changes in stock prices and dividends in the digital world. Based on these predictions, they decide whether to purchase stocks and the quantity to buy, in order to maximize their own utility. All agents independently form their own expectations and have the ability to learn, allowing them to adjust their decisions based on the success or failure of their previous predictions. Conversely, the decisions made by all traders determine the state of supply and demand competition and thereby determine the price of stocks.

As shown in Figure 3.18, the entire stock trading market constitutes a self-contained computational system. The model operates in discrete time steps, and the experiment can proceed indefinitely. At the beginning of each cycle, the current dividend is issued and perceived by all agents. Each agent uses current and historical data on dividends and prices to match with the "condition" part of their rules and forms expectations for the next period's price and dividend. Based on these expectations, they then optimize the allocation of their funds between stocks and bonds. A market maker calculates the new stock price according to the aggregate supply and demand imbalance of all agents and a market-clearing price equation. At the start of the next time step, agents update the accuracy of

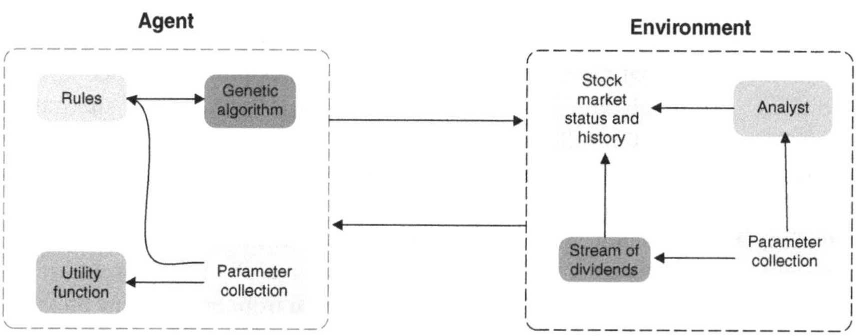

Figure 3.18 Structure of the interaction between agent and the stock market.

their prediction rules, and the new dividend is distributed. This process repeats until a termination condition is satisfied. The ASM offers a powerful metaphor for real-world stock markets. It can be used to test whether different expectations eventually evolve into identical rational expectations, thereby supporting the Efficient Market Hypothesis, or whether more complex behaviors emerge at both the individual and organizational levels, thus supporting the perspective of real investors and helping to explain real-world financial market phenomena. Subsequently, many researchers have improved the ASM model to conduct deeper analyses, such as the link between micro-level investor behavior and macro-level stock market dynamics, or the effectiveness of price limits.

3.4.2.2 Political Mechanism

Claudio Cioffi-Revilla from George Mason University proposed the MASON RebeLand model to analyze how the state of a polity or its political stability is influenced by internal (endogenous) or environmental (exogenous) processes, such as changes in the polity's economy, population, culture, natural environment (ecosystem), climate, or broader socio-natural stressors [76]. Figure 3.19 presents a "map" view of RebeLand as a polity or nation situated within a natural environment. The country consists of an island surrounded by water, thus omitting interactions with external or neighboring countries. The RebeLand environment includes terrain and a simplified weather system that can simulate climatic dynamics (e.g., prolonged droughts, climate change). The rebellion and political elements include a society and a government system that addresses public issues through policymaking. Initially, the government formulates policies to address social problems. In subsequent simulations, under certain conditions, the society may also give rise to rebels who interact with the government's military, along with other emergent phenomena.

There are two types of rebels. The main agents include the general population, cities, and countries. Cities represent local public administrative organizations, while the state represents the entire government system of the polity. Secondary agents include rebels, rebel groups, and government forces representing police and military units that arise from among the general population under various conditions. The rebels are supported by rebel groups, which fund alternative policies that oppose the official policies. Therefore, rebel groups represent the alternative political organizations that compete with the state. The government forces attempt to destroy the rebels by attacking them. This study presents three scenarios, showing stable, unstable, and failed political states, respectively. An important outcome is related to the overall resilience of the political system. Usually, the failure of a political system does not result from one or a few pressure issues to experience, but rather a large set of issues (such as inflation, rebellion, and environmental pressure). Such results may not immediately lead to actionable

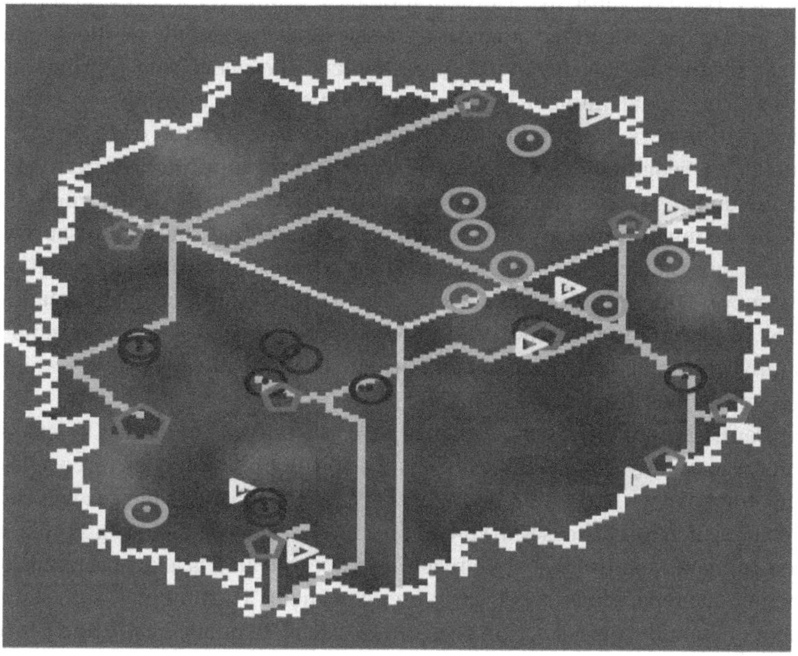

Figure 3.19 The map of RebeLand Island features its main natural and social characteristics. Cities are indicated in pentagons, natural resources in triangle, and the rebel forces and government forces in medium-dark gray circle and dark gray circle, respectively. Roads and provincial boundaries are represented by medium gray and light gray, respectively. The physical topography is displayed in green tones, and the island is surrounded by the ocean.

policy recommendations for specific projects, but at least, they could provide valuable new insights for researchers and policy analysts. This type of research is also used in election prediction and applied political science research [77].

3.4.2.3 Epidemic Spread

With the outbreaks of emerging infectious diseases such as SARS, H1N1, and COVID-19, epidemic transmission models have become a research hotspot. Traditional predictive models rely heavily on variable selection and expert knowledge, attempting to make optimal estimates based on uncertain information. However, the spread of infectious diseases is a dynamic and uncertain process, involving numerous influencing factors as well as many unknown aspects and unexpected events, making predictive research extremely challenging. Take the infection rate of a virus as an example—it depends not only on local social behavior, public health conditions, and political decisions but also on changes in response to evolving intervention measures. Computational experiments use artificial society models to achieve integrated analysis of basic data, modeling methods, and analytical results, mainly involving three aspects: the generation of epidemic dynamics (e.g., initial number of infections, contact rate, transmission speed, virus incubation period, mortality rate, recovery probability); the representation of spatial and geographical features (e.g., city type, transportation networks, population density, temperature, weather, age distribution, urban infrastructure); and the modeling of resources and governance capacity (e.g., medical resources, number of hospital beds, social organizations, control measures, information transparency). Different artificial society systems vary in their implementation methods, representation techniques, and levels of accuracy. Table 3.8 presents a comparative summary of the characteristics of several artificial society systems related to infectious disease research.

At present, computational experiments have become an important means of studying large-scale infectious disease transmission, mainly applied in three fields. (1) Trend forecasting: The development of transportation systems has made it easier for major infectious diseases to spread on a large scale. Analyzing and predicting their transmission trends before reaching a specific area is a key prerequisite for emergency preparedness. (2) Impact preassessment: The outbreak and evolution of an epidemic within a region pose direct threats to public health and cause secondary effects on the socioeconomic environment. Quantitative assessment of its impact serves as an important basis for emergency stockpiling and intervention decisions. (3) Intervention strategy optimization: A variety of intervention measures can be used in emergency prevention and control of major infectious diseases, each with different targets and intensities—for example, the choice between targeted immunization and herd immunity. In practice, selecting reasonable interventions and their intensities to form a combined intervention strategy—one that controls the epidemic while minimizing the cost of intervention—is a major challenge in emergency decision-making. However, no computational model exists that perfectly matches the real-world transmission process. All models are merely approximate abstractions of the actual transmission process and must continuously avoid and reduce uncertainty in data, modeling, computation, and result analysis in order to achieve simulation outcomes that increasingly approximate the real transmission process.

3.4.3 Parallel Optimization

The purpose of scientific research is to provide a causal hypothesis that can link existing facts. If the constructed artificial society model represents such a regularity, then the output behavior of the model can exert an effective impact on the real world. The parallel world, by establishing an artificial society model

Table 3.8 Comparison of Typical Characteristics of Large-Scale Propagation Simulation Systems.

Feature	BioWar [13]	EpiSimS [61]	GSAM [78]	CovidSim [79]	ASSOCC [80]	SIsaR [81]
Disease Type	Droplet transmission, physical contact	Smallpox, influenza	Not specific to any disease, uses H1N1 as an example	COVID-19, other respiratory viruses	COVID-19	COVID-19
Main Purpose	Impact assessment, strategy optimization	Impact assessment, strategy optimization	Study the spread and control of infectious diseases	Impact assessment, strategy optimization	Impact assessment, policy trade-off	Assess costs and benefits of different intervention policies
Simulation Scale	Medium-sized US city	Medium-sized US city	Global	National	National	National
Simulation Method	Multiagent	Multiagent	Multiagent	Geospatial Unit	Multiagent	Multiagent
Programming Language	C++	—	Java	C++	R language & NetLogo	NetLogo
Visualization	No	Yes	Yes	Yes	Yes	Yes
Open source	No	No	No	Yes	Yes	Yes (available online)

that is homomorphic to the real society, enables parallel execution and cyclic feedback between the two, thereby supporting the management and control of real complex systems.

3.4.3.1 Island Economy

Economic inequality is worsening globally and has drawn increasing attention due to its negative impacts on economic opportunity, health, and social welfare. Governments can use economic policies—particularly in taxation and redistribution—to improve social outcomes. However, the challenge lies in the fact that taxation tends to reduce productivity. Workers may choose to withdraw from labor if they are required to pay taxes on their income, thereby decreasing the utility of labor. This effect can be especially significant for highly skilled workers. Thus, a trade-off must be made between economic equality and productivity: Do redistributive interventions that allow wealth to be reallocated result in a reduction of total wealth available for redistribution? Because of the coupling relationship between taxation and labor, the key unresolved issue is how to mitigate economic inequality while maintaining productivity. In practice, such economic problems are difficult to study due to the lack of suitable economic data and the limited opportunities for experimentation.

To this end, Salesforce proposed a new research project called "The AI Economist." It conducts economic simulation through AI agents, where both individuals (agents) and tax strategies (planners) possess learning and adaptive capabilities [20]. This project introduces a two-level reinforcement learning (RL) framework involving citizens and tax policies, aiming to discover tax strategies that efficiently balance economic equality and productivity. This framework operates without relying on prior knowledge or modeling assumptions, enabling direct optimization for any socioeconomic objective and learning solely from observable data. Figure 3.20 presents the core implementation framework of The AI Economist, focusing on the evolutionary game mechanism between agents and the planner—that is, how RL is used to jointly optimize individual agent behavior and the tax policy employed in the economy.

In the inner loop, various types of worker agents gain experience by performing labor, earning income, and paying taxes. They adapt their behavior through trial-and-error learning to maximize their individual utility. In the outer loop, the planner agent (PA) adjusts the tax policy to optimize a social objective, which is based on a social welfare function from economic theory that considers the trade-off between income equality and productivity. The AI Economist offers a powerful framework that integrates economic theory with data-driven simulation, producing objective insights into the economic consequences of different tax policies. However, there are still some limitations—for example, it does not yet simulate human behavioral factors or interpersonal interactions,

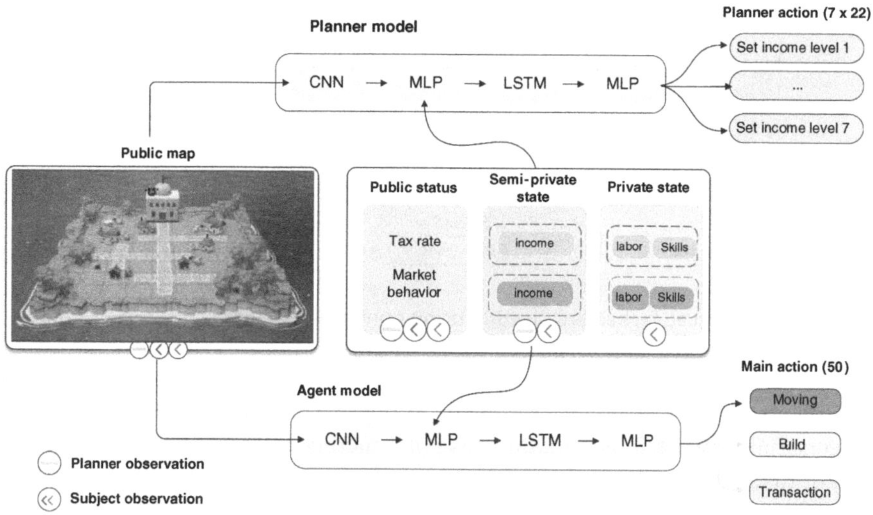

Figure 3.20 Island economy.

and it focuses on a relatively small economy. The developers at Salesforce hope that The AI Economist can address the complexities that traditional economic research struggles to handle and provide an objective tool for studying the real-world impact of policies. Researchers can further refine this framework for use in "what-if" experiments and counterfactual reasoning, which is often a key value-added feature of AI-driven simulators.

3.4.3.2 Virtual Taobao

Product search is the core business of Taobao, which is one of the largest retail platforms. Taobao's commercial objective is to increase sales revenue by optimizing the strategy of displaying PVs (page views) to customers. Since the feedback signal from customers depends on a sequence of PVs, the problem can be viewed as a multi-step decision-making problem rather than a single-step supervised learning task. The engine and the user serve as each other's environment. RL solutions excel at learning sequential decision-making and maximizing long-term rewards. However, one major obstacle in directly applying RL in these scenarios is that current RL algorithms typically require a large number of interactions with the environment, which incurs high physical costs—such as spending real money, taking days to months, poor user experiences, or even life-threatening risks in medical tasks.

To avoid physical costs, researchers use a simulator (i.e., Virtual Taobao) to conduct RL training, allowing policies to be trained offline in the simulator using any RL algorithm that maximizes long-term rewards [82]. However, simulating the behaviors of hundreds of millions of customers in a dynamic environment is

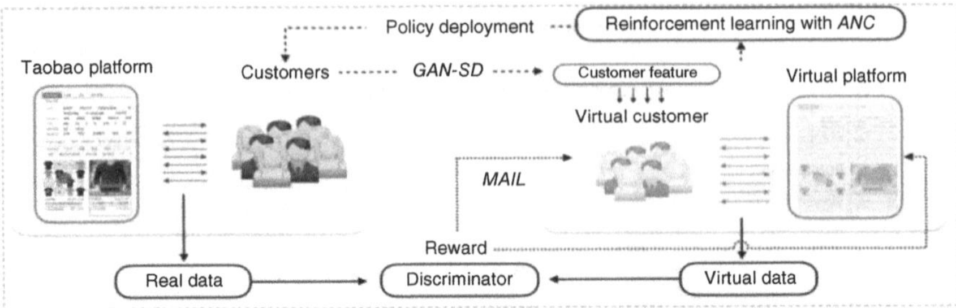

Figure 3.21 Virtual Taobao architecture using reinforcement learning.

more challenging. In this work, Virtual Taobao is constructed by generating both customers and interactions. As shown in Figure 3.21, the GAN-SD (generative adversarial network for simulating diversity) method is proposed to simulate different customers, including their requests; and a multiagent adversarial imitation learning (MAIL) method is proposed for interaction generation. MAIL simultaneously learns both customer and platform policies, following the idea of GAIL (generative adversarial imitation learning), using a generative adversarial framework. MAIL trains a discriminator to distinguish simulated interactions from real ones. The discrimination signal is used as a reward and fed back to the customer and platform policies to generate more realistic interactions. Once customer and interaction generation are complete, Virtual Taobao is established.

Experimental results show that Virtual Taobao successfully reconstructs attributes that closely resemble those of the real environment. Virtual Taobao can be used to train platform policies to maximize revenue. Compared to traditional supervised learning methods, the policies trained within Virtual Taobao achieved over a 2% increase in revenue when deployed in the real environment.

3.4.3.3 Autonomous Driving

Autonomous driving requires rigorous testing and validation of a vehicle's ability to understand complex traffic scenarios and make driving decisions—this is one of the major challenges in the field of AI. There are two main reasons for this. (1) Autonomous driving development demands vast amounts of data: According to an estimate by the RAND Corporation, a complete autonomous driving system must undergo at least 11 billion miles (approximately 17–18 billion km) of validation before reaching mass production readiness. (2) Extreme scenario testing is rare and hard to reproduce: Constructing real-world scenarios such as blizzards, heavy rain, or typhoons is extremely challenging, and creating, replicating, and iterating such environments are costly—almost impossible to accomplish in practice.

Therefore, training and testing autonomous driving strategies in virtual environments has become a feasible technical solution. It is controllable, repeatable, and safe. To ensure that simulation results in the virtual world closely match those in the real world, three levels of fidelity must be achieved: (1) geometric fidelity, which requires 3D scene simulation and sensor simulation, ensuring that both the environment and vehicle conditions mirror the real world; (2) logical fidelity, which involves simulating the decision-making and planning processes of the test vehicle in the virtual world; and (3) physical fidelity, which requires simulation of the vehicle's control behavior and the outcomes of its dynamics. Additionally, the simulation platform must support high concurrency, allowing for permutations and combinations of vehicle responses across all scenarios. Currently, parallel learning methods are increasingly applied in virtual scenario generation and intelligent testing of autonomous vehicles [38–41].

As shown in Figure 3.22, Tencent has built a cloud-based virtual city that runs in parallel with the real physical world, leveraging its simulation platform, high-definition mapping platform, and data cloud platform. Urban simulation encompasses not only static environmental information but also dynamic data such as traffic and pedestrian flow. It can also integrate virtual elements such as traffic flow information. This supports both the development and safety validation of autonomous driving, while also contributing to the construction of smart cities and intelligent transportation systems. To improve the utilization of road data and enrich testing scenarios, technologies similar to agent AI can be employed. By training traffic flow AI with large volumes of road-sensing data, highly realistic and interactive traffic scenarios can be generated for closed-loop simulation, enhancing testing efficiency and reducing data collection costs. For

Figure 3.22 Tencent TAD Sim simulation system scenario demonstration.

example, when an autonomous vehicle under test attempts to overtake, Agent AI can control nonplayer character (NPC) vehicles to perform realistic avoidance or other strategic behaviors consistent with real-world dynamics.

3.5 Chapter Summary

This chapter addresses the challenges of quantification, analysis, and regulation in social–ecological system governance. From the perspective of complexity science, it proposes a technical pathway that integrates computational experimentation with real-world social scenarios. This pathway includes three key component models: agent model, environment model, and rule model. Together, these components aim to provide technical support for building artificial society similar to real world. Finally, some classic cases of social simulation are given, covering three types: thought experiments, mechanism exploration, and parallel optimization.

References

1 Watts, D.J. and Strogatz, S.H. (1998). Collective dynamics of 'small-world' networks. *Nature* 393 (6684): 440.

2 Barabási, A.L. and Albert, R. (1999). Emergence of scaling in random networks. *Science* 286 (5439): 509–512.

3 Axelrod, R. and Hamilton, W.D. (1981). The evolution of cooperation. *Science* 211 (4489): 1390–1396.

4 Epstein, J.M. and Axtell, R.L. (1996). *Growing Artificial Societies: Social Science from the Bottom Up*. Brookings Institution.

5 Gardner, M. (1970). The fantastic combinations of John Conway's new solitaire game 'life'. *Scientific American* 223 (4): 120–123.

6 Schelling, T.C. (1971). Dynamic models of segregation. *Journal of Mathematical Sociology* 1 (2): 143–186.

7 Langton, C.G. Artificial life. In: *1991 Lectures in Complex Systems*, vol. 1992 (ed. L. Nadel and D.L. Stein), 189–241. Reading, MA: Addison-Wesley.

8 Builder, C.H. and Bankes, S.C. (1991). *Artificial Societies: A Concept for Basic Research on the Societal Impacts of Information Technology*. Santa Monica, CA: RAND Corporation.

9 Arthur, W.B., Holland, J.H., LeBaron, B. et al. (1997). Asset pricing under endogenous expectations in an artificial stock market. In: *The Economy as an Evolving Complex System II* (ed. W.B. Arthur, D. Lane, and S.N. Durlauf), 15–44. Addison-Wesley.

10 Gilbert, N. and Conte, R. (1995). *Artificial Societies, The Computer Simulation of Social Life*. University College London Press.

11 Basu, N., Pryor, R., and Quint, T. (1998). ASPEN: A microsimulation model of the economy. *Computational Economics* 12: 223–241.

12 Epstein, J.M. and Axtell, R. (1996). *Growing Artificial Societies: Social Science from the Bottom Up*. Brookings Institution Press.

13 Carley, K.M., Fridsma, D.B., Casman, E. et al. (2006). BioWar: scalable agent-based model of bioattacks. *IEEE Transactions on Systems, Man, and Cybernetics - Part A: Systems and Humans* 36 (2): 252–265.

14 Feiyue, W. (2004). The parallel system method and the management and control of complex systems. *Control and Decision* 19 (5): 485–489.

15 Wang, F.Y. (2007). Toward a paradigm shift in social computing: the ACP approach. *IEEE Intelligent Systems* 22 (5): 65–67.

16 North, M.J, Tatara, E., Collier, N.T. et al. (2007). Visual agent-based model development with repast simphony. Tech. rep., Argonne National Laboratory.

17 Squazzoni, F., Jager, W., and Edmonds, B. (2014). Social simulation in the social sciences: a brief overview. *Social Science Computer Review* 32 (3): 279–294.

18 Xiao, X. (2020). *Computational Experiment Methods for Complex Systems: Principles, Models, and Case Studies*. Science Press.

19 Zhang, W., Valencia, A., and Chang, N.B. (2021). Synergistic integration between machine learning and agent-based modeling: a multidisciplinary review. *IEEE Transactions on Neural Networks and Learning Systems* 34 (5): 2170–2190.

20 Zheng, S., Trott, A., Srinivasa, S. et al. (2022). The AI economist: taxation policy design via two-level deep multiagent reinforcement learning. *Science Advances* 8 (18): eabk2607.

21 Park, J.S., O'Brien, J., Cai, C.J. et al. (2023). Generative agents: interactive simulacra of human behavior. *Proceedings of the 36th Annual ACM Symposium on User Interface Software and Technology*, pp. 1–22.

22 Flache, A. and de Matos, F.C.A. (2021). Agent-based computational models. In: G. Manzo (Ed.), *Research Handbook on Analytical Sociology*, 453–473. Edward Elgar Publishing.

23 Axelrod, R. (1997). Advancing the art of simulation in the social sciences. In: *Simulating Social Phenomena*, 21–40. Berlin, Heidelberg: Springer Berlin Heidelberg.

24 Pinyol, I. and Sabater-Mir, J. (2013). Computational trust and reputation models for open multi-agent systems: a review. *Artificial Intelligence Review* 40 (1): 1–25.

25 Axtell, R.L. (2007). What economic agents do: How cognition and interaction lead to emergence and complexity. *The Review of Austrian Economics* 20 (2): 105–122.

26 Moss, S. and Edmonds, B. (2005). Towards good social science. *Journal of Artificial Societies and Social Simulation* 8 (4): 1–13.

27 De Marchi, S. and Page, S.E. (2014). Agent-based models. *Annual Review of Political Science* 17 (1): 1–20.

28 Bruch, E. and Atwell, J. (2015). Agent-based models in empirical social research. *Sociological Methods & Research* 44 (2): 186–221.

29 Roth, C. and Cointet, J.P. (2010). Social and semantic coevolution in knowledge networks. *Social Networks* 32 (1): 16–29.

30 Boero, F. (2010). The study of species in the era of biodiversity: a tale of stupidity. *Diversity* 2 (1): 115–126.

31 Wunder, M., Suri, S., and Watts, D.J. (2013). Empirical agent based models of cooperation in public goods games. *Proceedings of the Fourteenth ACM Conference on Electronic Commerce*, pp. 891–908.

32 Duffy, J. (2006). Agent-based models and human subject experiments. *Handbook of Computational Economics* 2: 949–1011.

33 Windrum, P., Fagiolo, G., and Moneta, A. (2007). Empirical validation of agent-based models: alternatives and prospects. *Journal of Artificial Societies and Social Simulation* 10 (2): 8.

34 Hägerstrand, T. (1965). A Monte Carlo approach to diffusion. *European Journal of Sociology/Archives Européennes de Sociologie* 6 (1): 43–67.

35 Janssen, M.A. and Jager, W. (2003). Simulating market dynamics: interactions between consumer psychology and social networks. *Artificial Life* 9 (4): 343–356.

36 Billari, F.C., Prskawetz, A., Aparicio Diaz, B. et al. (2007). The "wedding-ring" an agent-based marriage model based on social interaction. *Demographic Research* 17: 59–82.

37 Mäs, M. and Flache, A. (2013). Differentiation without distancing. explaining bi-polarization of opinions without negative influence. *PLoS One* 8 (11): e74516.

38 Dugundji, E.R. and Gulyás, L. (2008). Sociodynamic discrete choice on networks in space: impacts of agent heterogeneity on emergent outcomes. *Environment and Planning B: Planning and Design* 35 (6): 1028–1054.

39 DiMaggio, P. and Garip, F. (2011). How network externalities can exacerbate intergroup inequality. *American Journal of Sociology* 116 (6): 1887–1933.

40 Fountain, C. and Stovel, K. (2014). Turbulent careers: Social networks, employer hiring preferences, and job instability. *Analytical Sociology* 339–370.

41 Mare, R. and Bruch, E. (2011). Segregation dynamics. In: *The Oxford Handbook of Analytical Sociology* (ed. P. Hedström and P. Bearman). Oxford: Oxford University Press.

42 Lux, T. and Marchesi, M. (1999). Scaling and criticality in a stochastic multi-agent model of a financial market. *Nature* 397 (6719): 498–500.

43 Poletti, P., Ajelli, M., and Merler, S. (2011). The effect of risk perception on the 2009 H1N1 pandemic influenza dynamics. *PLoS One* 6 (2): e16460.

44 Frias-Martinez, E., Williamson, G., and Frias-Martinez, V. (2011). An agent-based model of epidemic spread using human mobility and social network information. *2011 IEEE Third International Conference on Privacy, Security, Risk and Trust and 2011 IEEE Third International Conference on Social Computing*, pp. 57–64. IEEE.

45 Sugden, R. (2000). Credible worlds: the status of theoretical models in economics. *Journal of Economic Methodology* 7 (1): 1–31.

46 Axtell, R.L., Epstein, J.M., Dean, J.S. et al. (2002). Population growth and collapse in a multiagent model of the Kayenta Anasazi in Long House Valley. *Proceedings of the National Academy of Sciences* 99 (suppl_3): 7275–7279.

47 Cederman, L.E. (2005). Computational models of social forms: Advancing generative process theory. *American Journal of Sociology* 110 (4): 864–893.

48 Tesfatsion, L. (2006). Agent-based computational economics: a constructive approach to economic theory. *Handbook of Computational Economics* 2: 831–880.

49 Diez Roux, A.V. (2015). The virtual epidemiologist—promise and peril. *American Journal of Epidemiology* 181 (2): 100–102.

50 Imbens, G.W. and Rubin, D.B. (2015). *Causal Inference for Statistics, Social, and Biomedical Sciences.* Cambridge: Cambridge University Press.

51 Morgan, S.L. and Winship, C. (2015). *Counterfactuals and Causal Inference: Methods and Principles for Social Research*, 2e. Cambridge: Cambridge University Press.

52 Grune-Yanoff, T. (2009). The Explanatory Potential of Artificial Societies. *Synthese* 169 (3): 539–555.

53 Elsenbroich, C. (2012). Explanation in agent-based modelling: functions, causality or mechanisms? *Journal of Artificial Societies and Social Simulation* 15 (3): 1. http://jasss.soc.surrey.ac.uk/15/3/1.html.

54 Macy, M. and Sato, Y. (2008). Reply to will and hegselmann. *Journal of Artificial Societies and Social Simulation* 11 (4): 11.

55 Casini, L. (2014). Not-so-minimal models: between isolation and imagination. *Philosophy of the Social Sciences* 44 (5): 646–672.

56 Ylikoski, P. and Aydinonat, N.E. (2014). Understanding with theoretical models. *Journal of Economic Methodology* 21 (1): 19–36.

57 Tornberg, A. (2019). Abstractions on steroids: a critical realist approach to computer simulations. *Journal for the Theory of Social Behaviour* 49 (1): 127–143.

58 Manzo, G. (2022). *Agent-based Models and Causal Inference.* John Wiley & Sons.

59 Russell, S. and Norvig, P. (1995). Intelligent agents. *Artificial Intelligence: A Modern Approach* 74: 46–47.

60 Tu, X. and Terzopoulos, D. (1994). Artificial fish: physics, locomotion, perception, behavior. *ACM Computer Graphics, Annual Conference Series, Proceedings of SIGGRAPH'94*, pp. 43–50.

61 Mniszewski, S.M. and Del Valle, S.Y. (2013). EpiSimS: large-scale agent-based modeling of the spread of disease. Los Alamos National Lab.(LANL), Los Alamos, NM (United States).

62 Wang, F.Y. (2022). The DAO to metacontrol for metasystems in metaverses: The system of parallel control systems for knowledge automation and control intelligence in CPSS. *IEEE/CAA Journal of Automatica Sinica* 9 (11): 1899–1908.

63 Grieves, M.W. (2005). Product lifecycle management: the new paradigm for enterprises. *International Journal of Product Development* 2 (1–2): 71–84.

64 Dorigo, M., Birattari, M., and Stutzle, T. (2006). Ant colony optimization. *IEEE Computational Intelligence Magazine* 1 (4): 28–39.

65 Kennedy, J. and Eberhart, R. (1995, November). Particle swarm optimization. *Proceedings of ICNN'95-international Conference on Neural Networks* (Vol. 4), pp. 1942–1948. IEEE.

66 Karaboga, D. and Akay, B. (2009). A comparative study of artificial bee colony algorithm. *Applied Mathematics and Computation* 214 (1): 108–132.

67 Dodds, P.S. and Watts, D.J. (2005). A generalized model of social and biological contagion. *Journal of Theoretical Biology* 232 (4): 587–604.

68 Chen, W., Yuan, Y., and Zhang, L. (2010). Scalable influence maximization in social networks under the linear threshold model. *2010 IEEE International Conference on Data Mining*, pp. 88–97. IEEE.

69 Hoppitt, W. and Laland, K.N. (2013). *Social Learning: An Introduction to Mechanisms, Methods, and Models.* Princeton University Press.

70 Reips, U.D. (2002). Standards for internet-based experimenting. *Experimental Psychology* 49 (4): 243.

71 Bask, M. and Bask, M. (2014). Social influence and the Matthew mechanism: the case of an artificial cultural market. *Physica A: Statistical Mechanics and its Applications* 412: 113–119.

72 Lavin A, Krakauer D, Zenil H, et al. (2021). Simulation intelligence: towards a new generation of scientific methods. arXiv preprint arXiv:2112.03235.

73 Markisic, S., Neumann, M., and Lotzmann, U. (2012). Simulation of ethnic conflicts in former Jugoslavia. ECMS. , pp. 37–43.

74 Ravi, B., Miodownik, D., and Nart, J. (2008). REsCape: an agent-based framework for modeling resources, ethnicity, and conflict. *The Journal of Artificial Societies and Social Simulation* 11 (27).

75 Axelrod, R. and Bennett, D.S. (1993). A landscape theory of aggregation. *British Journal of Political Science* 211–233.

76 Cioffi-Revilla, C. and Mark, R. (2010). Mason rebeland: an agent-based model of politics, environment, and insurgency. *International Studies Review* 12 (1): 31–52.

77 Laver, M. and Sergenti, E. (2012). *Party Competition: An Agent-based Model.* Princetion: Princeton University Press.

78 Epstein, J.M. (2009). Modelling to contain pandemics. *Nature* 460 (7256): 687–687.

79 Schneider, K.A., Ngwa, G.A., Schwehm, M. et al. (2020). The COVID-19 pandemic preparedness simulation tool: CovidSIM. *BMC Infectious Diseases* 20 (1): 1–11.

80 Ghorbani, A., Lorig, F., de Bruin, B. et al. (2020). The ASSOCC simulation model: a response to the community call for the COVID-19 pandemic. Review of Artificial Societies and Social Simulation. URL. https://rofasss.org/2020/04/25/the-assocc-simulation-model.

81 Pescarmona, G., Terna, P., Acquadro, A. et al. (2020) How can ABM models become part of the policy-making process in times of emergencies - the S.I.S.A.R. Epidemic model. Review of Artificial Societies and Social Simulation, 20th Oct 2020. https://rofasss.org/2020/10/20/sisar/.

82 Shi, J.C., Yu, Y., Da, Q. et al. (2019). Virtual-taobao: virtualizing real-world online retail environment for reinforcement learning. *Proceedings ofe AAAI Conference on Artificial Intelligence* 33 (01): 4902–4909.

4 Methodological Framework of Computational Experiments

This chapter details the computational experiment framework, comprising three core steps: (1) model docking between artificial society and the real world; (2) experimental design; and (3) causal analysis. To support the abovementioned theory, the MACE (multi-agent computational experiments) platform is introduced, which was independently developed by our team. It integrates the computational experimentation theories and technologies presented in Chapters 5–9, which can reveal causal relationships between individual behaviors and complex social phenomena. Furthermore, it enables dynamic optimization of decisions under complex and uncertain environments—thus realizing intelligent decision-making. This chapter can help readers gain a preliminary understanding of the entire computational experiment method, laying the foundation for the study of subsequent chapters.

4.1 Framework of Computational Experiment Method

The operation of complex social system involves coordination across multiple organizations or intelligent agents and requires an operational strategy that can adapt to various potential constraints. Computational experiments enhance system efficiency by identifying key factors, defining operational rules, and designing interventions, thereby offering a conceptual framework to explain, understand, guide, and regulate real-world social systems. As shown in Figure 4.1, Ref. [1] provides a general framework for computational experiment method, which can integrate agent-based modeling (ABM) logic and experimental techniques and support the customization of general method for different application fields. The framework of computational experiment method includes five steps: artificial society modeling, experimental system construction, computational experiment design, computational experiment analysis, and computational experiment verification, and forms a feedback closed loop. The key bridge connecting these

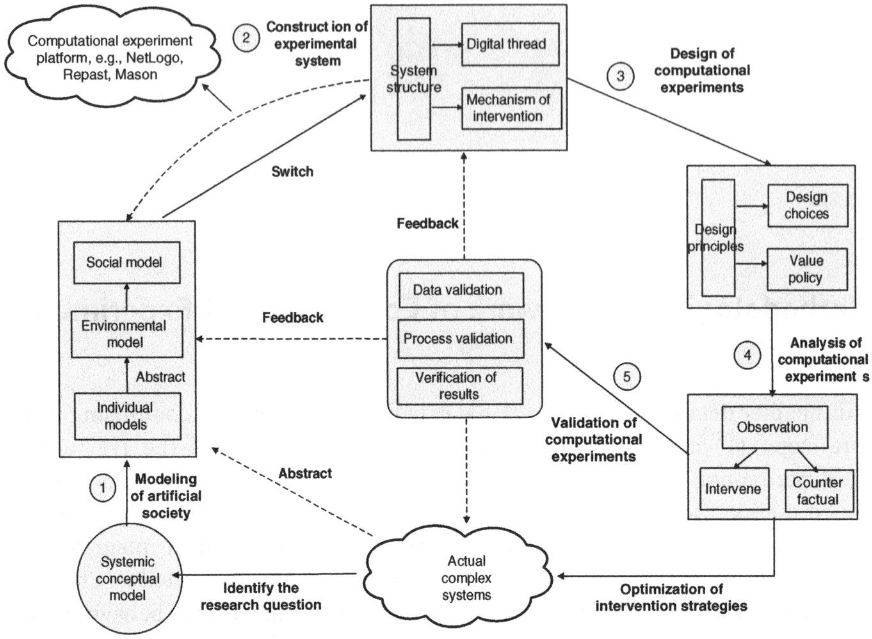

Figure 4.1 A technical framework for computational experiment method.

steps is also described in detail, including model docking (connecting artificial society with experiment system), scenario generation (connecting experiment system with experimental design), and causal analysis (connecting experimental design with experiment analysis). Details are described later.

4.1.1 Model Docking

This step mainly focuses on mapping different CPSS (cyber–physical–social system) scenarios to artificial society models. Due to the intricate nature of cyber–physical–social integration, CPSS's domain model often consists of multiple models, forming a model family. Key challenges include modeling autonomous/social individuals and capturing correlations between group behavior and algorithmic intervention. The details are as follows:

4.1.1.1 Social Space → Agent Model

In social space, the characteristics of CPSS are mainly reflected in the following. (1) The research object falls under the artificial society, which comprises various intelligent agents, such as humans, machines, organizations, and institutions.

Complex agents can be constructed by combining relatively simple agents. (2) All agents can be categorized into different classes, with each agent class possessing a unique structure, attributes, and decision-making mechanism that correspond to its functions.

In an artificial society, the individual agent model serves as a platform for integrating multidisciplinary knowledge that can be tailored to specific application scenarios. Typically, an individual agent's structure comprises four main components: (1) perception, (2) decision-making, (3) reaction, and (4) optimization, as detailed in [1]. The information control flow within the agent structure connects these components into a cohesive unit. The collective behavior of numerous agents contributes to the emergence of specific macro-level phenomena. The operation mode can be expressed by the formula as follows:

$$S_t = \{m_1(v_t), m_2(v_t), \ldots, m_n(v_t), Network(n)\} \tag{4.1}$$

$$v_t = f(Y_t) \mid_{\varepsilon(\alpha, v_t, E_t, Y_t) \Rightarrow <S_{t+1}, N, \Lambda>} \tag{4.2}$$

In Formula 4.1, S_t indicates that the nonlinear interaction between individuals draws forth the macro emergence characteristics of the system, such as population size, city size, and wealth size; $m_i(v_t)$ represents the dynamic mechanism of each individual; v_t can be described in Formula 4.2; and $Network(n)$ indicates the relationship between n agents. In Formula 4.2, $f(Y_t)$ represents the function mapping of decision Y_t to behavior v_t, where agent α completes the behavior v_t in the case of environment E_t and decision Y_t, and the state set of the next cycle S_{t+1} satisfies the constraint conditions N and the task criteria Λ of the agent.

4.1.1.2 Physical Space → Environment Model

In physical space, the characteristics of CPSS are mainly reflected in the following. (1) Agents can publish their resources, including IoT facilities, platforms, data, and algorithms, as services. These services can be customized and aggregated on-demand to meet user needs. (2) Social space affects system operation, requiring consideration of supply-side performance and pursuit of high efficiency (output/time). Additionally, it's crucial to consider demand-side impact and continuously provide valuable output to avoid invalid results.

In the artificial society, the environment model is the mapping of the actual service system in the computer. It is also the activity space for the agent to exist. To define the environment model more clearly, Formula 4.3 gives key elements required for environment model:

$$E = \{O_t, D_t, \Delta t, Input, Metrics\} \tag{4.3}$$

where E represents the service-oriented environment model; O_t represents datasets of all types of objects in the service system, including people, cars, roads, environment, and nonphysical objects; $D_t = \{(x, y)\}$ represents the perception space dimension, i.e., the coordinate set of individual objects; Δt represents the time dimension, i.e., the time period taken as per the perceived demand; *Input* represents the external influencing variables of the system; and *Metrics* represents the system evaluation indicators, that is, service performance (performance = valuable output/time used).

4.1.1.3 Cyber Space → Rules Model

CPSS in cyberspace evolves through a combination of top-down intervention and bottom-up emergence. Top-down intervention refers to the actions taken by policymakers, regulators, and other higher-level entities to intervene and control cyberspace through the creation of rules, policies, laws, and other measures. Bottom-up emergence refers to the self-organizing, self-adaptive phenomena of each individual agent and patterns that emerge within the network. Individual learning mechanisms are crucial drivers of system evolution, while adaptive adjustment of intervention algorithms can either accelerate or hinder the process. By identifying differences between actual performance and expected standards, intervention strategies can adjust individual guidance and achieve predetermined target status through cyclic feedback. In artificial society, the system's evolution mechanism is described mainly by rules models expressed by a formula:

$$Y_t(\alpha) = Y_{t-1}(\alpha) \oplus Fit(E_{t-1}(S)) \tag{4.4}$$

Formula 4.4 indicates the decision updating mechanism of agent α. Y_t is about how to optimize and update its own rule base gradually according to the feedback learning principle of "increasing benefit" so that they can take the behavior closer to the completion of the goal. *Fit*() represents the fitness function, which is regarded as the mapping relationship between policy and fitness. According to the concept of value engineering, the fitness value is generally defined as the ratio of outcome to cost. $\oplus$ represents that the decision-making mechanism Y_{t-1} is updated based on external perception $E_{t-1}(S)$, and S means the system state. Commonly used learning models are summarized into three categories: reinforcement learning (RL), evolutionary learning, and imitation learning [2, 3]. The example of an artificial stock market (ASM) can enable readers to better understand the evolution mechanisms of the rule model [4]. Artificial intelligence (AI) traders in this virtual market make predictions based on stock price and dividend fluctuations. They decide whether and how many stocks to buy based on these predictions. The collective decisions of all traders determine stock prices, creating a self-contained computing system. Traders can learn, adjust their decisions based on past outcomes, and generate innovation using genetic algorithms. For example, in the ASM, the agent (trader) makes decisions based on a set of rules. If a rule leads to a positive gain, its weight increases; if it results in a negative gain, its weight decreases. The probability of using each rule is proportional to its weight, favoring rules with higher weights. Traders can adjust their rules over time. The rule set evolves through a process of selecting, modifying, and recombining rules, allowing it to adapt to changing market conditions.

4.1.2 Scenario Generation

This step mainly focuses on constructing diverse experiment scenarios for the complexity of CPSS to evaluate the effects of various factors (such as geographical environment and characteristics of trigger event) on system performance. With the increasing complexity of CPSS, the number of influencing factors and their combinations are vast, and the experiment costs will also increase significantly. Generally, it is helpful to emphasize the factors that have a great influence on the system performance and study their influence degree and influence range. This section describes how to design computational experiments reasonably to obtain the ideal experiment results in the fastest and most economical way, which covers the design of input and output variables for CPSS.

As a typical complex system, CPSS involves many unknown variables or factors that can impact the system. Randomized controlled experiments are used to eliminate other possible confounding variables for the sake of seeking the causal relationship between variables. One factor at a time method is a widely used experiment strategy in practice, which is used to show how the response variable is affected by the changes of each factor, keeping other factors unchanged. However, this method has limitations, as it only allows for the study of one factor at a time. With a relatively large sample size, this can result in a less efficient experimental design.

Combinatorial testing (CT) involves cross-grouping values of various levels of multiple factors in the tested scenario to generate test cases. This enables accurate estimation of the main effect size of each experimental factor and the interaction between factors. The value of each factor is discrete and finite, and the multifactor (n-wise, $N > 2$) combination test can cover all value combinations of any N factors. The process of CT method for test case generation can be described as follows:

Assuming that there are k parameters in the test case, they constitute the element set $P^* = \left\{p_1^*, p_2^*, \dots p_k^*\right\}$, where the element p_i^* has n values, and the value set $V_i^* = \{v_{i1}, v_{i2}, \dots v_{in}\}$. As per the response combination coverage criteria, the test case set matrix TS is produced, where rows represent test cases and columns represent input parameters. The input parameter is the test scenario element, and the value is the importance of each element to the test scenario. One of the test cases is represented by $tc = (v_1, v_2, \dots v_k)\left(v_1 \in V_1^*, v_2 \in V_2^*, \dots, v_k \in V_k^*\right)$.

Experimenting with all possible values is neither practical nor necessary. Instead, a subset of factor values is typically used in the CT algorithm to fully cover all interactions between an arbitrary number of elements, thereby effectively solving the massive test cases in exhaustive testing (ET) methods. The key challenge in combination testing is determining the size of parameter sampling subset. If the subset is too large, the sampling significance will be lost; when the subset is too small, the results are unconvincing. For the design space composed of a K-dimensional unit cube defined by standard experiment input, the corresponding design methods mainly include Orthogonal Experimental Design, Latin Super Cube Experimental Design, Uniform Design, etc. [5].

The operation of CPSS is influenced by external variables, but obtaining specific sample data may be difficult due to privacy and legal restrictions. To address this, micro-individual data is generated based on macro statistical laws in the experimental design. The main challenge is devising a data generation strategy that closely resembles real-world situations. Synthetic datasets are crucial for testing intervention strategies over multiple historical periods without overfitting to a particular dataset. Two common approaches for generating synthetic datasets are resampling and Monte Carlo.

The first method is resampling. If we already have historical data for a certain period of time, new (unobserved) datasets can be generated by repeatedly sampling the observed dataset. Historical data is often limited, and it is easy to overfit, in order to deal with this situation, resampling technique is widely used. Resampling can be deterministic or random. Instances of deterministic resampling include jackknife (leave-one-out), cross-validation (onefold-out), and combinatorial cross-validation (permutation tests). Instances of random resampling include subsampling (random sampling without replacement) and bootstrap (random sampling with replacement). Subsampling relies on weaker assumptions; however, it is impractical when the observed dataset has limited size. Bootstrap can generate samples as large as the observed dataset, by drawing individual observations or blocks of them (hence preserving the serial dependence of the observations). The effectiveness of a bootstrap depends on the independence of the random samples, a requirement inherited from the central limit theorem.

The second approach to generating synthetic datasets is Monte Carlo. The Monte Carlo approach generates synthetic datasets by randomly sampling new datasets from an estimated population or data-generating process, rather than using observed data like a bootstrap method. It is a random simulation technique that uses probability density distribution to sample scenario parameters, generating random test cases that closely resemble real occurrences. The basic steps of the Monte Carlo method are as follows. (1) Describe the probability of scenario occurrence by representing scenario parameters as a probability distribution and constructing a probability model for these parameters. (2) Generate random values using various methods (e.g., importance sampling, accept–reject sampling,

Markov Monte Carlo sampling) to ensure that the sample set follows the real statistical distribution. (3) Utilize the sample set for simulation and derive estimates of scenario parameters.

4.1.3 Causal Analysis

This step mainly focuses on how to seek out the causal relationship between intervention variables and system emergence. The basic idea of computational experiments is to acknowledge that both the future outcome of events and the means of achieving such future outcomes are uncertain. The emphasis is on predicting what might happen in the future so that researchers can respond according to possible scenarios. To achieve this, it is essential to analyze the relationships between input and output variables in terms of correspondence and causality. Unlike traditional scenario analysis, which relies on artificial imagination and reasoning, computational experiment method introduces quantitative methods to construct the "scenario" of event occurrence, development, transformation, and evolution. The specific experiment analysis is carried out from three levels.

4.1.3.1 Black-Box Analysis (What Happened)

Black-box analysis is a mathematical abstraction of the computational experiment process, which is used to describe the changing rules between influencing factors and response variables by summarizing and analyzing the experimental data. Black-box analysis can be regarded as the secondary modeling of artificial society model. The most commonly used factorial model can be abstracted into the following form:

$$y = f(x) + \varepsilon \tag{4.5}$$

where y is the observation result of the experiment, namely, the response variable and $f(x)$ represents the mathematical relation implied by the experimental process, while ε represents the experiment error. As $f(x)$ tends to be unknown, it can be fitted by using a known mathematical function $g(x)$, called the factorial function model. Generally, the factorial function model refers to the mathematical abstraction of the experimental process, not limited to a linear model or statistical model. Given the differences between different experiments, the selection of factorial function is quite broad and changes with its development.

For linear characteristics, the regression model is often used. Assume that there is no interaction between the factors, only one factor should be changed at a time, while keeping other factors constant, and then the influence of different test levels on the response value can be studied. The commonly used effect model is as follows:

$$y_{ij} = \mu + \tau_i + \varepsilon_{ij} \tag{4.6}$$

where $i = 1, \dots, a; j = 1, \dots, n$. y_{ij} is the response variable of the experiment, μ is the average value of all variables, τ_i is the effect of the i-th factor, and ε_{ij} is the random variable representing the error. This model can use the completely random design as the experimental design strategy. The data can also be processed by single-factor variance analysis.

For nonlinear characteristics, response surface or neural network models are often used. The response surface method is based on multiple linear regression, including the first-order term, the second-order term, and the first-order interaction term between two factors of all significant factors. Neural network is a flexible nonlinear statistical model that can be trained using historical data and tested for accuracy using a test set.

4.1.3.2 Behavioral Analysis (How Does It Happen?)

Behavioral analysis describes the dynamic behavior of agents and analyzes the occurrence and development process of event chains. Assuming that the system complexity is represented by the comprehensive event C at the macro level, its presence or absence is usually related to a series of other events $\{X\}$. The conventional econometric regression method can only describe the statistical relationship between scenario 1 and scenario 2. In contrast, computational experiments can explain how scenario 1 evolves into scenario 2 through the behavior of micro individuals.

This kind of analysis is mostly based on agent behavior analysis: First, the agent behavior pattern is mined from massive agent behavior data; then, the agent starts self-learning (induction-probability model) and evolution (deduction-law model) behavior, they can optimize their behavior strategies and achieve evolution through manually specified learning rules, as well as through processes such as genetic learning and RL; finally, this evolving agent behavioral law is used to explain or predict future phenomena. This behavior-based scenario inference assumes that a simple agent behavior acts as an intermediary between source phenomenon 1 and predicted phenomenon 2. Specifically, the behavior chain between two phenomena can be explained through a chain of agents, which, in turn, can elucidate the correspondence between input and output variables. We consider the original phenomenon as the input variable, and the final phenomenon as the output variable.

4.1.3.3 Mechanism Analysis (Why It Happens?)

Mechanism analysis is used to identify causal relationships between scenarios and the system evolution process. While using behavioral analysis, if the micro-behavior assumption of the agent and the macroscopic features produced by the model are both similar to the real world, we generally think that this model can reasonably represent the laws of the real world. There are several problems with this logic: "A mechanism can produce a scenario" is far from the conclusion that "the mechanism is the cause of this scenario." It is possible for different agent models to generate the same macroscopic features. Therefore, determining the true underlying mechanism that generates these outputs is crucial for understanding causality.

In the computational experiment, "mechanisms" refer to the generation rules of agent behavior and the interaction rules between agents, as well as the internal mechanism of specific macro results. Mechanism-based behavior analysis is the process of analyzing the causality chain between input and output variables. The analysis focuses on the agent, especially the behavioral orientation of the agent. (1) Analyze the influence of hypothesis 1 on the behavioral orientation, including Belief-Desire-Intention (BDI), etc. [6], i.e., the causal chain between hypothesis 1 and the behavioral orientation. (2) Analyze the mechanism between the behavioral orientation of agents and other environmental factors, i.e., the generative rules of agent behaviors. (3) These rules will be manifested as the behavior of the agent in the experiment, and then the deduction results of hypothetical phenomenon will emerge.

The mechanism analysis aims to identify the causal relationship between experimental phenomena, that is, the "influence–action–emergence" link from macro phenomena to micro behaviors to macro phenomena. As a result, it can penetrate every aspect of the evolution of artificial society and restore the mechanism of intervention, rather than a statistical description between macro phenomena. Therefore, the what-if analysis is the core of the computational experiment. To further quantify the causal relationship between the two nodes, the probabilistic causal model (PCM) [7] can be used to formalize and define relevant operations.

4.2 Computational Experiment Platform Architecture

4.2.1 Realistic Background

If computational experimentation has a highest-priority task, it is to make simulations and experiments easier to build, run, and use—because this enables real-world decision-makers to use models more readily. Epstein envisioned a national center where decision-makers could access digital twins of every major city, along with precomputed models that incorporate various potential hazards. "If something actually happens—for example, the release of a toxic contaminant—we can select the most suitable model and perform near real-time calculations to determine the optimal combination of shelters and evacuation routes." Madhav Marathe from Virginia Tech has also considered this issue: When a Category 5 hurricane strikes, the mayor cannot afford to wait about a

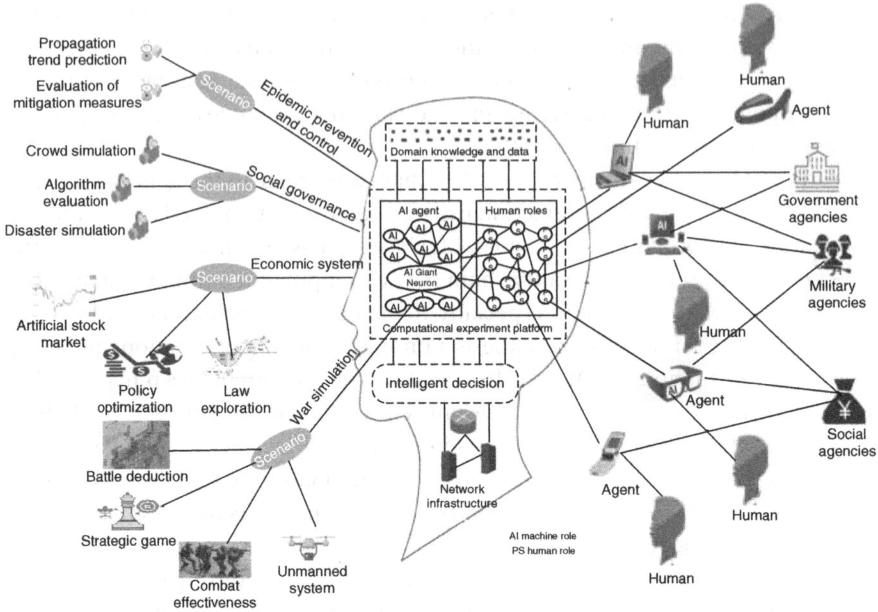

Figure 4.2 Computational experimentation platforms serve as central hubs connecting users with application scenarios. They provide standardized interfaces and tools to support the analysis, design, management, regulation, and integration of social systems.

week to receive an analysis of the storm's potential impact on the Puerto Rico power grid. She needs actionable information—"and that means models with simple user interfaces, running in the cloud, and capable of completing extremely complex analyses in a very short amount of time" [8] (Figure 4.2).

Marathe's laboratory spent four years developing and testing a software tool designed to provide personalized decision support for public health officials, enabling large-scale epidemic simulations and hypothetical scenario analyses—without the need to hire programmers. With just a few clicks, users can specify key variables such as the region of interest (ranging from a single city to the entire United States) and the type of disease—such as influenza, measles, Ebola, or newly emerging pathogens. By using the tool's built-in maps and graphical interfaces, users can observe the simulation process in real time and assess the potential outcomes of the recommended response strategies. The tool's underlying geographic models and synthetic populations are generalizable, making it applicable to other types of disasters as well, including chemical spills, hurricanes, and failures in power or energy networks. The ultimate goal of the tool is to help answer practical, personalized questions such as: "If I send Jimmy to school today, what is the probability he'll be infected with the Zika virus?"

Decision intelligence systems face common challenges: external environments are often characterized by fuzzy boundaries and uncertainty; internal factors are dynamic and self-organizing; and the interactions between internal and external elements display nonlinearity and emergent behavior. Computational experimentation adopts a more proactive and holistic perspective to simulate possible future scenarios and conduct situational forecasting in advance. To apply these forward-looking analyses across the full life cycle of decision-making, including formulation, analysis, implementation, and feedback, the dedicated computational experimentation platforms are required.

However, traditional ABM platforms are often tailored to specific domains or limited applications, and they struggle to integrate with the latest AI technologies (as shown in Table 4.1). The lack of a unified computational experimentation platform has become a significant bottleneck in the advancement of this field. More importantly, it limits the ability of many researchers—especially those interested in interdisciplinary simulation—to effectively apply computational experimentation in their work. Therefore, there is an urgent need for a foundational computational experimentation infrastructure that aligns well with the methodological framework of computational experiments, integrates social simulation logic with experimental design and analysis techniques, and supports interoperability with AI technologies such as machine learning, social networks, and large-scale models. Such a platform would enable applications across a wide range of social domains, including finance, economics, and politics.

4.2.2 Platform Architecture

The design goal of the MACE is to integrate multiagent modeling and simulation capabilities with counterfactual causal inference from experiments, providing tool support for the description, explanation, prediction, and optimization of complex social systems. The system does not predefine the individual behaviors of different agents; instead, it focuses on building a class library encompassing agent behavioral autonomy, learning, reasoning, interaction, collaboration, and game-theoretic mechanisms. On this platform, users can construct artificial society models tailored to various scenarios to represent real-world situations, design experiments and perform simulation runs according to research objectives, and carry out causal inference, pattern discovery, and decision optimization based on experimental data. The system's model library, algorithm library, and database are all extensible and can be continuously expanded and refined.

As shown in Figure 4.3, this chapter mainly introduces the four core components of the computational experiment platform. (1) Sandbox design module: Responsible for constructing AI agents and AI societies. AI agents represent intelligent entities and can be customized for specific applications, incorporating

Table 4.1 Introduction of Existing Social Simulation Tools.

Name	Advantage	Remarks
SWARM	Swarm originally emerged from the Santa Fe Institute, aiming to be used by researchers in various disciplines. The basic structure of Swarm is to simulate the aggregation of agents (multiple model-based agents) that interact with each other simultaneously.	http://www.swarm.org/
Repast	Repast was originally developed by the Social Science Computing Laboratory at the University of Chicago and is now managed by a nonprofit organization composed of members from government, education, and industry organizations. It supports modeling knowledge systems, agents, organizations, and institutions as recursive social structures. In 2007, Repast Simphony 1.0, abbreviated as Repast S, was officially released.	http://Repast.sourceforge.net/
NetLogo	NetLogo is a programming language and modeling environment specifically designed for agent-based modeling (ABM) simulations. It is widely used in fields such as complex systems research, education, and scientific research. Developed by the Center for Connected Learning and Computer-Based Modeling (CCL) at Northwestern University in the United States, it is characterized by being easy to learn, having strong visualization capabilities, and having flexible functions, making it suitable for both beginners and professional researchers.	http://ccl.northwestern.edu/netlogo/
StarLogo	StarLogo was developed by the MIT Media Lab in the early 1990s. It retains Logo's user-friendly simplicity (e.g., programming with "turtles" as basic agents) while enhancing support for multiagent parallel interaction, enabling users to easily simulate the dynamics of large-scale independent agents (e.g., ecosystems, social networks, swarm behavior).	https://www.slnova.org/
MASON	MASON is a Java-based ABM framework developed through collaboration between George Mason University (USA) and the University of Salerno (Italy). It emphasizes high-performance computing and low-level code control, making it ideal for researchers and developers requiring efficient, customizable models.	https://cs.gmu.edu/~eclab/projects/mason/

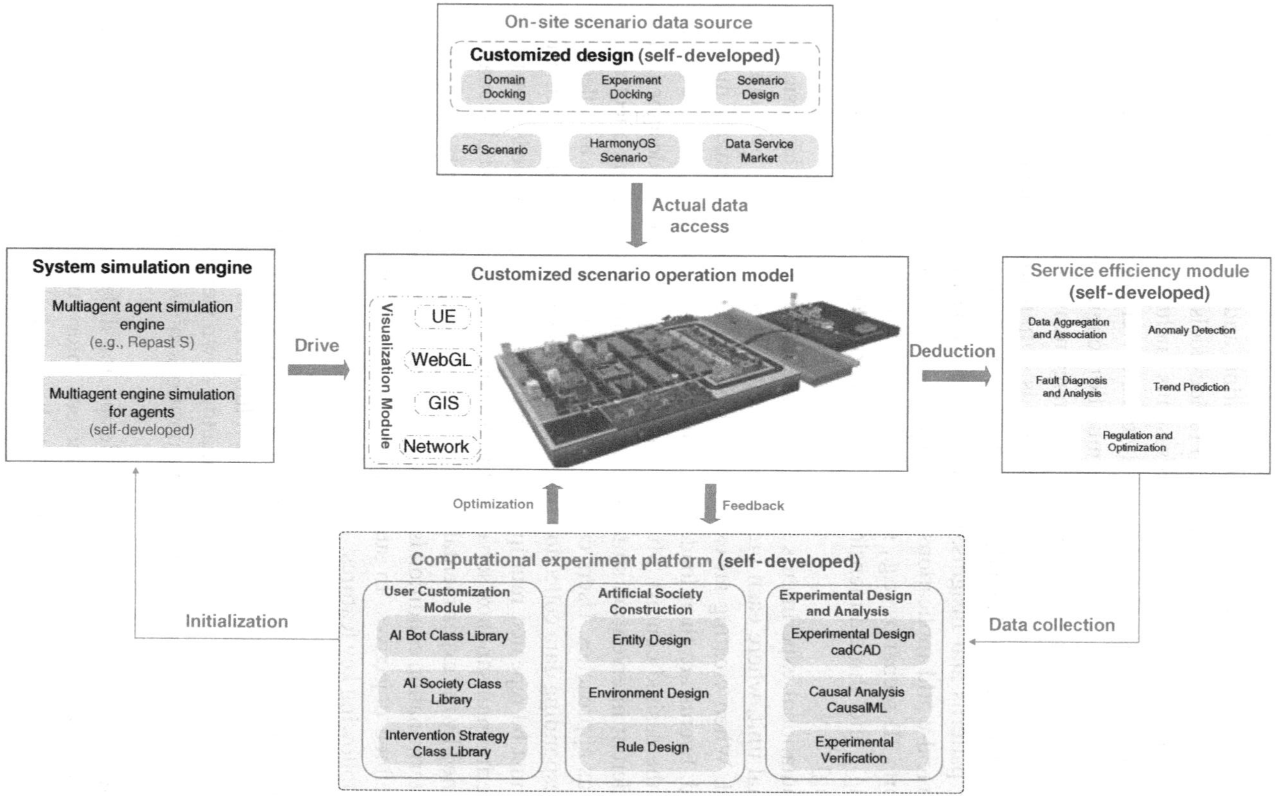

Figure 4.3 Technical framework of MACE.

features such as heterogeneity, bounded rationality, autonomous decision-making, and learning and evolution. AI societies correspond to social systems and include mechanisms for interaction, cooperation, and competition among agents. (2) Computational experiment module: Provides tools for the design, analysis, and optimization of experiments on social systems, including deterministic factorial experiments, stochastic factorial experiments, correlation-based causal analysis, and generative causal analysis. (3) Simulation engine module: Decouples the design of sandbox scenarios from the management of multiagent operations. Data generated from sandbox simulations is output to the data management module, after which it is analyzed by the computational experiment module to uncover patterns and insights. (4) Scenario docking module: Acts as a bridge between real-world applications and sandbox scenarios. Depending on the needs, it supports 2D scenarios (e.g., social networks), 2.5D scenarios (e.g., GIS-based modeling), and three-dimensional (3D) scenarios (e.g., Unity + WebGL-based modeling).

Computational experiments take "experimental projects" as the basic organizational unit, where each project represents a specific domain problem to be explored. When creating an experimental project, an initial scenario must be specified. Events are then loaded into this scenario to construct a sandbox that reflects the desired context. Multiple sets of experiments can be designed, each with different intervention strategies, enabling comparisons of the effectiveness of various approaches. Users can also define which specific data to collect and set corresponding data collection rules. To eliminate statistical noise, each experiment must be repeated multiple times with different random seeds. For users, the experimental sandbox serves as the research subject. The initial scenario is determined by the experimental setup, the intervention strategy acts as the input, and the termination condition is defined by either the runtime or the achievement of predefined objectives. In summary, the computational experiment platform has the following key features:

Visualization: The sandbox can be visualized via simple 2D interfaces or immersive 3D modeling, displaying system evolution from macro and micro perspectives. For example, a detailed map in a top-down view provides an overview of the environment, including real-time positions, movements, and interactions of each intelligent agent.

Customizability: Supports iterative design for different scenarios, including overall layout, environmental parameters, intervention strategies, and event sources. Beyond physical elements, it allows defining agent quantity, parameters, spatial distribution, relationships, interaction rules, and economic norms.

Openness: Enables flexible addition or removal of agents without compromising operational integrity. The external environment can be expanded by adding/removing sandbox entities, and different governance strategies can be loaded to influence agent behaviors and simulation results.

Persistence: While individual agent actions at each time step are transient, the overall organizational structure persists over time, allowing decisions and behaviors to accumulate and form a coherent causal trajectory.

Situational awareness: Experiments operate in a human-constructed environment where agents are aware of their spatial context, understanding their position and objects within their field of view, enhancing active environmental interaction.

Organization: Experiments run within predefined rules and constraints: agents interact with the environment in a limited action space, and environmental objects transition within a limited state space. These rules determine how the swarm intelligence system operates, ensuring coherence, understandability, and reflection of real-world complexity.

4.2.3 Development Trends

In 2005, researchers at Arizona State University proposed the concept of "Decision Theater," which is a high-end solution for group decision-making, characterized by a wide field of view and a high sense of immersion. It is fundamentally based on super-visualization computing, presenting all necessary information as vivid 3D visual images in front of the user. Decision-makers, scientists/professionals, planners, and design experts, as well as engineers/technicians, can immerse themselves in this virtual environment to explore, understand, and communicate data using methods that are difficult to implement in the real world. This immersive setting enables faster, more effective, and flexible decision-making, thereby achieving goals more efficiently.

The workflow of the Decision Theater is shown in Figure 4.4. (1) For the problem to be addressed, a large amount of relevant historical data is collected and organized according to different dimensions. (2) Government decision-makers and experts collaborate in a "what-if" hypothetical analysis. Based on various indicators, a problem model can be established, and simulation scenarios can be used to convert abstract expert knowledge into knowledge that is acceptable to the general public. (3) The simulation-based decision-making scenarios are visualized in 3D to provide decision-makers with multiple solution options and outcome predictions. (4) Finally, under a group decision-making environment where all stakeholders negotiate and reach consensus, a mutually satisfactory solution is selected to solve the problem.

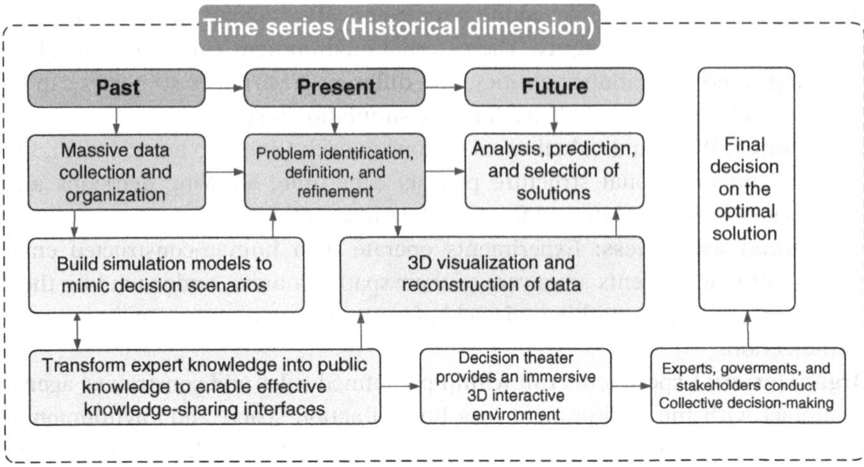

Figure 4.4 Workflow of the Decision Theater.

One representative application of the Decision Theater is the strategic planning case for sustainable water development in Phoenix, Arizona [9]. Phoenix, located in central Arizona, is one of the major cities in the western United States. Together with more than a dozen surrounding satellite cities, it forms a large metropolitan area with a population of around three million. Due to Arizona's typical tropical climate, long sunshine duration, and continuously growing population, water resource issues have become a common concern. The Decision Theater aims to integrate academic research with policymaking, leveraging advanced data visualization technologies to show the interrelations between climate change, water supply, and urbanization in Phoenix. This supports more effective communication between experts and policymakers and facilitates the development of strategic decisions that promote the city's future development.

The Decision Theater constructed a 3D model of the underground aquifer system named "WaterSim," which can display the impact of temporal changes, climate variations, population growth, urbanization, individual behaviors, and policy choices on groundwater development. It also allows users to adjust relevant data to simulate and estimate the potential outcomes of different management policies. An internal demonstration of the Decision Theater is shown in Figure 4.5, including groundwater beneath the city surface, population growth, geolocation, and water pumping energy consumption. Through clear visual representations, experts are able to predict the extent of drought impacts on Arizona and simulate the catastrophic consequences of hypothetical events. The creators of the Decision Theater hope to help decision-makers better "converse with the future," make

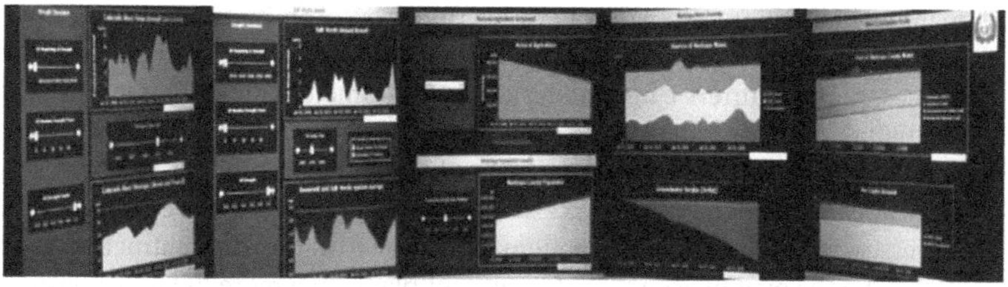

Figure 4.5 Demonstration of the changing trends of key water resource variables.

proactive response plans before major events occur, and thereby turn potentially disastrous outcomes into fully controllable results.

Unlike Decision Theater, the ambition of the metaverse is no longer limited to mere "simulation" of reality. It takes a step further by integrating the real and virtual worlds—or rather, by dissolving the boundary between them—ultimately aiming to extend and empower the physical world through the digital realm [10]. The metaverse is rapidly evolving into a virtual space that can both map the real world and exist independently of it. In this space, people in the real world can use digital identities to engage in entertainment, socializing, learning, and work, completely breaking the boundary between digital and physical experiences. At the same time, AI-powered digital humans are able to coexist and evolve alongside us, playing a crucial role in the social ecosystem of the metaverse and co-evolving with real humans [11]. As the metaverse continues to develop, its application scenarios are expanding rapidly—from gaming into a wide array of domains. The specific details are as follows:

Industrial design: The metaverse can create an interoperable collaborative space where architects, engineers, and designers co-design environments. For example, the metaverse will offer a social collaborative space for designing vehicles and a simulated environment for testing cars, especially autonomous vehicles. One day, you might visit a virtual showroom with your family and take a virtual test drive to decide which vehicle to purchase. Whether in architecture and design, media and entertainment, traditional manufacturing, or industries such as science, big data visualization, robotics, autonomous vehicles, and autonomous machines, it is possible to achieve physically realistic, photorealistic, or cinematic-grade 3D simulations. Before physical products exist in the real world, they can be digitized and virtually validated in the metaverse. Through virtual simulation, industrial production can undergo multiple rounds of testing and experimentation, thereby reducing costs, optimizing design, and improving efficiency.

Healthcare: In the healthcare sector, the metaverse will build a digital patient health ecosystem through augmented reality and virtual reality technologies, enabling bidirectional data flow between the physical entity and its digital twin. This brings immense benefits, for example, when developing new medical devices, manufacturers can conduct various parameter adjustment tests based on digital patients, identifying potential issues before actual production, significantly reducing R&D costs and enhancing final product performance and safety. In personalized diagnosis, doctors rely on customized simulation processes to track how each patient responds to different treatments, thus improving the overall accuracy of treatment plans. In a more long-term vision, digital patients not only create digital twins of important organs and tissues but also connect all these twins for comprehensive analysis—such as diagnosing chronic heart disease through signs like abnormal heart sounds, shortness of breath, and swollen ankles. Digital patients are not merely static digital records of health information but integrated spatiotemporal models that analyze every detail of the human body, establishing more accurate, personalized, and predictive long-term healthcare services. Large-scale digital patient integration also deepens understanding of population-level diseases and contributes to more insightful public health decision-making.

Military transformation: Military exercises are essential for combat readiness. Real battlefields are filled with uncertainties, requiring constant drills for full preparation. Traditional live-fire exercises are costly and often fail to simulate realistic combat scenarios—especially extreme ones. Pure computer-based simulations struggle to model battlefield uncertainty and human agency. A dedicated "synthetic training environment" combines the best of both real and virtual worlds. Computers can generate virtual military training grounds, equipment, missions, weapons, environments, and battlefields, where real soldiers can participate as avatars to practice combat skills. On these virtual battlefields, soldiers and strategic leaders can simulate imagined concepts, battle maneuvers, and formations beyond current limitations. Future metaverse-based war games will combine data capture, storage, and analysis technologies with synthetic environments, making it easier to analyze mistakes, failures, and successes.

Smart cities: The metaverse aims to integrate the physical world into a virtual one. Epic Games once built a virtual city spanning approximately 10–20 km^2, housing around 30,000 citizens with dynamic traffic and buildings. Vehicles have their own behaviors: They stop at traffic lights and stop when seeing pedestrians, and every citizen differed in appearance, behavior, and clothing. Using such a generated digital city as a platform, social management personnel can conduct experiments that are otherwise difficult in the real world. In the metaverse, real subjects can integrate with virtual identities, greatly enhancing the credibility of social experiment outcomes. For example, in emergency disaster relief, fire departments can rely on fire spread models to forecast fire progression and improve current prediction methods. With the metaverse, firefighters can also recreate past fire scenarios using real data, providing immersive digital twin environments for responders, operators, and engineers to evaluate firefighting effectiveness.

4.3 Operation Steps of the Experimental Platform

Based on the abovementioned characteristic requirements, Figure 4.6 shows the operational process of MACE, including artificial society modeling (agent design, environment design, and rule design), domain knowledge docking (knowledge graph construction and domain knowledge injection), experimental design (selecting different influencing factors, response variables, and experimental method), experimental operation (transforming operation results into intuitive images), experimental analysis (data analysis and causal inference), and report generation. This section explains how to use the computational experiment platform to implement each specific step.

4.3.1 Step 1: Domain Docking

4.3.1.1 Knowledge Import

As shown in Figure 4.7, users can import knowledge base files into the system as domain knowledge for the operation of the artificial society. Supported file formats include Word, Excel, and databases. The system's built-in large language model can help analyze the content of the knowledge documents and generate a knowledge graph, which serves as the basis for subsequent outputs of the large language model. The knowledge base generated from the imported documents will be stored in the system, and users can directly view its contents by clicking on the existing knowledge base on the right-hand side.

4.3.1.2 Domain Docking

As shown in Figure 4.8, domain knowledge is represented in the form of a knowledge graph, which can integrate and analyze large amounts of fragmented information, clearly present domain knowledge, and provide strong support for future decision-making and analysis. After the knowledge file is converted into a knowledge base, users can view the constructed knowledge graph on the agent docking interface. The system also allows users to filter and view detailed information of each node. For incorrect knowledge nodes or problematic relationships between nodes, the system supports user-driven modification.

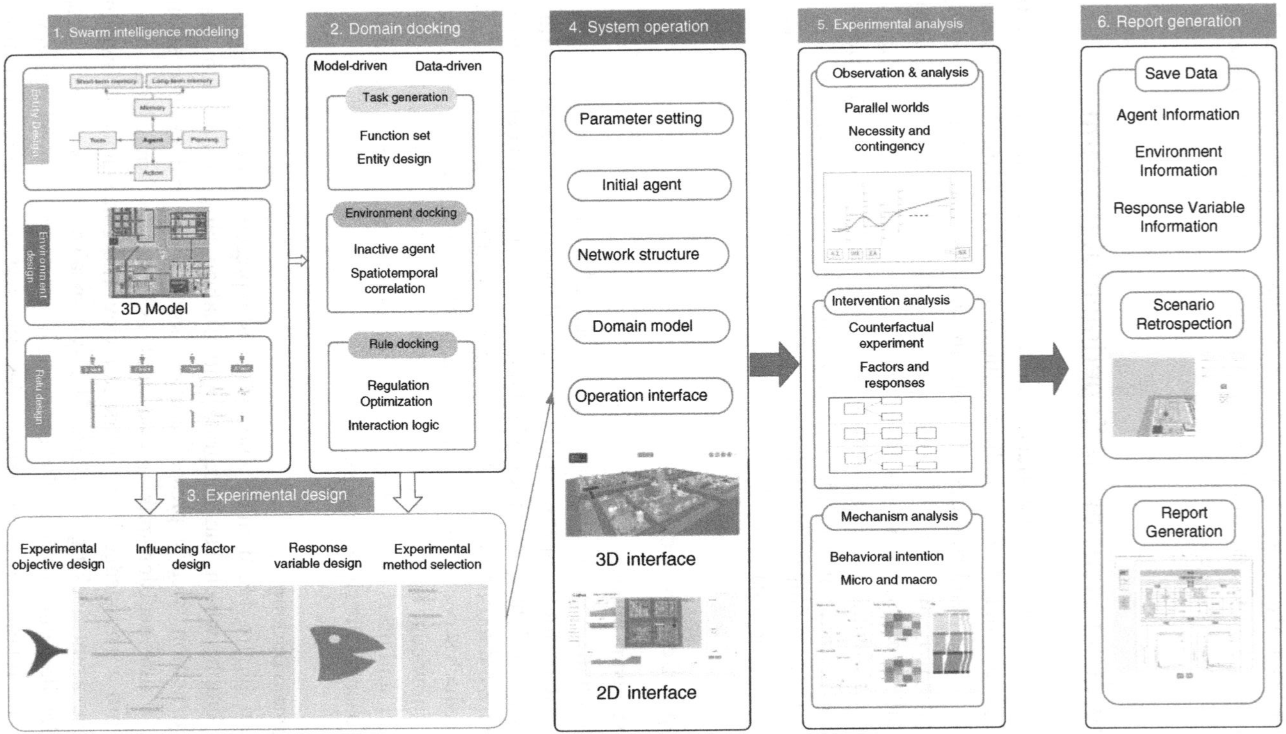

Figure 4.6 Operational process of MACE.

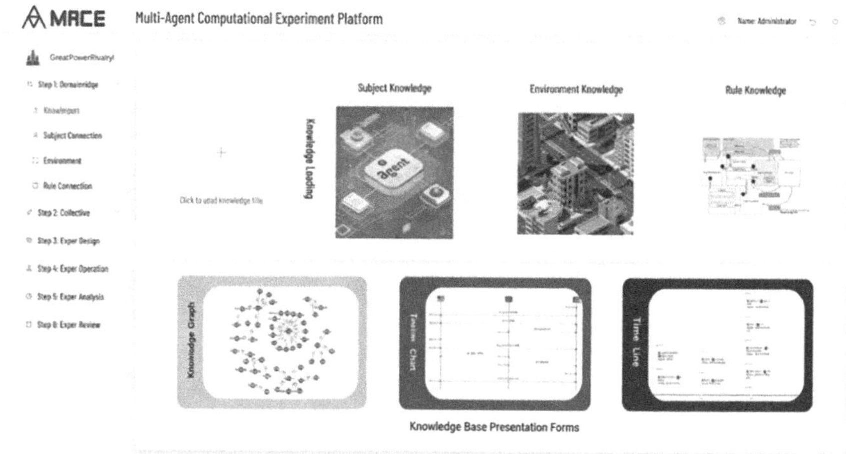

Figure 4.7 Knowledge import interface.

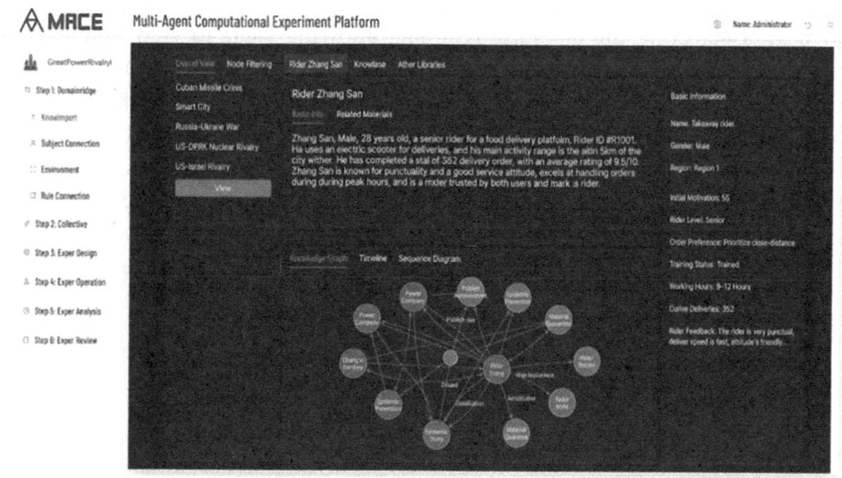

Figure 4.8 Knowledge graph in domain docking.

4.3.1.3 Knowledge Management

The system supports continuous extraction and integration of knowledge to ensure that the knowledge graph is dynamically updated. As shown in Figure 4.9, a knowledge base management module is added on the front end, allowing administrator users to perform deletion, modification, and query operations on each piece of data. Since the system includes multiple domain knowledge bases, such as the "Cuban Missile Crisis Knowledge Base" and the "Large Language

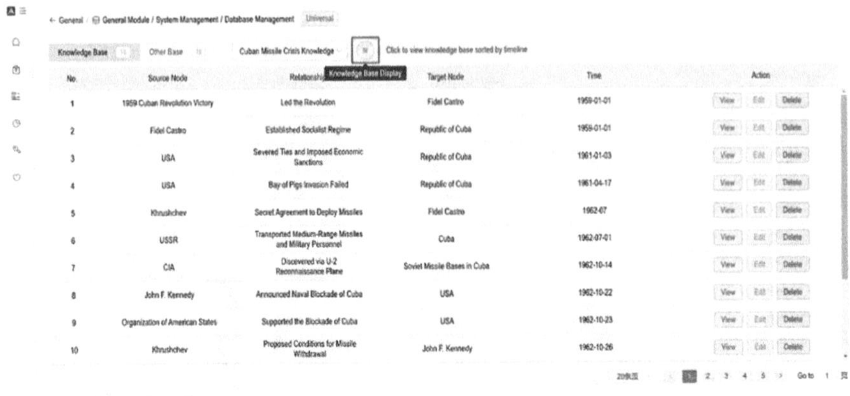

Figure 4.9 Database management interface.

Model Prompt Knowledge Base," a filter box for selecting specific databases is added after selecting the database type tab, allowing users to view and operate only within the same database. It also provides dynamic visualization functions, enabling users to explore and understand complex relationships within the data.

At present, most of the knowledge entries in the system contain time stamps. The majority of the knowledge follows a pattern like "In xxxx (xx month xx day), 'xxx' had an impact on 'xxx'." As shown in Figure 4.10, in order to visually highlight the importance of time in knowledge, the knowledge base is additionally displayed by year. Users can click the "Expand" button at the top of the

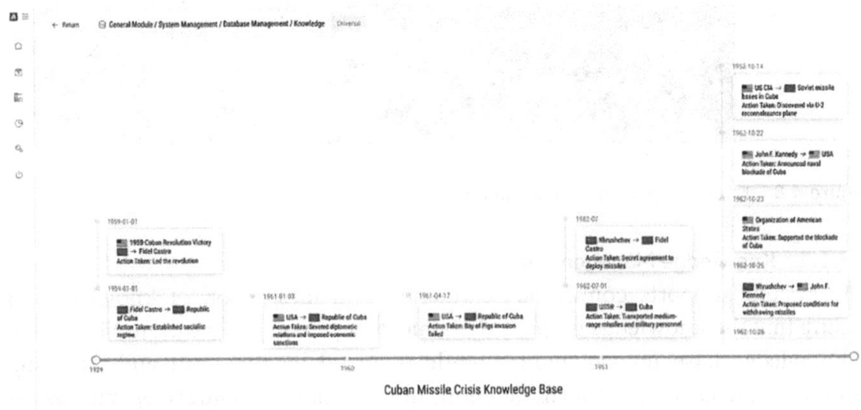

Figure 4.10 Knowledge base interface.

knowledge base to enter the "Knowledge Base Display" interface and view the time-sequenced presentation of the knowledge base. To further emphasize the chronological order of events in the "Knowledge Base Display" interface, users can drag the "start" and "end" nodes on the timeline below. The cards displayed above will show the event relationships that occurred between the selected start and end time points.

4.3.2 Step 2: Swarm Intelligence Modeling

4.3.2.1 Agent Design

The platform supports two types of agent design: one is the traditional ABM style, and the other is the large-model agent style. As shown in Figure 4.11, in the traditional agent design interface, users need to configure parameters separately for the agent's feature set, perception module, cognition module, and behavior module.

As shown in Figure 4.12, the large-model agent requires the design of the agent's role information, mechanism configuration, and extended parameters.

As shown in Figure 4.13, after users complete and submit all agent parameters, they can preview the generated agents. By clicking the "Agent Preview" button on the "Agent Design" interface, users can access the agent details page. This page presents all designed agents in a tabular format, including each agent's name, type, affiliated organization, work efficiency, and connections between agents. The page also supports fuzzy search functionality to help users locate specific agents. For example, entering the query "C" will return results such as "C_Robot_1, C_Robot_2, … , C_People_4."

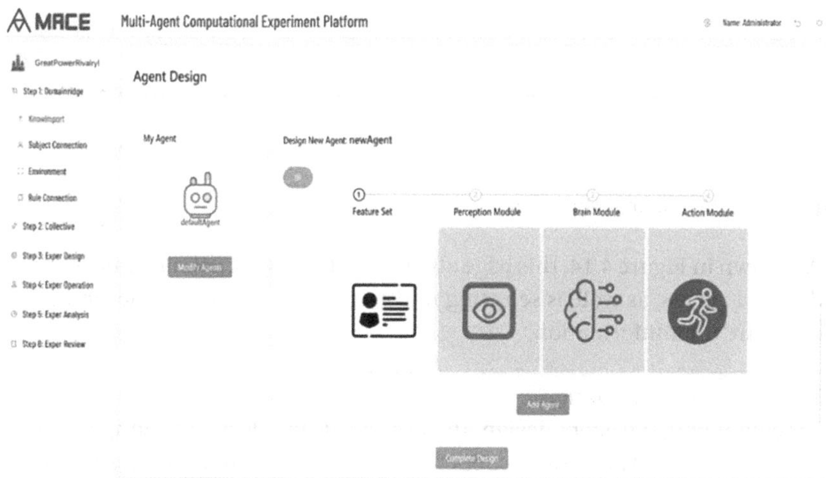

Figure 4.11 Agent design interface.

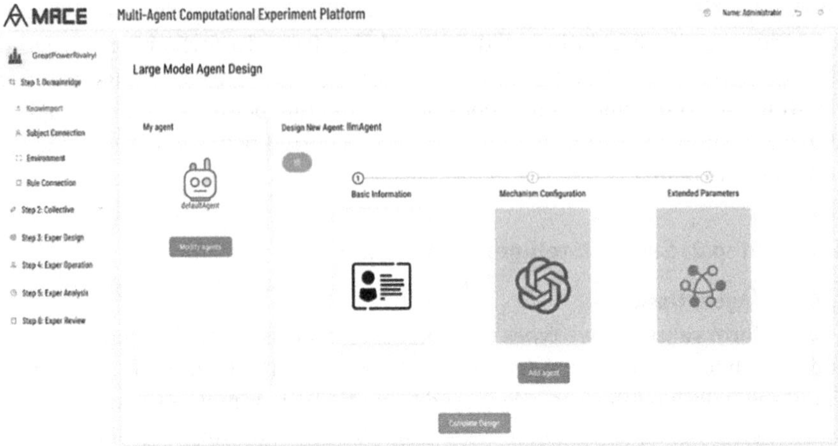

Figure 4.12 LLM agent interface.

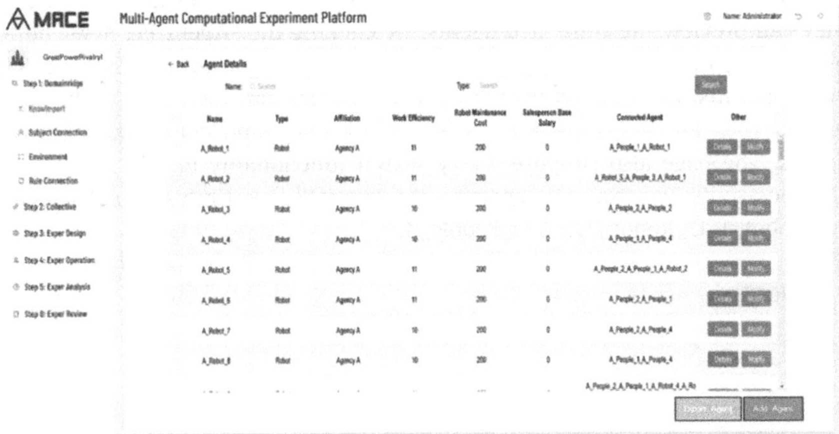

Figure 4.13 Agent preview interface.

As shown in Figure 4.14, this page also supports viewing detailed information of individual agents, as well as selecting multiple agents via checkboxes in the table to compare their information.

4.3.2.2 Environment Design

After completing the agent design, the user needs to select and configure the system environment. The system supports displaying the runtime interface (social environment) using one of the following map types: 2D grid map, 3D grid map, or GIS map. Once the social environment is selected, the user is required to customize

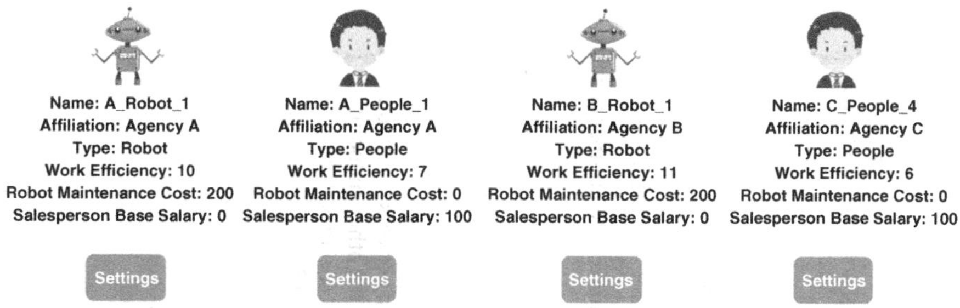

Figure 4.14 Agent comparison diagram.

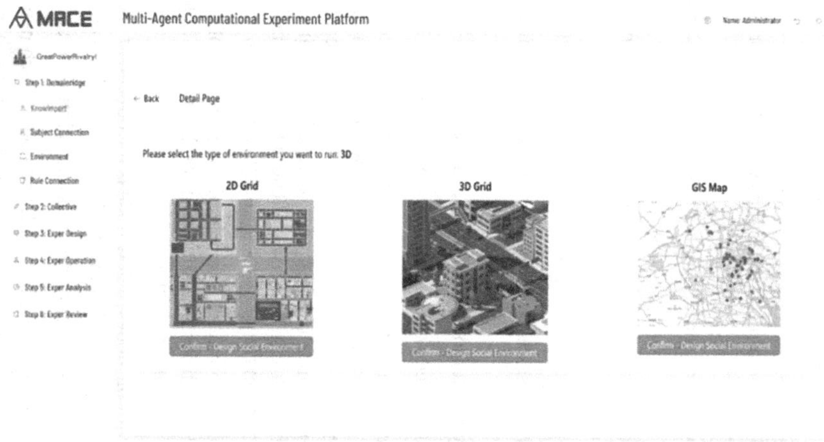

Figure 4.15 Environment model design.

the area length, area width, and the activity range of the agents. Both area length and width are numerical values. To prevent invalid user input, a slider combined with a numeric input display is used, which not only facilitates user adjustments but also ensures input validity. The values for area length and width range from 1 to 100. To help users define the activity range of agents more easily, the map is divided into different independent regions. Users can specify where each agent can operate by selecting the corresponding activity region(s) through a selection box (Figure 4.15).

Besides the physical environment, agents also exist within social environments. As shown in Figure 4.16, the platform offers 10 types of network structures for configuration, including 5 types of regular networks, 2 types of random networks, 2 types of small-world networks, and 1 type of scale-free network. Table 4.2 provides descriptions of the different network types.

Table 4.2 Description of Different Network Types.

Network Type	Description
Isolated node network	Each agent in the network has no direct connections with other agents, thus cannot collaborate and must handle orders independently.
Globally coupled network	Each agent is connected with every other agent, allowing collaboration with any agent. The network forms a complete graph.
Star network	The key central agent is connected to all other agents, but those agents can only collaborate with the central agent, not with each other.
One-dimensional ring	Agents form a ring network, where each agent collaborates only with its two adjacent neighbors.
Two-dimensional lattice	Agents can collaborate with their four neighbors in up, down, left, and right directions. The number of agents must be a perfect square.
Random network $G(N, L)$	N agents are connected by L randomly placed links, where $0 \leq L \leq \frac{N(N-1)}{2}$. Initially, L is set as specified $\frac{N(N-1)}{8}$.
Random network $G(N, P)$	Each pair of N nodes is connected with probability P, where $0 \leq P \leq 1$. The initial value of P is 0.25.
WS small-world network	Consider a nearest-neighbor coupled network with N nodes arranged in a ring. Each node is connected to its $K/2$ nearest neighbors on both the left and right sides, where K is an even number. The parameters satisfy $N \gg K \gg \ln(N) \gg 1$ (a K-nearest neighbor regular network). Then, each edge in the network is randomly rewired with probability p, meaning one endpoint of the edge remains fixed while the other endpoint is randomly selected from the nodes in the network. The small-world network model specifies that no multiple edges or self-loops are allowed in the network. Initially, K is set to 4, and p is set to 0.3.
NW small-world network	In a K-nearest neighbor regular network, a randomly selected pair of agent nodes are connected with an edge with probability p. The edge addition operation ensures that no multiple edges or self-loops occur in the network. Initially, K is set to 4, and p is set to 0.3.
BA scale-free network	Starting from a network with m_0 nodes, a new node is introduced at each step and connected to m existing nodes, where $\leq m_0$. The probability π that the new node connects to an existing node v_i is proportional to the degree k_i of node v_i, i.e., $\pi_i = \frac{k_i}{\sum_j k_j}$. After t steps, this algorithm generates a network with $N = t + m_0$ nodes and mt edges. Initially, the network is configured with two nodes. Each newly added node connects to two existing nodes in the network according to the preferential attachment mechanism.

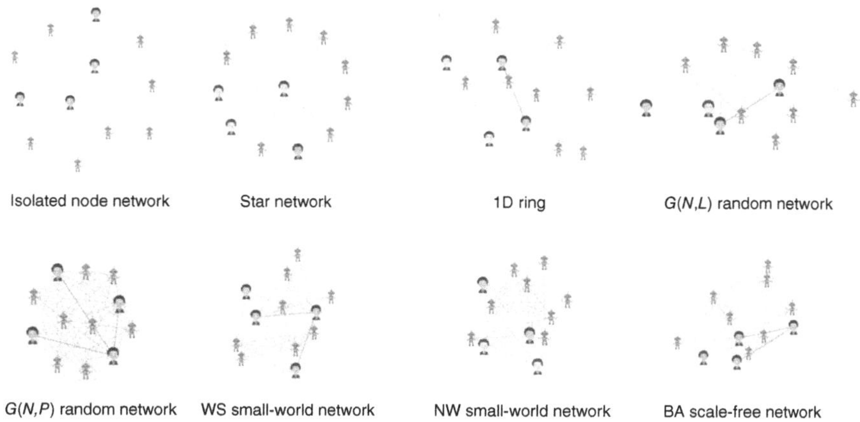

Figure 4.16 Example of network diagram.

4.3.2.3 Rule Design

After completing the design of agents and the environment, users need to define the system's internal interaction rules. As shown in Figure 4.17, users can define the interaction logic between agents and the environment, as well as between agents themselves, through the rule design interface. By clicking the "Agent Interaction" connection line, users can configure agent interactions, including interaction type, interaction content, and triggering conditions.

The core of agent–environment interaction lies in establishing precise behavior-triggering mechanisms, thereby constructing a cognitive closed loop

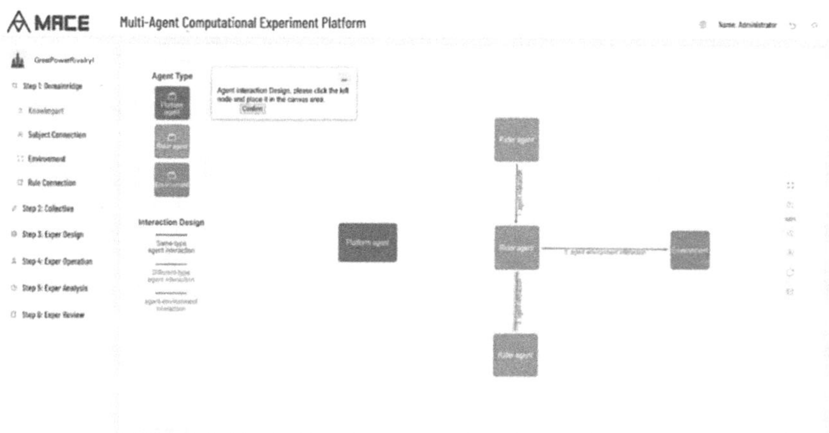

Figure 4.17 Design of the rule model.

of "perception–decision–response." The core of agent–agent interaction focuses on information sharing and transmission, forming a social network between agents; or on task allocation and negotiation to ensure that multiple agents can collaboratively complete complex tasks according to the preset cooperation modes.

The rule design process makes full use of the knowledge graph constructed earlier, which serves as the core support for agents to understand and interact with their environment. Unlike traditional preset rules or simple data matching, agents can query and utilize knowledge within the knowledge graph while perceiving environmental data. When an agent detects changes in the environment, it can match and associate the perceived information with existing entities (such as devices, locations, and events) and relationships (such as "located in," "belongs to," and "affects") in the knowledge graph, thereby gaining a deeper understanding of the current situation and enabling higher-level intelligent reasoning.

At the same time, the knowledge graph provides rich contextual information for agent decision optimization. Agents can access comprehensive information about the capabilities and historical performance of other agents, allowing them to choose suitable collaborators and cooperative strategies.

4.3.3 Step 3: Experimental Design

Each computational experiment has multiple possible design schemes. To select the most appropriate one, a cause-and-effect diagram is used here as a tool to organize information. As shown in Figure 4.18, the response variable is drawn along

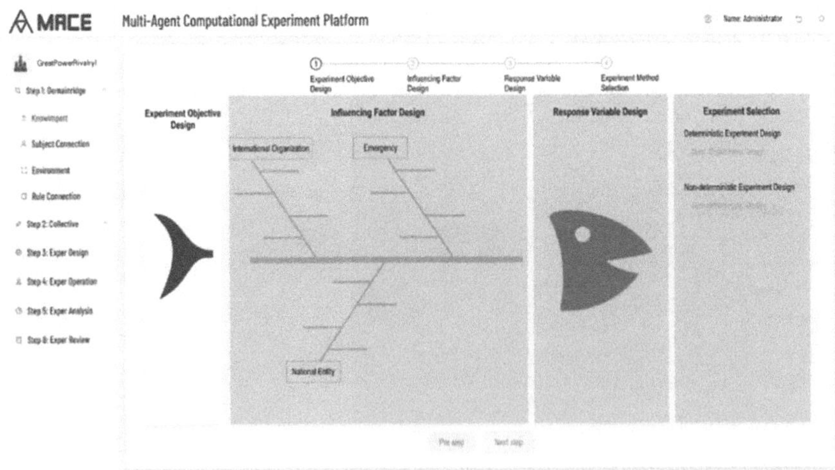

Figure 4.18 Experimental design interface.

the spine of the diagram, while potential influencing factors are arranged along a series of ribs. Based on this, the selection of the experimental design is carried out, including clarifying the experimental objective, considering the sample size (number of repetitions), determining an appropriate experimental order, and deciding whether to divide into blocks or involve other randomization constraints. The first four steps are reflected in the step bar at the top of the page and can be controlled using the "next step" and "previous step" buttons below. Different colored overlays are used to indicate the step currently in progress.

4.3.3.1 Experimental Objectives

Objective design is the core part of an experimental design scheme. It clearly states the goals the experiment aims to achieve, defining precisely what phenomenon is to be explained, what hypothesis is to be tested, or what problem is to be explored. Experimental objectives fall into four categories: descriptive research, explanatory research, predictive research, and parallel optimization. Descriptive research focuses on reproducing processes related to past or present situations, without involving causality or prediction. Explanatory research identifies and estimates causal relationships through counterfactual experiments, emphasizing the discovery of underlying patterns. Predictive research focuses on intelligent decision-making, i.e., "If something happens in the future, what would be the best response?" Parallel optimization aims to empower the real world through computational experiments (Figure 4.19).

4.3.3.2 Influencing Factors and Response Variables

In experimental design, it is necessary to plan and organize various influencing factors that may affect the experimental results. The purpose is to systematically

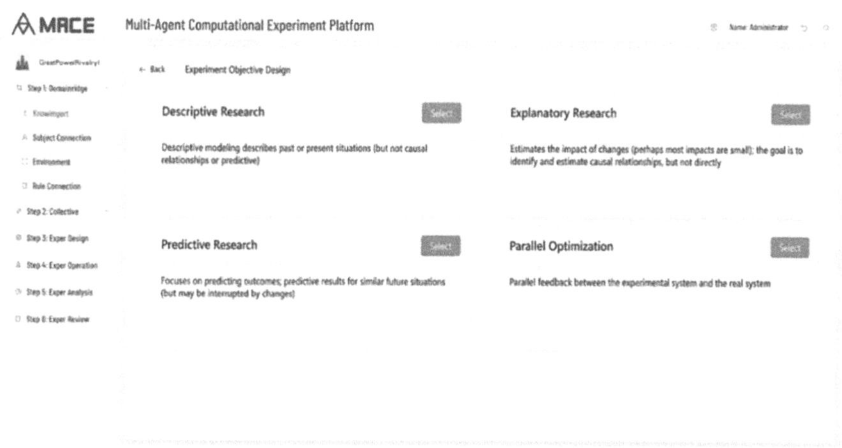

Figure 4.19 Experimental objective design interface.

study how these factors, either individually or in combination, impact the experimental subjects and thereby affect the dependent variables. For example, in an experiment investigating material strength, factors such as material composition, processing techniques, and environmental temperature may all influence the strength of the material. The design of influencing factors aims to determine how to study the effects of these factors. In multiagent systems, influencing factors refer to various elements that affect the overall performance and behavior of the system. These factors may include the types of agents involved (e.g., large language model [LLM] agents, RL agents), environmental conditions (e.g., intervention measures), and communication mechanisms between agents (e.g., network topology). The design of influencing factors is located in the "fishbone" section of Figure 4.18. By clicking the node at the end of a "fishbone," a dialog box will pop up, allowing users to independently design that influencing factor.

The design of response variables mainly serves to determine the variables used to measure the effects or outcomes of an experiment. Response variables, also known as dependent variables, are those variables that experimenters focus on and that change in response to variations in independent variables (experimental factors) during the experiment.

In multiagent systems, response variables change with variations in agents' strategies, environmental factors, and their communications. The design of response variables corresponds to the "fish head" section in Figure 4.18. Clicking on the "fish head" opens the response variable dialog box. To support users in customizing response variables, a formula editor component is introduced here (as shown in Figure 4.20), allowing users to edit the calculation rules of response variables using the formula editor.

Figure 4.20 Formula editor interface.

4.3.3.3 Experimental Design Plan Generation

The experimental design method is the core of the entire process. Choosing different experimental method means using various approaches to observe and analyze how different combinations of influencing factors affect the response variables. To visually present the relationships among the algorithms, the system displays all methods and their hierarchical relationships in a tree diagram. Users can click the "Expand" button to the right of the algorithm selection box to view the tree diagram. The tree diagram supports collapsing and expanding of nonleaf nodes, allowing users to better understand the relationships. For example, in Figure 4.21, the branch nodes represent collapsed nodes, while the leaf nodes represent fully expanded nodes.

After completing the first four steps of the experimental design, the overall experimental design plan will be displayed based on the user's selections in those steps (Figure 4.22). The overall plan is divided into four parts, each corresponding to one of the first four steps of the experimental design. The top-rightmost text boxes indicate the options selected by the user. The two columns on the left represent the influencing factors, the two middle columns contain fixed content, and the far right column shows the response variables. Below, the gray and orange text boxes represent the experimental objectives and experimental method, respectively.

4.3.4 Step 4: Experiment Execution

After users complete agent modeling and experimental plan design, they can proceed to the system execution interface to run and observe the experiment. During

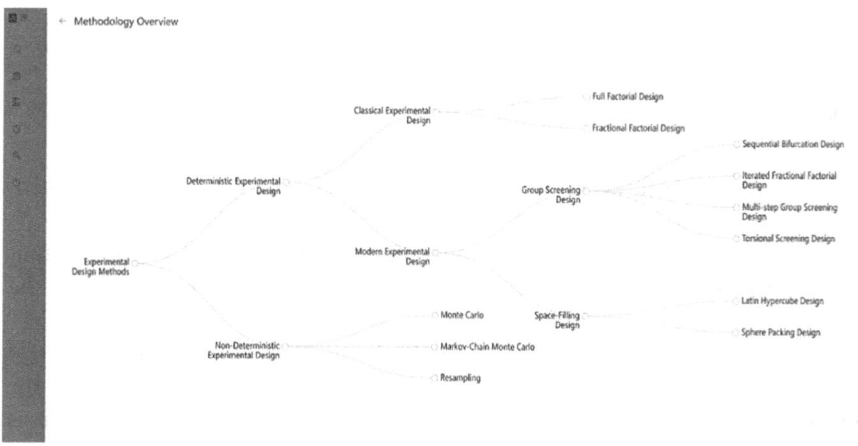

Figure 4.21 Preview interface of experimental design method.

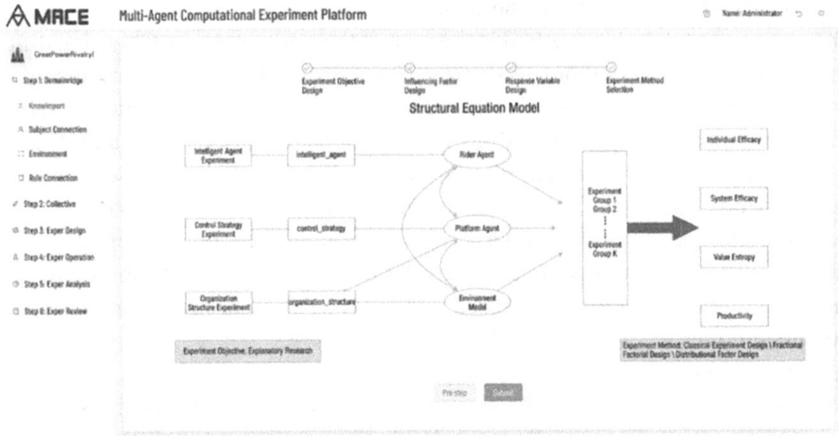

Figure 4.22 Overall experimental design plan.

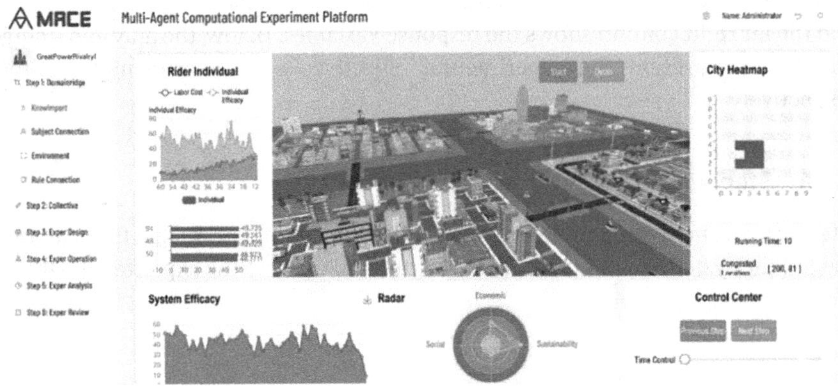

Figure 4.23 Main execution interface.

execution, agents operate according to the parameters set by the user and the generated experimental plan.

As shown in Figure 4.23, a large number of agents run within a 3D city environment. Users can drag, pan, zoom, or rotate the 3D city to observe the status of the agents.

During execution, the left side of the 3D city displays real-time properties of some agents; the right side shows a heatmap of agent movements throughout the city along with real-time monitoring data for any anomalies. Below the 3D city is overall system data analysis, including system efficiency and several key system metrics represented by radar charts.

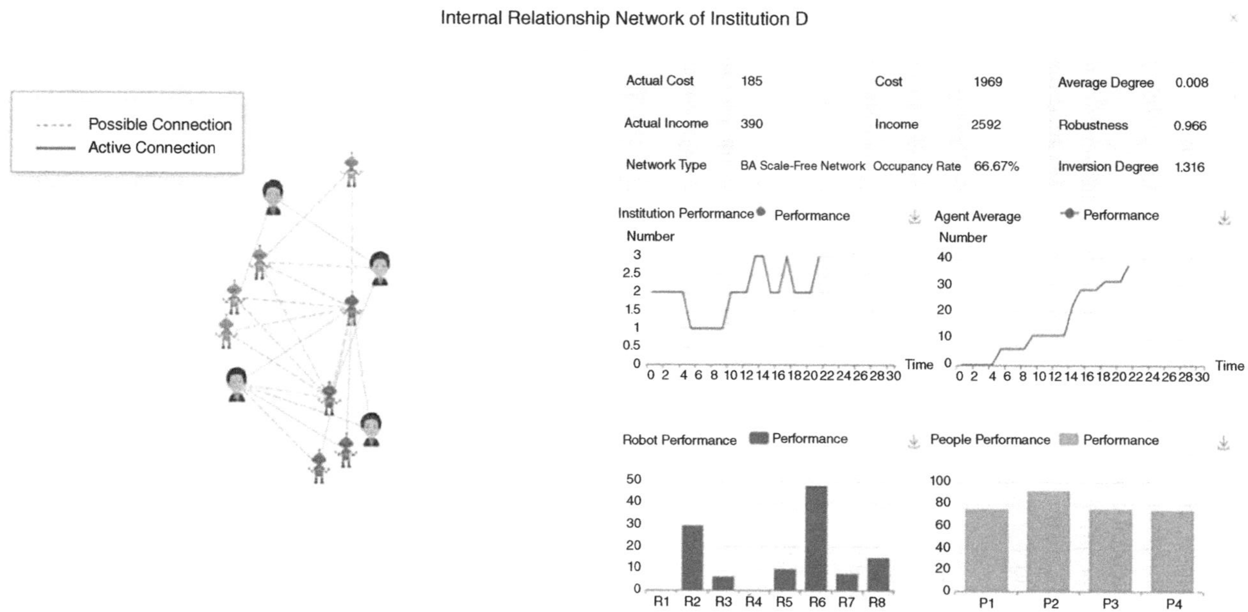

Figure 4.24 Agent organizational network graph data display.

At the bottom right is the system control center, supporting single-step execution and time rewind features during system operation.

To provide users with an intuitive way to explore the activities and data of agents within the organization, the execution interface also supports clicking on an agent to enter its internal network graph. In the network graph, dashed lines between agents indicate potential collaboration relationships, while solid lines represent agents actively cooperating to handle the same order. The network graph uses different colors to indicate the agents' idle or busy status: dark shading indicates idle and bright solid bold signifies that the agent is currently working (Figure 4.24).

4.3.5 Step 5: Experimental Analysis

4.3.5.1 Experimental Analysis

The experimental analysis module is primarily responsible for anomaly detection in the experimental results. The system outputs data at both the system level and the individual agent level. Users can view data at these two levels separately and perform experiment process tracing and intervention adjustments. Figure 4.25 shows the interface for system-level anomaly detection, while Figure 4.26 displays the interface for individual-level anomaly detection. The system offers multiple anomaly detection methods for users to choose from.

After each run of the system, the main data of that run will be output on the experimental analysis interface (Figure 4.27), including the experiment round number, input conditions for the current round, system runtime for the round, and overall system performance. Additionally, results from different experiment rounds are displayed in a scrolling marquee format to facilitate more intuitive comparison by users.

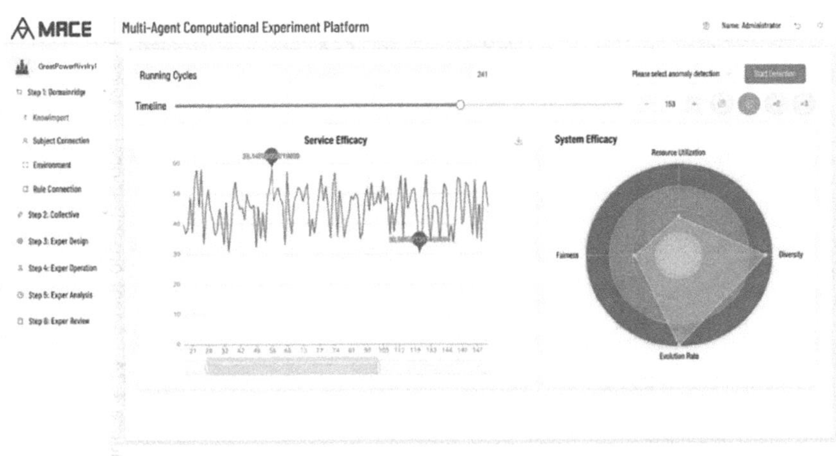

Figure 4.25 System anomaly detection.

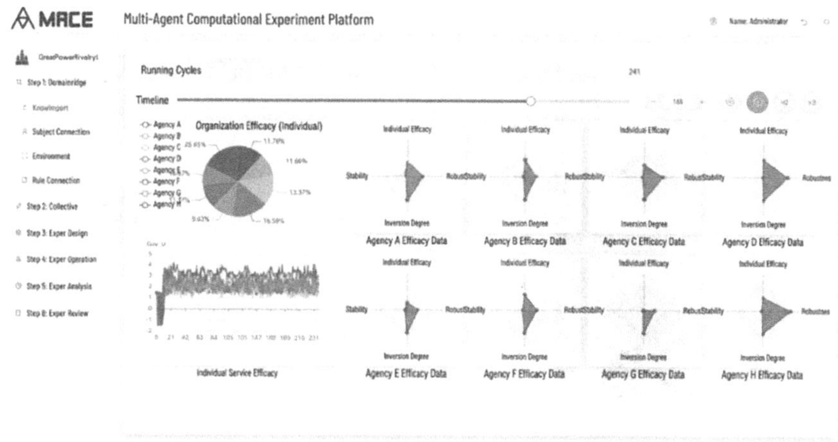

Figure 4.26 Individual anomaly detection.

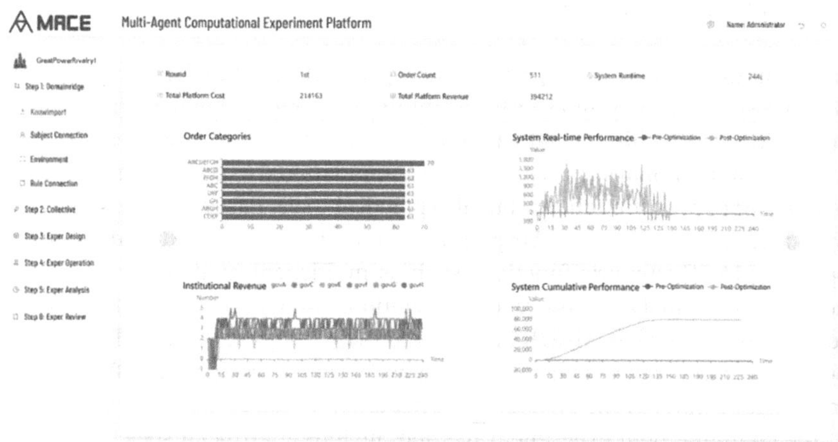

Figure 4.27 Experimental analysis interface.

4.3.5.2 Report Generation

After all steps in the system are completed, an experimental report is automatically generated. The report includes the design of experimental parameters (such as the number of different types of agents, control methods, system runtime duration, agent vision range, and order generation patterns), evaluation metrics (covering both individual- and system-level indicators), and heatmaps of agent movement trajectories. This comprehensive report helps users better analyze the experimental results (Figure 4.28).

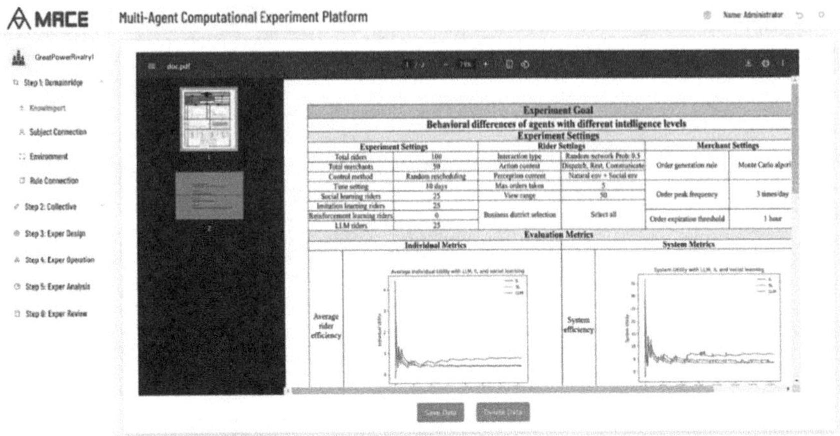

Figure 4.28 Experimental report generation interface.

4.4 Chapter Summary

Computational experiments provide a new perspective for analyzing and studying complex systems. They allow us to test intuition through inference, trial and error, and estimation within virtual environments. To become a mainstream research tool in computational social science, the computational experiment methodology requires not only sound theory but also practical tools. Based on this, this chapter introduces the computational experiment support platform **MACE**, developed by our lab. MACE supports functions such as multiagent modeling, experimental design, causal analysis, and report generation. After years of development and refinement by our lab, MACE has become increasingly mature and is currently being tested in various application domains. We look forward to its official release in the near future. From Chapter 5 to Chapter 8, we have elaborated on each step of the computational experiment method, including artificial society modeling, experimental system construction, experimental design, experimental analysis, and experimental validation.

References

1 Xue, X., Yu, X., Peng, C. et al. (2023). Computational experiments: past, present and perspective. *Acta Automatica Sinica* 49 (2): 246–271.

2 Xue, X., Wang, S., Zhang, L. et al. (2018). Social learning evolution (SLE): computational experiment-based modeling framework of social manufacturing. *IEEE Transactions on Industrial Informatics* 15 (6): 3343–3355.

3 Zhou, D., Xue, X., and Zhou, Z. (2022). SLE2: the improved social learning evolution model of cloud manufacturing service ecosystem. *IEEE Transactions on Industrial Informatics* 18 (12): 9017–9026.

4 Arthur, W.B., Ermoliev, Y.M., and Kaniovski, Y.M. (1987). Path-dependent processes and the emergence of macro-structure. *European Journal of Operational Research* 30 (3): 294–303.

5 Montgomery, D.C. (2017). *Design and Analysis of Experiments*. Hoboken: John Wiley & Sons.

6 Rao, A.S. and Georgeff, M.P. (1995). BDI agents: from theory to practice. Icmas. Vol. 95.

7 Pearl, J. (2009). Causal inference in statistics: an overview. *Statistics Surveys* 3: 96–146.

8 Marathe, M. and Vullikanti, A.K.S. (2013). Computational epidemiology the challenge of developing and using computer models to understand and control the diffusion of disease through populations. *Artificial Intelligence and Machine Learning* 56 (7): 88–96.

9 White, D.D., Wutich, A.Y., Larson, K.L., and Lant, T. (2015). Water management decision makers' evaluations of uncertainty in a decision support system: the case of WaterSim in the Decision Theater. *Journal of Environmental Planning and Management* 58 (4): 616–630.

10 Peng, C., Xiangning, Y., Ma, W. et al. (2023). Computational experiments: virtual production and governance tool for metaverse. *International Journal of Crowd Science* 7 (4): 158–167.

11 Goodfellow, I.J., Pouget-Abadie, J., Mirza, M. et al. (2014). Generative adversarial networks. arXiv preprint arXiv:1406.2661.

5

The First Step – Building an Artificial Society

In sandbox games like *The Sims*, numerous AI agents form an artificial society via self-development and continuous evolution, with unimaginable energy and vitality. On the one hand, the emergence of an artificial society is a bottom-up, self-organizing, and self-growing process, where the overall state of the system depends on the spontaneous interactions among different agents. On the other hand, an artificial society is also a product of top-down planning, and the evolutionary paths of social ecology can be changed by using macro-control strategies. Behind each AI agent, there may be a program, a real human, or a large language model (LLM)-based agent. As shown in Figure 5.1, modeling artificial societies is the core component of the computational experiment method, which needs to perfectly map the complex behaviors at all levels of real social systems, including individual autonomy, learning and evolution, competitive-cooperative games, social emergence, etc. [1]. To address this, this chapter analyzes the model characteristics of artificial societies first, including agent heterogeneity, bounded rationality, learning and evolution, and interactive behaviors. Then, this chapter proposes a hierarchical modeling framework for artificial societies, including four layers: autonomous characteristics, evolutionary characteristics, interactive characteristics, and emergent characteristics. Next, corresponding implementations of modeling technology are given for each layer. Finally, the intelligent logistics system is used as a case to demonstrate the entire process of customized modeling of artificial societies.

5.1 Analysis of Characteristics of Artificial Society

The agent model is the core of artificial society modeling and also a container for different intelligent knowledge to function. At present, researchers primarily manage the decision-making processes of agents by defining some simple deductive rules for each agent, thus generating macroscopic emergent phenomena.

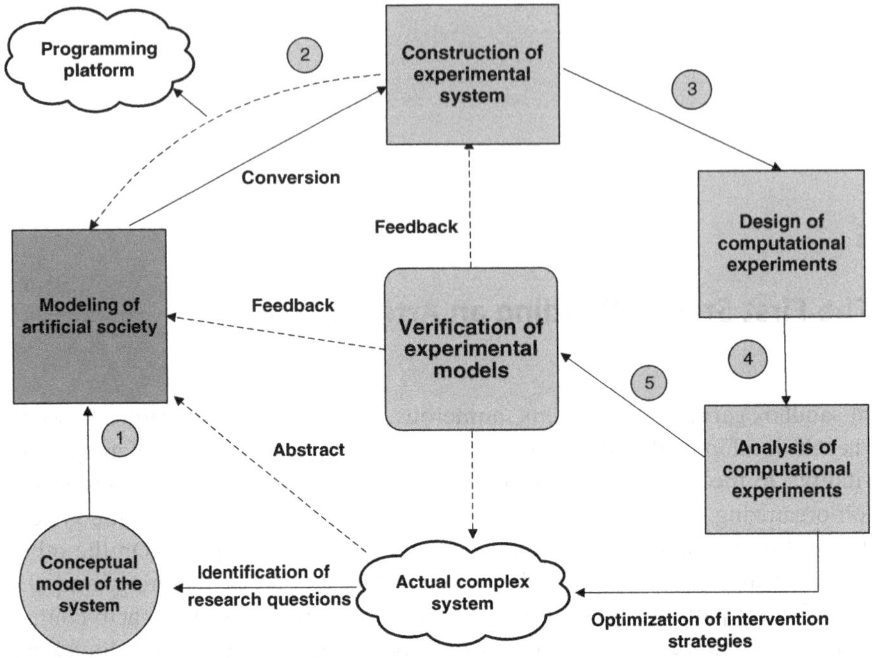

Figure 5.1 The role of artificial society modeling within computational experiment method.

However, it is generally believed that such models often fail to fully reflect the true decision-making patterns of agents. A good agent model should have human-like characteristics, and the factors influencing its decision-making are diverse, including internal states, its environment, and corresponding dynamic rewards and risks. It may even involve complex social factors, different cultural backgrounds, and the learning curves of agents. Therefore, how to use various AI technologies to enhance the design of agent models, especially how to model the anthropomorphic characteristics of agents (including bounded rationality, heterogeneity, memory, learning and evolution, interaction, etc.), has become a key challenge for the ongoing development of this field.

5.1.1 Heterogeneity of Agents

The heterogeneity of agents includes two aspects: (1) attribute heterogeneity, which is usually manifested in differences in specific parameter values (mostly exogenously fixed) and (2) behavioral heterogeneity, including the finite-type model (N-type model), where agents may switch between different types of strategies, but the strategies are always fixed and no new strategies are created [2], and the autonomous agent model, which employs heterogeneous strategies and can create new strategies through learning [3]. According to the "KISS" modeling principle, before introducing heterogeneity into the model, it is necessary to clarify whether the research question requires considering heterogeneity or not to avoid over-engineering.

In general, the decision to introduce heterogeneity is mainly based on the following aspects: (1) whether heterogeneity affects the operation of the system and whether the degree of heterogeneity determines the magnitude of the impact. For example, in transportation systems, agents have heterogeneous beliefs about whether a certain road will be congested in the coming period, which may directly affect traffic flow. (2) Whether characterizing heterogeneity helps in analyzing the design process of a policy or mechanism. For instance, tax policies have significantly different impacts on different groups, with the poor being more sensitive to them, so the study of issues related to social equity needs to consider heterogeneity. (3) Heterogeneity can help in "solving" complex dynamic game problems. Solving dynamic game problems is often difficult because it essentially involves solving a dynamic optimization problem with multiple objective functions. If there are a large number of individuals and information asymmetry exists, the model is largely unsolvable. By introducing heterogeneity, a dynamic game process of multiple agents can be "simulated." Through the design of agents' decision-making methods and learning behaviors, it can be observed that each agent continuously adjusts its strategies during the game, jointly driving the global system toward a certain "equilibrium." Although an "optimal solution" may not be obtained in the end, a "reasonable solution" can be achieved to a certain extent.

5.1.2 Bounded Rationality of Agents

Traditional economics has always been premised on perfect rationality, assuming that agents can access all information and choose the option that maximizes utility from multiple alternatives. Although these models have proven effective in some applications, obstacles remain when applied to many real-world problems. As Rosenfeld and Kraus [4] summarized, these obstacles include three aspects: (1) decision-makers often do not adhere to perfectly rational behavior, even when an optimal expected utility policy is specified; (2) decision-makers often lack complete information and thus may not know the quantitative structure of their environment; and (3) even if people aim to act rationally and have complete information about a given problem, they may fail to compute the optimal solution due to the intractable computational complexity of many real-world issues. Therefore, Herbert A. Simon argued that it is necessary to consider the basic physiological and psychological limitations of humans and the resulting cognitive constraints, leading to the proposal of the concept of "bounded rationality" [5].

"Bounded rationality" generally involves two types of constraints: (1) incomplete information, where agents lack sufficient information and knowledge to enumerate all alternatives, and they face uncertainty regarding the payoffs of each option and (2) imperfect rationality, where agents do not necessarily make decisions in a purely self-interested manner, nor do they always possess sufficient computational capacity to obtain the "optimal solution," often only able to choose a fundamentally reasonable "satisfactory solution." German psychologist Gerd Gigerenzer termed the decision rules of agents as "fast and frugal heuristics" [6]. To model human-like behavior in agents, bounded rationality approaches can be applied instead of perfect rationality methods.

As shown in Figure 5.2, considering the bounded rationality of agents in the agent-based model (ABM) provides an interface for introducing the research conclusions of psychology and behavioral economics. Prospect theory studies how people make decisions under uncertainty, clarifies the decision-making criteria that most bounded rational decision-makers may adopt, and provides valuable theoretical support for the bounded rationality modeling of agents [7]. Bounded

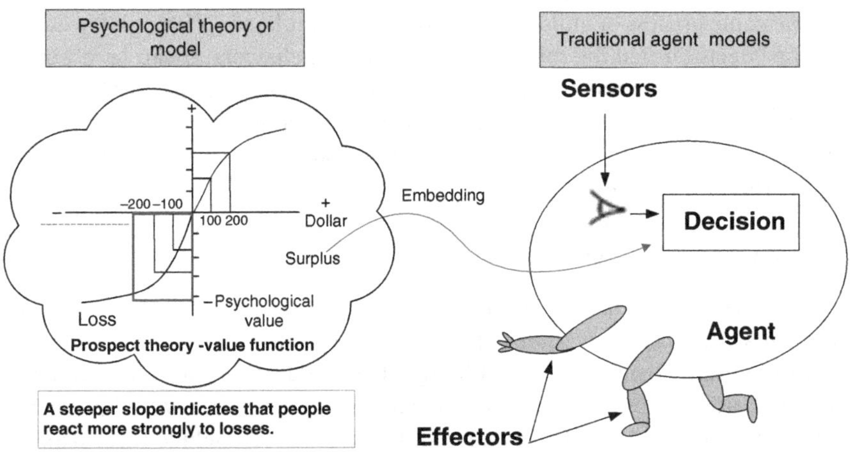

Steps for embedding psychological models	
	• Step 1: Select and identify a psychological theory or model. • Step 2: Systematically simulate the psychological theory or model. Use the structural equation method to determine the structure of the simulation model, and use the psychometric method to determine the initial values of the model variables. • Step 3: Embedding. Solve the time synchronization and data communication problems between simulated psychological models and traditional computing models.

Figure 5.2 Psychological models embedded in ABM.

rationality lies between "complete irrationality" and "perfect rationality," representing a trade-off between modeling flexibility and result rationality. It is necessary to ensure that the "modeling flexibility" is sufficient to characterize the key dynamic mechanisms of the system and to also ensure the "result rationality" of agent behaviors and models so that the final conclusions have reference value. This raises another question: If agents do not make decisions based on perfect rationality from an optimization perspective, then to what specific extent is the so-called "bounded rationality-based decision-making," and can it ensure the rationality (nonoptimality) of agent decisions and the rationality of model results?

5.1.3 Learning Evolution of Agents

"Bounded rationality" focuses on the "cognitive and behavioral system" of agents, i.e., how decisions are made; "learning behavior" focuses on how agents iterate their cognitive and behavioral systems, i.e., how to adjust decisions. To enhance the flexibility of agent decision-making and interaction, we can model agent decisions based on bounded rationality. However, determining the degree of bounded rationality becomes a challenge, as it may even affect the rationality of model results. In this case, we can introduce a learning mechanism to adjust the bounded rational behavior of agents, seeking a more appropriate balance between "modeling flexibility" and "result rationality." This is analogous to parameter fitting in deep neural networks. As is well known, deep neural networks have the "potential" to fit various complex functions, but only through the process of "parameter tuning" can this potential be converted into practical capacity. Similarly, using bounded rationality to characterize agent behavior provides the "potential" for models to approximate reality. However, if the system and model are too complex, the potential can only be realized through a "parameter tuning" process of "learning."

When agents have learning capabilities and behaviors, they can drive the system to exhibit certain patterns at a macro level and over longer time scales, transforming a "complex system" into a "complex adaptive system." This state may be a "convergent steady state," essentially achieving a dynamic game equilibrium through a "learning phase"; it may enter a stable oscillatory trajectory; or it may be a "chaotic" state with no discernible pattern [8]. The following example of "wolf pack foraging evolution" illustrates this.

For instance, suppose a group of wolves live on a plain, each independently searching for food. When two wolves discover food simultaneously, each wolf has two strategies: the "dove strategy" and the "hawk strategy" (Table 5.1). Assume each wolf adopts the hawk strategy with probability P and the dove strategy with probability $1 - P$. Meanwhile, all wolves continuously learn and adjust the value of P based on experience. Finally, the macro indicator "average P value of all wolves"

Table 5.1 Strategies That Wolves Can Adopt in the Hypothetical Experiment.

Wolf A / Wolf B	**Hawk Strategy** (Must monopolize food, fight until both sides are injured)	**Dove Strategy** (Can share, give up if the other party refuses)
Hawk strategy (must monopolize food, fight until both sides are injured)	P, P	$P, 1-P$
Dove strategy (can share, give up if the other party refuses)	$1-P, P$	$1-P, 1-P$

is defined as the toughness of the entire wolf pack. Suppose initially, all wolves have $P = 1$, indicating extremely high toughness. Because every food competition leads to mutual harm, some wolves may quietly reduce their P values to improve their survival chances, causing the macro-level toughness of the pack to decline. As more wolves do this, some may find that adopting more hawk strategies allows them to eat more, so they start to quietly increase their P values, and the pack's toughness gradually rises again. In this deduction process, the pack's toughness fluctuates continuously up and down in value. This macro-level pattern over a longer time scale is driven by the micro-level individual learning over shorter time scales.

In this example, the internal game of the system leads to individual learning and drives forward the dynamic evolution of the system. In other scenarios, changes in the external environment of the system may also lead to dynamic evolution of the system driven by individual learning, such as the adaptation of ant colonies to environmental changes and the adaptation of viral populations to drugs. In the example of the wolf pack, no matter how the agents learn, they can only modify the existing functions and parameters in their cognitive and behavioral system and cannot create something out of nothing. However, in reality, through learning, people can not only modify their own strategies but even create new strategies, which may trigger abrupt changes in the entire system. Therefore, it is generally believed that in order to capture the complexity of human intelligence and psychological characteristics, it is necessary to combine agent models with technologies such as uncertainty reasoning, machine learning (ML), and data mining.

5.1.4 The Interaction Behavior of Agents

Agent interaction is the fundamental mechanism leading to "emergence." At the micro level, agents make decisions and interact according to local rules, but their impacts diffuse, rebound, and oscillate across the entire system, eventually forming emergent phenomena at the macro level. Take epidemic transmission as an example: If each person is regarded as an agent, the local interaction between agents depends on individual behaviors, contact frequency, external constraints, etc., while the emergent global phenomenon is the epidemic distribution across the entire region. Macro-level social phenomena are not determined by individual behaviors alone, but by the complex network behaviors embedded in individuals, communities, and markets. To understand these complex social processes, it is essential to focus on individual behaviors, social network structures, and their dynamics.

Social network analysis (SNA) uses formally defined graph-theoretic methods to rigorously quantify relational patterns among social entities, aiding the understanding of various social phenomena [9]. ABM has also proven to be a valuable approach for analyzing interactions between individuals or groups [10] and is capable of capturing feedback relationships between heterogeneous agents' behaviors and their surrounding environments. Since interactions between agents can be mapped to nodes and links in network science, these two methods can be easily integrated. Embedding networks into ABM allows for the definition of agent sets with spatial interactions or social connections.

Because individual behaviors and network structures are deeply intertwined, social systems often exhibit nonlinear and unpredictable behaviors. Integrating social networks into ABM helps in understanding these processes. Additionally, ABM can address the sampling biases in network structures commonly found in empirical methods of social sciences [11]. As complete network data is rarely available, a comprehensive understanding of all nodal relationships is usually impossible. Computational modeling can serve as a "virtual laboratory" to explore systems in space and time and test hypotheses about causal relationships [12]. Therefore, combining theory-driven approaches with empirically driven network science helps bridge the gap between the two methods and unlocks numerous research possibilities for the study of human behavior. Application cases of ABM integrated with social networks include epidemic transmission [13], innovation diffusion [14], social integration [15], etc.

5.2 The Modeling Framework of Artificial Society

5.2.1 Hierarchical Modeling of Artificial Society

In modeling an artificial society, it is necessary to characterize characteristics at different levels and also needs to incorporate evolutionary processes and spatial relationships at multiple scales. In this way, it can reveal how macro-level dependency relationships are gradually formed through cross-level minor changes,

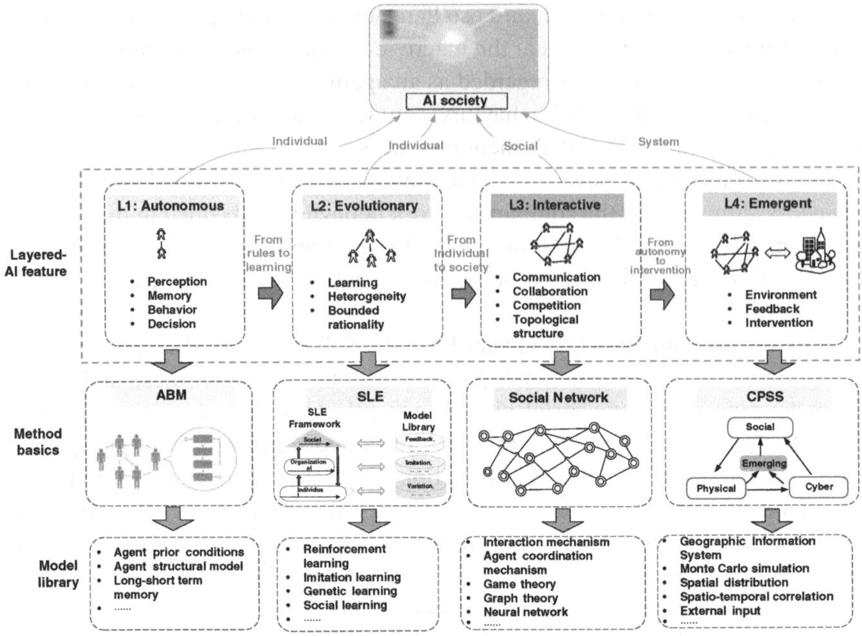

Figure 5.3 The hierarchical modeling framework of artificial society.

which is highly consistent with the view of generative causal relationships. The interdependencies mentioned here include the interdependencies between entities at the same level, the "downward" dependency of the behavior of entities at a given level on the behavior of entities at a lower level, and the "upward" dependency of the behavior of entities at a given level on entities at a higher level. For artificial societies with multi-scale characteristics, agents are endowed with a variety of complex features and behaviors and can independently adjust their behaviors, strategies, and decisions according to environmental changes and acquired information. To adapt to different application fields and specific problems, a hierarchical modeling framework for artificial societies is presented in Figure 5.3. Ranging from micro-level agents to the macro-level AI society, this framework is divided into four layers that span the system across different scales. Each layer has corresponding modeling theories and technologies, which can be expanded as needed in practice. Specific descriptions are as follows:

5.2.1.1 L1: Autonomous Characteristics of Agents

Agents make decisions based on their perception capabilities, memory mechanisms, and behavior sets. Autonomous characteristics enable agents' behaviors to exhibit human-like traits. Agent template serves as a container for integrating knowledge from different domains to meet specific needs. The autonomous characteristics of agents can be described by the following formula:

$$L_t(Agent) =< R, S_t, E_t, V_t, M_t, D_t > \tag{5.1}$$

where R represents the characteristic attributes of the agent, which remain relatively stable; S_t is the dynamic attribute of the agent, subject to changes over time; E_t denotes the agent's perception capability, including the perception range; V_t is the agent's behavioral capability, encompassing spontaneous behaviors and all actions triggered by external events; M_t refers to the agent's memory capability, including short- and long-term memory and the duration of memory retention; and D_t is the agent's decision-making capability, including task decomposition and reasoning abilities.

5.2.1.2 L2: Evolutionary Characteristics of Agents

Based on L1, L2 begins to emphasize the learning and evolutionary characteristics of agents. Diversity is introduced by endowing agents with distinct personalities, goals, and value orientations, thereby giving rise to agent heterogeneity and uniqueness. The evolutionary characteristics of agents can be described by the following formula:

$$L_2(Agent) = \langle F, Y_t \rangle \tag{5.2}$$

where F represents the prior conditions for agent learning and evolution, including heterogeneity and bounded rationality, and Y_t is the agent's learning capability, which enables the agent to learn from environmental feedback or other agents.

5.2.1.3 L3: Interactive Characteristics Among Agents

The diversity and interactions of agents ultimately form a complex and dynamic AI society. Based on the autonomous characteristics of L1 and the learning characteristics of L2, L3 emphasizes the interaction, competition, and cooperation mechanisms among agents, constituting the organizational structure of the AI society. Interactive characteristics can be described by the following formula:

$$L_3(Society) = \langle C_t, G_t, P_t, TS \rangle \tag{5.3}$$

where C_t are the interaction rules among agents, including the interaction range; G_t are the cooperation rules among agents, including cooperation conditions; P_t are the competition rules among agents, including resource constraints; and TS represents the topological structure among agents, such as tree-like, star-shaped, and grid-based topological structure.

5.2.1.4 L4: Emergent Characteristics of the AI Society

In the AI society, agents not only exhibit competition and cooperation but also exhibit time-varying correlations and nonlinear interactions among agent groups and between agents and the environment, which serve as the fundamental driving forces for system emergence. Emergent characteristics can be described by the following formula:

$$L_4(System) = \langle Env, Q_t, N_t \rangle \tag{5.4}$$

where Env represents the interaction between the environment and agents; Q_t represents the feedback mechanism from the system to agents, through which agents can extract knowledge and improve decision-making capabilities; and N_t denotes the intervention strategies adopted by the system for sustainable development.

5.2.2 Modeling Technologies for Evolutionary Characteristics

At the L2 layer, ABM developers have long faced challenges in formulating behavioral rules for agents. ML-based reasoning models can encode agents' behavioral rules into a predefined priori models, providing an alternative solution to constructing agent decision-making mechanisms. The integration of ABM and ML is primarily reflected in two layers:

5.2.2.1 Policy Agent (Macro Level)

In complex systems such as economic and transportation systems, the behaviors of micro-level agents are influenced by policies at the macro level. Similarly, macro-level decision-makers can formulate certain policies to direct the behaviors of micro-level agents or achieve specific goals. For example, governments may guide economic actors to maintain low unemployment or stable price levels; transportation regulators can adjust traffic light settings to alleviate congestion in crowded cities [16, 17]. As shown in Figure 5.4, policy agents represent macro-regulatory entities that collect the aggregated states of all micro-agents, use ML to predict unknown macro-variables or detect macro-patterns/features (termed as "emergence"), and then refine strategies (regulatory rules, incentive policies, etc.). Role agents are regulated entities representing micro-agents in social ecology, endowed with subjective initiative to adjust their behaviors based on regulatory strategies in order to maximize their individual effectiveness. A cyclic feedback process exists between policy agents and role agents: Policy agents learn to adjust governance strategies based on indicator changes, influencing the cognition, decision-making, and behaviors of role agents. Under the impact of current governance strategies, role agents evolve through learning, affecting system effectiveness indicators. Through repeated iterative cycles, a balanced state is ultimately achieved.

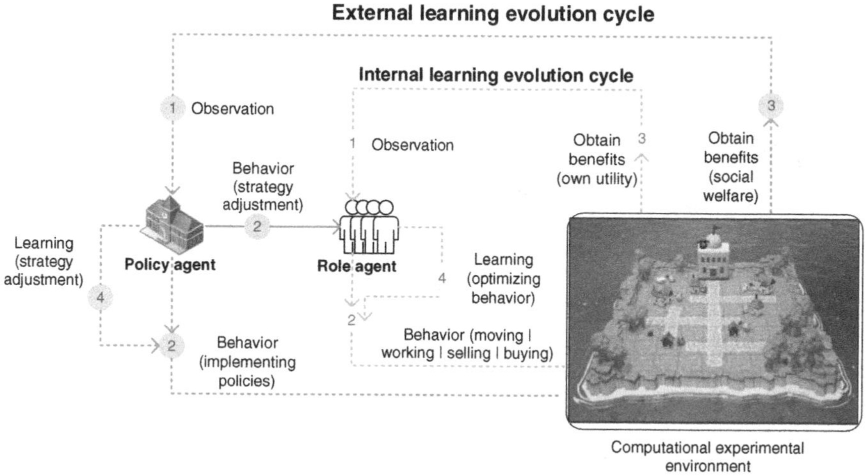

Figure 5.4 Application of ML in ABM at the macro level (policy regulation).

5.2.2.2 Role Agent (Micro Level)

Role agents are more like active entities in a simulation environment. They can use their own observation values or historical data to predict unknown variables or mine certain patterns and then feed the predicted patterns back to their optimization module to generate optimal behaviors [18]. Taking the path planning problem as an example, each vehicle in the transportation system is regarded as an independent agent. By predicting the congestion level of candidate routes, each vehicle agent decides the route to take [19]. As shown in Figure 5.5, when a vehicle agent intends to move from point 1 to point 2, it must decide which route to choose. In this case, the traffic flow in the physical environment is crucial for the agent's road selection, which is transformed into a prediction problem. In such cases, supervised ML techniques can be adopted to establish a model describing the decision-making process of vehicle agents. Supervised ML uses collected data to train the model, and the provision of more data for model training endows ABM with the adaptation and learning capabilities. It should be noted that there is a cyclic interaction between agent prediction and the environment; that is, the behavior of agents will significantly affect the environment, and the environment, in turn, will affect the behavior of agents. This may affect the prediction effect of agents, so modeling should be carried out with caution.

Agents not only possess behavioral characteristics but also, in most cases, have specific goals to achieve. Therefore, agents can adopt various learning methods to optimize their decision-making mechanisms. Although traditional rule-based models can improve agent behavior to a certain extent, these rule-based approaches cannot adapt to the constant changes in the environment

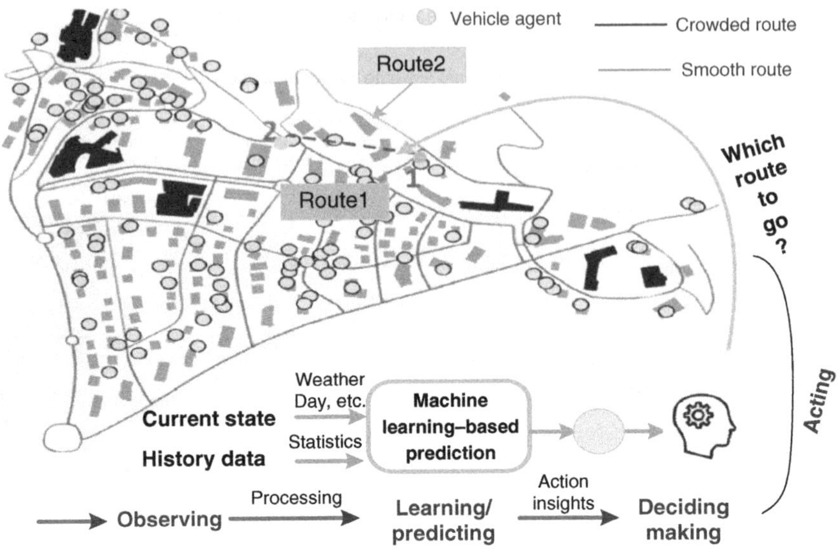

Figure 5.5 Application of ML in ABM at the micro level (route planning).

where agents operate. When an agent operates in a dynamic environment, it gains more knowledge and experience, which are stored in the form of data. This data can be used to refine the agent's decision-making model, enabling the agent to better achieve its goals. ABM allows agents to evolve and adjust their strategies through three main methods: supervised ML methods, which primarily use empirical data for learning; reinforcement learning (RL) methods, which can learn from online-generated streaming data [20]; and evolutionary learning methods, which seek optimal strategies through reproduction, crossover, and mutation.

5.2.3 Modeling Techniques for Interaction Characteristics

At the L3 level, the integration of ABM and social networks can be categorized into three scenarios: (1) Figure 5.6(a) illustrates an exogenous network, focusing on how fixed network structures affect agent states and system dynamics. Social networks enable adequate representation of social interactions between agents. The focal agent (marked by a light bulb) changes its own state (from black to gray) based on the current states of the connected agents. During simulation, model dynamics are determined by the interactions of agents connected through the exogenous network. (2) Figure 5.6(b) represents an endogenous network, focusing on how network structures evolve and what types of structures that

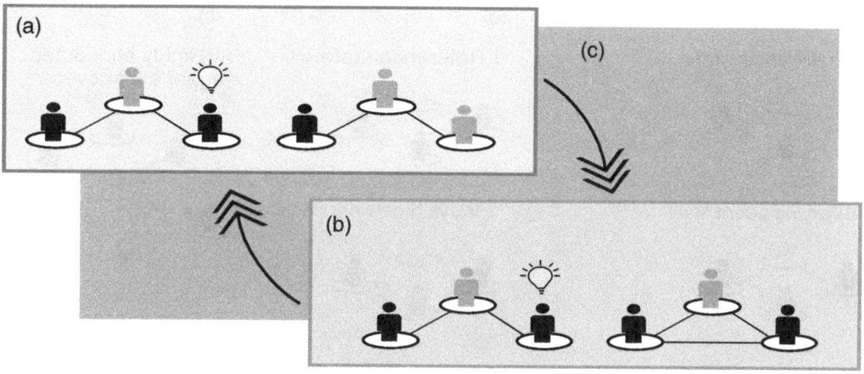

Figure 5.6 The integration of ABM and social networks encompasses three types: exogenous, endogenous, and co-evolutionary networks. (a) Dynamics on **exogenously** imposed network. (b) **Endogenous** network formation. (c) **Co-evolutionary network:** Dynamics on network affects network formation and vice versa. *Source:* [21] / with permission of The Royal Society.

may emerge in specific scenarios. ABM incorporates individual behaviors and environmental impacts into the network formation model. Connections between agents may appear or disappear, but agent states do not necessarily change. (3) Figure 5.6(c) represents a co-evolutionary network, which simultaneously considers endogenous network formation and dynamic updates of agent states. Agents may change their states due to network connections, and they can also adjust network connections based on their own states [22]. However, co-evolutionary networks have rarely been applied in ABM to date.

Comprehending the aggregate patterns resulting from individual-level assumptions and model rules is crucial for interpreting ABM results. In ABM, there are three main methods for evaluating the role of social networks. (1) Agent-centric approach: While networks play a significant role in interactions between agents, but different model outcomes can be obtained by modifying input parameters unrelated to the network itself (e.g., agent states (**1–2 in** Figure 5.7(a))), represented here by agent colors, or external influences such as policies (**3 in** Figure 5.7(a)). (2) Network-centric approach: The agent states remain unchanged, which focuses on the impact of network-level modifications on model outcomes—for example, altering the initial network density (**1–2 in** Figure 5.7(b)) or size (**3 in** Figure 5.7(b)). (3) Structure-explicit analysis: Model outcomes depend not only on agent attributes or network attributes but also on the specific positions of agents within the network. This example illustrates the initial conditions for innovation diffusion (marked in gray) in two scenarios: a highly connected agent as the innovator (**1 in** Figure 5.7(c)) and a loosely connected agent as the innovator (**2 in** Figure 5.7(c)).

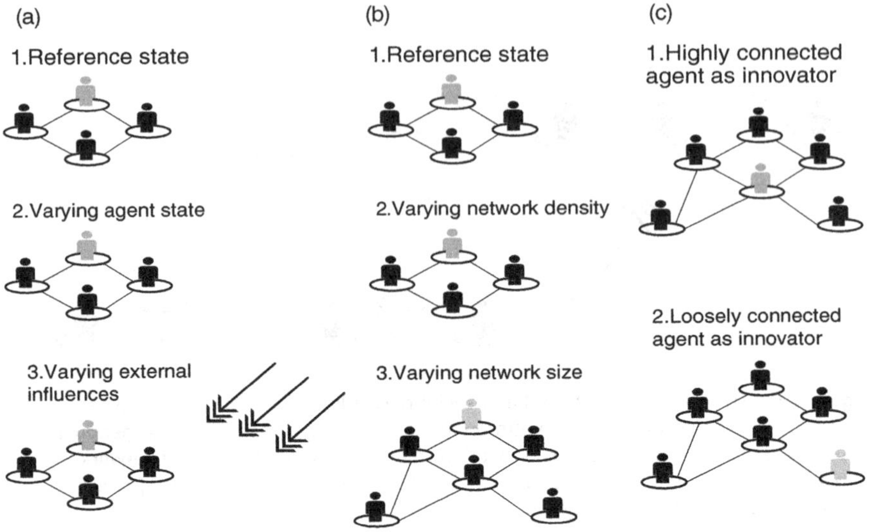

Figure 5.7 Agent-centric analysis, network-centric analysis, and structure-explicit analysis. (a) Agent-centric. (b) Network-centric. (c) Structurally explicit.

5.3 The Modeling Framework of Artificial Society

5.3.1 L1: Autonomous Characteristics of Agents

To simulate real human behavior, the autonomous characteristics of agents are primarily manifested in the ability to reason, plan, and act upon itself, other agents, and the environment [23]. As shown in Figure 5.8, perception (via sensors) provides agents with key information about the environment. The decision-making module of agents then uses rules or goal-based reasoning patterns to develop targeted task plans and take appropriate actions. Knowledge in memory plays an important role in the planning process. The pseudocode detailing the implementation logic of autonomous characteristics is presented in Table 5.2.

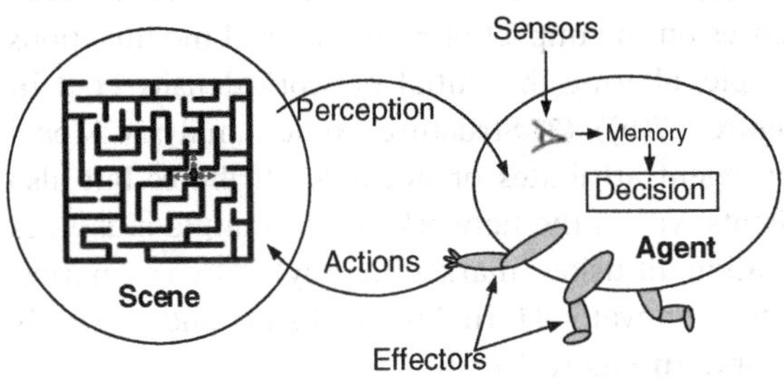

Figure 5.8 The implementation logic of individual autonomous characteristics.

Table 5.2 The Implementation Workflow of Individual Autonomous Characteristics.

Input: Agents with their initial attributes and knowledge.

Output: Agents with their updated attributes and knowledge.

Step 1: Information Perception and Storage

Agents acquire their status information (e.g., position) and environmental information (e.g., other agents and scene information within their perception radius).

Step 2: Memory Mechanism of Agents

Agents determine the types of information they need to store and retrieve, as well as their memory requirements, and save memory information as "state–behavior" pairs. The generated historical data is stored within their memory and continuously updated. Agents can selectively retain, update, and output their memories, thus modeling long- and short-term memories by controlling forget gates, input gates, and output gates. This memory mechanism allows agents to capture critical information and long-term dependencies when processing sequential data, thereby facilitating better decision-making and prediction [19]. The memory mechanism of agents is formally represented by the following formula:

$$M_t = M_{t-1} \odot f_t + A_t \odot i_t$$

where M_t is an agent's memory; M_{t-1} is the agent's previous memory; f_t is the forget vector, determining the features and their weights retained in memory; A_t is the new memory obtained from inputs; i_t is the update vector, determining the feature weights in new memory; and $\odot$ denotes element-wise multiplication.

Step 3: Decision-Making and Action of Agents

The agent's decision-making mechanism refers to the internal logic and algorithms that determine the agent's intelligence level and behavior. According to the agent's intelligence, decision-making is divided into three levels. (1) "Zero intelligence" mode: Low-intelligence agents randomly make decisions from the action set. (2) "Observation-action" mode: Medium-intelligence agents make decisions based on the current state, with the decision mechanism composed of a set of rules in the form of "if ... then ..." statements. The formula is as follows:

$$E_t * S_t * M_t \rightarrow D_t.Rules * V_t \rightarrow S_{(t+1)}, \pi(v \mid s) \rightarrow Q(s, v), \quad Q(s, v) \rightarrow \pi(v \mid s)$$

where E_t represents perception, S_t represents the state, and D_t. *Rules* are the rules set in the decision-making mechanism. $\pi(v \mid s)$ is the policy function, denoting the probability of choosing action v given the state s. $Q(s, v)$ represents the expected cumulative reward for a given specific state and action.

(3) "Observation-prediction" mode: High-intelligence agents make decisions based on reasoning, using a loss function to represent the effectiveness of previous decisions. Better strategies with smaller loss function values are assigned greater weights. This decision-making is expressed by the following formula:

$$D_t = W_h * [D_{t-1}, S_t] + b_h$$

where D_t and D_{t-1} are the weights of decision information at time t and $t-1$, respectively; W_h is the decision weight vector; and b_h is the bias vector.

Step 4: Update the Agent's Information List for the Next Round of Iteration

5.3.2 L2: Evolutionary Characteristics of Agents

The evolutionary characteristics of agents are primarily reflected in their learning capabilities. Agents with different intelligence levels exhibit distinct cognitive and behavioral patterns, as well as varying methods for updating their strategy sets. As illustrated in Figure 5.9, bounded rationality and heterogeneity serve as preconditions for agent learning, enabling agents to make near-optimal decisions within limited resources and time without exhaustive searching or analysis. On this basis, agents continuously refine their decision-making through self-reflection, imitation of others, experience absorption from scenarios, etc. [24]. The bounded rational decision-making of agents is expressed as $P(a_i|s_i)$, which represents the probability that agent_i takes action a_i in state s_i, rather than a deterministic manner. This provides an interface for introducing research conclusions in psychology and behavioral economics, such as prospect theory and acquired bias models. Table 5.3 presents the implementation logic of evolutionary characteristics.

5.3.3 L3: Interaction Characteristics among Agents

The interaction characteristics of agents is mainly reflected in the collaboration among them, which is a process that combines individual task-solving capabilities with communication abilities. In complex coordination tasks, differences in goals or intentions among agents may arise, posing difficulties for collaboration. As shown in Figure 5.10, in the agent social network, interaction serves as the foundation for both collaboration and competition. Agents can achieve interaction through direct peer-to-peer communication, indirect broadcasting, or transmission via intermediaries to enhance mutual understanding and decision-making capabilities. They can share common goals and reasonably allocate tasks based

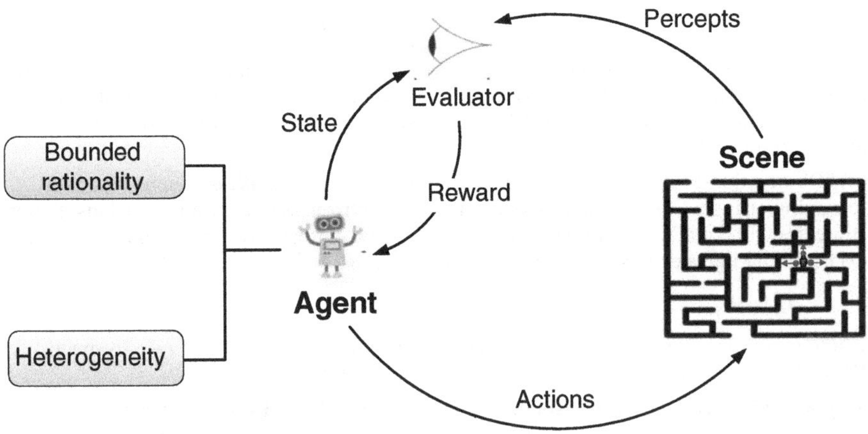

Figure 5.9 The implementation logic of individual evolutionary characteristics.

Table 5.3 The Realization Logic of Evolutionary Characteristics.

Input: Agents with their initial attributes and knowledge.

Output: Agents with their updated attributes and knowledge.

Step 1: Genetic Evolution Modeling of Agents

Identify the characteristic attributes of a single agent and represent them as a matrix. During the evolution process, a random value between 0 and 1 is generated. If this value is less than the mutation threshold, the agent undergoes mutation. A random weight variable in the weight matrix is then replaced with a random value (fluctuating within 10% of the initial value). This induces minor changes in the agent, potentially leading to a more adaptive one. Additionally, a smaller fluctuation range avoids catastrophic failures in the neural network.

Genetic evolution of an agent involves learning from its predecessors. When agents in the simulation environment can produce offspring, the offspring inherits certain attributes from its parents using the formula:

$$Offspring = parent_1 * a + parent_2 * (1 - a)$$

where *Offspring* is the attribute vector of the offspring, a is a randomly generated value between 0 and 1, and $parent_1$ and $parent_2$ denote the weight matrices of the attributes.

Step 2: Imitation Learning Modeling of Agents

Agents can learn from information acquired by neighboring agents. Each agent's capability attribute serves as a knowledge point. Let S_{ij} be the i-th knowledge point of the j-th agent, and $S_{r1}, S_{r2}, \ldots, S_{rk}$ represent the knowledge points of other agents in the organization. Each knowledge point is decomposed into multiple slices (e.g., the agent's location, stored order count). After comparing the normalized attribute values of all agents, the optimal slices from different agents are selected. By combining these optimal slices, a final learning reference is formed. The agent then compares its own slices with the corresponding slices of this virtual learning object to retain superior knowledge points.

Step 3: Social Learning Modeling of Agents

Agents judge their game strategies based on the accepted social culture and their historical strategies during the game. Social learning is modeled as follows. First, define the learning society by identifying all agents within the entire or a partial region as members of the learning society. This region is defined by

$$If\ |X_d - X_1| = Rand(\square) * (X_2 - X_1),\quad then\ X_d \in this\ society$$

where X_d is a point in the region, X_1 is a reference point, X_2 is the central point, and *Rand*() is a random value in [0,1]. Then, calculate the viability of all agents:

$$Viability_i = \sum_{j=1}^{n} w_j * Attribute_j$$

where $Viability_i$ is the viability value of the i-th agent; w_j is the weight of the j-th attribute; and $Attribute_j$ is the value of the j-th attribute. After that, calculate the normalized fitness value of all agents:

$$Fitness_i = \frac{viability_i}{\frac{1}{n}\sum_{j=1}^{n} viability_j}$$

where $Fitness_i$ is the normalized fitness value of the i-th agent. Finally, select the top N agents with the highest fitness values over the previous cycles in the region. Broadcast the knowledge points of these elite agents for others to observe. Upon receiving the broadcast, agents compare the recommended knowledge points with their own and decide whether to imitate and update their knowledge based on their strategy-switching rules.

Step 4: Update the Agents' Information List for the Next Round of Iteration

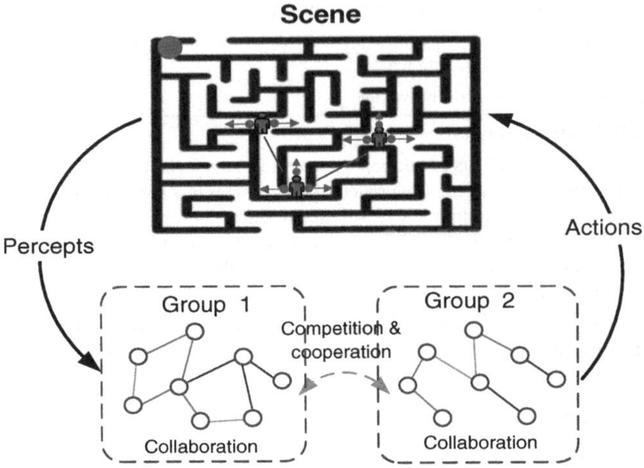

Figure 5.10 Modeling logic of interaction characteristics among agents.

on individual skills and resources availability. Through effective task division and cooperation, agents can improve work efficiency. Table 5.4 presents the implementation logic of interaction characteristics.

5.3.4 L4: Emergent Characteristics of AI Society

As shown in Figure 5.11, if the individual behavior of agents is regarded as the causal variable (input), then emergence serves as the resulting phenomenon

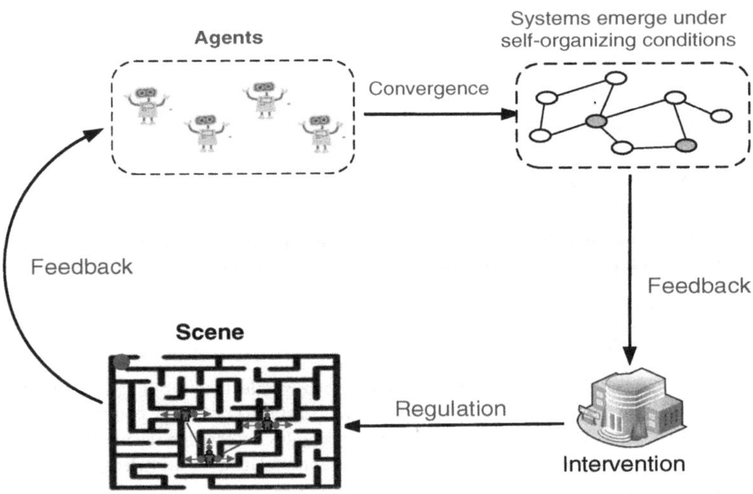

Figure 5.11 The modeling logic of emergent characteristics.

Table 5.4 The Implementation Logic of Interaction Characteristics.

Input: Agents with their initial attributes and knowledge.

Output: Agents with their updated attributes and knowledge.

Step 1: Design the Interaction Network Topology of Agents in the System

Interactions among agents can be represented by a network structure. In addition to the fixed topology in the system, changes in network topology—including node failure, edge failure, edge addition, and other processes—can be used to study unexpected events in society. This work, combined with mechanism design, will ultimately constitute an optimal topology design problem.

Step 2: Record Agents' Connections and Acquire Agents' Relationship Lists

Agents construct their own relationship networks through the number of collaborations, which can be expressed by the following formula [25]:

$$P_{ij} = \frac{N_i + N_j}{2N} + \frac{n_{ij}}{T}$$

where P_{ij} is the strength of the relationship between $Agent_i$ and $Agent_j$; N_i and N_j are the numbers of connection objects in the strong connection relationship lists of $Agent_i$ and $Agent_j$, respectively; N is the total number of agents in the network; n_{ij} is the actual number of encounters between $Agent_i$ and $Agent_j$; and T is the total number of repeated operations.

Step 3: Modeling of Collaboration Process

The adjustment of an agent's collaboration intention depends on the following formula:

$$(v_{i\_co} \geq v_{i\_in})?Evolve(no):\ Evolve(yes)$$

where v_{i_co} and v_{i_in} represent the profits obtained by the i-th agent when choosing a collaborative strategy and an independent operation strategy, respectively. An agent tends to join a new collaborative alliance only when the above formula is not satisfied. The profit distribution vector is calculated according to the following formula:

$$Profit(i) = w_a * Input_{will} * Output_p * Profit_{total}$$

where $Profit(i)$ is the profit of $Agent_i$ after participating in collaboration; $Input_{will}$ represents the intensity of investment willingness, calculated as the cost invested by $Agent_i$ in collaboration divided by the total cost to complete the task, i.e., $Cost(i)/Cost_{total}$; $Output_p$ represents the value contribution degree, calculated as the effective value created by $Agent_i$ in collaboration divided by the total effective value created after completing the task, i.e., $Value(i)/Value_{total}$; w_a is the profit distribution coefficient; and $Profit_{total}$ is the total profit obtained when completing the task. Negotiations among agents often achieve a win–win outcome, leading to the emergence of Nash equilibrium in the system. The negotiation mechanism is expressed by the following formula:

$$C_i = f(coordinate(H_i))$$

where C_i is the collaborative strategy; $f()$ represents the coordination function; and H_i represents the experience of $Agent_i$.

Step 4: Modeling of the Organizational Structure of the System or Group

According to the group's evolutionary pattern, organizational structures include centralized star topologies, autonomously interactive mesh topologies, hierarchical topologies, etc.

Step 5: Update the Agents' Information List for the Next Round of Iteration

(output). The regulatory mechanism measures the macroscopic emergent phenomena and then exerts a regulatory effect on the microscopic individual behaviors. Specific regulatory measures need to be designed for specific scenarios, including intervention and induction. Individual behaviors will update their decision-making mechanisms upon receiving system feedback, thereby generating new behaviors. The system will then produce second-order emergence based on the original first-order emergence. In this way, a closed loop is formed between the overall system and individual behaviors, that is, a two-way feedback mechanism (feedforward and feedback), which may lead to positive feedback loops and mutual reinforcement. In this case, a small fluctuation deviating from equilibrium can lead to exaggerated and amplified effects. Table 5.5 presents the implementation logic of emergent characteristics.

5.4 Case Study: Intelligent Logistics System

The intelligent logistics system can provide personalized solutions for complex transportation demands. This section constructs experimental scenarios (cases 1–4) of the intelligent logistics system based on the characteristics of L1–L4 layers, respectively, to verify the effectiveness of the previously proposed hierarchical modeling framework for artificial society.

5.4.1 Construction of Experimental System

The intelligent logistics system is composed of various logistics resources, including mechanical vehicle agents, manual vehicle agents, data, algorithms, and facilities, jointly forming an "Agentic Service System." As shown in Figure 5.12, the operation of the intelligent logistics system is based on supply–demand matching [27], mainly including three modules. (1) Demand side: Customer orders are regarded as system inputs to represent fluctuations in the external market environment. (2) Supply side: Composed of two types of agents, namely, mechanical vehicle agents and manual vehicle agents. Among them, as adaptive agents, manual vehicle agents decide to expand their scale or exit the logistics system according to their own profits. (3) Operation side: The platform agent monitors, analyzes, intervenes in, and regulates the evolutionary process of the system. Previous agent modeling simulations only focused on certain characteristics of agents or systems, while the hierarchical modeling framework covers more complex features, as specifically shown in Table 5.6.

Four experimental scenarios are customized here, requiring modeling of intelligent logistics systems with different complexities to understand their operation process. The **middle** part of Figure 5.12 shows the operation interface of the

Table 5.5 The Implementation Logic of Emergent Characteristics.

Input: Agents with their initial attributes and knowledge.

Output: Agents with updated attributes and knowledge, and updated environmental information.

Step 1: Define the Impact of the External Environment on Agents

The impact of the external environment perceived by agents is expressed by the following formula:

$$E =< S_e, X, \delta_e >, S_e * X \rightarrow \delta_e$$

where S_e is the set of environmental states; X is the set of external events; and δ_e represents the new environmental state based on the current state.

Step 2: Model the Adaptation Mechanism of Agents

Agents select actions based on their current state, interact with the environment, and obtain immediate rewards and feedback on the next state. Based on this feedback, agents update their value functions and strategies to make better decisions in the future. For instance, specific algorithms and methods can vary according to problem characteristics and research needs [26], such as Q-learning, Deep Q-Network (DQN) and policy gradient.

(1) The feedback information received by agents from the environment is represented using a reward function as follows: $R(t) = r(S_t, a(t))$, where $R(t)$ is the immediate reward obtained by the agent when selecting an action based on the current state at time t; $r()$ is the reward function; S_t is the environmental state at time t; and $a(t)$ is the action selected by the agent at time t.

(2) The formula for the agent's value update function is expressed as:

$$V(s(t)) = V(s(t)) + \alpha * (R(t) + \gamma * V(s(t+1)) - V(s(t))),$$

$$Q(s(t), a(t)) = Q(s(t), a(t)) + \alpha * (R(t) + \gamma * Q(s(t+1), a(t+1)) - Q(s(t), a(t)))$$

where $V(s(t))$ is the state value estimate at time t, i.e., the expected value of cumulative rewards the agent can obtain in state s; $Q(s(t), a(t))$ is the action value estimate at time t, i.e., the expected value of cumulative rewards the agent can obtain after selecting action a in state s; $s(t+1)$ is the next state the agent enters after selecting action $a(t)$ at time t; α is the learning rate; and γ is the discount factor measuring the importance of future rewards.

(3) The formula for updating the strategy based on value function estimation is as follows:

$$\pi(t+1) = argmax\, Q(s(t), a(t))$$

where $\pi(t+1)$ is the strategy at time $t+1$ and *argmax*() operation denotes selecting the action that maximizes $Q(s(t), a(t))$.

Step 3: Model the System Intervention Mechanism

System operators continuously adjust strategies according to the system's goals and states, which is expressed by the following formula:

$$Intervention(R_t, J_t) = < S_{System}, S_t, Goal_{System} >$$

where $Intervention(R_t, J_t)$ is the intervention mechanism; R_t and J_t are the system's incentive strategy and intervention strategy, respectively; S_{System} is the system state under the influence of agents; S_t is the environmental state at time t; and $Goal_{System}$ is the system's goal.

Step 4: After a certain number of cycles, steps 2 and 3 are repeated to achieve continuous evolution.

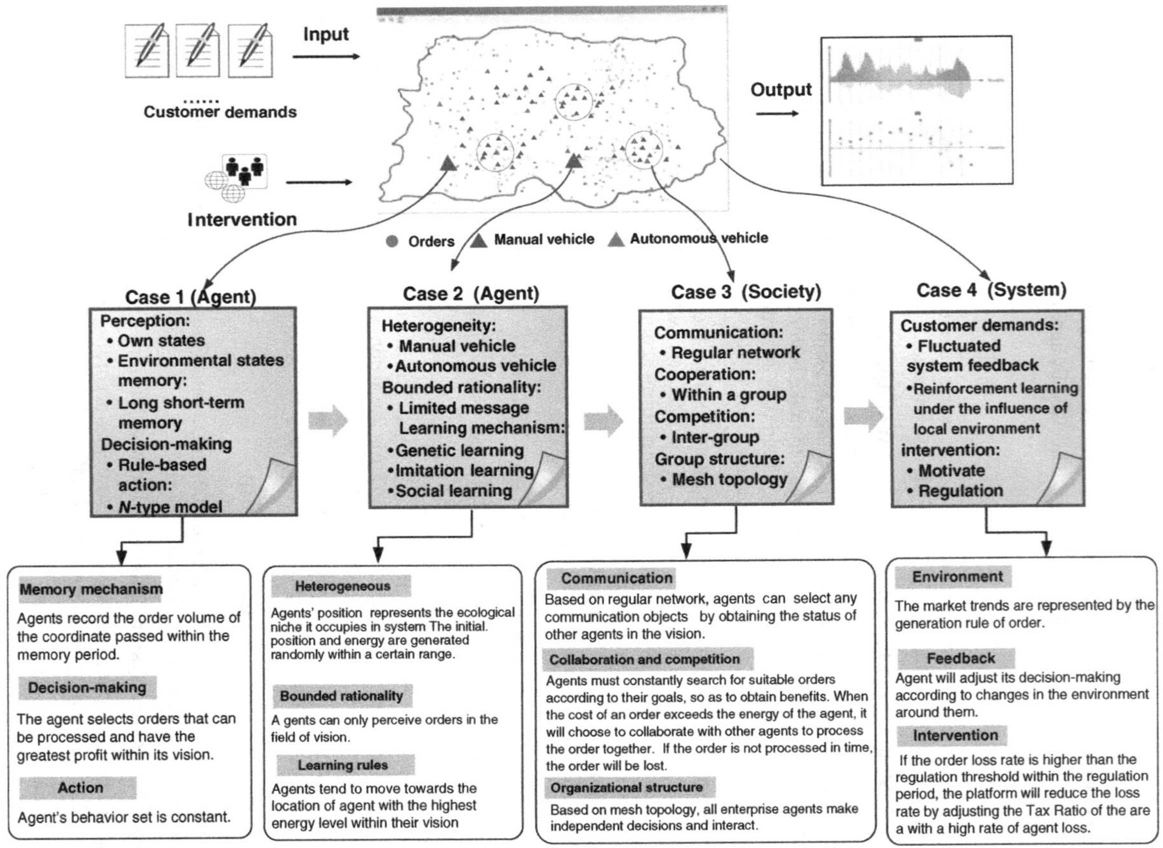

Figure 5.12 The overall framework of the experimental system.

Table 5.6 Comparison and Explanation of Existing Work.

	L1				L2					L3			L4	
Models	**F1**	**F2**	**F3**	**F4**	**F5**	**F6**	**F7**	**F8**	**F9**	**F10**	**F11**	**F12**	**F13**	**F14**
ABM + SN	✓		✓	✓				✓		✓	✓			
ABM + ML	✓		✓	✓						✓		✓	✓	
ABM + LLM	✓	✓	✓	✓	✓			✓	✓	✓				
SLE [28]	✓		✓	✓			✓	✓	✓			✓	✓	✓
SLE2 [29]	✓		✓	✓			✓	✓	✓	✓	✓	✓	✓	✓
Ours	✓	✓	✓	✓	✓	✓	✓	✓	✓	✓	✓	✓	✓	✓

Notes: F1: perception; F2: memory; F3: decision-making; F4: behavior; F5: heterogeneity; F6: bounded rationality; F7: genetic evolution; F8: imitation learning; F9: social learning; F10: communication; F11: cooperation; F12: competition; F13: environmental effect; F14: intervention mechanism.

experimental system, representing the geographical activity space of agents. Due to variations in economic development levels, consumption habits, etc., the demand markets in different spatial locations differ significantly, which are reflected in the number of orders, order distribution locations, and other aspects. In the initial state, supply-side agents and orders are randomly distributed in different locations. The design of complexity characteristics of the experimental system is shown in the **lower** part of Figure 5.12. Scenario 1 is designed from an individual perspective, including only manual vehicle agents, mainly considering the impact of agents' autonomous decision-making and behavior on system simulation. Scenario 2 is also designed from an individual perspective, including mechanical vehicle agents and manual vehicle agents, mainly considering the impact of the two types of agents' learning mechanisms on system simulation under limited information conditions. Scenario 3 is also designed from a social perspective, mainly considering the impact of collaboration mechanisms under interactive networks on system simulation. Scenario 4 is also designed from the perspective of the entire system, considering the impact of external intervention mechanisms on system simulation in a fluctuating market environment.

The performance indicators for experimental evaluation include (1) the number of manual vehicle agents, reflecting the scale of manual vehicle agents; (2) the average assets of manual vehicle agents, reflecting the development status of individual agents; (3) order loss rate: a metric for evaluating system anomalies, calculated as the ratio of lost orders to generated orders within a specific time period; and (4) platform total revenue: the platform's revenue derived from manual vehicle agents' service value, reflecting the operational status of the system. Table 5.7 presents the experimental schemes for the four scenarios.

Table 5.7 Experimental Schemes and Research Objectives.

Case	Scheme Design	Experimental Objectives
Scenario 1	All vehicle agents have the same perception mechanism, memory, decision-making mechanism, and behavior set. Vehicle agents differ in initial positions and assets.	As a control group, analyze the impact of autonomous characteristics on system operation.
Scenario 2	Based on scenario 1, manual vehicle agents add a learning mechanism; mechanical vehicle agents have different initial positions, and can perceive and process low-difficulty orders within a radius.	When both manual and mechanical vehicle agents are present, analyze the impact of learning characteristics on system operation.
Scenario 3	Based on scenario 2, add collaboration mechanisms among agents (including collaboration between manual vehicle agents and between manual and mechanical vehicle agents) to enable agents to complete orders beyond their individual capabilities.	When collaboration exists among agents, analyze the impact of collaboration mechanisms on system operation.
Scenario 4	Based on scenario 3, in addition to manual and mechanical vehicle agents, add a platform operational agent that regulates the system through external intervention.	Compared with other cases, study the impact of intervention and regulation mechanisms on system operation.

5.4.2 Design of the Experimental System

As shown in Figure 5.13, manual vehicle agents search for specific orders in the environment and consume a certain amount of assets during the search process. Conversely, when a manual vehicle agent captures and processes a demand, its assets increase accordingly. When its assets reach a threshold, it can generate offspring agents; when its assets fall below the death threshold, the agent exits and disappears. The characteristic attributes, state attributes, perceived events, behaviors, and decision-making mechanisms of manual vehicle agents are specified in **Tables A.5.1–A.5.5**, respectively.

Mechanical vehicle agent is a type of agent with relatively low intelligence, which can only handle low-difficulty orders within a narrow business scope and can also be invoked by manual vehicle agents to improve order processing efficiency. Additionally, compared with manual vehicle agents, mechanical vehicle agents often have lower service costs and higher processing speeds. The main settings of mechanical vehicle agents are shown in **Table A.5.6**.

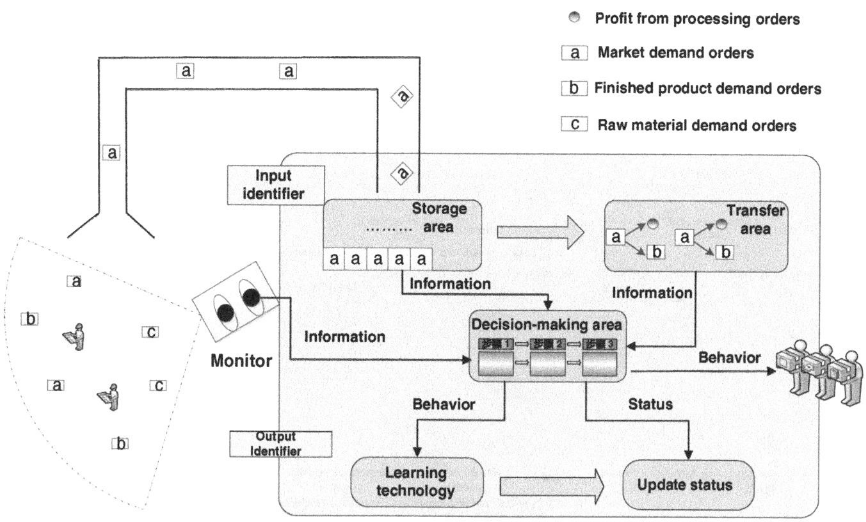

Figure 5.13 The structure of manual vehicle agent.

The vehicle agent (including manual vehicle agents and mechanical vehicle agents), as the main body of the system, has their behavioral process shown in **Table A.5.7**.

Manual vehicle agents switch among three states—normal, reproduction, or death—based on their own asset status. To maintain survival and reproduction, they need to continuously make decisions that help improve their survival assets. In this process, environmental settings and platform strategies have significant impacts on them. The main settings of the operational-side platform agent are shown in **Table A.5.8**.

The agents on the demand side are embodied as orders in the system, and their specific settings are shown in **Table A.5.9**.

5.4.3 Analysis of Experimental Results

During the operation of the experimental system, the evolution process of the intelligent logistics system is shown in Figure 5.14. In scenario 1, manual vehicle agents only rely on perception to obtain order information within their own vision range and gain profits after processing orders to maintain survival. Due to limited information, a large number of manual vehicle agents died due to insufficient orders (Figure 5.14(a)). In scenario 2, agents learned location knowledge from those with abundant orders in the system, leading to the aggregation of agents in areas with more orders (Figure 5.14(b)). In scenario 3, manual vehicle agents could process orders beyond their own processing capacity through collaboration. Compared

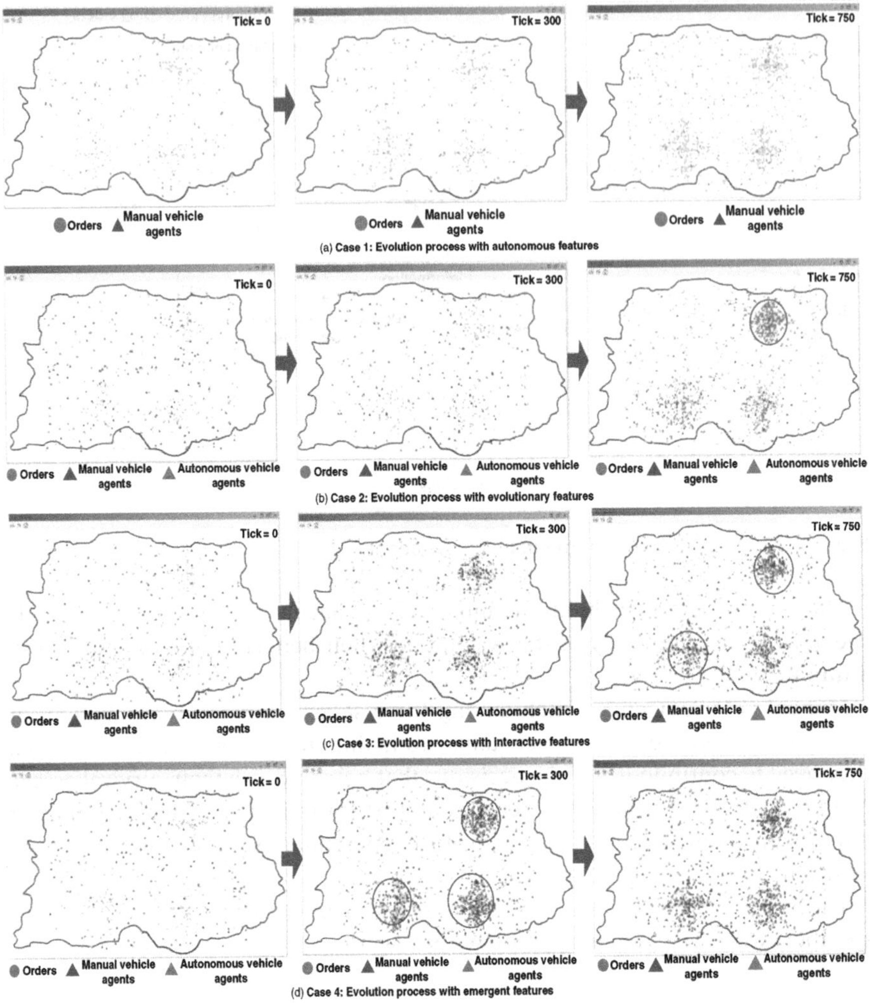

Figure 5.14 Evolution processes of intelligent logistics systems with different characteristics.

with scenario 2, the number of agent nodes in scenario 3 increased (Figure 5.14(c)). In scenario 4, influenced by the tax reduction strategy, some agents chose to carry out business in areas with high-order loss rates, making the location distribution of agents more uniform compared with previous scenarios (Figure 5.14(d)).

Figure 5.15(a) and (b) present individual-level analyses. In scenario 1, agents overly relied on individual exploration, leading to delays in order acquisition and thus significant assets losses. Due to insufficient assets replenishment, a large

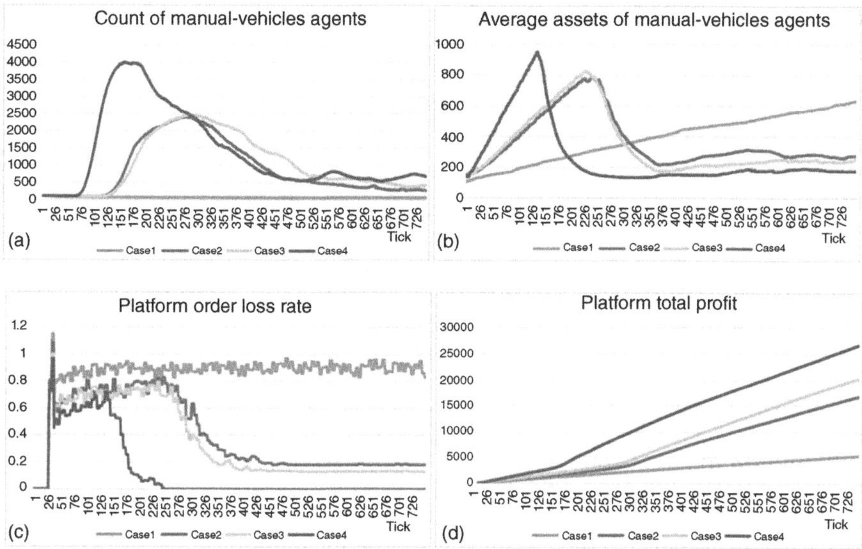

Figure 5.15 Experimental result analysis of intelligent logistics system. (a) Count of manual-vehicles agents. (b) Average assets of manual-vehicles agents. (c) Platform order loss rate. (d) Platform total profit.

number of agents died, leaving only a few highly adaptable agents to survive, with the number of agents remaining at a low level. However, this also resulted in sufficient resources and less competition for the surviving agents, maintaining consistently high average profits. In scenarios 2–4, due to the improved intelligence of agents (via learning, collaboration, and intervention mechanisms), the number of agents and average profits rose rapidly in the early stage and entered a dynamic balance in the later stage due to limited social resources. Among them, scenario 3 benefited from agent collaboration, with more order acquisition opportunities, and the agent scale and average profits were slightly higher than those in scenario 2. Scenario 4, through the intervention inducement mechanism, could process more orders, ultimately achieving a stable state with the largest agent scale and higher individual average profits. Figure 5.15(c) and (d) provide system-level analyses. In scenario 1, due to the limited visibility of agents, the number of processed orders was very small, and the order loss rate was extremely high. Therefore, the platform's profits remained low and grew slowly due to the limited number of completed transactions. In scenarios 2 and 3, agents could improve their behaviors by learning from the experience of other agents, thereby better understanding demand and processing orders more effectively. Compared with scenario 2, the collaboration of agents in scenario 3 further improved efficiency, the order loss rate slightly decreased, and the platform's revenue

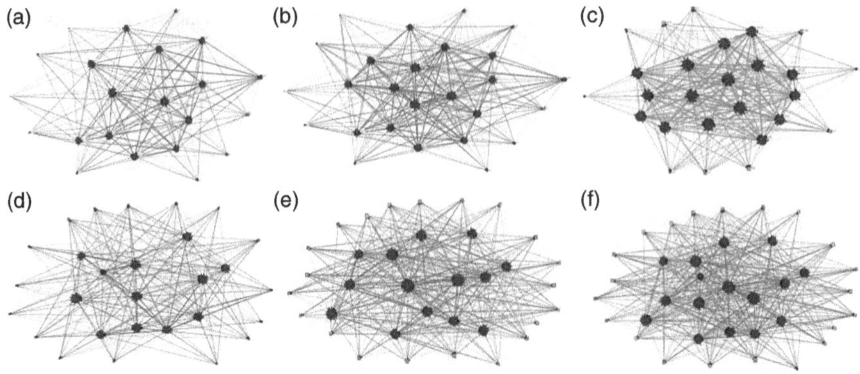

Figure 5.16 Evolution network analysis of intelligent logistics system. (a) Case 3: Early stage. (b) Case 3: Medium stage. (c) Case 3: Later stage. (d) Case 4: Early stage. (e) Case 4: Medium stage. (f) Case 4: Later stage.

increased. In scenario 4, the platform's regulation strategy significantly reduced the order loss rate, and the increase in transaction volume greatly improved the platform's total profits.

In the four groups of experiments, agent collaboration existed in scenarios 3 and 4, including two types of collaboration: between manual vehicle agents and between manual vehicle agents and mechanical vehicle agents. Figure 5.16 is a collaborative network analysis diagram constructed from partial agent collaboration data obtained through uniform sampling of experiments, reflecting the evolution of collaboration between agents, where dark gray dots represent human-driven vehicle agents and light gray squares represent machine vehicle agents. The greater the influence weight of a node in the social network, the larger the corresponding node size. The upper part (a, b, c) of Figure 5.16 is the collaboration network diagram of scenario 3. When there is no platform intervention, the scale of agent collaboration is small, and it is mainly concentrated on collaboration between human-driven vehicle agents. The lower part (d, e, f) of Figure 5.16 is the collaboration network diagram between agents in scenario 4. With the intervention of the platform, the scale of agent collaboration and the frequency of collaborative behaviors significantly improved, especially the increased collaboration between manual vehicle agents and mechanical vehicle agents, confirming the concept that human–machine collaboration advocated in today's society improves production efficiency.

This experiment presents the usage process of the hierarchical modeling framework, and the constructed artificial societies exhibit autonomous, learning, collaborative characteristics and emergent phenomena; simultaneously, the experimental results are consistent with our intuitive experience. This verifies the effectiveness of the hierarchical modeling framework to a certain extent and provides a specific customizable reference path for the model design of AI agents and AI societies.

5.5 Chapter Summary

This chapter deeply discusses the modeling difficulties of artificial societies, including heterogeneity, bounded rationality, learning and evolution, and interactive behaviors. To effectively build a bridge between complex feature modeling and AI technologies, we propose a hierarchical artificial society modeling framework. Based on the increasing complexity of agent and systems, it details the behavioral patterns and decision-making processes of agents across four layers, namely, autonomous characteristics (L1), evolutionary characteristics (L2), interactive characteristics (L3), and emergent characteristics (L4). Among them, L1 emphasizes the perception, memory, behavior, and decision-making capabilities of agents; L2 focuses on the learning mechanism of agents based on heterogeneity and bounded rationality; L3 analyzes the communication, collaboration, competition, and network topology between agents; and L4 examines the environmental impact, feedback loops, and intervention mechanisms of the system from a macro perspective. This framework not only helps standardize the modeling process but also greatly reduces the threshold for applying the computational experiment method in different fields.

References

1 Xiao, X., Xiang-Ning, Y., De-Yu, Z. et al. (2023). Computational experiments: past, present and perspective. *Acta Automatica Sinica* 49 (2): 246–271.

2 Brock, W.A. and Hommes, C.H. (1997). A rational route to randomness. *Econometrica: Journal of the Econometric Society* 65: 1059–1095.

3 Chen, S.H. and Yeh, C.H. (2001). Evolving traders and the business school with genetic programming: a new architecture of the agent-based artificial stock market. *Journal of Economic Dynamics and Control* 25 (3–4): 363–393.

4 Rosenfeld, A. and Kraus, S. (2009). Modeling agents through bounded rationality theories. IJCAI (vol. 9), pp. 264–271.

5 Simon, H.A. (1990). Bounded rationality. In: *Utility and Probability*, 15–18. London: Palgrave Macmillan.

6 Gigerenzer, G. (2004). Fast and frugal heuristics: the tools of bounded rationality. In: *Blackwell Handbook of Judgment and Decision Making*, vol. 62 (ed. D.J. Koehler and N. Harvey), 88.

7 Kahneman, D. and Tversky, A. (2013). Prospect theory: an analysis of decision under risk. In: *Handbook of the Fundamentals of Financial Decision Making: Part I* (ed. L. C. MacLean and W. T. Ziemba), 99–127. Hackensack: World Scientific Publishing.

8 Weibull, J.W. (1997). *Evolutionary Game Theory*. MIT press.

9 Scott, J. (2011). Social network analysis: developments, advances, and prospects. *Social Network Analysis and Mining* 1 (1): 21–26.

10 Gilbert, N. (2019). *Agent-based Models*. Sage Publications.

11 Costenbader, E. and Valente, T.W. (2003). The stability of centrality measures when networks are sampled. *Social Networks* 25 (4): 283–307.

12 Carley, K.M. (2009). Computational modeling for reasoning about the social behavior of humans. *Computational and Mathematical Organization Theory* 15 (1): 47–59.

13 Verelst, F., Willem, L., and Beutels, P. (2016). Behavioural change models for infectious disease transmission: a systematic review (2010–2015). *Journal of the Royal Society Interface* 13 (125): 20160820.

14 Kiesling, E., Günther, M., Stummer, C., and Wakolbinger, L.M. (2012). Agent-based simulation of innovation diffusion: a review. *Central European Journal of Operations Research* 20 (2): 183–230.

15 Rockenbauch, T. and Sakdapolrak, P. (2017). Social networks and the resilience of rural communities in the Global South: a critical review and conceptual reflections. *Ecology and Society* 22 (1): 10.

16 Balaji, P.G., German, X., and Srinivasan, D. (2010). Urban traffic signal control using reinforcement learning agents. *IET Intelligent Transport Systems* 4 (3): 177–188.

17 Bone, C., Dragicevic, S., and White, R. (2011). Modeling-in-the-middle: bridging the gap between agent-based modeling and multi-objective decision-making for land use change. *International Journal of Geographical Information Science* 25 (5): 717–737.

18 Augustijn, E.W., Kounadi, O., Kuznecova, T. et al. (2019). Teaching agent-based modelling and machine learning in an integrated way. GeoComputation 2019.

19 Gehrke, J.D. and Wojtusiak, J. (2008). Traffic prediction for agent route planning. In: *International Conference on Computational Science*, 692–701. Berlin, Heidelberg: Springer Berlin Heidelberg.

20 van der Hoog, S. (2017). Deep learning in (and of) agent-based models: a prospectus. arXiv preprint arXiv:1706.06302.

21 Gross, T. and Blasius, B. (2008). Adaptive coevolutionary networks: a review. *Journal of the Royal Society Interface* 5 (20): 259–271.

22 Will, M., Groeneveld, J., Frank, K., and Müller, B. (2020). Combining social network analysis and agent-based modelling to explore dynamics of human interaction: a review. *Socio-Environmental Systems Modelling* 2: 16325–16325.

23 Wang, F.Y., Guo, J., Bu, G., and Zhang, J.J. (2022). Mutually trustworthy human-machine knowledge automation and hybrid augmented intelligence: mechanisms and applications of cognition, management, and control for complex systems. *Frontiers of Information Technology & Electronic Engineering* 23 (8): 1142–1157.

24 Malashenko, Y.E. and Nazarova, I.A. (2020). Analysis of critical damage in the communication network: I. Model and computational experiment. *Journal of Computer and Systems Sciences International* 59 (5): 745–754.

25 Kapucu, N. and Garayev, V. (2013). Designing, managing, and sustaining functionally collaborative emergency management networks. *The American Review of Public Administration* 43 (3): 312–330.

26 Xue, X., Wang, S., Gui, B. et al. (2016). A computational experiment-based evaluation method for context-aware services in complicated environment. *Information Sciences* 373: 269–286.

27 Li, H., Chen, Y., Li, K. et al. (2023). Transportation internet: a sustainable solution for intelligent transportation systems. *IEEE Transactions on Intelligent Transportation Systems* 24 (12): 15818–15829.

28 Xue, X., Wang, S., Zhang, L. et al. (2018). Social learning evolution (SLE): computational experiment-based modeling framework of social manufacturing. *IEEE Transactions on Industrial Informatics* 15 (6): 3343–3355.

29 Zhou, D., Xue, X., and Zhou, Z. (2022). SLE2: the improved social learning evolution model of cloud manufacturing service ecosystem. *IEEE Transactions on Industrial Informatics* 18 (12): 9017–9026.

6
The Second Step – Constructing the Experimental System

To analyze the dynamics of complex social systems via computational experiments, we transform artificial society models into social simulators. As an integrated whole, social simulators represent the intersection of general modeling technologies and diverse domain scenarios. However, there currently exists no standardized or systematic approach for domain model docking. Consequently, researchers often rebuild models from scratch, wasting significant time and resources. To address this deficiency, this chapter first presents the developmental trajectory of social simulators and examines the progressive integration of virtual and real worlds. Subsequently, using the "Corrupted Blood" incident in *World of Warcraft* as a case study, we demonstrate the potential of online games as virtual laboratories. The chapter then provides detailed specifications for an integration framework connecting domain scenarios with artificial society models, including interface protocols for agent models, environment models, and rule models. Finally, through a case study focused on coal mine disaster emergency management, we illustrate the construction process of computational experiment systems (Figure 6.1).

6.1 From Game Worlds to Virtual Societies

Against the technological backdrop of different eras, virtual worlds have taken various forms. In recent decades, the development of virtual worlds has significantly overlapped with the evolution of online games. Since the birth of online gaming, humans have constantly dreamed of transcending real-world activities to experience vibrant virtual lives. Starting from the latter half of the 20th century, this dream gradually became tangible as virtual worlds became increasingly accessible. As outlined in John David N. Dionisio's paper "3D Virtual Worlds and the Metaverse: Current Status and Future Possibilities" [1], the developmental

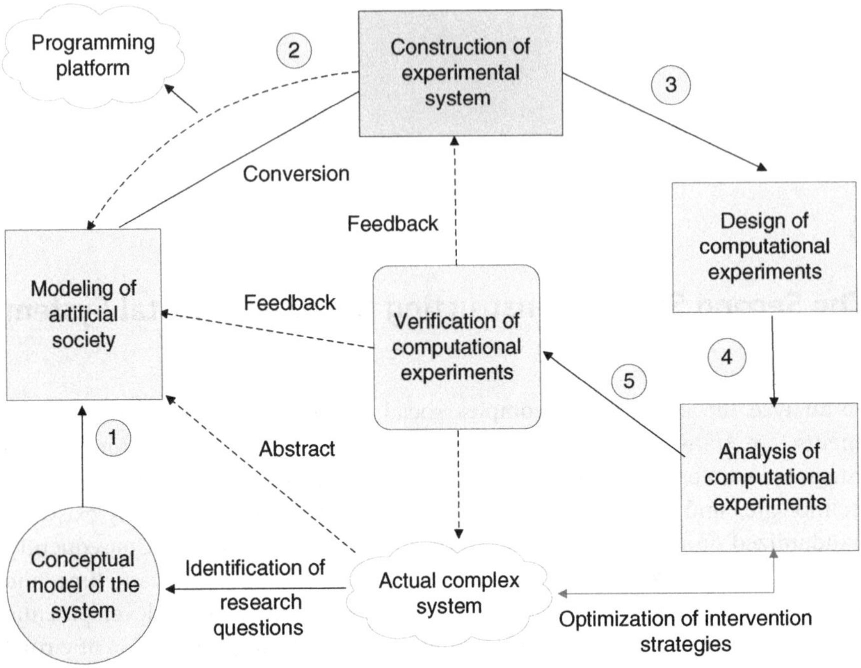

Figure 6.1 The role of the experimental system within the computational experiment method.

trajectory of virtual worlds can be systematically traced through the lens of digital games.

6.1.1 The First Phase 3D Virtual Worlds

Since the 1970s, computer technology has advanced rapidly. Computers were no longer giant machines confined to laboratories; ordinary people could now own personal computers, laying the foundation for the popularization of digital games. Concurrently, the emergence of network technology enabled players in different locations to interact and compete within the same virtual world. These two breakthroughs formed the bedrock for the subsequent development of virtual worlds.

In 1978, Roy Trubshaw wrote a brief program in a computer lab and symbolically named it MUD (Multi-User Dungeon/Dimension/Domain). This marked the formal dawn of digital virtual worlds. MUD was a purely text-based world where players performed virtual actions by entering commands, enabling them to interact with others in real time, level up their characters, complete quests, and explore the environment. It became the earliest authentic example of real-time multiplayer online gaming.

Though simple in gameplay, these Role-Playing-Games (RPGs) already contained critical elements of virtual worlds: chat functions, character creation, identity systems, missions, and open-world exploration. Players engaged in a fictional setting, selecting races and factions, and forming teams to fulfill tasks through collaborative interactions. This can be considered the prototype of "Script killing (the murder role-playing game)" and has laid the narrative foundation for later online games such as Neverwinter Nights, Diablo, and *World of Warcraft*. Stories and life experiences that were once impossible could now be realized in the virtual world. This was the earliest embryonic form of virtual worlds. Table 6.1 lists influential works that significantly shaped the evolution of virtual worlds.

With further advancements in computer technology, graphical games gradually replaced text-based games (as shown in Figure 6.2). The immersive experience of

Table 6.1 Influential Works on the Development of Virtual Worlds.

Time	Work	Impact
1955/1977	*The Lord of the Rings* *The Silmarillion*	Tolkien's works, hailed as the greatest fantasy world-building achievements of the 20th century, inspired countless subsequent games and virtual worlds.
1974	*Gungeons & Dragons*	Created by Gary Gygax and Dave Arneson, this work marked the birth of modern RPGs.
1981	*True Names*	One of the earliest cyberpunk novels, offering a comprehensive depiction of virtual spaces.
1984	*Neuromancer*	A cyberpunk classic that profoundly influenced later virtual world concepts.
1992	*Snow Crash*	Introduced and detailed the concept of the metaverse, laying foundational ideas for future virtual worlds.

Figure 6.2 Interface of a 3D game. *Source:* YuanShan / https://news.sohu.com/a/562521040_121210895 / last accessed on February 14, 2026.

colorful visual graphics far exceeded the textual narratives imagined by players, leading to the gradual decline of MUD-style games. The first major graphical virtual world project was Habitat, a pioneering MMORPG (Massively Multiplayer Online Role-Playing Game) and also the first commercially successful virtual world. Habitat's groundbreaking contributions were threefold. (1) It constructed the first large-scale virtual world using a graphical interface. (2) It introduced the concept of the avatar, a digital extension of humans. (3) It established a fully digital society where Habitat's evolution was driven by players' imagination and creativity. This laid the trajectory for future development of virtual worlds, marking the divergence between virtual worlds and traditional online games.

In 1995, Worlds Inc. launched the first fully 3D virtual world. This open world had no predefined protagonist—every player was a real individual who collectively shaped the world's evolution. Social interactions within this virtual space mirrored the complexity of human behaviors in the real world, resembling an alternative network paradigm that is parallel to the Internet. The same year, Alpha World rebranded its 3D web browser as Active Worlds, a virtual world inspired by the metaverse vision in *Snow Crash*. It rapidly became one of the most significant 3D social platforms, attracting tens of thousands of users. In 1996, OnLive! Traveler debuted as the first virtual world to support voice chat and directional audio cues, further blurring the boundaries between the virtual and physical realms.

As "Active Worlds" and "OnLive! Traveler" continued to evolve, virtual worlds were maturing significantly. The paper "Active Worlds: Geography and Social Interaction in Virtual Reality" [2] summarized the key characteristics of such virtual platforms. (1) Digital citizenship: Users were categorized into tourists and paid citizens. Tourists could only access public spaces and engage in limited community activities, while paid citizens unlocked broader privileges. (2) Digital property ownership: Active Worlds featured two types of spaces: public spaces, which were open to all players for constructing personal buildings, and private spaces, reserved for specific paid groups and granting ownership rights to the structures—a concept that can be seen as a direct precursor to NFTs (Non-Fungible Tokens, NFTs). (3) User stratification: Users exhibited diverse attributes, with distinctions between real players and nonplayer characters (NPCs). Moreover, the world was not singular. Active Worlds hosted more than 800 parallel subworlds. Each subworld had unique rules and themes, akin to distinct game instances, allowing users to choose environments that align with their preferences.

By this stage, virtual worlds had nearly achieved parity with the physical realm in terms of technical infrastructure: 3D environments, social systems, user-generated content (UGC), and digital assets formed a robust framework. However, a critical gap remained: the lack of a sustainable economic system.

While the tokenization of in-game assets (first implemented in Habitat) served as a foundation, true viability required interoperability between virtual and real-world economies. Without this bridge, virtual assets remained mere data, devoid of any tangible value.

6.1.2 The Second Phase: Development of Virtual Economic Systems

From a simulation perspective, virtual worlds gradually approached the real world in two dimensions: one being the physical dimension, where virtual worlds transitioned from text to graphics and then to 3D, synchronizing real-world physics into virtual environments; the other being the social dimension, where they transplanted interpersonal social relationships to establish broader social networks. Thus, beyond visual technology, virtual worlds required the introduction of more real-world social mechanisms, which led to the emergence of virtual economic systems. Even within virtual worlds, players' energy and time were limited, necessitating choices between tasks such as resource collection, skill learning, and entertainment activities. Consequently, trading behaviors involving virtual goods emerged.

Initially, trading activities were driven by players' needs for items such as food, equipment, and resources, with item values determined by practicality and rarity. Time or money invested in acquiring these items also added to their value. This basic supply and demand relationship is very similar to the real world. *Second Life*, released in 2002, revolutionized this simple transaction model. Like previous virtual worlds, it allowed users to create items that could be traded using its virtual currency, the Linden Dollar. Uniquely, players could exchange these in-game tokens for US dollars on markets operated by residents, at an exchange rate that fluctuated around 270 Linden Dollars per USD by 2007. As transactions increased, arbitrage became commonplace, which solidified the necessity of a robust economic system within virtual worlds.

In *Second Life*, players could not only enjoy gameplay but also apply their skills to create virtual assets such as buildings, landscapes, vehicles, furniture, and machinery. These assets could be used personally or traded, forming the core of various virtual economic activities. Players could even upload more intricate creations, which, upon being uploaded to *Second Life*, were protected under digital asset ownership laws. This allowed creators to earn substantial incomes, ranging from several hundred to several thousand dollars per month, and achieve financial independence. This phenomenon further fueled user creativity, thereby promoting the development of UGC. Virtual economics serves as one of the most crucial topics in both virtual worlds and the metaverse ecosystem, encompassing areas like cryptocurrencies, economic models, DeFi, GameFi, and NFTFi.

6.1.3 The Third Phase: Toward Decentralized Virtual Worlds

Since 2007, virtual worlds have progressively moved toward decentralization, with the contours of the metaverse becoming increasingly clear (refer to **Table A.6.1**). A decentralized network eliminates distinctions between user terminals and servers, treating each user as an independent server node that is capable of both accessing and processing data. Without centralized servers, virtual worlds become collectively owned by all users, liberating creativity free from central rule constraints. Decentralization's greater significance lies in interoperability across platforms through public and open-source protocols that enable the sharing of computing power and data, which integrated disparate platforms seamlessly. Essentially, a true virtual world is the aggregate of all subvirtual worlds, where individual subworlds remain autonomous yet interconnected through unified standards such as user identity, digital assets, and interaction data. Data remains consistent across different worlds, which is akin to game instances, enabling players to traverse between them.

During the Web 1.0 era, the Internet itself served as a prototype framework for virtual worlds, with data communicated between web pages via TCP/IP and HTTP/HTTPS protocols. However, this form of connectivity primarily benefited platforms and developers rather than end-users. In the Web 2.0 era, mobile Internet emerged, facilitating interconnectivity among social platforms through OpenID or OAuth authentication methods. Yet, centralized platforms wielded excessive control, often blocking external links to maintain traffic supremacy, thereby limiting true interoperability as a result of commercial interests. By contrast, the Web 3.0 era simplifies everything with tools like MetaMask, enabling seamless navigation across applications on the Ethereum (ETH) network while ensuring the interoperability of digital assets and personal information.

Projects like Roblox, The Sandbox, and even Meta's metaverse initiatives and the ETH network within the metaverse realm leverage a large number of collaborative users and server nodes to form the foundational layer of decentralized virtual worlds. The sharing of data among different nodes ensures the stable operation, while open-source systems facilitate the continuous improvement of virtual worlds. At this stage, virtual worlds have reached a self-evolving phase where all users collectively own and create the virtual world, which marks an ongoing evolutionary process.

6.1.4 The Fourth Phase: Social Simulators and AI NPCs

When constructing virtual societies, humanity has long harbored a dream similar to that in the science fiction television series "Westworld": creating NPCs with human-like qualities—naturalistic behaviors that are free from mechanical

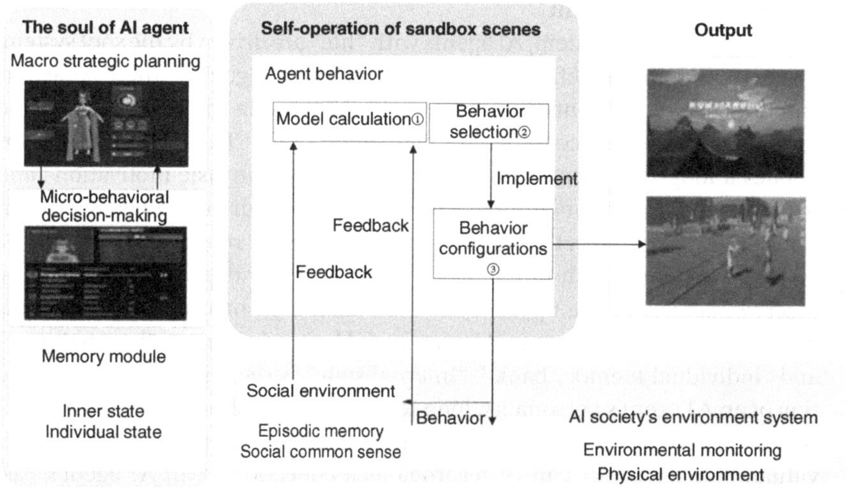

Figure 6.3 Interface of "Living Chang'an City" game.

rigidity. Each NPC should possess unique personality traits, goals, and value orientations, exhibiting diversity and individuality in their interactions with one another. Ultimately, when these AI-driven NPCs interact at scale, they will form a complex artificial society with dynamic emergence. In this paradigm, human participants can engage actively, co-creating narratives and shaping social dynamics. This vision represents the ultimate objective of social simulators—to replicate the richness of human society through computational means (Figure 6.3).

In 2024, the company Parametrix.ai released a city demo called "Living Chang'an City" [3]. This demo is a prototype of an "AI society" capable of continuous operation, where AI NPCs with different personalities live. The AI NPCs connect and influence each other in social life, leading to the constant emergence of new events in Chang'an City, which gradually evolving into a broad blueprint like "Westworld."

The key point of "Living Chang'an City" is to create AI NPCs with "life" and a "feedback mechanism" that enables AI NPCs to interact with one another. Among them, the Soul Module creates AI NPCs with "life," equipping them with the ability to constantly change their thinking and behavior in response to environmental changes; the Environmental System interacts with and provides feedback to numerous AI NPCs, and drives the self-operation of the entire AI NPC ecosystem. These two components interact with each other through the feedback mechanism of the system engine, exerting influence on each other to enable the functioning of AI society.

(1) **AI Agent's Soul System**
In the experimental system, AI agents with "life" are driven by the soul system. The Soul System can define the role, personality, and goals of these AI agents and record various events that occur to the AI agent in the virtual scene. The soul system is mainly composed of two model layers. The upper layer incorporates a long-sequence strategy model based on "intrinsic motivation" and is responsible for formulating macro-strategies that drive the behavior decisions of the lower layer; the lower layer is a multi-task model represented by a symbolic system, which formulates specific behavior decisions based on the macro-strategies of the upper layer. The "memory" module of AI agents is used to integrate the upper and lower model layers, including "internal state" and "individual memory bank." "Internal state" is the result of the combination of an AI agent's personality, long-term goals, and short-term goals, which is reflected in changes in numerical values such as decision intention; "individual memory bank" can be regarded as a collection of an AI agent's past experiences, formed through interactions between the AI agent and the Environmental System. AI agents will exhibit a variety of behavior patterns based on their distinct personalities and past experiences.

(2) **AI Agent's Environmental System**
The Environmental System consists of the Physical Environment and the Social Environment. The Social Environment enables AI agents to interact with one another and provide feedback and records information such as language, behavior, and relationship data generated from these interactions and feedback processes, forming a virtual scene memory bank. This memory bank also contains some knowledge and rules of human society, which will continue to affect the subsequent behavior decisions of AI agents. This is the key to establishing the feedback mechanism. The physical environment includes the virtual scene itself (such as buildings) and the physical presence information of other AI agents. Due to the existence of the time dimension, during the evolution of the experimental sandbox environment, the Environmental System will change under the influence of AI agents' behavior; conversely, changes in the Environmental System will also act on AI agents' soul system, thereby affecting the AI agents' behavior decisions. However, there is a premise for the "mutual causality" between AI agents and the environment; that is, AI agents have "life" and can make decisions that conform to their own logic and take corresponding actions.

(3) **Self-Operation of the Sandbox Scene**
AI agents calculate and execute their behavior in the sandbox scene based on inputs from their own Soul System and state values provided by the Environmental System, and output the calculation results to the behavior selector for the compound configuration of agent behaviors, while correcting the model

calculation at the next moment through feedback. The System Engine updates the state of the virtual environment, as well as the internal state values and memory systems of the agents, in each cycle before proceeding to the cyclic calculation of the next moment. When the agents' configuration behavior is executed or the Environment System is updated, information output is formed and displayed in real time. In this process, each step is multi-node and non-linear. Each AI agent has N possible choices, leading to the emergence of countless storylines. In the experimental sandbox, as long as an AI agent takes action, the entire social ecosystem is activated. Massive numbers of AI agents interact and provide feedback all the time, resulting in the constant emergence of content. Because of the existence of the time dimension, a huge memory bank is formed, which, in turn, influences the future decision-making of AI agents, ultimately evolving into an independently operable AI Society.

6.2 Virtual Experiments with Humans Outside the Loop

6.2.1 The Corrupted Blood Incident in World of Warcraft

On September 13, 2005, Blizzard Entertainment released patch 1.7 for *World of Warcraft*, introducing a new dungeon called Zul' Gurub. The dungeon's boss, Hakkar the Soulflayer (also known as the Blood God), could inflict players with a harmful spell (known in gaming terms as a "debuff") during combat. This debuff, called Corrupted Blood, dealt 250–300 damage every few seconds and spread to nearby characters. Low-level characters could die within seconds, while high-level characters could survive thanks to their stronger health pools and healing capabilities.

Normally, this plague remained confined to the dungeon, dissipating over time or upon death. However, players discovered a loophole: by infecting their pets (player-controlled companions), dismissing them within five seconds of infection, and later recalling them in crowded zones outside the dungeon, players could cause the plague to persist. Dismissed pets retained the infection, and the debuff timer would pause until they were summoned again. Exploiting this, players spread the plague globally, causing mass devastation (Figure 6.4). Within days, Corrupted Blood became *World of Warcraft*'s "Black Death"—cities became uninhabitable, forcing players to either avoid crowded areas or abandon major cities entirely.

Blizzard attempted to mitigate the crisis through quarantines but ultimately "cured" the plague by altering Hakkar's magic from a disease into a red lightning bolt. The outbreak (infecting over half the characters on many servers) garnered

Figure 6.4 A city during the Corrupted Blood outbreak in *World of Warcraft*.

significant real-world media attention, highlighting such games as virtual laboratories.

In-game players with healing abilities mimicked real-world healthcare workers, forming volunteer rescue teams to heal and evacuate affected players. Some warned players at city gates to avoid infected zones. However, others intentionally spread the plague across the map. Hakkar's initial transmission resembled zoonotic spillover (animal-to-human), followed by human-to-human spread. Dense urban areas acted as epicenters, with mages and warlocks—capable of teleportation—spreading the plague to remote regions as "super-spreaders." Player behavior mirrored that in real-world pandemics, offering epidemiologists with unique research material.

6.2.2 Public Health Insights from the Corrupted Blood Incident

In animal disease dynamics research, epidemiologists rely on observational and retrospective studies, as controlled experiments on infectious diseases in animals are economically unfeasible and ethically unacceptable. Computer simulations can bypass these constraints by modeling virtual populations, but they often lack internal variability and unexpected outcomes driven by host behaviors in real populations. To address this, researchers developed complex artificial societies like TRANSIMS (Transportation Analysis Simulation System, Los Alamos National Laboratory) and EpiSimS (Epidemiological Simulation System,

Virginia Bioinformatics Institute). In these virtual societies, the agents' behavior trajectories are programmed based on real-world data.

While games can be viewed as agent-based epidemiological models, games that use human-controlled characters better approximates real-world pandemic scenarios than those relying on random algorithms. Human role-playing, where virtual agents are controlled by human players, bridges the gap between real-world epidemiology and large-scale simulations. By comparing human-driven and purely computational simulations for the same disease, researchers can study differences in the impact of behavior on disease spread. Game simulations face two limitations: (1) human actions in virtual worlds may diverge from real-world responses to pandemics under similar conditions and (2) the limited number of human players constrains the experimental validity of games as models.

In the Corrupted Blood case, healers rushed to infected zones to assist others, potentially prolonging the outbreak by remaining in infected areas. Their actions might either increase survival time of infected individuals—thereby raising transmission rates—or reduce mortality. Such nuanced behavioral patterns are difficult to replicate in computational models. Human responses are unpredictable and require experiments with voluntary participants to determine the ratio of different population reactions. Similarly, quarantine failures cannot be accurately predicted by numerical methods alone, as quarantine failures depend on human decision-making.

World of Warcraft players invest heavily in the game, paying monthly fees and spending hours daily. Many craft believable avatars, embracing responsibilities and social expectations. This commitment aligns virtual reactions with real-world behaviors. Future experiments could adjust disease transmissibility and mortality rates to simulate diverse pathogens, testing human responses to hypothetical scenarios that are impossible to replicate in real life.

In March 2007, Israeli epidemiologist Ran D. Balicer published a paper linking the Corrupted Blood plague to Severe Acute Respiratory Syndrome (SARS) and avian influenza, arguing that role-playing games could model the spread of infectious diseases [4]. In August 2007, Tufts University's Fefferman and University of North Carolina (UNC's) Lofgren co-authored an article in *The Lancet Infectious Diseases* discussing the event's epidemiological significance [5]. Additionally, the U.S. Centers for Disease Control and Prevention (CDC) contacted Blizzard Entertainment for outbreak data to study real-world pandemic response strategies.

6.2.3 The Potential of Online Games as Laboratories

MMORPGs hold immense potential as virtual laboratories. These games often host massive player bases (*World of Warcraft* once had 6.5 million players), enabling the conduct of controlled epidemic simulations. However, MMORPGs remain games at their core, prioritizing enjoyment and entertainment. Thus, motivating players to participate in epidemiological experiments can be challenging. Yet, plagues and epidemics already occupy significant narrative roles in games, as disease transmission—whether intentional or accidental—has become a pivotal plot element in several major titles. Researchers must seamlessly integrate epidemiological experiments into game design, ensuring players do not perceive themselves as participating in overtly deadly simulations but instead immerse themselves in coherent, logical environments where in-game death is framed as a critical risk. Such efforts may require collaboration with the gaming industry.

Using game systems for research avoids the ethical and privacy constraints inherent in studies involving human patients. Research unfeasible in real populations can be conducted as in-game events, with data collected in an unbiased manner since server logs record all in-game actions. Finally, repeating experiments across different player groups (or servers) can establish detailed, reproducible, and standardized frameworks for epidemiological research, enabling the verification of results and the conduct of alternative analyses.

While online games open new experimental avenues, external validation is still necessary before these insights can be applied to real-world contexts. Virtual scenarios can be calibrated to closely mimic real historical outbreaks and validated by comparing simulation results to actual outbreak data. However, like many simulation technologies, even without external validation, having access to such an experimental framework provides critical insights into systemic dynamics.

Considering the needs of public health research, designing and implementing virtual outbreaks shows great promise for bridging the gaps between traditional epidemiology and computational simulations. In such controlled environments with known disease parameters, neither approach alone suffices. If epidemic designs are seamlessly integrated into game worlds as part of expected player experiences, human responses to disease in real-world scenarios can be observed and captured within computational models. These experiments could then offer reproducible empirical analysis, deepening understanding of individual and collective behaviors during outbreaks. Leveraging games as experimental frameworks allows for deeper exploration of epidemiological complexity within social groups.

6.3 Model Integration of Social Simulators

By drawing on online gaming frameworks, we can extract the core components of social simulators: digital agents, digital objects, and a Runtime Engine. Figure 6.5 presents an integrated architecture for social simulators, capable of integrating

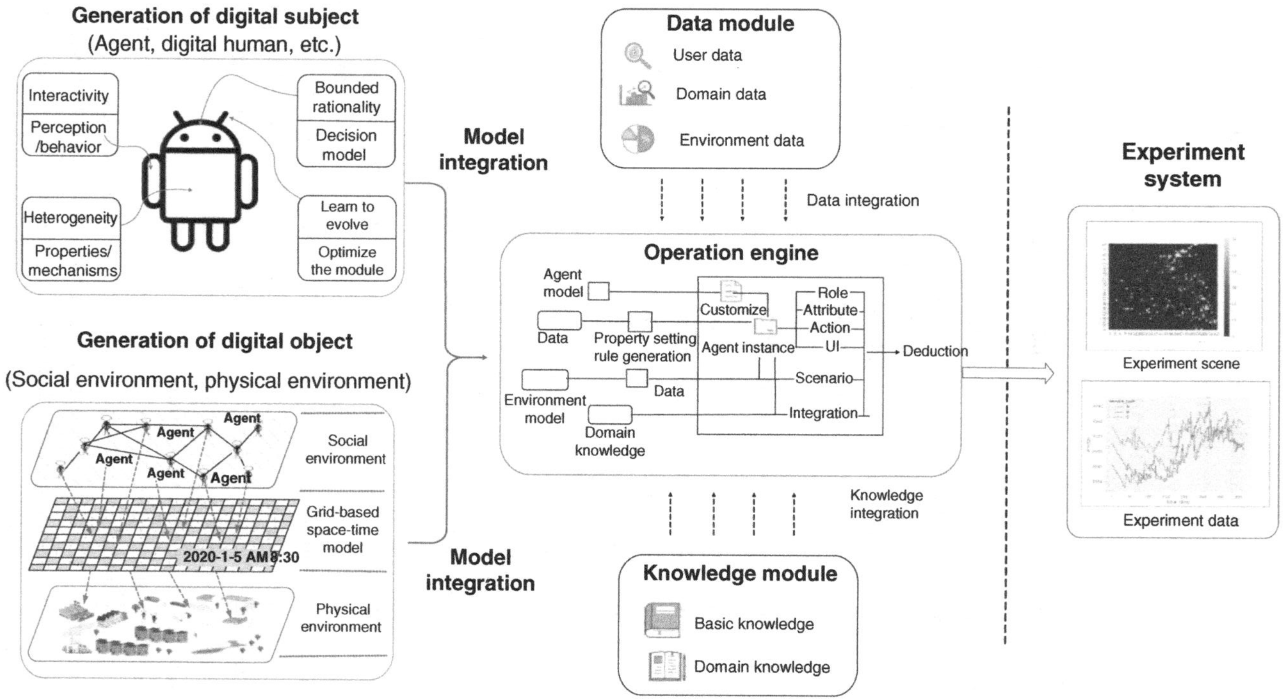

Figure 6.5 Integrated architecture of a social simulator.

multilevel models and injecting domain-specific scenarios, knowledge, and data [6]. Based on specific requirements, this framework can generate a variety of computational experiment scenarios. Digital agents can be either autonomous agents or human-controlled roles. Agent modeling emphasizes anthropomorphic features such as heterogeneity, bounded rationality, interactivity, and learning/evolution, while human roles require deep integration of "human cognition" and "role behavior." Digital objects include physical environments and social environments, both of which can be easily reused or extended according to experimental needs. The runtime engine drives the operations of the artificial society, governing the interactions and update rules between agents and environments.

This section focuses on the interface framework between domain knowledge and artificial society models. Based on the fundamental elements of artificial society models, the integration specifications are categorized into three types—agent models, environment models, and rule models—to lower development barriers and establish a robust foundation for widespread application [7].

6.3.1 Interface Framework for Domain Models

If computational experiments aim to address existing or potential real-world societal issues, high fidelity between artificial societies and real-world systems becomes critical. As a bridge between the problem space (left side of Figure 6.6) and the solution space (right side), the domain interface framework (central part) maps the studied social system (i.e., the domain model) to a multiagent simulation system (i.e., the artificial society model) in virtual space. Before model integration, application scenarios (i.e., problem domains) must be analyzed to identify the required domain knowledge and data. Domain knowledge describes the characteristics of the target social system, and the analysis of such knowledge follows these steps:

- **Identifying system boundaries:** In social systems, factors are diverse, relationships are tightly coupled, and information flows are complex. Initially, research objectives and scope must be clearly defined to avoid unbounded expansion. The target social system can be treated as a "black box," where the box represents the system boundary, which defines its functional scope: Agents that compose the system are placed inside the box, while external interventions affecting the system are positioned outside.
- **Defining system agents:** In social simulators, agents are fundamental components. Human-controlled agents represent individuals characterized by profiles, personalities, and tasks contextualized to specific application scenarios. Autonomous agents are dynamically distributed across end-devices, edge nodes, pipelines, and cloud platforms, providing services via interfaces. These agents operate on platforms and undergo continuous optimization during system runtime. All agents are uniquely identified by a unique ID.

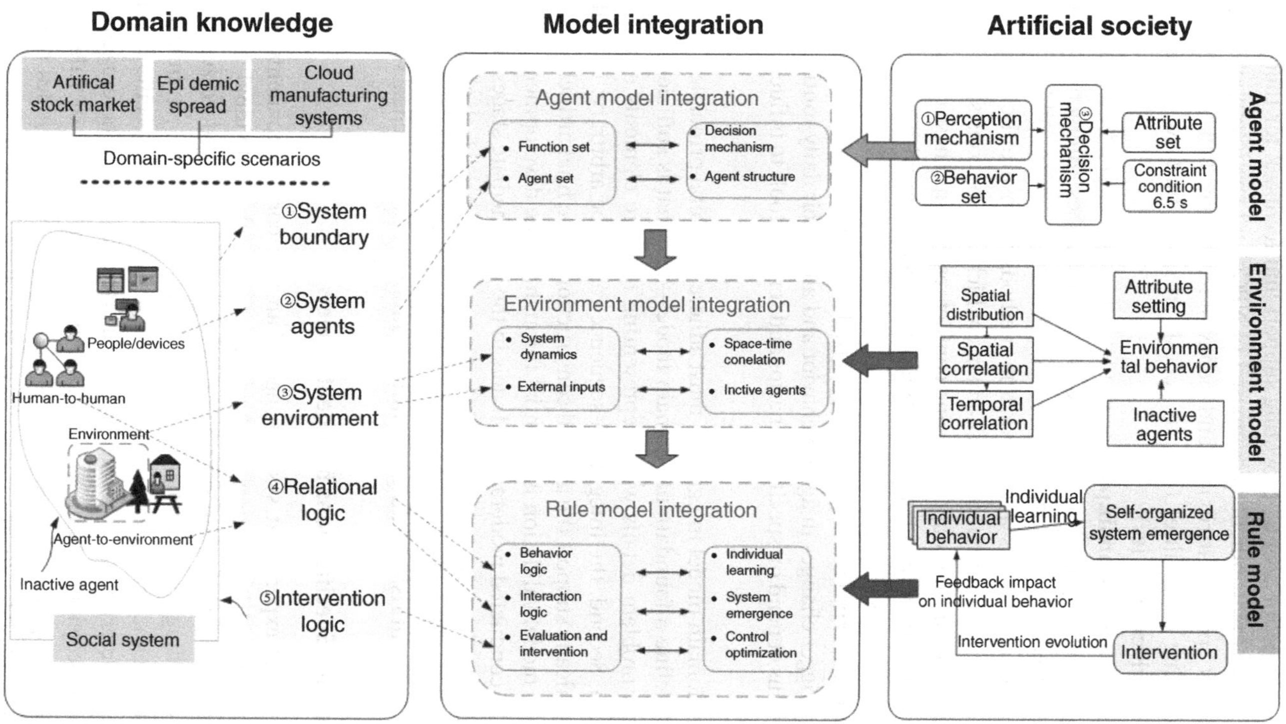

Figure 6.6 Interface framework between domain models and artificial society models.

- **Recognizing system environments:** The physical environment serves as the activity space for agents, providing support and feedback for their behaviors. The social environment is composed of NPCs and spatiotemporal logic. NPCs possess specific attributes and behaviors but may lack independent decision-making capabilities (e.g., in pedestrian or traffic flow simulations, NPCs lack such capabilities).
- **Mapping agent relationships:** Agent interactions are defined through three key components: topological structure, which describes interaction modes (e.g., centralized, decentralized structures); interaction rules, which impose constraints on agent roles and interaction methods to ensure the correctness of behaviors; and coordination mechanisms, which ensure consistent actions across agents, including cooperation and conflict resolution. These components collectively govern how agents communicate, collaborate with one another, and resolve conflicts within the system.
- **Identifying external interventions:** External interventions are imposed measures that trigger changes in agent behavior and alter their daily routines. Trigger conditions depend on specific scenarios and serve as tools to influence agent behaviors.

Generally, domain model integration must adhere to several principles. First, it is critical to define a foundational domain model—one that is sufficiently clear to capture all problem details while representing the core elements of the reference system. Second, the integration sequence cannot be arbitrary; instead, it must systematically support the gradual construction of the final artificial society model. Third, model validation is essential throughout the development process but should be conducted moderately. Models in the intermediate stages are valuable but immature, and their value should not be negated due to a lack of sufficient empirical evidence. Fourth, it is unnecessary to incorporate all domain details, as excessive refinement risks leading to infinite iteration without practical value.

6.3.2 Interface Specifications for Agent Models

In computational experiment systems, agents are the core components of the model. Agent model integration involves customizing the individual modules of an agent based on domain-specific knowledge. With a standardized agent structure (including perception, action, decision-making, and optimization) as a template, researchers apply various technologies to meet specific requirements, such as determining whether agents are homogeneous or heterogeneous, whether their decision-making mechanisms are zero-intelligent or rule-based, and whether they possess learning capabilities. If the system includes multiple agent types, each type must be integrated separately based on the research question. Table 6.2 outlines the interface specifications for agent models.

Table 6.2 Interface Specifications for Agent Models.

Option	Domain Knowledge	Artificial Society Model
Model Specification	Clarify the research objectives and questions, then define system boundaries and functional scope. Determine the types of agents required in the system based on domain characteristics.	Based on identified agents in the domain model, set two prior conditions: (1) whether agents are homogeneous or heterogeneous and (2) whether agents exhibit bounded rationality or full rationality.
Attribute Configuration	Define unique attribute values for each autonomous agent in the system, including static attributes (e.g., ID codes, gender) and dynamic attributes (e.g., current location, asset accumulation).	Determine feature and state attributes for agents in the artificial society based on key attributes identified in the domain model.
Perception Module	The perception module acts as the agent's "eyes," enabling it to observe its surroundings. Define the agent's perceptual range and the objects/environments it can sense.	Implement the perception module using two components: (1) encoding self-perception information (self-state, environmental feedback) and (2) encoding interaction information received from other agents (e.g., task instructions, shared data).
Memory Module	Functions as the agent's "storage," allowing past experiences to influence current decisions and future outcomes. Define events to be recorded, memory capacity, and forgetting mechanisms: (1) Memory fading over time; (2) memory replacement when capacity is exceeded; and (3) prioritized forgetting of low-importance memories.	Implement the memory module as two types of lists: individual memory stores personal experiences, while system memory stores observed events. Retrieval methods include (1) uniform sampling; (2) time-series sampling (recent memories prioritized via time decay); and (3) priority-based sampling (combining relevance, timeliness, and importance).
Decision-Making Module	Define utility functions to guide decision-making. Agents evaluate expected benefits (via utility functions) to maximize their gains, balancing short-term and long-term goals.	Implement decision mechanisms categorized as (1) zero-intelligence: random behavior; (2) observe-act: explicit rule-based "if...then..." logic; and (3) observe-predict-act: knowledge/reasoning-driven decisions.
Action Module	Define the action spaces for agents in response to external stimuli and decompose them into executable action sequences.	Implement action mechanisms via two components: (1) proactive actions taken by the agent and (2) reactive behaviors triggered by external events.
Learning Module	Define intelligence differences among agent types, reflecting individual learning capabilities and macro-level system evolution. This explains how "complex systems" transition to "complex adaptive systems."	Adaptive behavior rules include (1) N-type models (switching fixed strategies) and (2) self-discovery of new strategies through learning (e.g., evolutionary learning, imitation learning, reinforcement learning).

6.3.3 Interface Specifications for Environmental Models

In complex social systems, human behavior patterns are closely linked to the environment. This is a critical factor in studying issues such as pandemic control, transportation planning, and public opinion monitoring. For example, researchers have long collected daily commuting data from urban residents through surveys and crowd sensing to establish transportation flow prediction models. In simulators, the environment acts as the backdrop or stage for agent actions and interactions. Depending on the problem domain, environmental models are generally abstracted into three layers. (1) First layer: Describes the physical environment, including ecosystems, buildings, roads, transportation infrastructure, and climate. (2) Second layer: Focuses on geographic spatial information, establishing a spatial coordinate system to define entity locations. (3) Third layer: Describes human interaction networks and external interventions. As shown in Figure 6.7, common environmental models include network environments, map environments, digital twin environments, and grid environments. Grid environments divide space into regular cells where agents move and interact. These environments are simple to implement, particularly suitable for problems such as land-use planning or spatial diffusion analysis, and facilitate proximity analysis due to clearly defined neighbors (e.g., adjacent or diagonal cells).

Environmental model integration maps domain knowledge into a standardized geospatial framework to capture spatial distributions and spatiotemporal correlations. In thought experiments, most environmental models are abstract, serving only as operational backdrops for simulations. In application scenarios, the environment model must align with the details of real-world systems. Table 6.3 outlines the interface specifications for environmental models.

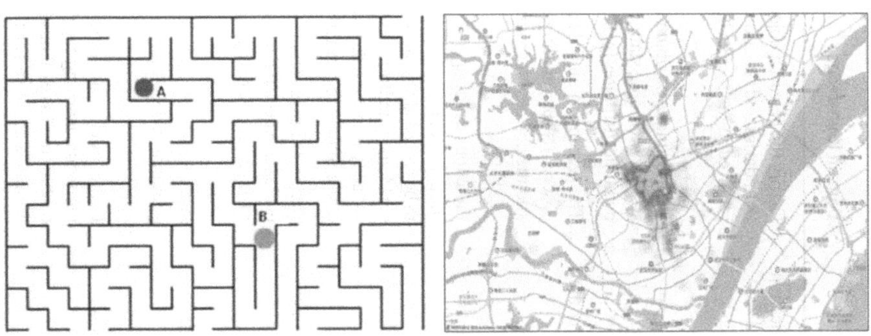

Figure 6.7 Common physical environmental models.

Table 6.3 Interface Specifications for Environmental Models.

Category	Item	Environmental Feature	Type
Physical Environment	Abstract environment	Uses discrete grids to represent space, enabling the execution of simulations under limited computational resources. However, high abstraction limits simulation scale and the complexity of agent–environment interactions.	Early artificial society models often adopted grid models, such as the Sugarscape model and the segregation model.
	2D map	A 2D environmental system, more cost-effective than 3D simulations. Abstracts real-world elements (e.g., buildings, roads, climate) into physical models.	Many typical artificial society cases use this type of mode, such as EpiSimS.
	Digital twin	Fully maps real geographic environments into a 3D virtual space, including both static and dynamic data (e.g., traffic flow, passenger flow).	Tencent built a parallel virtual city system mirroring the real physical world, using NPCs (nonplayer characters) to simulate real-time traffic conditions.
	Metaverse	Combines virtual environment with real humans to test human reactions in unprecedented scenarios. It is useful for military exercises, strategic simulations, and emergency management.	Applied in fields such as military exercises, strategic simulations, and emergency management.
Social Environment	Grid neighborhood	Grid distance reflects the influence between an agent and its neighbors. Neighbor selection rules are critical (e.g., random nearest neighbors, Von Neumann neighborhoods, Moore neighborhoods).	Widely used in grid-based simulations.

(Continued)

Table 6.3 (Continued)

Category	Item	Environmental Feature	Type
	Social network	Integrates social relationship networks (including strong/weak ties) into agent behavior models. Key algorithms include random edge addition, overlapping node linking, and weighted network generation.	Used in networked artificial societies (e.g., disease/rumor propagation models). Network types include random, regular, and scale-free networks.
	External factors	Includes trigger events (natural/social/accidental) and external interventions. The type, intensity, and timing of interventions influence the progression of events.	For example, in [8], an S-shaped growth function was used to model user/developer growth in an APP ecosystem, with the model calibrated using Apple Store data for validation.
Correlation Relationships	Spatial distribution	Collect spatial data to define the distributions of agent/environment entities (including total quantity, types, and population adaptation characteristics).	Design initialization algorithms ensuring (1) the agent count matches real-world statistics and (2) the distribution of environmental entities aligns with reality.
	Spatial correlation	Collect spatial migration data to establish one-to-many mappings between agents and activity locations.	Design agent movement rules that match real-world human behavior patterns.
	Spatiotemporal correlation	Collect spatiotemporal data to define the relationships between agent, their activities, and time.	Design daily behavior rules that integrate regularity and randomness in individual actions.

Table 6.4 Interface Specifications for Rule Models.

Option	Domain Knowledge	Artificial Society Model
Individual Learning	Determine whether agent decision rules evolve over their life cycle. If agents possess learning mechanisms, they iteratively optimize their rule sets to refine their decision-making processes.	Based on the level of an agent's awareness (or rationality), adaptive behavior rules for agents include N-type models (selecting strategies among N fixed options) or learning models (discovering new strategies through learning). Implementation mechanisms include reinforcement learning, genetic learning, imitation learning, and social learning.
System Emergence	Collect interaction data (e.g., cooperation, transactions) between agents. Determine whether dynamic interactions lead to system-level phenomena or cultural traits and whether these phenomena further retroactively influence individual decisions.	Initialize social relationships between agents based on interaction data. Design individual agent interaction rules (i.e., how to select interaction partners under constraints). Finally, describe emergent system-level phenomena using topological structures and organizational rules. These elements allow numerous combinations, each corresponding to a distinct organizational type.
Intervention Evolution	External intervention strategies are adjusted based on discrepancies between system-level outcomes and predetermined target states. These strategies either accelerate or hinder system evolution (e.g., platform algorithms such as pricing incentives or governance policies such as taxation).	Design an interactive closed-loop between the adjustment of intervention strategies and the evolution of individual learning. Interventions are dynamically adjusted (e.g., via reinforcement learning) based on feedback, influencing the daily behaviors of agents. Utility functions prioritize satisficing solutions over optimal solutions. Additionally, initial intervention settings (e.g., intensity, timing) can be altered to evaluate their impact on intervention effectiveness.

6.3.4 Interface Specifications for Rule Models

In artificial societies, rule models serve as the core of system's cyclic mechanisms. They define interaction protocols between agents, between environment, and for agent–environment relationships. These rules can be set based on real-world principles or hypothetical constructs that researchers design to explore social behaviors and dynamics under specific scenarios. In artificial societies, agent learning is a key driver of system evolution. During evolution, agents generally follow the feedback learning principle of "increasing returns" to iteratively optimize their decision-making mechanisms, so as to take actions that bring them closer to achieving their goals. Behavioral interactions between agents generate system-level emergent phenomena, which, in turn, influences individuals, leading to complex second-order emergent phenomena. Additionally, during the construction of an artificial society, external intervention strategies must be implemented to enable controlled evolution of the system. Table 6.4 outlines the interface specifications for rule models.

6.4 Case Study

This section presents the construction process of a computational experiment system using a coal mine disaster emergency management case, illustrating the role of the proposed model integration framework in detail [9, 10].

6.4.1 Scenario Description of Coal Mine Emergency

In modern society, unexpected crisis events occur from time to time. With urbanization and globalization, the impacts of these events have expanded significantly, leading to increasingly severe economic and social losses, and even threatening public safety. The question of how to respond to emergencies promptly and effectively has attracted widespread attention from both society and academia. Emergency response challenges manifest in two aspects. (1) Unpredictable escalation: Incidents may develop to destructive scales unpredictably. (2) Complex consequences: Incidents trigger "domino effects" and cascading impacts. Traditional measures often fail to handle such emergencies. Computational experiments can simulate crisis scenarios, enabling risk analysis, early warning, and proactive preparedness to minimize losses.

Here, coal mine emergency management is used as a case study to demonstrate the construction workflow of the computational experiment system. Given the diversity and variability of coal mine accidents, along with their economic, legal, and ethical ramifications, testing emergency strategies in real environments is infeasible, especially for coal mine accidents such as gas explosions and outbursts. To address this challenge, we have designed various "experiments" using a 3D virtual mine simulation platform to accurately replicate different accident processes. This enables the analysis and evaluation of emergency response measures under diverse scenarios to determine their effectiveness. Details of the computational experiment system construction are as follows.

6.4.2 Model Library Integration and Operations

As shown in Figure 6.8, domain model analysis clarifies behavioral details and influencing factors during emergency evacuation. Based on the model library, emergency-related entities such as disaster sources, miners, and rescue teams are defined as agents, with their domain-specific features mapped to the state attributes of agents. Using the knowledge base and database, prototypes of various agents are established and initialized. Below, we detail the model library (**Table A.6.2**), knowledge base (**Table A.6.3**), and database (**Table A.6.4**) of the experimental system [11].

The Agent Generation Module creates agents, including Disaster Source, Miner, and Rescue agents. Environmental variables include basic geographic data from the Geographic Information System (GIS) and related constraints (e.g., tunnel conditions, disaster source distributions, miner distributions). Baseline data is sourced from the Database Module and Agent/Environment Parameter Configuration Module. The model integration and data processing workflow of the coal mine disaster emergency experiment system is illustrated in Figure 6.9.

(1) **Disaster Source Agent Operations**
Figure 6.9 (top-right) shows the Disaster Source agent's behavior execution process at time t. At time t, the disaster source agent first calculates disaster evolution trends based on parameter configurations. It then updates environmental variables such as disaster coverage area and diffusion direction. The results are output to the Simulation Module. Subsequently, the disaster source agent enters the "$t+1$" cycle.

(2) **Rescue Agent Operations**
Figure 6.9 (top-left) illustrates the Rescue agent's behavior execution process. After receiving task assignments, the Rescue agent first collects data, including information on evacuation targets, rescue team capabilities, and environmental conditions in evacuation zones. Once data is collected, the Rescue agent formulates an evacuation plan. Computational results from optimization models and drill simulations in the Simulation Module provide technical support for evacuation planning. After finalizing the plan, action plans are designed, evacuation orders are issued, and actions are executed. Results are

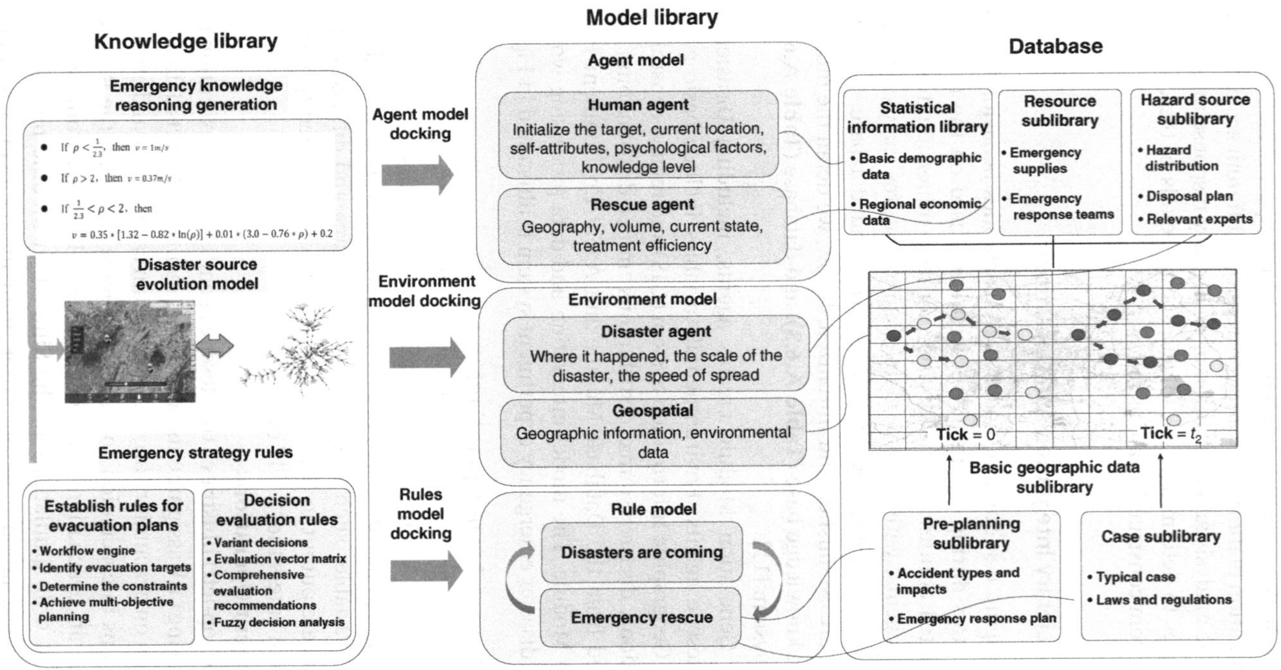

Figure 6.8 Coal mine disaster emergency flowchart.

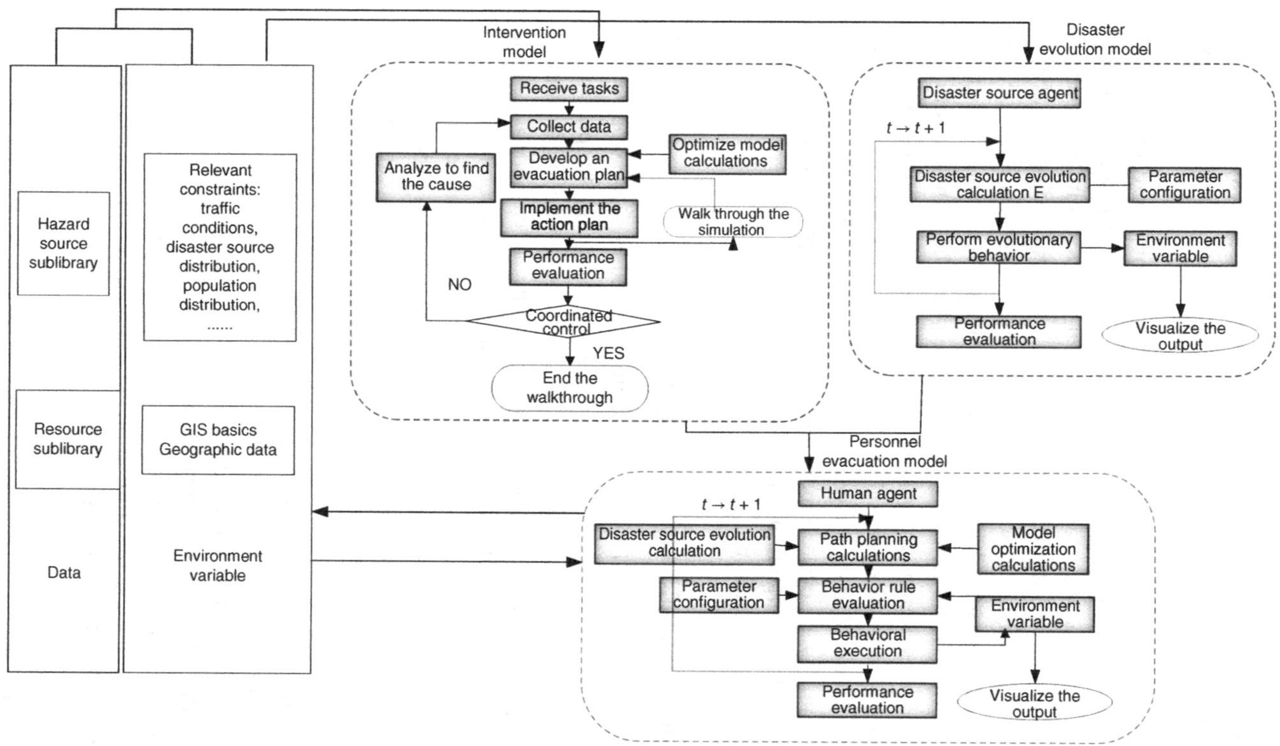

Figure 6.9 Model integration of the coal mine disaster emergency experiment system.

output to the virtual simulation environment. If the proposed evacuation plan meets requirements, the cycle ends; otherwise, a new evacuation drill begins. When the Emergency Decision Module is not involved in system simulation, the results represent unguided natural emergency evacuation processes.

(3) **Miner Agent Operations**
Figure 6.9 (bottom) details the Miner agent's behavior execution process at time *t*. The Miner agent first uses the evacuation path-planning model from the knowledge base to calculate evacuation routes. Simultaneously, multi-objective resource optimization scheduling rules and decision analysis rules from the Knowledge Base are applied for path optimization. The Disaster Source agent model calculates disaster evolution to provide constraints for path planning. After determining the route plan, the Miner agent computes behavior rules based on parameter configurations, completing behavior execution and environmental updates step by step. Results are output to the Simulation Module for performance evaluation. The Miner agent then enters the "$t+1$" cycle. Upon completing their plans, individual Miner agents exhibit emergent group behaviors, such as Miner agent interactions, competition, and queuing.

6.4.3 Simulator Configuration and Operation

The development of social simulators primarily employs two approaches: (1) self-programming to achieve greater modeling flexibility and (2) platform-based development to prioritize efficiency. Currently, its development remains in its early stages, where development efficiency outweighs modeling flexibility. Researchers typically select a development platform to automatically generate framework code and manually implement specific functionalities, reducing programming effort. If the platform is reliable, its auto-generated code is trustworthy. Platform-based simulator development has become mainstream, with popular tools including Swarm, Repast, Mason, and NetLogo.

As shown in Figure 6.10, the social simulator adopts a distributed and concurrent multiagent system architecture. It comprises five modules: Data Collection, Parameter Configuration, Preprocessing, Plan Execution, and Output. The Infrastructure Module integrates knowledge rules from the Knowledge Base and models from the Model Library. The Preprocessing Module generates agent behaviors in the simulation environment. The Execution Module dynamically updates initial agent state attributes and the simulation environment. Finally, results are output to the Output Module. Details are as follows:

(1) **Data Collection**
The Data Collection Module gathers data required by other modules. Sub-databases (e.g., Event Source Database, Agent Database, and Basic

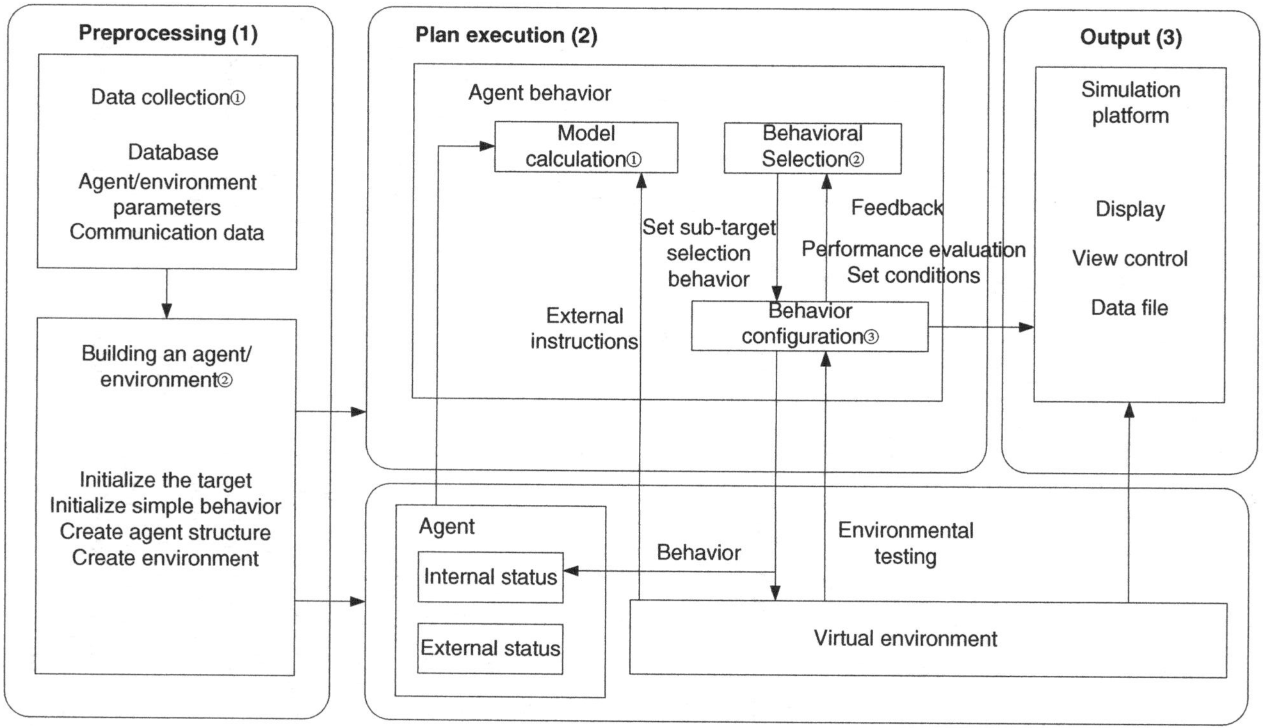

Figure 6.10 Architecture of the social simulator.

Geographic Database) provide foundational support for system modules, while statistical information databases, operational information databases, and case databases supply data for experimental analysis, guiding the design of intervention strategies. The Data Management Module performs two functions: storing, updating, and modifying database records via Communication Module inputs; and integrating multidimensional data from the sub-databases.

(2) **Parameter Configuration**
The Parameter Configuration Module accepts user-defined parameters, which are submitted to the Agent Generation Module and Simulation Environment Module to customize event scenarios. Agent parameters include behavioral entity parameters of disaster sources (e.g., event location, disaster diffusion speed) and attributes (e.g., quantity, type, spatial distribution), while environment parameters define agent activity zones and event-related constraints (e.g., road conditions, disaster source distribution, and agent distribution).

(3) **Constructing Agents and Environments**
Upon completing data collection, the Agent and Environment Construction Module generates customized agent entities and virtual environments. Agent structures are created by the Agent Generation Module, while the Simulation Module constructs the virtual environment. After construction, agent objectives and simple behaviors are initialized. The environment creation includes GIS-based data and constraints (e.g., road conditions, disaster/agent distributions).

(4) **Plan Execution Module**
The Plan Execution Module computes and executes agent behaviors in the simulation environment by using preprocessed agent state values, external state values, commands from the virtual environment, as well as knowledge rules and models from the Knowledge Base and Model Library. Computed results are output to the Behavior Selector, setting subgoals for agents at the next time step. After behavior selection, behavioral experience learning rules from the knowledge base configure composite agent behaviors based on selection results and environmental data, with behavioral feedback adjusting behavior selection parameters. Agent internal states are updated, and behavior status is transmitted to the virtual environment via the Simulation Module. The virtual environment is dynamically updated using environment update rules from the knowledge base before the system proceeds to the next simulation cycle.

(5) **Output Module**
Information is output upon the completion of agent behavior execution or environment updates, enabling the refresh of virtual environment. Results are

displayed on the system interface through 2D/3D visual scenes, images, or text, visualizing system evolution at both macroscopic and microscopic scales.

6.4.4 Experimental System Interface

As shown in Figure 6.11, based on geographic and monitoring data from a state-owned coal mine, a 3D coal mine simulation environment was constructed, which integrates modules of environmental systems (e.g., the network structure of mine tunnels, current excavation positions, ventilation directions), airflow dynamics (e.g., real-time gas concentrations, CO and other toxic gas concentrations, wind speed), working faces (e.g., quantities of mining equipment, compositions, and distributions of mining equipment in each underground zone), and equipment (e.g., main fans, local fans, dampers, shearers, conveyors, pumps, vehicles, air compressors).

This simulation platform can replicate diverse disaster scenarios, such as gas explosions, coal dust explosions, and flooding incidents. By loading emergency intervention strategies, the platform simulates miners' realistic emergency responses and evacuation behaviors within the virtual environment. At the end of each execution cycle, agent and environmental states are updated and visualized in the 3D virtual coal mine to intuitively demonstrate the effectiveness of the applied emergency interventions.

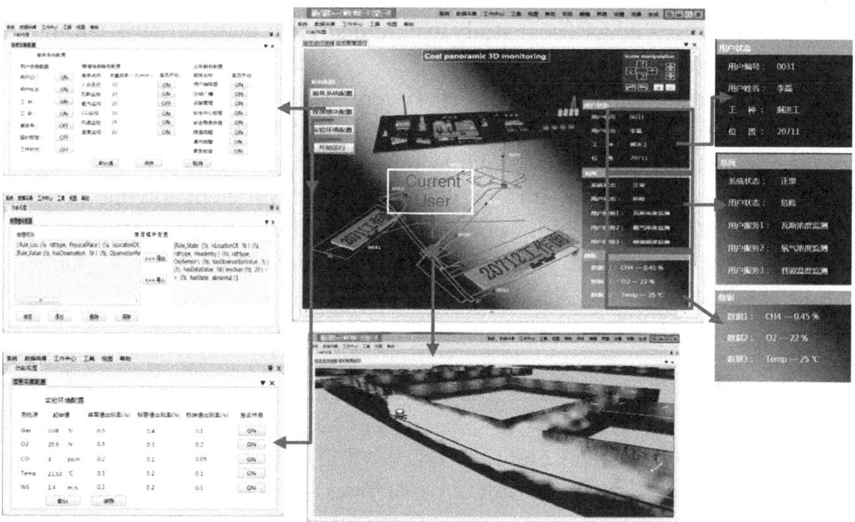

Figure 6.11 Configuration and operation interface of the virtual coal mine simulation platform.

A cyclical feedback learning process is established between the virtual environment and the real-world system. First, artificial society models are constructed in the virtual space through data and knowledge correlation analysis, enabling the conduct of computational experiments. Various stress conditions and extreme scenarios from the real world are replicated in the virtual environment. Second, intervention strategies are iteratively optimized via machine learning during the feedback loop, which can identify potential risks, conflicts, and hazards and provide "references," "predictions," and "guidelines" for practical applications. Finally, the validity of these strategies is verified by guiding real-world system operations and detecting implementation errors. Additionally, dynamic real-world data inputs enhance the realism and accuracy of the 3D coal mine simulation platform.

6.5 Chapter Summary

Computational experiments provide a novel perspective for analyzing complex systems, allowing intuitions to be validated through virtual system logical inference, trial and error, and estimation. To achieve predefined objectives, static artificial society models must evolve into dynamic computational experiment systems. This chapter presents an integrated design framework for constructing computational experiment systems, with an emphasis on integrating domain-specific knowledge into artificial society models. A coal mine disaster emergency simulation system is presented as a case study to demonstrate the workflow of computational experiments.

The maturation of computational experiment systems is an ongoing iterative process. Minor adjustments to computational models may cause significant changes in system performance, while exponential increases in the number of parameter combinations further complicate the problem-solving process. Therefore, identifying critical influencing factors and uncovering corresponding rules constitute another core technology in computational experiments—namely, experimental design.

References

1 Dionisio, J.D.N., Iii, W.G.B., and Gilbert, R. (2013). 3D virtual worlds and the metaverse: current status and future possibilities. *ACM Computing Surveys* 45 (3): 1–38.

2 Schroeder, R., Huxor, A., and Smith, A. (2001). Activeworlds: geography and social interaction in virtual reality. *Futures* 33 (7): 569–587.

3 Chaocanshu Technology. (2023). GAEA: Bringing "living" virtual scenes to reality. Retrieved from https://www.chaocanshu.cn/news/detail?id=80.

4 Balicer, R.D. (2007). Modeling infectious diseases dissemination through online role-playing games. *Epidemiology* 18 (2): 260–261.

5 Lofgren, E.T. and Fefferman, N.H. (2007). The untapped potential of virtual game worlds to shed light on real world epidemics. *The Lancet Infectious Diseases* 7 (9): 625–629.

6 Xue, X., Xiangning, Y., Zhou, D. et al. (2024). Computational experiments for complex social systems: integrated design of experiment system. *IEEE/CAA Journal of Automatica Sinica* 11 (5): 1175–1189.

7 Xue, X., Xiang-ning, Y., Zhou, D. et al. (2024). Computational experiments for complex social systems, part III: the docking of domain models. *IEEE Transactions on Computational Social Systems* 11 (2): 1766–1780.

8 Lim, S.L. and Bentley, P.J. (2012). How to be a successful app developer: lessons from the simulation of an app ecosystem. *ACM Sigevolution* 6 (1): 2–15.

9 Xue, X., Wang, S., Gui, B., and Hou, Z. (2016). A computational experiment-based evaluation method for context-aware services in complicated environment. *Information Sciences* 373: 269–286.

10 Xue, X., Chang, J., and Liu, Z. (2014). Context-aware intelligent service system for coal mine industry. *Computers in Industry* 65 (2): 291–305.

11 Lu, J. (2002). Mathematical models of staff evacuation and escape from building. *Journal of Wuhan University (Engineering Edition)* 35 (2): 68–70.

7

The Third Step—Experimental Design and Generative Explanation

In the 1990s, the Santa Fe Institute (SFI) pioneered the use of computer simulation to address the challenge of economic system modeling, laying the foundation for research tools in complex economics. Policy simulation extends beyond agent-based modeling (ABM) and reproducing economic systems' disequilibrium states. Instead, it integrates deeper economic insights into simulation models, with the aims of evaluating policy effects through "stress tests," analyzing influencing factors, conducting causal inference of policy impacts, and exploring solutions for preventing manipulative behaviors. Although social simulation has advanced across applications, researchers often favor deduction over experimentation. While these methods capture system evolution, they rarely offer causal insights into operational mechanisms or emergent laws.

The computational experiment method consists of generative deduction and generative experimentation. Figure 7.1 illustrates the role of experimental design in computational experiment methods. Similar to ABM, generative deduction employs deductive simulation to reproduce real-world settings or phenomena, or to construct and observe potential alternative worlds, in order to identify the influencing factors (independent variables) that drive changes in output responses (dependent variables) [1]. Generative experiments are primarily implemented through controlled experimental settings. By modifying the rules, parameters, and external interventions of the experimental system, various combinations of these elements can be tested in an "artificial laboratory," generating data well-suited for studying causal relationships and enabling the construction of structural causal models between influencing factors and response variables [2]. This chapter outlines the computational experiment design framework, comprising three core modules: descriptive modeling of artificial societies, explanatory experimental design, and predictive meta-modeling. Subsequently, the technical details of

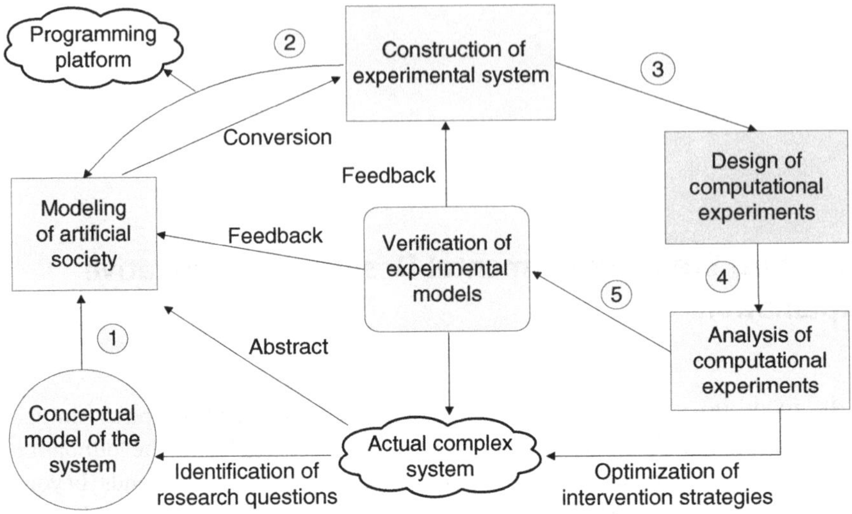

Figure 7.1 The role of experimental design within computational experiment method.

generating experimental scenarios based on experimental design are explained. Finally, to validate the effectiveness of the proposed method, a case study of the API service market is presented.

7.1 Comprehensive Framework of Computational Experiment Design

7.1.1 Hierarchical Relationships of Experimental Design

The initially proposed classical experiment design is a mathematical theory and method based on probability theory and mathematical statistics. It takes real physical systems as the research object to design experiments efficiently and enable effective analysis of experimental results. By contrast, unlike the research objects of traditional experiments, computational experiments are conducted on artificial social models.

In computational experiments, several key concepts are defined as follows:

- Experimental indicators measure the quality of experimental results and are usually associated with experimental outputs (responses).
- Experimental factors refer to input variables or parameters that may influence these indicators.
- Factor levels are the possible values that experimental factors can take.

- An experimental point is formed by one set of values of all factors, and a collection of such points constitutes an experimental scheme.
- The experimental domain is the space composed of the value ranges or discrete value sets of all experimental factors.

Figure 7.2 shows the three layers of computational experiment design: the experiment layer, the scenario layer, and the operation layer. A computational experiment consists of multiple experimental scenarios, and each scenario contains several operations. In this structure, an experimental scenario corresponds to an experimental point, while multiple experimental operations are conducted to reduce the impact of random uncertainty on results. According to the implementation stages, computational experiment design can be divided into three stages: screening experimental design, formal experiment design, and sampling experimental design. At the experiment layer, factors are

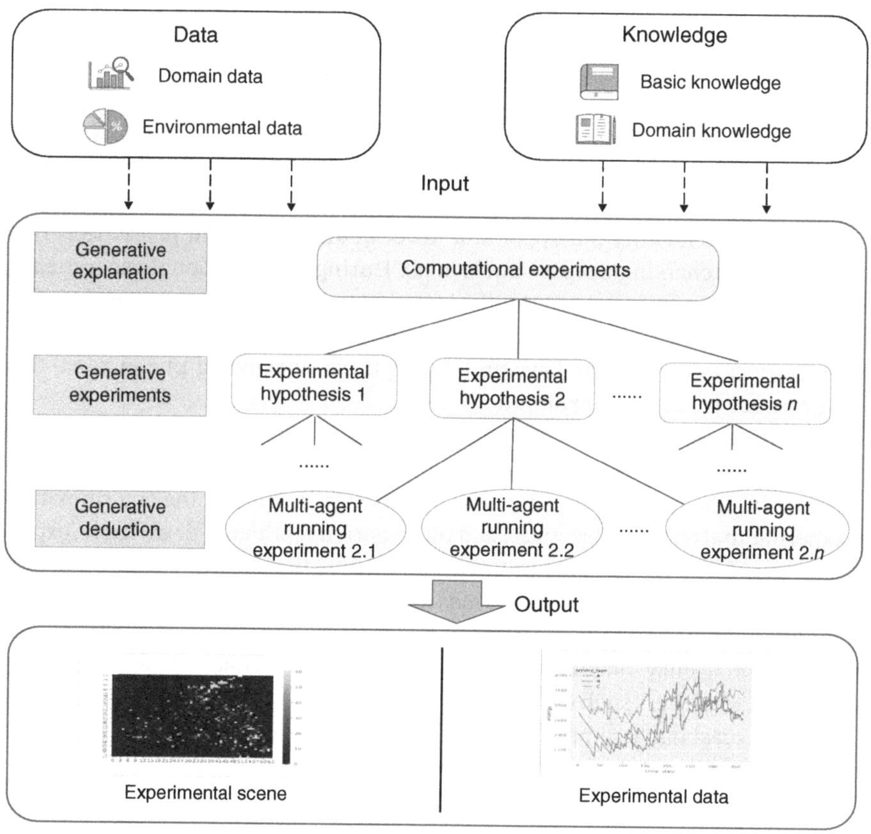

Figure 7.2 Design hierarchy of computational experiments.

identified from experimental objectives for use in the scenario layer. At the scenario layer, a screening design filters out critical factors, followed by a formal design for more detailed experimentation. At the operation layer, a sampling design is applied to reduce the impact of randomness with fewer simulation runs.

At the hypothesis layer, one must consider all contexts, including historical real-world contexts (for model calibration and optimization) and unobserved potential contexts. In this way, a series of experimental results can be generated, and past events can be regarded as experimental outcomes. These experiments relate not only to key factors but also to the parameter settings of the contexts. If these parameters are incorporated as variables into influencing factors, the characteristics of path dependence and context dependence can be effectively explained. By analyzing the system evolution process in experiments, the underlying laws of system states and properties can be derived as parameters gradually change.

At the operation layer, results across different contexts may be unstable. In some contexts, results may lack repeatability. If results do not conform to any probability distribution function, this indicates that the context is located precisely in the critical region of a system phase transition. In fact, criticality is a normal state in complex systems. At this point, the concept of nonequilibrium phase transition should be adopted to study the context. Alternatively, parameters can be fine-tuned to identify the upper and lower bounds of critical phase transition states, thereby enabling deeper conclusions. During computational experiments, phenomena that completely contradict intuitive expectations may emerge. Rather than simply doubting the results, parameters should be fine-tuned within the given context to explore the laws underlying these unexplored phenomena; this is the essence of scientific exploration.

As system complexity increases, the number of influencing factor combinations grows exponentially. Therefore, the successful application of computational experiments is highly dependent on experimental design, including the determination of experimental objects, objectives, and intervention methods. For example, experiments may focus on variance or treatment effects, the experimental population may be homogeneous or heterogeneous, and the factors under study may be single or multiple. Additionally, these human and social factors involve elements of randomness and agency, making traditional Design of Experiments (DOE) methods insufficient to address such unique and complex challenges (Table 7.1). Therefore, achieving a reasonable computational experiment design remains a core challenge in the field.

Table 7.1 Comparison of Experimental Characteristics.

No.	Traditional Experiments	Computational Experiments
1	Few influential factors	Numerous influential factors
2	Linear or low-dimensional responses	Nonlinear or multidimensional complex responses
3	Single-factor effects	Multiple qualitative and quantitative effects
4	Negligible high-order interactions	Numerous high-order interactions
5	Homogeneous errors	Heterogeneous errors
6	Normally distributed errors	Diverse error distributions
7	Black-box models	Models integrating large amounts of domain knowledge for interpretability
8	Univariate responses	Multidimensional indicators and complex behavioral characteristics

7.1.2 Design Framework of Computational Experiment Method

Compared with traditional experiment approaches, the computational experiment method offers a unique way to integrate descriptive, explanatory, and predictive models through generative deduction and generative experiments [3]. Its core idea is to explore the generative mechanisms of social phenomena rather than to explain them through general social laws or to merely identify statistically relevant factors. As shown in Figure 7.3, the comprehensive framework of computational experiment design consists of three modules: (1) *descriptive module* (upper part of Figure 7.3), which identifies influencing factors and response variables of dynamic system operations; (2) *explanatory module* (both sides of Figure 7.3), which observes the behavior of artificial societies under interventions (such as new strategies or parameter changes) through factorial experiment design [4]; and (3) *predictive module* (lower part of Figure 7.3), which constructs a meta-model to provide deeper insights into the evolution of artificial societies. Details are as follows:

- **Descriptive Module: Identification of Influencing Factors and Response Variables**
  Computational experiments employ computer simulation technology to address the generative explanation of complex social systems. The key question is: How do distributed local interactions among heterogeneous agents generate

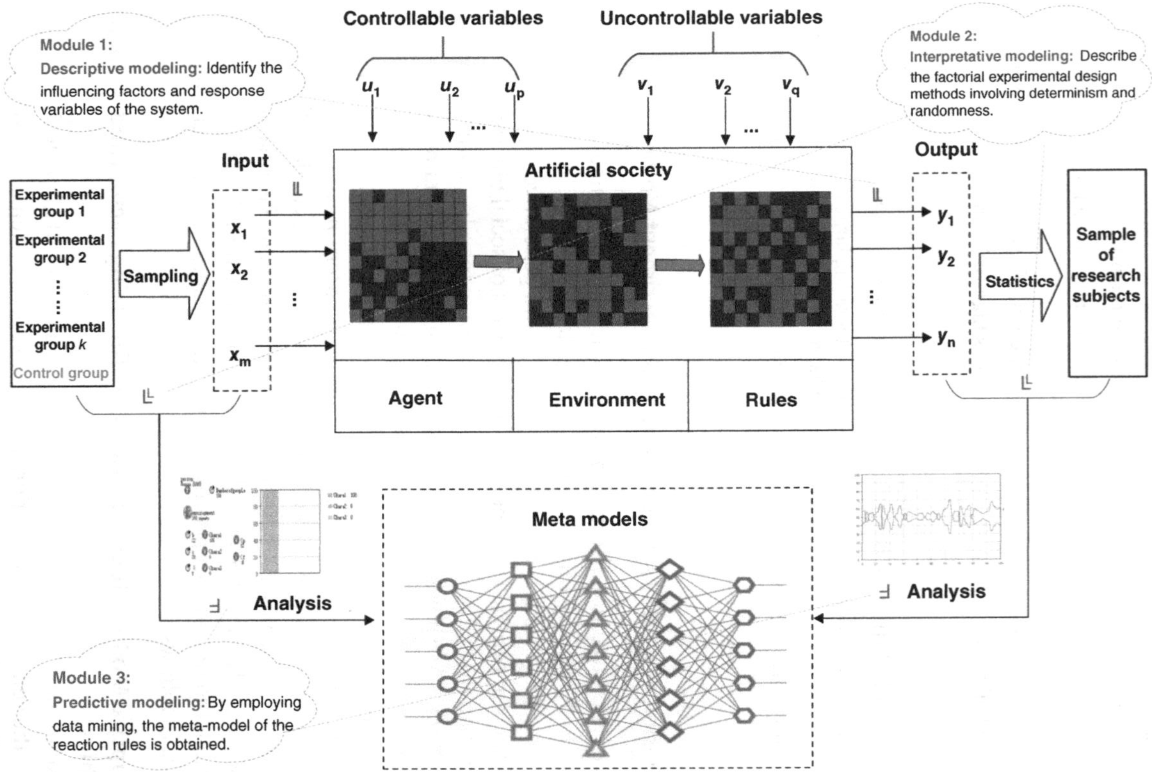

Figure 7.3 The comprehensive modeling framework based on computational experiment design.

predetermined macro-level rules from the bottom up? Generative problems focus on "how" rather than "why," and they are addressed by transforming generative processes into agent-based artificial societies. The key in this process is whether the artificial society can generate meaningful macro-level outcomes, otherwise the explanation fails. Generative sufficiency is a prerequisite for explaining the emergence of macro-level outcomes.

Artificial societies adhere to the principle of "simple consistency," offering holistic descriptions of real-world complex systems [5, 6]. Essentially, an artificial society is a formal logical model that investigates the relationships among a limited set of variables while controlling for others. As more variables are considered, accurately depicting the dynamics of an artificial society becomes increasingly challenging. On the one hand, events often result from the combined effects of multiple factors; on the other hand, the complexity and uncertainty of these factors complicate the characterization of their interrelationships. To facilitate subsequent explanatory and predictive modeling, the following steps should be followed: (1) Identify influencing factors: Which factors affect the system's state and behavior? (2) Analyze the paths and modes of influence: How do these factors exert their effects on the system? (3) Explore underlying factors: What influences these factors?

In general, unknown contexts can be categorized into several scenarios: (1) abrupt environmental changes, especially those caused by major disturbances; (2) changes in the network topology of agent interactions, such as fully connected graphs, random graphs, scale-free complex networks, and small-world complex networks; (3) system changes under intervention strategies, which are particularly common in social systems—for example, the introduction of new technologies, institutions, or cultures often requires ex ante evaluation; (4) changes in agent types, attributes, or behaviors—if the system is open, new agents may emerge, leading to contextual shifts; (5) contextual variations due to changes in system noise; and (6) miscellaneous cases. As mentioned, regardless of the scenario, we must consider not only the contexts themselves but also their underlying causes and associated probabilities before conducting experiments. Once contexts are defined, researchers should not focus excessively on whether experimental conclusions align with prior experience. Many researchers make a mistake at this stage: When encountering results inconsistent with prior experience, they prematurely conclude that the experiment is invalid and abandon further analysis of these anomalous phenomena. Thus, they adhere rigidly to "heredity" while neglecting "variation." Yet heredity bears little relevance to innovation, whereas variation constitutes its very essence.

- **Explanatory Module: Factorial Experiment Design**
  As shown in Figure 7.3, the experimental process can be viewed as a transformation of multiple input combinations into outputs with one or more observable response variables. Here, $x_1, x_2, \ldots, x_m$ denote the inputs of the artificial social system; $y_1, y_2, \ldots, y_n$ represent the outputs; $u_1, u_2, \ldots, u_p$ indicate controllable factors or decisions; and $v_1, v_2, \ldots, v_q$ correspond to uncontrollable factors or events. The objectives of experimental design are to (1) determine the set of factors influencing system outputs through computational experiments; (2) identify the most effective controllable factors u_2 that can steer outputs toward optimal levels; and (3) select controllable factors u_2 that minimize the impact of uncontrollable factors or events v_2 on the system.

  The most basic method to uncover relationships between inputs and outputs is manual model exploration guided by analysts or experts. Although this process is straightforward and requires no advanced technical expertise, it often introduces biases and may overlook potentially insightful unexplored regions of the input space. Therefore, a more systematic approach is needed to uncover input–output relationships. (1) For physical systems with deterministic elements and finite value domains, classical experiment design methods (e.g., random sampling, full or partial factorial designs, central composite designs) and modern methods (e.g., group screening designs, space-filling designs) are typically employed [7]. (2) For complex social systems with numerous uncertainties and random value spaces, Monte Carlo and resampling methods play crucial roles.

  To develop computational experiment schemes and effectively guide experiments, in-depth research on computational experiment design is required. In comparison with traditional experiments, computational experiments exhibit the following characteristics. (1) All experimental factors can be easily modified, enabling experiments with a large number of factors and levels. (2) They can explore richer and more valuable insights within a larger experimental domain. (3) Pseudo-random numbers can be employed, and experiments remain controllable, thereby eliminating the need for randomization and blocking. (4) They place greater emphasis on sequential design, in which subsequent experimental points are determined based on the previous results so as to effectively reduce sample size, avoid oversampling, and improve simulation efficiency. Relative to general simulation experiments, computational experiments on complex systems are also characterized by a greater number of factors, more factor levels, and longer experiment runtimes.
- **Predictive Module: Meta-Model of Artificial Society**
  After collecting input and output data, in-depth analysis can extract the underlying rules of these relationships. The primary goal of this analysis is to improve the capacity to predict and control these relationships effectively. To

overcome the limitations of manual exploration and experimental design-based techniques, the machine learning (ML) field provides a broad set of methods. These methods construct nonlinear mappings from artificial society inputs to outputs. The learned (potentially nonlinear) function (i.e., meta-model) serves as an approximate representation of the input–output relationship of the original simulation model [8].

Meta-model methods aim to directly predict outcomes of interest without explicitly focusing on identifying causal relationships. Traditional predictive models assume that out-of-sample data follow the same statistical distribution. In contrast, the focus here is on improving the applicability of predictive models across various contexts, such as uncontrollable factors or intentional interventions. It also encompasses extreme scenarios in entirely new situations where input features are set to unprecedented values. By comparing meta-model predictions with patterns observed in artificial society evolution, the accuracy and reliability of predictions can be reliably validated. Additionally, integrating optimization techniques (e.g., exact or heuristic methods) with meta-modeling offers a promising approach for identifying input parameter combinations that maximize or minimize specific characteristics of the original model outputs.

7.2 Introduction to Experimental Design Methods

In computational experiments, the artificial society serves as a descriptive modeling tool and forms the foundation for explanatory and predictive modeling. It is used to define, measure, and analyze the relationships between variables of interest. This section addresses three key questions: (1) How can influencing factors and response variables be identified, especially when involving agents' intelligent characteristics? (2) How can the model's response to factor changes and the robustness of outputs to parameter uncertainties be explored, thereby enhancing understanding of model behavior and structure? (3) How can correlations between system inputs and outputs be identified to uncover inherent system patterns that provide valuable insights for decision-makers?

7.2.1 Descriptive Modeling: Identification of Influencing Factors

Computational experiments act as generators of simulated data, capable of generating scenarios that might not arise in reality and producing large volumes of simulated data for specific problems [9]. The state and behavior of a system are influenced by multiple factors, which evolve dynamically over time. The Pareto principle (i.e., the 80/20 rule) [10] indicates that only a small proportion of factors are truly significant. To comprehensively understand the research problem

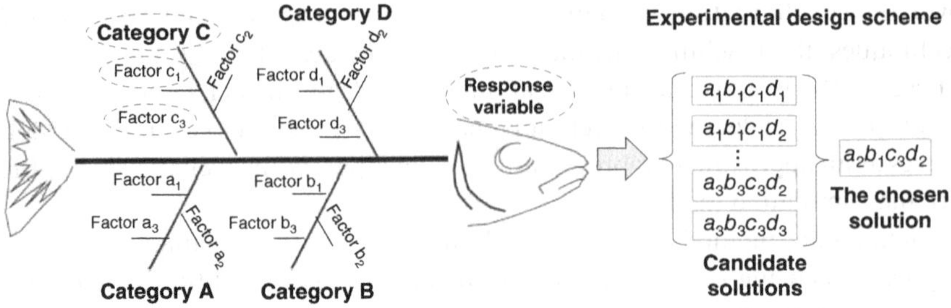

Figure 7.4 The cause-and-effect diagram of computational experiments.

and reduce the computational burden, it is essential to identify key influencing factors and response variables before experimental design. The multi-factor analysis method is a commonly used tool, analyzing the impact of each factor according to the pattern of "how the system's state and behavior will be affected if a certain factor changes." This method facilitates insights into the relationships and structures of complex problems, thereby supporting better adaptation to uncertainties and changes.

Each computational experiment has multiple possible design schemes. To select the most appropriate one, a cause-and-effect diagram is often used as an organizational tool [11]. As shown in Figure 7.4, the cause-and-effect diagram, also known as a fishbone diagram, places the effect or response variable of interest along the central "spine." Potential causes or design factors are arranged along the "ribs." Based on this, experimental design decisions include clarifying objectives, considering sample size (number of repetitions), determining the appropriate experimental order, and deciding whether to apply blocking or other randomization constraints. The specific steps are as follows:

- **Problem identification and formulation**
  Researchers must first establish a clear understanding of the problem under study and how to collect data, with at least a qualitative understanding of data analysis. Clearly stating the problem often facilitates a deeper understanding of the phenomenon under study and its eventual solution. An experiment may have multiple goals, including identification (which factors cause differences and estimating the magnitude of response changes), discovery (what happens if we design new strategies, variables, operating conditions, etc.), and verification (what factors lead to differences between two response variables). Since real-world systems are often highly complex, an experiment cannot answer all key questions. A sequential approach is often more effective, where the entire study can consist of a series of smaller experiments, each with a specific goal, whose results are then aggregated to form the overall conclusion.

- **Selection of response variables**
  Before conducting an experiment, it is crucial to define the response variables and their measurement methods. The design of response variables requires considering three key aspects. (1) Inherent measurement bias: Social systems are complex and changeable, and measurements are abstract quantitative indicators used to describe and compare system performance. In a sense, measurement results are inherently limited, reflecting only part of the truth. (2) Context-specific biases in the measurement process: Response variables are designed to achieve system goals, but in practice, the selection of response indicators often conflicts with goal optimization. (3) Context-dependent data interpretation: The same measurement data can yield completely different results through different interpretations, which may lead to highly misleading conclusions.
- **Selection of factors**
  Identifying factors affecting a system is a complex yet critical process that requires in-depth thinking and comprehensive consideration of multiple aspects. The key is to systematically enumerate all factors that may influence the system's state and behavior. Different factors may have interrelationships, with some exerting direct impacts while others exert indirect ones. A change in one factor can trigger a chain reaction, leading to broader changes in the system's state. Additionally, influencing factors may interact to form complex impact networks. Identifying these factors also requires considering their weights and priorities—some factors are more influential than others, so priority should be given to the most critical ones. Meanwhile, some factors are difficult to predict and control, such as individual cognitive abilities or external regulatory strategies, but they may nonetheless exert significant impacts on problem-solving.

  Based on this, computational experiments classify influencing factors into three categories: design factors, control factors, and nuisance factors. Design factors are those actually selected for study in the experiment. Control factors are variables that may affect responses but are not of current interest, so they are maintained at specific levels. Nuisance factors (also known as noise factors) are variables that naturally vary and cannot be controlled during the experiment; their influence is typically mitigated through randomization. To estimate the combined effects of multiple factors, a sequential branching method clusters factors into groups, tests their significance, and then conducts branching tests on the significant groups. The estimated effect of a factor group with r repetitions is expressed as:

$$\widehat{\beta}_{i-j;r} = \frac{Y(j); r - Y(i-1); r}{2} \tag{7.1}$$

where $\widehat{\beta}_{i-j;r}$ denotes the aggregated main effects from factor i to j (averaged over r repetitions); $Y(j)$; r is the response value when factors from x_1 to x_j are at high level (+1) and others at low level (−1); and $Y(i-1)$; r is the response value when factors from x_1 to x_{i-1} are at high level (+1) and others at low level (−1). Notably, blocking the effects of different realizations of nuisance factors can significantly reduce their impact on experimental results.

- **Selection of levels and ranges**
Once design factors are selected, researchers must determine their variation ranges and specific levels, as well as how to control and measure these values. They must also decide the expected variation range for each factor and the number of levels for each variable. In experimental design, factor-level selection is based on historical observations and corresponding outputs. Specifically, x_{N+1} is determined by $(x_1, \cdots, x_N)$ and corresponding response variables $(y_1, \cdots, y_N)$. The formal representation of the random approximation method is

$$x_{N+1} = x_N - a_N(y_N - \alpha) \tag{7.2}$$

where x_{N+1} is the $(N+1)$-th factor variable in the sequence, x_N is the N-th factor variable, y_N is the response variable corresponding to x_N, a_N is a predefined sequence of constants, and α is a predefined constant. Generally, the initial range of factor variation is usually set wide; as understanding deepens regarding which variables are important and which levels yield optimal effects, the range is typically narrowed.

- **Factorial experiment design**
The one-factor-at-a-time method is a commonly adopted experimental strategy. It involves selecting an initial point or baseline combination for each factor, then continuously varying each factor within its allowable range while keeping others at baseline levels, to examine how single-factor changes affect response variables. The main drawback of this strategy is that it ignores potential interactions between factors, which can make it difficult to yield consistent outcomes when combining different levels of factors. In reality, complex systems often exhibit nonlinear relationships, and the assumption of no interaction effects frequently does not hold.

  If interactions exist, an appropriate approach to handling multiple factors is conducting a factorial experiment, where all factors vary simultaneously instead of varying them individually. Assuming two factors each with two levels, a 2^2-factorial experiment enables researchers to study the individual effects of each factor and determine whether interactions exist. Generally, for k factors each with two levels, a factorial design requires 2^k runs. A key feature of factorial designs is their efficient use of experimental data. For systems involving four, five, or more factors, it is usually unnecessary to test all possible combinations of factor levels. Fractional factorial experiments, as variants of basic factorial designs, examine only a subset of all possible combinations.

- **How do these factors affect system state and behavior?**
  Understanding how each factor affects the system requires analyzing its pathways of influence (direct or indirect), the nature of influence (e.g., positive or negative correlation), the magnitude of influence (degree of impact), feedback loops (relationships between factors leading to complex feedback), and time delays (influences that do not manifest immediately but take time to appear). This analysis aids in deepening the understanding of each factor's operational mechanism. Through in-depth analysis, the focus can be traced to further links in the impact chain—i.e., considering not only direct influencing factors but also those that influence them, and so on. Extending the impact chain helps identify related or underlying factors that drive changes in influencing factors, thereby revealing the root causes of system changes.

7.2.2 Explanatory Modeling: Factorial Experiment Design

By exploring the model's response to factor changes and the robustness of outputs under parameter uncertainties, we can better understand the model's behavior and structure. Factorial experiment analysis can be divided into two approaches: local analysis and global analysis. Local analysis involves changing the value of a single input variable and observing the resulting changes in the model output. This helps in understanding the model's response to individual parameters and identifying the parameters with the greatest impact on the output, i.e., the main effects of factors. Global analysis helps to understand the model's response to multiple parameters, identify the parameter combinations with the greatest impact on the output, and account for interactions among parameters, i.e., factor interactions. In computational experiments, experimental design methods are categorized into two types: deterministic factorial experiment design and stochastic factorial experiment design [12].

Deterministic factorial experiment design refers to experiments in which factors and factor levels are known and can be controlled by adjusting key factors. In a fully deterministic factorial experiment design, factor levels are discrete, meaning the values of experimental factors are partitioned into several segments. For example, a parameter can take discrete values such as $a_1, a_2, a_3, \ldots$. Continuous factor levels, on the other hand, mean experimental factors have continuous values, such as $[b_1, b_2]$. This is considered a special case of discrete multilevel research objects, allowing more precise experimental analysis of factors but with enormous computational costs. As shown in Figure 7.5, deterministic experimental design methods are divided into the following two types according to the complexity of research objects. (1) Classical design methods: These methods are suited to cases

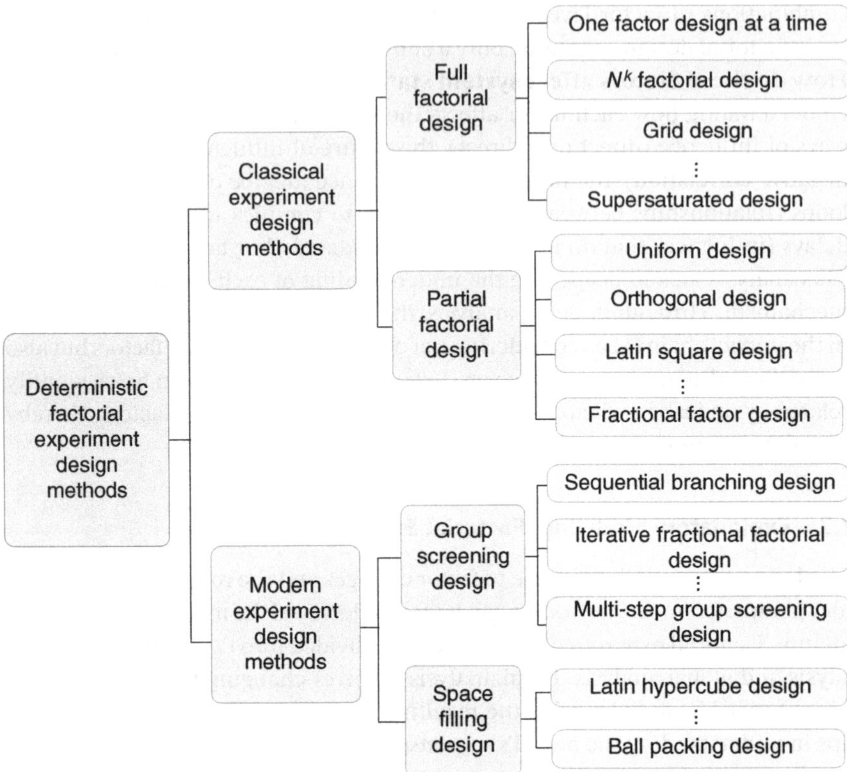

Figure 7.5 Selection of computational experiments design scheme.

involving a few factors and a few levels. Input–output data from such simulation experiments are typically analyzed using linear regression (meta) models represented by low-order polynomials. (2) Modern design methods: These methods are suitable for experimental objects involving discrete multi-factor multilevel, multi-factor few-level, few-factor multilevel, and continuous factor-level scenarios. The influencing factors in these types are often complex, leading to combinatorial explosions that cannot be effectively addressed through simple permutations, combinations, or fractional factorial designs. Compared with physical experiments on continuous experimental objects, computational experiments are more flexible in execution, allowing the formulation of strategies for the next stage based on existing experimental results.

In classical design methods, the experimental design methods for each type of research object are further divided into fractional factorial design and full-factorial experiment design based on whether all possible levels of all factors are analyzed, as specifically shown in **Table A.7.1.**

In modern design methods, they are divided into two categories: group screening design and space filling design. The group screening design adopts the method of sequential experiments, that is, the selection of experimental factor levels is determined according to the previous level selection and the corresponding output. Space filling design is suitable for experimental occasions where there is no random error, but the focus is on controlling system deviation. When the concern about random error is far less than that about the model itself, attention should be paid to the system deviation, that is, the difference between the approximate model and the real mathematical function. The goal of space filling design is to limit the system deviation, and its test scheme can obtain an ideal coverage, so as to lay a foundation for obtaining information-rich experimental result data to provide decision support, as specifically shown in **Table A.7.2**.

When the input of an experimental system involves numerous uncertainties, deterministic analysis methods may fail to effectively find solutions. In such cases, it is necessary to obtain statistical insights into the experimental system through repeated experiment runs. Probabilistic sampling plays a crucial role in addressing the output responses of experimental systems. Common random inputs are used to characterize the stochastic properties of artificial societies. In scenarios with multiple completely random factors, simulation datasets are also employed as outputs to observe differences in system responses. Determining the random data generation strategy that produces datasets closest to real-world conditions becomes a key problem to solve. Ideally, the data generation strategy should possess strong generalization capabilities, ensuring that the results of computational experiments not only perform well in test scenarios but also provide reliable predictions for unknown scenarios.

When the distribution of factor levels is known, the Monte Carlo method is typically used to generate new datasets that follow the known distribution. Unlike bootstrapping (which samples from observed datasets), the Monte Carlo method randomly draws new (unobserved) datasets from the estimation population or data generation process. It utilizes the probability density distribution of scenario parameters for sampling, generating random scenario test cases that closely approximate real occurrence probabilities. The basic steps of the Monte Carlo method are as follows. (1) Randomize the construction or description of scenario occurrence probabilities, representing scenario parameters as probability distributions. (2) After constructing a probability model for scenario parameters based on real scenario data, sample from known probability distributions, including importance sampling, acceptance–rejection sampling, MCMC sampling, and other complex distribution sampling methods. (3) Perform simulation modeling after sampling from known probability distributions to establish estimators for scenario parameters.

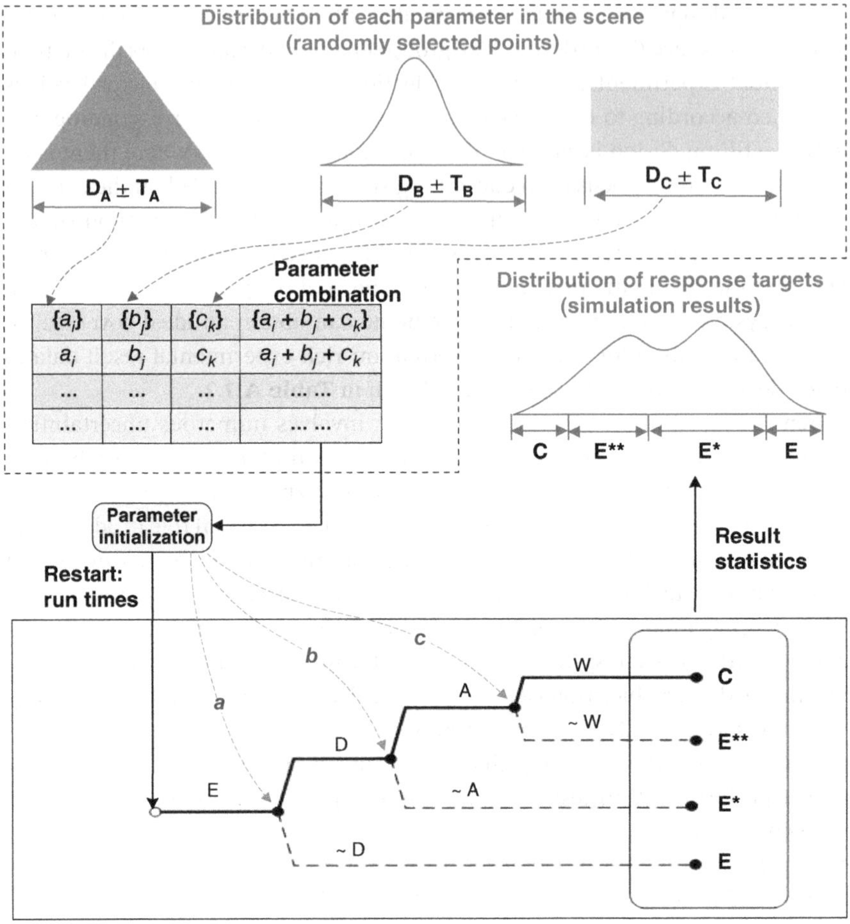

Figure 7.6 Monte Carlo simulation process of artificial society.

As shown in Figure 7.6, a certain scenario is defined by three parameters: Parameter A follows a triangular distribution, parameter B follows a normal distribution, and parameter C follows a uniform distribution. When using the Monte Carlo method, it is necessary to randomly sample values from parameters A, B, and C, denoted as $\{a_i\}$, $\{b_j\}$, $\{c_k\}$, respectively, and perform numerous simulations to calculate the response target values $\{a_i + b_j + c_k\}$. After the N-th simulation, the distribution of possible scenarios can be obtained [13]. Below Figure 7.6, event E occurs in a specific environment at the initial time point r_0. At the subsequent time point r_1, individual agents can decide their adaptive changes based on bounded rationality to address challenges in the current environment, i.e., event D. If no change is made (i.e., event ~D), the result remains E, and this logic continues sequentially. In the sequential logic model, the occurrence of event C is interpreted as a branching process in the sample space Ω that undergoes multiple games and resolutions, representing one of the possible events. The emergence of event C requires at least four necessary sequential conditions, with their probabilities being significantly low, while the probabilities of other outcomes (failing to satisfy E, E*, and E**) are relatively high. In the conditional logic model, when system complexity manifests as a comprehensive event, it can be analyzed based on the current research context to explain its necessary or sufficient conditions. Generally, system complexity C occurs in a dual causal mode: either depending on the simultaneous occurrence of necessary conditions (events connected by "AND") or relying on the occurrence of any sufficient condition (events connected by "OR").

7.2.3 Predictive Modeling: Meta-Model and Law Mining

Many real-world systems are extremely complex, making direct simulation and computation difficult or even infeasible. To address this challenge, "surrogate models" emerge as simplified representations to replace complex system models. By approximating and simplifying complex models, surrogate models significantly improve computational efficiency while maintaining acceptable accuracy. They find wide applications in industrial design, financial risk assessment, medical diagnosis, and other fields. For example, earth system models (ESMs) involve vast spatial–temporal scales and complex modeling processes, leading to enormous computational costs. Supervised ML tools can train ESM surrogates on selected expensive simulation samples. The goal of a surrogate model (or statistical simulation) is to replace simulator code with an ML model, enabling more efficient inference of simulator outputs than running the full simulator. Thus, a surrogate model is a statistical model (a model of the model) for the simulator, while the simulator itself is a mechanistic model of the world.

A fundamental goal of computational experiments is to reveal internal patterns and laws of systems for prediction and control. In computational experiments, macro-level emergence is determined by the behaviors of all micro-agents, meaning all micro-agent parameters or (initial) conditions must be considered. As the number of agents increases, the number of parameters grows linearly, and parameter combinations increase exponentially, leading to the so-called "curse of dimensionality." One solution is to establish a mapping structure between parameters and macro-emergent outputs. As a surrogate model for artificial societies, the meta-model describes the operational laws between input factors and response variables. Figure 7.7 illustrates the process of simulating an artificial society by constructing a meta-model. (1) Input various parameter combinations into the artificial society model, generating corresponding outputs through

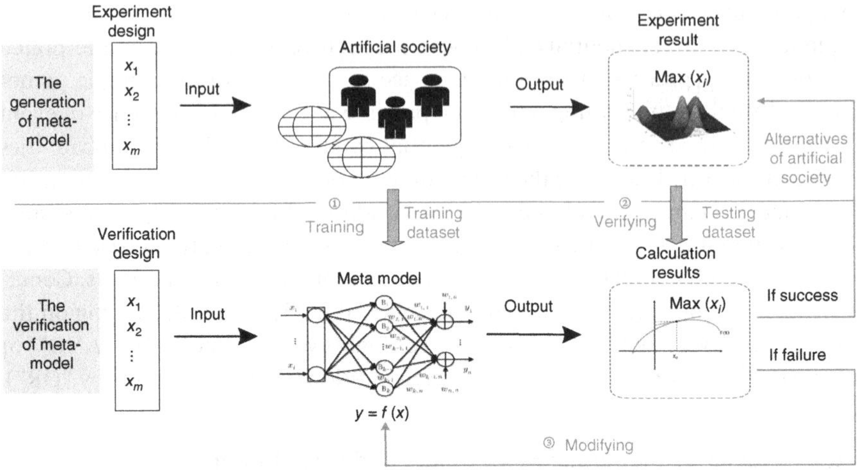

Figure 7.7 The verification and adjustment process of meta-model.

generative inference. Within the specified experimental domain, simulate specific sample points and fit these results using a simplified analytical expression (meta-model). (2) Verify the meta-model through extremum point comparison. If verification succeeds, the results from the analytical expression are considered equivalent to simulation results. If not, corrective measures such as increasing the number of experimental points or narrowing the fitting range are adopted. By identifying correlations between system inputs and outputs, inherent patterns in the system can be discovered, providing valuable insights for decision-makers to solve practical problems.

With the increasing complexity of artificial society models, meta-models require higher mathematical representation capabilities. Spatial statistical models and ML methods can be applied to learn the functional relationship from inputs to outputs of simulation models, such as artificial neural networks (NNs), Kriging, support vector regression (SVR), random forests (RFs), multivariate adaptive regression splines, response surface methods, and linear regression models [14, 15]. Below are some common meta-model forms for representing the operational rules extracted from the system.

- **Statistical model**
  Statistical models excel at identifying correlations between variables and are particularly suitable for the following scenarios: (1) scenarios with low levels of uncertainty, as they typically satisfy model assumptions (such as confidence intervals and hypothesis testing) during construction; (2) medium-sized datasets with fewer attributes but sufficient observations; (3) scenarios focused on isolating the effects of a limited number of variables; (4) scenarios where the

marginal error level in predictions is acceptable; (5) cases where interactions between independent variables can be predetermined to be minimal; and (6) instances where interpretability takes precedence over prediction accuracy. Typically, statistical models implement parametric methods such as linear and logistic regression.

The Kriging method [16, 17], also known as spatial correlation modeling, can be used to handle complex response surface structures. This model assumes that there is some spatial correlation among points in the multidimensional input space and exploits these correlations to make predictions. The basic Kriging model can be expressed as follows:

$$f(d) = \mu + \delta\,(d) \tag{7.3}$$

where μ is the mean in the entire experimental area and $\delta\,(d)$ is the additional noise, which constitutes a covariance stationary process with mean 0. The main disadvantage of Kriging model is that the huge computational amount of parameter acquisition and the assumed spatial correlation are difficult to verify. The parameter-free regression model does not rely on specific data distribution assumptions and is suitable for situations where the data distribution is unknown. The formal representation of the one-dimensional parameter-free regression model is as follows:

$$Y = m(X) + \varepsilon \tag{7.4}$$

where ε is a random error, and $m(X) = E(Y\,|\,X)$ is called the regression function, here only assume that the function is a continuous function without introducing other parameters, compared to polynomial regression, the advantage of parameter-free regression model lies in its continuity.

- **Machine learning model**
  Compared with statistical methods, ML methods have higher requirements on data. Compared with statistical methods, the main purpose of ML is to train models that can provide consistent predictions, often at the cost of interpretability. These models require large amounts of data, including numerous attributes and observations, and are particularly suitable for (1) results with low randomness; (2) extensive training through numerous precise repetitions; (3) treating models as "black boxes" when interpretability is secondary to achieving accurate results; and (4) scenarios where isolating the effects of specific variables is unnecessary. ML primarily employs nonparametric methods, such as K-nearest neighbors, decision trees, RFs, gradient boosting methods, and support vector machines (SVMs). Some studies have compared the different performances of these techniques based on criteria such as accuracy, robustness, transparency (i.e., interpretability), and efficiency (i.e., runtime) [18, 19]. These studies show that SVR typically produces the most accurate meta-models. Take the SVR as an

example. The goal of SVR is to find a hyperplane by constructing a regression model to fit the training sample as much as possible and try to avoid overfitting while keeping the prediction error minimal. The model function can be expressed as follows:

$$f(x) = \omega^T x + b \tag{7.5}$$

where $\omega = (\omega_1; \omega_2; \cdots; \omega_d)$ is the weight vector of the input feature and b is the bias term. An "interval zone" r is created on both sides of the linear function, with a distance of ε (i.e., the tolerance setting), which is usually determined through cross-validation, and no loss is calculated for all samples that fall into the interval zone. The original problem of the SVR optimization target is not only to maximize the interval band r but also to minimize the loss to determine the parameters w, b. That is shown as follows:

$$\min_{w,b} \frac{1}{2}\|\omega\|^2 + C\sum_{i=1}^{n}(\xi_i + \hat{\xi}_i) \tag{7.6}$$

$$\text{s.t.}(x_i) - y_i \leq \varepsilon + \xi_i,$$

$$y_i - f(x_i) \leq \varepsilon + \hat{\xi}_i$$

$$\xi_i \geq 0, (\hat{\xi}_i) \geq 0, \quad i = 1, 2, \ldots, n$$

where ξ_i and $\hat{\xi}_i$ are relaxation variables used to handle cases where data points fall within the boundary interval; C is the penalty parameter used to control the weight of the error term; and n is the number of samples.

Although SVR uses kernel functions to approximate highly nonlinear systems and generate accurate meta-models, this reduces the interpretability of prediction functions, making SVM a "black-box" technique [20]. While prediction functions are precise for unlabeled instances, they do not offer direct interpretability. To enhance transparency, techniques for extracting understandable rules (such as IF-THEN rules) from prediction functions can be used. However, rule extraction adds an abstraction layer that may lead to information loss. RF is another viable technique for modeling input–output relationships in artificial societies. As an ensemble model composed of numerous decision trees, it is represented by IF-THEN rules and grows using the CART algorithm [21]. Its training involves different bootstrap sets and random feature subsets split at each node. These attributes reduce variance, making RF competitive in classification and regression tasks. The rule selection algorithm in RF directly captures accurate and comprehensive rule sets without additional abstraction, maintaining interpretability despite the model's complexity.

- **Neural network model**

  As function approximation machines, NNs are also well-suited for surrogate model construction tasks: A feedforward network defines a mapping $y = f(x; \theta)$

and learns the values of parameters θ to achieve optimal function approximation. The universal approximation theorem states that a sufficiently large NN can approximate any nonlinear function with finite parameters. Although capable of representing any function, NN layers may be excessively large, leading to incorrect learning and poor generalization.

7.3 Design of Scenario Generation Algorithm

This section first presents the overall framework of the scenario generation algorithm, followed by a detailed explanation of each module. Specifically, the environment agent (EA) generates extreme scenarios, while the social agent (SA) constructs a collaborative network of agent interactions. Finally, the planner agent (PA) executes rule generation and experimental scheme calibration.

7.3.1 Problem Description of Scenario Generation

The scenarios of a service ecosystem can be conceptualized as an integrated outcome of the interaction between the demand environment and service entities within a time interval T. This framework systematically defines its constituent elements, including demand environment characteristics, operational statuses of service supply-side entities, interaction networks among stakeholders, and response variables reflecting system effectiveness. Within the time period T (i.e., the dimension of scenarios) and in a known probability space, the parameter vector is composed of random variables following a finite discrete distribution. Thus, $\tau = (\varepsilon_1, \varepsilon_2, \cdots, \varepsilon_T)$ represents a sequence of T-dimensional random vectors. A realization of τ, denoted as $\tau^k = (\varepsilon_1, \varepsilon_2, \cdots, \varepsilon_{T_k})$, is referred to as a scenario.

Scenario generation methods are used to produce a set S containing n scenarios $S(S = \tau^1, \tau^2, \cdots, \tau^n)$, where each scenario represents a plausible real-world situation. Under the same parameter settings in the experimental system, a scenario is associated with an occurrence probability p_s; [22, 23] the higher the probability, the more likely its occurrence, and the probabilities sum to 1. Considering both environmental and agent-related factors as well as response variables, the parameter vector ε_t at time t is defined as:

$$\varepsilon_t(e_t, a_t, u_t) = (E(e_t), I(a_t), U(u_t))^T, \tag{7.7}$$

$$\tau^c = min\{G_{Scenes}(S) \mid e, a, u\}, \tag{7.8}$$

$$X \to S, \ \ \forall \tau^k \in S, \ \ \tau^c \subset S \tag{7.9}$$

where $E()$ represents the normalization of the vector e composed of environmental elements at time t; $I()$ represents the normalization of the vector a composed of

Agent collaborative network elements; $U()$ is the representation method of the system effectiveness u; τ^c represents the scenario set finally generated by the scenario generation algorithm, which contains c scenarios, and the sum of the probabilities of all scenarios is less than or equal to 1; $G_{Scenes}(S)$ represents generating the minimum number of scenarios from the solution set S; X represents the sample space used to generate the corresponding elements in the parameter vector ε_t; and $\rightarrow$ indicates that the n-dimensional vector ε_t of the state of each scenario τ^k in the scenario set S at each moment is obtained by sampling from the n value distributions in the set X, respectively.

The system effectiveness u_t at time period t, as the response variable of the system, reflects the performance indices of interest to researchers and represents the overall state of the system at time t. According to Ref. [24], value entropy is adopted to measure the overall health of the system. Assuming that the total number of individuals in the system is N_{total}, the system effectiveness u_t at time t is defined as

$$u_t = e^{1-\left|\frac{-\sum_{j=1}^{m} p_j \log_2 p_j - \log_2 \sqrt{N_{total}}}{\log_2 \sqrt{N_{total}}}\right|} \tag{7.10}$$

$$p_j = \frac{N_j}{N_{total}} \tag{7.11}$$

where m represents the number of niches divided in the service ecosystem; N_j is the number of individuals in niche j; $\log_2 \sqrt{N_{total}}$ is the optimal entropy of the system; and $-\sum_{j=1}^{m} p_j \log_2 p_j$ is the current entropy of the system. The relative distance between the feature vector ε_t of the generated scenario and the original scenario features measures the extent δ to which the generated scenario τ retains the initial scenario features. The formula is

$$\delta = \frac{\sum_{l=1}^{L} \left|\tau_t^i - Es(t)\right|}{\sum_{l=1}^{L} \left|Es(t)\right|} \times 100\% \tag{7.12}$$

where L is the dimension of the scenario parameter vector ε and $Es(t)$ ($t = 1, 2, \ldots, T$) is the scenario composed of the expected values of historical scenarios at T time points.

As shown in Figure 7.8, based on tasks, domain data, and knowledge, the scenario generation algorithm operates in a closed-loop architecture through three large language model (LLM)-driven agents. The PA initiates the process by integrating domain knowledge and synthesizing system constraints. The EA then generates extreme scenarios with long-tail distributions through adversarial semantic modeling. Concurrently, the SA constructs heterogeneous agent populations by deriving the backbone structure of social networks. These components collaboratively generate the multi-period system state ε_t, which is fed into iterative experimental calibration via a feedback mechanism. This process ultimately achieves the self-optimized generation of the scenario set $S = \{\tau^1, \tau^2, \ldots, \tau^n\}$ for autonomous experiments in service ecosystems.

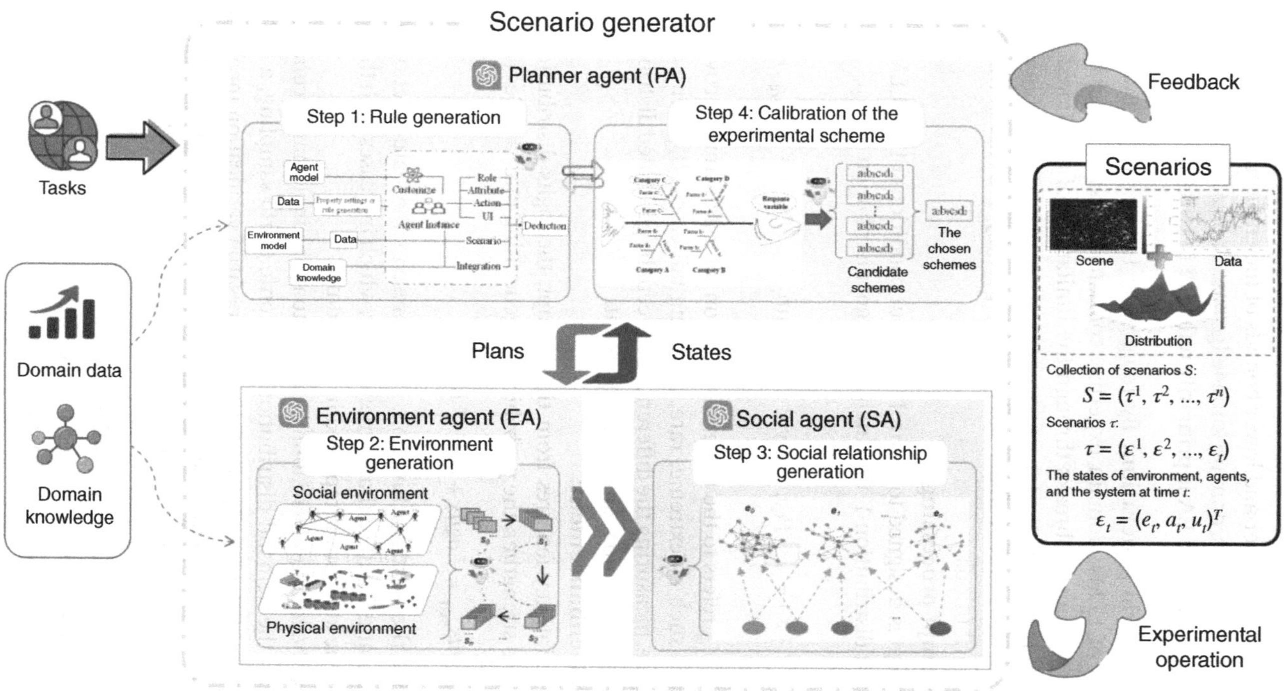

Figure 7.8 The implementation logic of autonomous individual characteristics.

7.3.2 Agent External Environment Generation Algorithm

The EA calculates the upper and lower bounds of the environmental distribution based on historical datasets, domain knowledge bases, and adversarial prompt sets [25]. The innovation of the EA is primarily reflected in two aspects:

1) **Diverse scenario generation:** By integrating the semantic understanding capabilities of LLMs, domain expertise contained in knowledge bases, and adversarial prompts, it breaks through the limitations of traditional methods that solely rely on statistical features of historical data. This approach generates more diverse and realistic scenarios, especially extreme scenarios with long-tail distributions.
   - **LLM prompt processing:**
     The prompt is designed to control the input conditions of the LLM, guiding the output distribution Y to match the target distribution Y_{target}.

$$P(Y \mid X) = \prod_{t-1}^{T} P(y_t \mid X, y < t), argmin_X D_{KL}(P(Y \mid X) \| Y_{target}) \tag{7.13}$$

     where Y_{target} refers to the target distribution after adjustment, covering various types such as extreme, rare, and long-tail cases. D_{KL} denotes the KL divergence, measuring the difference between the generated distribution and the target distribution, ensuring alignment.
   - **Long-tail enhancement:**
     By extracting key features from the prompt, the logits distribution can be adjusted to strengthen the probability of generating rare samples [26]:

$$Logits_{new} = Logits + \delta \cdot I\,(f \in F_{tail}) \tag{7.14}$$

     where I is the indicator function; F_{tail} represents the feature set of long-tail demands; and δ is the enhancement factor, which increases with the number of times the keyword is repeated in the prompt.
2) To address the temporal variability and fluctuations in demand scenarios generated from the demand distribution via Monte Carlo sampling, a reinforcement learning (RL) framework with an adversarial verification mechanism is introduced. This mechanism evaluates and filters the credibility of the generated scenarios, thereby ensuring realism and reliability. The process includes the following key point:
   - **Computation of the credibility score r_s:**
     A credibility score is calculated for each demand scenario to assess whether it deviates excessively from realistic logic.

$$r_s = w_p \cdot T_p(e_k) + w_c \cdot T_c(e_k), w_p + w_c = 1 \tag{7.15}$$

The basic probability-to-credibility conversion function is defined as [27]

$$T_p(e_k) = \frac{In(p_k + \alpha)}{ln(max(p) + \alpha)} \tag{7.16}$$

where p_k is the probability of occurrence for demands, normalized to [0,1], and α is a smoothing factor.
Environmental feature consistency is defined as [28]

$$T_c(e_k) = exp\left(-\beta \cdot \|e_k - \mu\|^2\right) \tag{7.17}$$

where μ represents the mean vector of all demand environment embeddings and β is a sensitivity coefficient for consistency.

- **Determination of the high-credibility threshold θ_{high}:**
Based on the output of a binary classification model assessing environmental credibility, we determine the optimal classification threshold θ_{high} that maximizes classification accuracy while minimizing potential risk loss [29, 30]. That is:

$$\theta_{high} = \underset{\theta}{argmax}(F1score(\theta) + \lambda \cdot Cost_r(\theta)) \tag{7.18}$$

where θ is the candidate threshold, initially set to the average of all r_s values; ***F1score*** is the combined metric for evaluating classification performance (based on Precision and Recall); and λ is a trade-off parameter.
The risk cost function ***Cost$_r$()*** [31] is defined as

$$Cost_r(\theta) = \sum_{Misjudged} Severity(\theta, e_k) \cdot e^{\alpha.RD} \tag{7.19}$$

where each ***Misjudged*** scenario refers to an unrealistic or unreliable environment mistakenly accepted as a credible demand scenario; ***Severity*** reflects the economic loss severity caused by such misjudgment; and α is a sensitivity factor: the greater the α, the higher the penalty for misclassification.
As a dynamic penalty coefficient for risk cost, the system updates the penalty weight of λ through an RL mechanism [32]:

$$\lambda_{t+1} = \lambda_t + \eta \cdot \left(\frac{C_a}{C_p} - 1\right), \quad \eta \in (0.01, 0.1) \tag{7.20}$$

where η is the learning rate; C_a is the actual loss of the system caused by misclassification, and C_p is the predicted cost from the model. If $C_a > C_p$, it means underestimation of risk and the model should increase the penalty degree λ.

The specific details of the demand environment modeling and extreme environment generation algorithm executed by the EA are shown in Algorithm 1 of **Table A.7.3**.

7.3.3 Collaboration Network Generation Algorithm

The SA takes effective social data D, filtered and summarized from domain data and knowledge, as input, aiming to generate the backbone network S that reflects the structure relationship among subjects. The innovation of the SA is mainly reflected in two aspects:

1) **Feature extraction enhances the identification of implicit relationships:** In the feature extraction stage, multi-modal data such as text (news, annual reports, etc.), images (statistical charts), and statistical results of datasets are integrated. The semantic understanding and analysis capabilities of LLMs are utilized to identify implicit relationships (such as potential competition or cooperation chains) and extract potential features of supply-side agents and groups [33]. It includes the following key points:
   - **Input representation (embedding)**
     The semantic understanding of LLM is achieved by converting text (such as descriptions of supply-side agents, behavior logs, interaction records) into high-dimensional vector representations.

$$x_i = Tokenizer(text_i), \quad h_i = LLM_{encoder}(x_i) \tag{7.21}$$

     where x_i is the text token sequence of the i-th agent; $h_i \in \mathbb{R}^d$ is the feature vector (embedding) of this agent in the semantic space; and d is the vector dimension.
   - **Semantic feature extraction**
     The multi-head attention mechanism of the LLM is used to extract key semantics [34], yielding the final contextual representation:

$$Ate(Q, K, V) = softmax\left(\frac{(Q \cdot K)^T}{\sqrt{d_k}}\right)V \tag{7.22}$$

     where Q is the "query" vector, representing the element features that need to be focused on currently; K is the "key" vector, representing the element features to be retrieved or matched; and V is the "value" vector, representing the information that is actually to be extracted or combined. These three vectors all come from the linear transformation of the input embedding. $(Q \cdot K)^T/\sqrt{d_k}$ calculates the similarity (dot product) between Q and K and then divides by $\sqrt{d_k}$ to prevent the value from being too large, and the result is an attention weight matrix. $softmax(\cdot)$ normalizes the weights to [0,1], and the sum is 1, representing the attention proportion of each element. For group features, the representations of multiple agents can be concatenated or averaged and then input into LLM for cross-entity attention calculation.
   - **Latent feature mapping**
     Project the semantic vector output by LLM to the "latent feature space" [35]:

$$z_i = f_\theta(h_i) = h_i + \sigma(W \cdot Ate + b) \tag{7.23}$$

where z_i is the latent feature vector of the agent; W and b are trainable parameters; and $\sigma(\cdot)$ is a nonlinear activation function.

- **Group feature aggregation**

  Combine the latent feature vectors $z_1, z_2, \cdots, z_n$ of multiple agents in the group into an overall group feature vector g to extract group features:

$$g = AGG(\{z_1, z_2, \cdots, z_n\}) \tag{7.24}$$

  $AGG()$ is an aggregation function based on graph neural networks, which realizes the aggregation of information received by agents from neighbors and is used to update the agent's own feature representation and selection tendency [36].

$$AGG = \sum_{j \in \mathcal{N}(i)} \alpha_{ij} z_j \tag{7.25}$$

  where the neighbor set $N(i)$ represents the direct upstream or downstream calling objects of agent i; the attention weight α_{ij} is determined by the relationship and feature similarity between APIs, normalized to [0,1], and the sum of the weights of all neighbors is 1.

2) In the process of relationship network construction, it comprehensively considers the relationships among individuals, between individuals and groups, and among groups. By setting a reasonable relationship strength threshold, it can accurately screen out effective relationship connections, making the constructed relationship network more in line with the complex structure of the actual social system [37].

   The strength of the relationship between individuals s_{ij} reflects the number of times agent i and agent j are invoked simultaneously within a certain time period.

   Regarding the threshold $\mathcal{M}_{ij}$ set for s_{ij}:

$$\mathcal{M}_{ij} = \omega_{ij} * \left(\frac{\sum_{k=1}^{N_{Total}} C_k}{N_{Total}} \right) \tag{7.26}$$

   where ω_{ij} is the weight coefficient; C_k is the total number of invocations of agent k within a year; and N_{Total} is the total number of agents.

   The strength of the relationship between an individual and a group s_{ig} is:

$$s_{ig} = \frac{C_{ig}}{\sum_{k \in I} C_{kg} + \varepsilon} \tag{7.27}$$

   where C_{ig} represents the number of times individual i invoked in group g; C_{kg} is the total number of invocations of all individuals $k \in I$ by group g; and ε represents a smoothing coefficient ($\varepsilon = 1$).

Regarding the threshold $\mathcal{M}_{ig}$ set for s_{ig}:

$$\mathcal{M}_{ig} = \omega_{ig} * \left(\frac{\sum_{k=1}^{N_{Total}} \sum_{l=1}^{N_g} s_{ig_l}}{N_{Total} * N_g} \right) \tag{7.28}$$

where ω_{ig} is the weight coefficient, and N_g is the total number of collaborations. The strength of the relationship between groups $s_{g_1g_2}$ is

$$s_{g_1g_2} = \omega_{g_1g_2} * \frac{|I_{g_1} \cap I_{g_2}|}{\sqrt{|I_{g_1}| * |I_{g_2}|}} \tag{7.29}$$

where $I_{g_1} \cap I_{g_2}$ is the set of individuals invoked jointly by groups g_1 and g_2 and I_{g_1} and I_{g_2} are the total numbers of agents invoked by groups g_1 and g_2, respectively. Regarding the threshold $\mathcal{M}_{g_1g_2}$ set for $s_{g_1g_2}$:

$$\mathcal{M}_{g_1g_2} = \omega_{g_1g_2} * \left(\frac{\sum_{l=1}^{N_g} s_{g_1g_2}}{N_g} \right) \tag{7.30}$$

where $\omega_{g_1g_2}$ is the weight coefficient; N_g is the number of groups.

The specific details of the social relationship network generation algorithm executed by SA are shown as Algorithm 2 in **Table A.7.4**.

7.3.4 Scenario Generation Algorithm

PA generates executable logical expressions and an optimized experiment plan based on industry norms and domain knowledge, by integrating the generated external demand sample set and internal subject sample set. The execution steps of PA are as follows:

1) **Rule generation:** According to the experimental purpose, Program-aided Language (PAL) model is adopted to automatically compile natural language rules into a set of executable logical expressions $\boldsymbol{L}$ [38, 39], guiding the collaboration among various modules of the three types of intelligent agents.

$$L = \{\lambda_i = f_c(r_i, G) \mid r_i \in R\} \tag{7.31}$$

where each element in $R = \{r_1, r_2, \cdots, r_m\}$ is a natural language rule (e.g., "If the market demand shrinkage rate >30%, then reduce the production rate!"), and the natural language rule r_i and the experimental target G are transformed into a logical expression through the compilation function $f_c()$. $\lambda_i : X \to \{0, 1\}$ represents the input state X and outputs a Boolean value (whether the condition is satisfied). The form is as follows:

$$\lambda_i(x) = \left\{ \begin{cases} 1, & if cond_i(x) \Rightarrow act_i(x) \\ 0, & otherwise \end{cases} \right. \tag{7.32}$$

2) **Experimental scheme adjustment:** Measure the degree of matching between the scheme and the experimental target as well as the degree of satisfying constraints, and use the gradient descent method to simplify the experimental plans generated from the above distribution, so as to reduce scenarios with high repetition. It includes the following key points:
   - **Generate initial schemes**
     Starting from the current distribution state $S^{(t)}$ and the logical rule set L, generate a batch of initial candidate experimental plans P_{raw} based on the Monte Carlo method.

$$P_{raw} = g_{MC}(S^{(t)}, L) \tag{7.33}$$

     For each plan p, evaluate its degree of matching with the target G and the constraint conditions (I, V) [40]:

$$M(p) = \alpha \cdot GoalFit(p, G) + \beta \cdot ConsSfy(p, I, V) \tag{7.34}$$

     where $GoalFit(p, G)$ is used to measure whether the plan conforms to the experimental target; $ConsSfy(p, I, V)$ judges whether the plan meets the input constraints (domain specifications I, environmental boundaries V). α and β are weights, which are used to balance the relative importance of target matching and constraint satisfaction.
   - **Update of the scheme set**
     The experimental plan set is denoted as $\{p_1, p_2, \cdots, p_n\}$ and is updated using the gradient descent method.

$$P^{(t+1)} = P^{(t)} - \eta \cdot \nabla_P \mathcal{L}(P^{(t)}) \tag{7.35}$$

     where η is the learning rate; $\nabla_P \mathcal{L}(P^{(t)})$ is the gradient of the loss function with respect to the plan set, representing the optimization direction; iterations and updates are performed continuously until the convergence condition is satisfied:

$$||\nabla_P \mathcal{L}(P)|| < \varepsilon \tag{7.36}$$

     The loss function $\mathcal{L}(P)$ is defined as follows [41]:

$$\mathcal{L}(P) = \frac{1}{|P|} \sum_{p \in P} [1 - M(p)] + \lambda \cdot Ry(P) \tag{7.37}$$

     where λ is used to control the intensity of redundant punishment and $Ry()$ is a redundant punishment determination function. If there are many similar or repeated plans in the plan set, the punishment will be increased.

The specific details of the social relationship network generation algorithm executed by PA are shown in Algorithm 3 of Table 7.2.

Table 7.2 Algorithm 3: Rule Generation Algorithm Implemented by Planning Agent (PA).

Input: Domain specification documents I, upper and lower bounds of environmental distribution V_{max} and V_{min}, and annual backbone relationship sets N;
Output: Executable logical expressions L, optimized experimental scheme P_{opt};

Initialize an empty system constraint set $\boldsymbol{C}$ for storing subsequently generated system constraints;
for each document d in I **do** // Generate system constraints
  Based on d, use LLM combined with RAG technology to generate constraint condition constraint, and add it to the system constraint set C;
  Combine PAL technology with the LLMs, convert the constraint condition c into an executable logical expression (denoted as "expression") according to Formula (7.31), formalize the expression, and then add it to the set of executable logical expressions L;
end for
Initialize the experimental plan P_{raw} according to Formula (7.33);
while not converged **do**
  Calculate the gradient value ∇J based on current environmental parameters and backbone relationship parameters to measure the scheme matching degree and constraint satisfaction degree;
  According to the calculated ∇J, adjust the experimental scheme using the formula $Ev^{(k+1)} \leftarrow Ev^{(k)} - \eta \nabla J$, where η is the learning rate controlling the step size of each scheme adjustment;
  if the norm of the gradient $\| \nabla J \| < \varepsilon$ (ε is the preset convergence threshold) **then**
    Break
  end if
end while
while not converged **do**
  Execute the current experimental scheme P to obtain experimental *results*;
  Analyze the experimental *results*, compare with the network relationships of each year combined with N, and obtain feedback information *feedback*;
  Adjust the experimental scheme $\boldsymbol{P}$ and the executable logical expression set $\boldsymbol{L}$ based on the feedback information;
  if *isConverged*(feedback) **then** // The feedback meets the convergence condition
    Set the convergence flag converged to **true**;
end while
return The generated executable logical expression set L and optimized experimental scheme $P_{opt} \leftarrow P^{(t)}$.

7.4 Case Study: API Service Market

Currently, the development scale of Web services is unprecedented in scale. As of March 2025, the number of developers invoking services on the world's largest public API marketplace, RapidAPI, has exceeded four million, with over 40,000 APIs available. RapidAPI adopts a revenue-sharing governance strategy to achieve rapid expansion [42]. In contrast, the ProgrammableWeb ecosystem, which employed an indirect profit strategy, was ultimately forced to cease operations, highlighting the critical importance of service ecosystem governance. Many researchers are interested in conducting in-depth studies on its governance methods but face challenges in designing experimental schemes and inefficient experimental processes. Motivated by these demands, this section conducts extensive experiments using computational experiment methods to address the following three questions regarding the scenario generation algorithm proposed in this paper:

Question 1: Can it account for extreme scenarios in service ecosystem governance?

Question 2: Can it effectively reflect the collaborative characteristics among individuals within the service ecosystem?

Question 3: Are the generated scenarios more accurate and efficient?

7.4.1 Construction of Experimental System

ProgrammableWeb is an authoritative online platform for the API ecosystem, established in 2005. In the 2010s, with the rapid expansion of the API economy (e.g., open APIs from Twitter, Google Maps), ProgrammableWeb became an industry benchmark. However, as market competition intensified (e.g., the establishment of RapidAPI in 2014), it faced challenges such as user attrition and declining revenue, rendering it unsustainable to cover maintenance costs. Consequently, it ceased operations in 2022. Based on the ProgrammableWeb dataset and the RapidAPI official data, we designed experiments to analyze the evolutionary process of the ProgrammableWeb service ecosystem within a fluctuating social environment throughout its life cycle, demonstrating the experimental design process proposed in this paper and verifying its effectiveness.

As shown in Figure 7.9, we statistically analyzed the ProgrammableWeb dataset and API call data from RapidAPI to extract environmental features that reflect social demand fluctuations. In addition, we derived the relational structure among entities from the invocation relationships between APIs and Mashups in the ProgrammableWeb dataset, thus forming the initial distribution of characteristic parameters from the raw data.

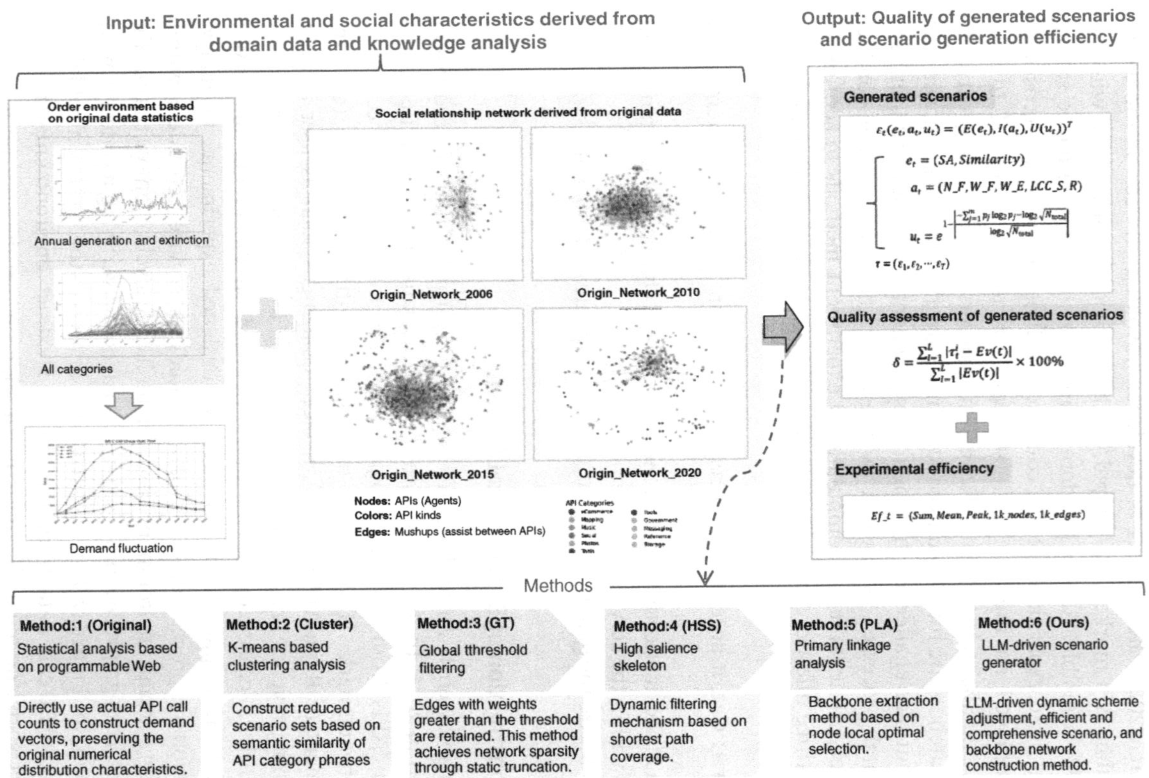

Figure 7.9 Computational experiment design framework. Using the environmental and social characteristics derived from the dataset and knowledge analysis as inputs, six different experimental design methods are applied to generate scenario sets and evaluation indicators, including assessments of scenario quality and experimental efficiency.

Table 7.3 Methods Compared in the Experiment.

No.	Method Name	Notes
Method 1	Baseline method	Rules are established based on the statistical analysis of raw data, and the system evolves in accordance with these rules;
Method 2	Clustering [43]	The K-means algorithm is used to partition the initial distribution into clusters. The cosine similarity between samples and cluster centers is calculated, and adaptive thresholding is then used to remove outliers and noise, producing the final experimental plan.
Method 3	Global threshold filtering [44]	Through static truncation, edges with weights exceeding a predefined threshold are retained, enabling sparsification of samples under given environmental states and narrowing the distribution of generated scenarios.
Method 4	High Salience Skeleton [45]	Significant service-dependency relationships are preserved based on statistical hypothesis testing. A single-source Dijkstra's algorithm is applied to generate a shortest path tree for each node. The significance score of each edge is calculated according to path coverage frequency, and edges with scores above the threshold are retained.
Method 5	Principal link analysis [46]	For each element or node, the adjacent edge with the maximum weight is selected, and only bidirectional maximum-weight connections are retained to form an undirected network.
Method 6	Scenario generation algorithm	LLM-driven collaboration among three functional agents to generate and optimize experimental schemes.

Drawing on our analysis of the current research landscape, we categorized the service ecosystem governance methods compared in this experiment into the following groups (see Table 7.3).

In the scenario generation algorithm, according to Formula (7.7), the state representation of the scenario vector $\varepsilon_t(e_t, a_t, u_t)$ in the API service market at time t is shown in **Table A.7.5.**

We used the methods outlined in Table 7.3 to generate diverse experimental scenarios, and the outcomes were evaluated in terms of both quality and efficiency to assess the effectiveness of different experimental design approaches. The specific evaluation metrics used for this assessment are detailed in **Table A.7.6.**

7.4.2 Experimental Design

The experimental design aims to investigate the impact of different governance strategies on the evolution of service ecosystems under varying social demand fluctuations. The environmental design is outlined as follows:

- **Design of request environmental scenarios (EA):**
  The environmental scheme models the fluctuation of social demand within the research period. This is represented by a fluctuation function with time as the independent variable. For experimental design purposes, APIs are categorized into four types: infrastructure, life services, enterprise management, and social entertainment. The number of Web service calls in the ProgrammableWeb dataset (2006–2022) is counted and converted into semantic features. The domain knowledge base K is built from collected knowledge on the development history of Web services, such as open APIs from Twitter, Google Maps, and the rise of the API economy around 2013. This is then converted into a text format suitable for LLM input to generate event logic rule sets. The generated adversarial prompt set P includes the following information:
  ① Extract economic variables strongly correlated with market fluctuations from the trigger model (including demand, price, and policy).
  ② Generate market performances under different economic environments (such as recession periods and policy stimulus periods).
  ③ Ensure that the generated scenarios are logically consistent with historical data and economic theories.
  ④ Reveal the chain reaction mechanism of risk factors (e.g., price fluctuations → corporate profit → expenditure ratio → system stability).
  In Algorithm 1 of Section 7.3.2, the environment-related parameter design for the API service market is shown in **Table A.7.7**.
- **Design of subject collaborative relationships (SA):**
  Use API and Mashup data from D (2006–2022) as input for Algorithm 2. Count the API and Mashup sets, and use them as LLM input to derive the statistical characteristics of individuals (I_f) and groups (G_f). Algorithm 2 in Section 7.3.3 shows the parameter design for generating individual collaborative relationships in the API service market, as outlined in **Table A.7.8**.
- **Design of experimental schemes generation process (PA):**
  Extract industry specification documents from the domain knowledge base, including relevant documents such as intradomain rules, standards, and business processes. Then, generate system operation constraint rules using the fused information. Initialize the environmental scheme within the given boundary conditions of environmental distribution, and continuously optimize the environmental scheme using iterative gradient descent optimization until

convergence. Algorithm 3 in Section 7.3.3 shows the parameter design for the scenario generation process in **Table A.7.9**.

7.4.3 Analysis of Experimental Results

Each experimental design method generates a final plan, which, when executed, produces a corresponding scenario set. The experimental results are statistically analyzed to answer the three research questions as follows:

1) **Analysis of the generated demand environment (question 1):** We collected data on active APIs from ProgrammableWeb and RapidAPI (2005–2022) and mapped discrete API categories (e.g., e-commerce, social) into a continuous semantic space. An LLM agent merged functionally similar categories into four main semantic communities. APIs were classified into four categories: infrastructure, lifestyle services, enterprise management, and social entertainment. For each group, the number of Mashup calls was counted to construct a demand time-series simulating market changes.

   The EA in the scenario generator fits the market demand's evolutionary trend. The LLM uses online information and inferences to present patterns like growth, saturation, and oscillation. Impact weights across time scales are quantified through data analysis. Figure 7.10 shows that our method determines extreme boundaries of environmental fluctuations, significantly improving environmental coverage (Figure 7.11).
2) **Analysis of generated social relationships (question 2):** The interaction relationships within the ProgrammableWeb ecosystem are presented through

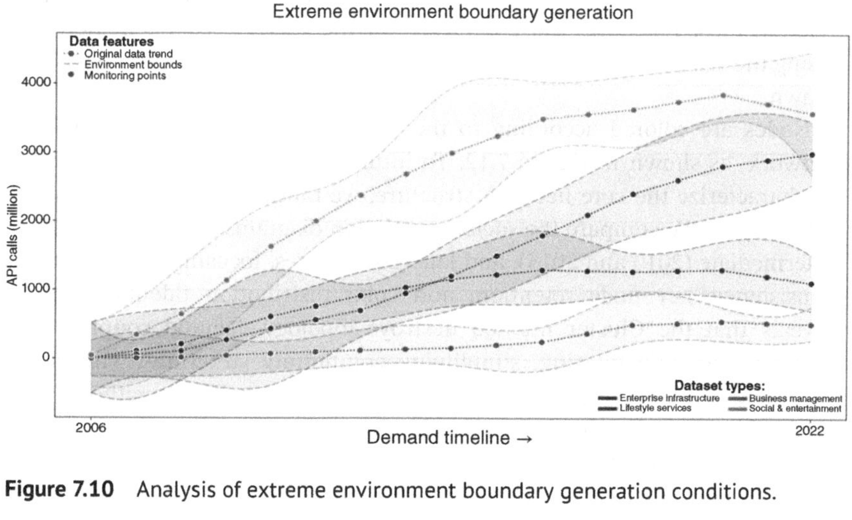

Figure 7.10 Analysis of extreme environment boundary generation conditions.

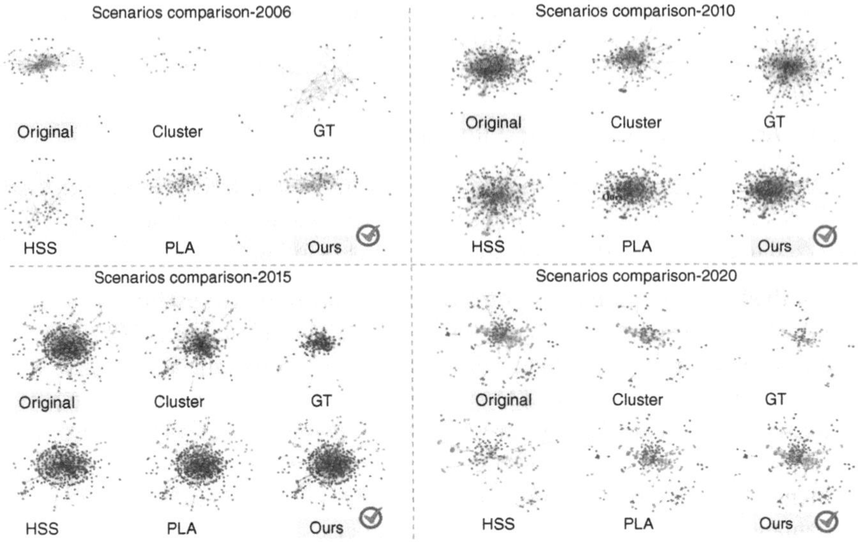

Figure 7.11 By analyzing environmental and social characteristics from datasets and knowledge, six experimental methods are used to process and output a scenario set with evaluation indicators (including scenario quality and experimental efficiency evaluations).

an API–API co-occurrence network. Mashups serve as service composition units, while APIs act as basic functional individuals. The co-occurrence frequency of each API in Mashups is counted, and weighted edges are constructed to generate a relational network diagram for 2005–2020. In the original state, numerous API nodes have dense connections and complex relationships, reflecting the complexity of the service ecosystem network. Using the six methods, an annual API–API co-occurrence network diagram is drawn.

Nodes are colored according to the community division results of each network, as shown in Figure 7.12. To intuitively verify the methods' ability to characterize the core network structure, we randomly select one scenario for analysis. We compare the social association diagrams of the initial (2006), intermediate (2010 and 2015), and late (2020) states, focusing on community consistency, key node retention, and color distribution fidelity. Analysis reveals that the Cluster method destroys topology–demand coupling due to strict semantic filtering, simplifying community structures. The global threshold filtering (GT) and high-significance skeleton (HSS) methods, constrained by static thresholds, sacrifice key connections to reflect cross-category collaboration. The tree-like constraints of PLA cause excessive filtering, leading to deviations in community numbers and color distribution, making

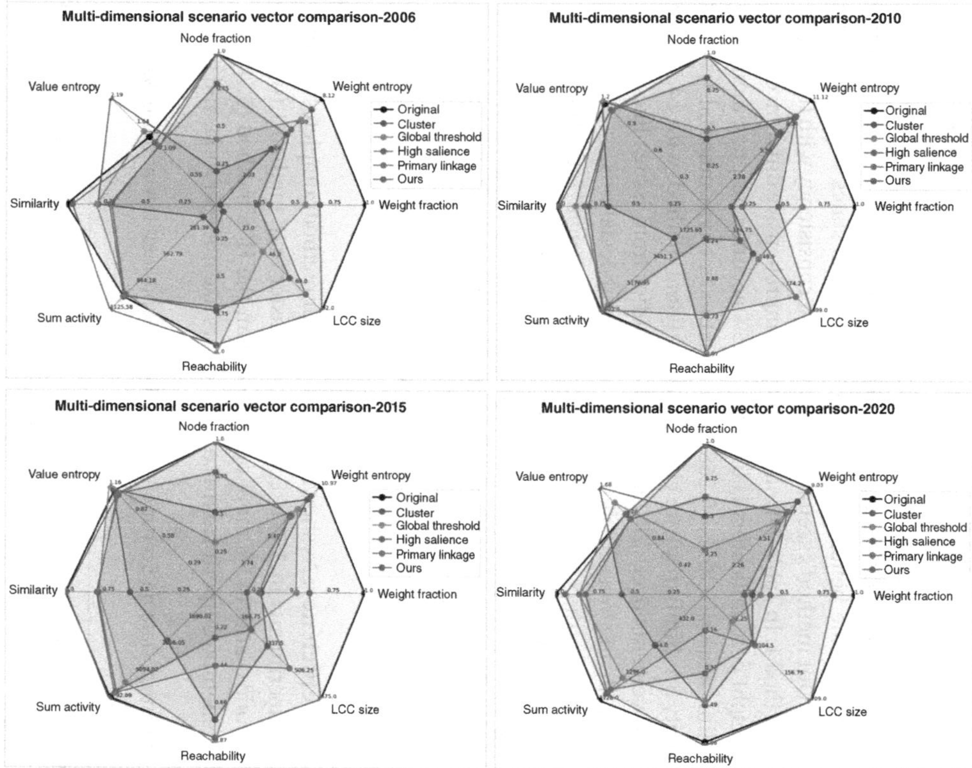

Figure 7.12 Effect of the scenario generation algorithm. The figure compares 8D scenario vectors generated by six methods (Original, Cluster, GT, HSS, PLA, and our method). Our method is closest to the original scenario's comprehensive features.

it hard to reflect real-world mesh-like interactions. In contrast, our method maintains community structures and color distributions consistent with the original state, retains cross-community connections, and reflects real-world collaboration models.

3) **Quality and efficiency analysis of the scenario generator (question 3):** Through multidimensional analysis, we comprehensively evaluated the differences between the proposed scenario generation algorithm and other methods.

Performance evaluation: Based on the calculation and comparison of the 8D scenario vector characteristic parameters, as shown in Table 7.4, method 1 (original) was used as the benchmark, comparing scenario states across multiple time periods. Our method's scenarios show the best consistency with those generated by method 1. In Table 7.4, our method's comprehensive difference degree (δ) across eight core indicators is 0.0738, a 72.5% accuracy improvement over principal link analysis (PLA) (0.2685). Our method's environmental features (total activity SA, Similarity) are closer to the original method, with 14.5% and 19.4% improvements over GT. Other methods suffer from issues like missing key APIs, sacrificing long-tail demands, or structural deviations.

In social relationship generation (node ratio NF, weight ratio WF, weight entropy WE, Size of the Largest Connected Component [LCCS], Reachability), our method closely resembles method 1 in network integrity, global reachability, and weight distribution diversity. In contrast, methods like GT, Cluster, HSS, and PLA exhibit varying degrees of adaptive defects or connectivity losses. In overall system state evaluation (VE), our method's deviation from method 1 performs best.

Experimental efficiency: As shown in Table 7.5, our method completes multiple rounds of experiments in 3.171 seconds, 75.7% faster than method 1 (original) (13,037 seconds) and 44.5% more optimized than the HSS method (5,709 seconds). Our method outperforms in core indicators like time per thousand nodes,

Table 7.4 Statistical Data on Scenario Vector Indicators and Total Deviations of Various Methods.

Method	SA	Similarity	NF	WF	WE	LCC_S	Reachability	VE (u)	Deviation (δ)
Original	0.9776	1	1	1	1	1	0.9712	0.8608	—
Cluster	0.3587	0.622	0.4643	0.226	0.8272	0.418	0.7886	0.9865	0.4329
GT	0.8578	0.8179	0.3942	0.5397	0.7869	0.3894	0.9427	0.9093	0.2972
HSS	0.92287	0.7753	0.7816	0.2214	0.7009	0.4214	0.3885	0.8114	0.3556
PLA	0.9448	0.7987	1	0.2997	0.7087	0.674	0.4826	0.8049	0.2685
Ours	0.9817	0.9767	0.9978	0.6335	0.8973	1	0.9755	0.824	**0.0783**

Table 7.5 Experimental Efficiency Time Statistics of Different Methods.

Method	Sum (ks)	Mean (ks)	Peak (ks)	/1kn (ks)	/1ke (ks)
Original	13.037	0.869	10.142	25.775	4.166
Cluster	18.977	1.265	18.142	76.665	22.430
GT	6.181	0.412	5.671	31.862	8.813
HSS	5.709	0.381	5.314	14.329	14.884
PLA	5.937	0.396	5.393	11.739	12.701
Ours	**3.171**	**0.211**	**1.585**	**6.279**	**1.907**

time per thousand edges, and peak time. It shows excellent stability, with average timings significantly better than comparative methods. The growth rate of time per thousand nodes/edges in our method is slower than the growth rate of network scale, highlighting its potential in handling large-scale data and providing methodological support for service ecosystem governance research.

In summary, through dynamic weight adjustment and spatiotemporal coupling modeling, our method identifies inefficiencies in traditional experiment approaches. It optimizes environmental distribution and social relationship integrity while significantly improving experimental efficiency. This introduces a new paradigm for service ecosystem governance research.

7.5 Chapter Summary

The computational experiment method offers a unique approach to analyzing complex social systems through generative inference and experiments. Generative inference deduces macro-level rules from agent configurations, while generative experiments test and select rule sets. However, there is no systematic approach for abductive analysis of human and social factors in cyber–physical–social systems. Based on this, this chapter introduces the computational experiment design framework, comprising descriptive modeling of artificial societies, explanatory modeling of experimental design, and predictive modeling of meta-models. Furthermore, the chapter elaborates on the process of generating experimental scenarios and validates the framework through a case study of the API service market.

Despite advancements, several challenges remain. Meta-models, while useful for predictive modeling, suffer from issues related to reproducibility, applicability, generalizability, and interpretability. There is also a gap between descriptive and predictive modeling, which could be bridged by causal inference. Additionally, the proposed scenario generation algorithm faces limitations such as semantic biases in LLMs and scalability concerns for large, complex scenarios. Future research should focus on integrating RL, enhancing validation modules, and incorporating causal inference to address these challenges, ultimately improving governance tools for dynamic service ecosystems.

References

1 Xue, X., Li, G., Zhou, D. et al. (2022). The research roadmap of service ecosystem: a crowd intelligence perspective. *International Journal of Crowd Science* 6 (4): 195–222.

2 Kahng, M., Thorat, N., Chau, D.H. et al. (2018). Gan lab: understanding complex deep generative models using interactive visual experimentation. *IEEE Transactions on Visualization and Computer Graphics* 25 (1): 310–320.

3 Xue, X., Zhou, D., Yu, X. et al. (2024). Computational experiments for complex social systems: experiment design and generative explanation. *IEEE/CAA Journal of Automatica Sinica* 11 (4): 1022–1038.

4 Salganik, M.J., Dodds, P.S., and Watts, D.J. (2006). Experimental study of inequality and unpredictability in an artificial cultural market. *Science* 311 (5762): 854–856.

5 Xue, X., Wang, S., Zhang, L. et al. (2018). Social learning evolution (SLE): computational experiment-based modeling framework of social manufacturing. *IEEE Transactions on Industrial Informatics* 15 (6): 3343–3355.

6 Peng, C., Yu, X., Ma, W. et al. (2023). Computational experiments: virtual production and governance tool for metaverse. *International Journal of Crowd Science* 7 (4): 158–167.

7 Politis, S., Colombo, P., Colombo, G. et al. (2017). Design of experiments (DoE) in pharmaceutical development. *Drug Development and Industrial Pharmacy* 43 (6): 889–901.

8 Ferber, J. and Gutknecht, O. (1998). A meta-model for the analysis and design of organizations in multi-agent systems. *Proceedings International Conference on Multi Agent Systems (Cat. No. 98EX160)*, pp. 128–135. IEEE.

9 Lu, M., Chen, S., Xue, X. et al. (2021). Computational experiments for complex social systems—part II: the evaluation of computational models. *IEEE Transactions on Computational Social Systems* 9 (4): 1224–1236.

10 Sanders, R. (1987). The Pareto principle: its use and abuse. *Journal of Services Marketing* 1 (2): 37–40.

11 Ullman, J.B. and Bentler, P.M. (2012). Structural equation modeling. In: *Handbook of Psychology*, 2e, 2.

12 Kleiner (2010). *Design and Analysis of Simulation Experiments*. Beijing: Electronic Industry Press.

13 Yu, C.H. (2002). Resampling methods: concepts, applications, and justification. *Practical Assessment, Research & Evaluation* 8 (1): 19.

14 Salle, I. and Yıldızoğlu, M. (2014). Efficient sampling and meta-modeling for computational economic models. *Computational Economics* 44 (4): 507–536.

15 Edali, M. and Yucel, G. (2018). Automated analysis of regularities between model parameters and output using support vector regression in conjunction with decision trees. *Jasss-The Journal of Artificial Societies and Social Simulation* 21 (4): 1.

16 Cui, K., Zheng, X., Wen, D. et al. (2013). Researches and applications of computational experiments. *Acta Automatica Sinica* 39 (8): 1157–1169.

17 Kleijnen, J.P.C., Van Beers, W., and Van Nieuwenhuyse, I. (2012). Expected improvement in efficient global optimization through bootstrapped kriging. *Journal of Global Optimization* 54 (1): 59–73.

18 Clarke, S.M., Griebsch, J.H., and Simpson, T.W. (2005). Analysis of support vector regression for approximation of complex engineering analyses. *Journal of Mechanical Design* 127 (6): 1077–1087.

19 Li, Y.F., Ng, S.H., Xie, M., and Goh, T.N. (2010). A systematic comparison of metamodeling techniques for simulation optimization in decision support systems. *Applied Soft Computing* 10 (4): 1257–1273.

20 Barakat, N. and Bradley, A.P. (2010). Rule extraction from support vector machines: a review. *Neurocomputing* 74 (1–3): 178–190.

21 Breiman, L., Friedman, J., Stone, C.J., and Olshen, R.A. (1984). *Classification and Regression Trees*. CRC Press.

22 Xie, S. (2013). Research on scenario generation and reduction techniques in stochastic unit commitment problems. Ph.D. dissertation. Hunan University.

23 Mesquita, C.M., Valle, C.A., and Pereira, A.C.M. (2024). Scenario generation for financial data with a machine learning approach based on realized volatility and copulas. *Computational Economics* 63 (5): 1879–1919.

24 Yu, X., Xue, X., Zhou, D. et al. (2024). Beyond traditional metrics: the power of value entropy in multidimensional evaluation of the service ecosystem. *2024 IEEE International Conference on Web Services (ICWS)*, pp. 611–621. IEEE.

25 Hsieh, C.J., Si, S., Yu, F.X. et al. (2023). Automatic engineering of long prompts. arXiv preprint arXiv:2311.10117.

26 Li, H., Ning, Y., Liao, Z. et al. (2023). In search of the long-tail: systematic generation of long-tail knowledge via logical rule guided search. arXiv preprint arXiv:2311.07237.

27 Guo, C., Pleiss, G., Sun, Y. et al. (2017) On calibration of modern neural networks. *International Conference on Machine Learning*, pp. 1321–1330. PMLR.

28 Goldberger, J., Hinton, G.E., Roweis, S. et al. (2004). Neighbourhood components analysis. In: *Advances in Neural Information Processing Systems*, vol. 17 (ed. L.K. Saul, Y. Weiss, and L. Bottou).

29 Lipton, Z.C., Elkan, C., and Naryanaswamy, B. (2014). Optimal thresholding of classifiers to maximize F1 measure. In: *Joint European Conference on Machine Learning and Knowledge Discovery in Databases*, 225–239. Berlin, Heidelberg: Springer Berlin Heidelberg.

30 Jansche, M. (2005). Maximum expected F-measure training of logistic regression models. *Proceedings of Human Language Technology Conference and Conference on Empirical Methods in Natural Language Processing*, pp. 692–699.

31 Zhou, Z. and Liu, X. (2006). Training cost-sensitive neural networks with methods addressing the class imbalance problem. *IEEE Transactions on Knowledge and Data Engineering* 18 (1): 63–77.

32 Sutton, R.S. and Barto, A.G. (1998). *Reinforcement Learning: An Introduction.* Cambridge: MIT press.

33 Chi, T.C., Fan, T.H., Rudnicky, A.I. (2023). Attention alignment and flexible positional embeddings improve transformer length extrapolation. arXiv preprint arXiv:2311.00684.

34 Ampazis, N. and Sakketou, F. (2024). Diversifying multi-head attention in the transformer model. *Machine Learning and Knowledge Extraction* 6 (4): 2618–2638.

35 Bengio, Y., Courville, A., and Vincent, P. (2013). Representation learning: a review and new perspectives. *IEEE Transactions on Pattern Analysis and Machine Intelligence* 35 (8): 1798–1828.

36 Veličković, P., Cucurull, G., Casanova, A. et al. (2017). Graph attention networks. arXiv preprint arXiv:1710.10903.

37 Wasserman, S. and Faust, K. (1994). *Social Network Analysis: Methods and Applications.* Cambridge University Press.

38 Gao, L., Madaan, A., Zhou, S. et al. (2023) Pal: program-aided language models. *International Conference on Machine Learning*, pp. 10764–10799. PMLR.

39 Fang, M., Deng, S., Zhang, Y. et al. (2024). Large language models are neurosymbolic reasoners. *Proceedings of the AAAI Conference on Artificial Intelligence* 38 (16): 17985–17993.

40 Kalyanmoy, D. (2001). *Multi-Objective Optimization Using Evolutionary Algorithms*, 497. John Wiley & Sons.

41 Tanabe, R. and Ishibuchi, H. (2019). A review of evolutionary multimodal multiobjective optimization. *IEEE Transactions on Evolutionary Computation* 24 (1): 193–200.

42 Smith, J. and Lee, K. Governance strategies in API ecosystems: a case study of RapidAPI. *Journal of Digital Economy* 12 (3): 45–60. https://doi.org/10.1000/jde.2023.012.

43 Ikotun, A.M., Ezugwu, A.E., Abualigah, L. et al. (2023). K-means clustering algorithms: a comprehensive review, variants analysis, and advances in the era of big data. *Information Sciences* 622: 178–210.

44 Tian, Z., Wang, Y., Li, X. et al. (2024). Typical daily scenario extraction method based on key features to promote building renewable energy system optimization efficiency. *Renewable Energy* 236: 121420.

45 Yang, C., Indurkhya, B., See, J. et al. (2024). Skeleton ground truth extraction: methodology, annotation tool and benchmarks. *International Journal of Computer Vision* 132 (4): 1219–1241.

46 Greenacre, M., Groenen, P.J.F., Hastie, T. et al. (2022). Principal component analysis. *Nature Reviews Methods Primers* 2 (1): 100.

8

The Fourth Step – Experimental Analysis and Causal Inference

The study of system complexity primarily has two objectives: to explore underlying patterns and to develop theoretical explanations. Pattern exploration seeks to clarify the mechanisms behind the emergence of system complexity, while theoretical explanations aim to identify the fundamental causes of this complexity. Laws are generally defined as mappings between variables, whereas theories offer causal explanations of system behavior [1, 2]. Agent-based modeling (ABM) is an important approach for studying complex systems, but it tends to emphasize *simulation* over *experimentation*. As a result, ABM often struggles to deeply uncover the governing operational principles. Unlike conventional scenario analysis that relies on human reasoning, computational experiments emphasize *counterfactual experiments*—that is, creating parallel worlds that simulate alternative "evolutionary paths" of real-world events [3]. By systematically adjusting input variables and observing the resulting changes in output variables, computational experiments provide a robust tool for causal inference, thereby addressing the limitations of traditional ABM. Together, these methods offer causal insights into the dynamic evolution of systems. This chapter first outlines the explanatory theory of system complexity. Then, from the dual perspective of *horizontal* and *vertical* causality, it presents a detailed causal analysis framework for computational experiment method, including observational analysis, interventional analysis, and mechanism analysis. Finally, the effectiveness of this framework is validated through real-world case studies (Figure 8.1).

8.1 The Laws of System Complexity

Complexity is one of the most intricate concepts in modern science, and no unified definition has yet been established; perhaps, such a definition may never exist. The different layers of the real world—physical, biological, social,

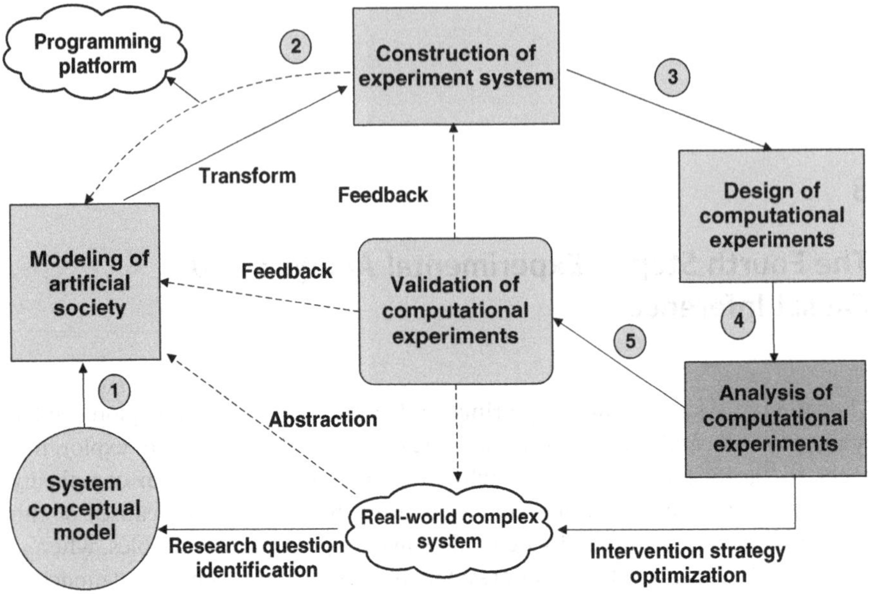

Figure 8.1 Role of experimental analysis within computational experiment method.

and conscious layers—each exhibit distinct forms of complexity that should not be conflated. One cannot substitute lower-level complexity for higher-level complexity, nor dismiss lower-level complexity on the basis of higher-level complexity. Complexity can be understood from various perspectives and at different levels. This section first provides a categorical overview of the causes of complexity, followed by a detailed discussion using structural complexity and distributional complexity as examples. The formulas presented here are mainly drawn from Claudio Cioffi-Revilla's work *Introduction to Computational Social Science: Principles and Applications*. For more details, refer to the original text [4].

8.1.1 Emergent Phenomena in Complex Systems

To explore the causal mechanisms of complex systems, it is first necessary to understand their emergent phenomena. Jochen Fromm's classification [5] offers profound insights into these emergent behaviors. As illustrated in Figure 8.2, emergent phenomena can be categorized into four types based on their causal relationships.

The first type is simple emergence, which involves unidirectional, bottom-up causality and is typically reflected in basic statistical properties. In contrast, weak emergence is more intriguing because it reflects top-down causality, where

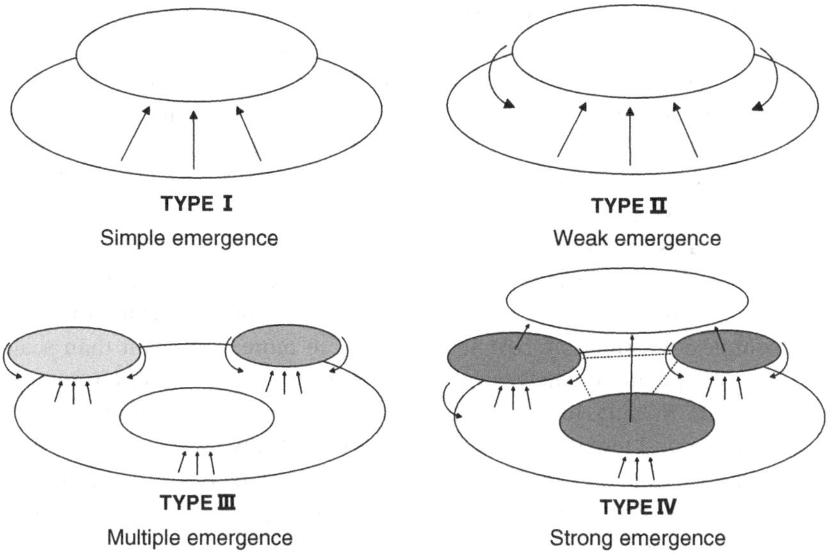

Figure 8.2 Fromm's classification of emergence.

individual behaviors are influenced by collective rules, leading to expressions of collective intelligence. Multiple emergence captures the intricacies of interactions between different causal layers, as seen in simulations like the "Game of Life," where static and dynamic patterns of various sizes and shapes coexist, demonstrating the interplay of multilevel causal relationships. The most complex form, strong emergence, encompasses three levels: micro, mesa, and macro.

Understanding these types of emergence is crucial for causal inference in complex systems. While macro-level phenomena originate from micro-level mechanisms, they can effectively substitute for micro-level descriptions in causal reasoning. This is the core idea of causal emergence. Macro-level behavioral descriptions are often more intuitive, whereas micro-level mechanisms contain the full complexity of the system. However, how micro-level causal relationships relate to those at the macro-level remains an open question, requiring further exploration and also addressing the following challenges:

Challenge 1: Exploring the Interrelationship Between Multiple Influencing Factors and Complexity

Complex systems involve a series of complex factors and their interwoven relationships, including disturbances from both external environmental factors and internal fluctuations or human-induced errors. Such noise obscures experimental data, making accurate causal inference difficult. Noise hides not only true causal relationships but also risks, leading to incorrect conclusions.

Complexity arising from system scale: The number of system components serves as an indicator of the system's scale. The formation of complexity requires a sufficiently large system size. A vast scale poses difficulties for description and processing, rendering traditional methods ineffective. Simple systems do not exhibit complexity. A sufficiently large scale is a necessary condition for the generation of complexity, but it is not a sufficient condition. Even a giant system is not necessarily a complex system.

Complexity arising from system structure: The diversity and heterogeneity of system components lead to a variety of interrelationships and disparities, which are fundamental drivers for the emergence of complexity. In terms of complexity formation, structural effects are more significant than scale effects. The greater the heterogeneity of components, the harder it is to integrate them. Notably, hierarchical structures are one of the main sources of complexity, and researchers in the field of complexity almost unanimously emphasize this point. Nonhierarchical structures, which are relatively simple, have only two levels: elements and the whole. Hierarchical structures involve intermediate layers between elements and the system as a whole. Without navigating and integrating different levels, the final system cannot be fully constructed. The more layers there are, the more likely complexity will emerge.

Complexity arising from openness (environment): Closed systems do not exhibit complexity; complexity must arise in open systems. When external environmental factors no longer act as disturbances and the system's inherent structure instead becomes dominant, closed systems face two issues: They cannot resist external shocks, and black-box methods fail to capture their dynamics. Consequently, such systems inevitably show certain types of complexity. Openness is a major contributor for the emergence of complexity, and the interaction between complex systems and their environments is a critical manifestation of systemic complexity.

Challenge 2: Agent Learning and Adaptability Increases the Complexity of Causal Identification

Complexity arising from proactivity and dynamism: In simple systems, actors are passive, and causes and effects are clearly distinguishable. In contrast, in complex systems, different components interact and co-evolve with their environments, forming networks of causal interdependencies that can lead to complexity, especially when agents possess a certain level of autonomy and continuously adapt to their environment. In this process, the system as a whole may exhibit emergent behavioral complexity. A core tenet of research at SFI research is that adaptivity gives rise to complexity. All adaptive systems (i.e., complex adaptive systems, CAS) emerge through continuous adaptation to dynamic environments.

Complexity arising from individual intelligence: Even if a system composed of nonintelligent individuals generates complexity through self-organization, it is usually relatively rudimentary and low-level. Intelligent agent systems (such as those studied by the SFI's CAS research) can sense their surrounding environments, predict future system states, learn from experience, and develop behavioral rules, thereby achieving adaptive transformations. Thus, they are inherently complex. The higher the intelligence of individual agents, the greater the system's complexity. Agent intelligence is one of the major sources of complexity. For example, land combat systems typically involve a limited number of agents but are still complex due to the high intelligence of each agent, resulting in unpredictable and emergent system behaviors. These systems are, therefore, classified as open complex systems.

Complexity arising from human rationality: In systems where humans are essential components, system behavior is influenced by human rationality. In particular, in competitive systems such as games, the rationality of participants (e.g., strategic thinking, decision-making logic, risk assessment) is a significant source of complexity. If we assume complete rationality (i.e., unlimited reasoning), the root of complexity may be eliminated, and game players would always adopt the "max-min strategy." Such systems are simplified and can be analyzed using standard game theory. However, bounded rationality (i.e., limited reasoning) leads to the emergence of true complexity.

Challenge 3: Emergence from Micro to Macro Levels Masks the Underlying Micro Mechanisms

Complexity arising from nonlinearity: Philosophers have long pointed out that the ultimate cause for the development and change of things is interactions, but interactions can be either linear or nonlinear. Linearity implies additivity, uniformity, invariance. It fails to meet the conditions for generating complexity. Linear systems are usually simple; linear interactions produce predictable outcomes and cannot give rise to complexity. Nonlinearity, by contrast, implies rich diversity, heterogeneity, variability, asymmetry, irregularity, singularity, and creativity. Nonlinear interactions between elements or subsystems are the internal mechanisms behind complexity. However, linear systems themselves can also contain significant heterogeneity. Weak nonlinearity may not necessarily lead to complexity, and it is often treated as a disturbance term. In such cases, linear models supplemented with perturbation methods may still describe the system effectively. Only strong nonlinearity, especially intrinsic nonlinearity, can truly give rise to complexity.

Complexity arising from uncertainty: Randomness is a key form of uncertainty. Simple systems exhibit stable random processes, which can be modeled statistically. Physical systems with well-behaved random processes often conform to the law of large numbers and fall within the scope of statistical analysis. However, in living systems and social systems—where components are intelligent and interact in complex ways—relying solely on statistical laws cannot capture the system's fundamental characteristics. Current statistical approaches are inadequate to handle such randomness. For example, it is impossible to deduce the overall behavior of a biological system solely from individual genetic markers, nor can the overall characteristics of a social system be statistically inferred from features of individual members. Another major source of uncertainty is fuzziness. Fuzziness is both a cause of complexity and a result of its expression. Fuzzy set theory was introduced specifically to deal with such complexity. It holds that when system complexity surpasses a certain threshold, precision in description and the value of interpretation are often incompatible; the two cannot coexist.

8.1.2 System Complexity Theory: Explanation

A fundamental function of theory is to explain phenomena. Therefore, a theory must account for observed events by drawing on initial conditions or driving factors, and its arguments must conform to the model of scientific explanation. A defining feature of scientific theories is that they must present a narrative linking antecedents (causes) with consequences (effects)—a causal story. System complexity theory explains the emergence of complexity in systems through causal relationships. Its distinguishing feature lies in its ability to interpret observed facts or empirical patterns using the available data (Table 8.1).

Given that system complexity emerges as a result of individual-level games (as opposed to most natural state outcomes), the modeling and explanation of the emergence of system complexity depend on the outcomes of these individual strategic interactions. Based on whether causal events occur within the sample space, the emergence of system complexity can be interpreted through the causal logic of event occurrence. That is, causal events must occur in a nonrandom manner in order for system complexity to manifest. For a collective behavior to emerge, certain key combinations of causal events must occur in a specific configuration; otherwise, the collective behavior will not emerge. For instance, the formation of an international alliance requires that events related to national strategy, leadership capacity, interest alignment, and coalition recruitment occur in a particular configuration, or in an equivalently effective alternative configuration.

Formally, the event function can precisely represent the causal logic and explain how a composite event arises. However, this raises several questions: What types of event functions exist? How do different event functions explain the emergence of composite events? How can the probability of a composite event

Table 8.1 Definitions of Related Concepts in System Complexity.

Concept Name	Concept Explanation	Remarks
System Complexity Emergence	Suppose system complexity is represented as a composite event C at the macro level. It originates from the combination of a set of basic events (sample points in the sample space Ω) and the coupling process among these basic events.	The composite event C consists of two components: 1) a series of basic observational events (defined by coupling outcomes and relevant natural state structures) and 2) a rule of operations that links these events together.
Event Function	When the emergence of a complex event Y is related to the occurrence or sequence of other events X, the mapping $\varphi:(X)\to Y$ is referred to as the event function of Y.	The event function φ() defines causal explanation and essentially corresponds to a functional mapping for causal inference. Thus, the event Y is expressed as $Y=\varphi(X)$.
Emergence Event Function of System Complexity	When the emergence of system complexity, represented by a composite event C, is related to the occurrence or sequence of other events X, the mapping $\varphi:(X)\to C$ is referred to as the emergence event function of C.	Therefore, the event C is expressed as $C=\varphi\{X\}$.

be determined from an event function? To address these and similar questions, we must examine the logic of system complexity at the micro level and identify two causal explanatory models based on sequential logic and conditional logic. According to the conceptual framework of Simon's theory [6]—including environmental complexity, goal-seeking behavior, bounded rationality, adaptiveness, and near-decomposability—the sequential tree of forward causal logic enables the construction of models of adaptation and system complexity. The model shown in Figure 8.3 provides a first-order representation of Simon's theory. The key insight is that various outcomes within the sample space Ω can be generated through combinations. Specifically, the emergence of system complexity C requires at least four necessary sequential conditions, which collectively have a notably low probability; without them, the successful emergence of system complexity C is impossible. In contrast, alternative outcomes (failures E, E^*, and E^{**}) occur with relatively higher probabilities, making them more likely to manifest overall.

At certain initial time points r_0, event E occurs within a specific environment. At subsequent time points r_1, in response to environmental challenges, individuals can decide whether to make adaptive changes based on bounded rationality,

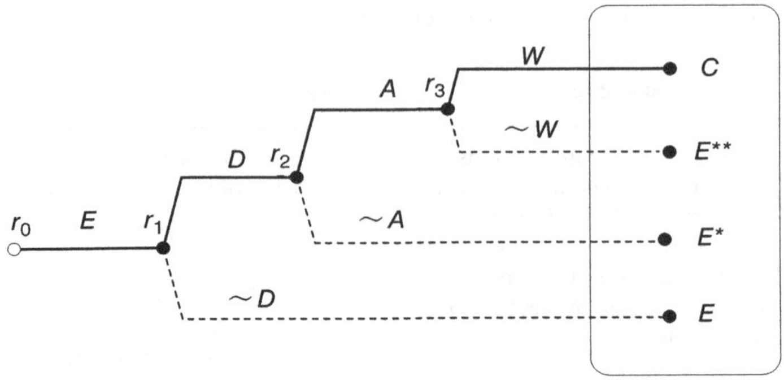

Figure 8.3 Forward sequential causal logic tree of Simon's adaptation theory and the emergence of system complexity.

i.e., event D. If they choose not to change (i.e., event $\sim D$), they will continue to endure the same environmental impacts as before, regardless of the outcome (result E).

If they decide to adapt, they then face the choice of whether to actually implement the decision and carry out adaptive measures—this is event A. If they fail to execute this action (i.e., event $\sim A$), they must still endure the environmental impact; however, time has passed, leading to outcome E^*. It can be argued that $E \approx E^*$.

If individuals have taken action, at some point in time, behavioral feedback may or may not be effective. If the feedback is effective (event W), the result is success with increased complexity, since a more complex system must now be maintained (result C). If the feedback is ineffective (event $\sim W$), the result is failure with continued exposure to environmental impact (result E^{**}). It can be argued that $S(E^{**}) > S(E)$, where $S(X)$ denotes the impact or inefficacy associated with event X.

8.1.3 Modeling Explanation Based on Sequential Logic

In the sequential logic model [7], the emergence of system complexity as a composite event can be explained by the sequence of events or the preceding event path. In this framework, the emergence of system complexity C is interpreted as the outcome of a branching process P involving multiple agent games and decisions, one of the possible outcomes within the sample space Ω. Sequential logic typically emphasizes process-oriented causal reasoning. Among many possible outcomes, the composite event (system complexity C) that occurs according to sequential logic is generally interpreted as a possible result, not a necessary one.

Sequential logic gives rise to extensive-form games, in which causal processes are illustrated using sequential event trees. The initial event I marks the beginning of the sequential process $P_N(I \to C)$, and after the N-th branching node, a certain event C may occur. The probability of C is very low unless a specific series of conditions are met. The branching nodes between the initial event I and the outcome are determined by the results of games—shaped by individual choices, natural states, and game-theoretic outcomes. Therefore, according to sequential logic, system complexity can emerge from the possible event paths that evolve from the initial state (Table 8.2).

In many cases, intuition tends to place greater emphasis on the cardinality of causal relationships than on their probabilities. In fact, researchers studying the prior causal events leading to C (system complexity) are often more concerned with individual occurrences rather than their total frequency, and altering the probability of these prior events is often more impactful than changing their total frequency. Without proper analytical reasoning, such conclusions are not intuitively obvious and must be validated through computational methods.

8.1.4 Model Interpretation Based on Conditional Logic

Under the conditional logic model, system complexity is treated as a composite event C, whose emergence (i.e., the emergence of system complexity) can be explained by identifying necessary and/or sufficient conditions. Conditional logic emphasizes logical structure over causal process. Ultimately, this model portrays the emergence of complex events (i.e., system complexity C) as logically deducible or probabilistically inferable, rather than as unpredictable or accidental outcomes (Table 8.3).

The two preceding types of causal relationships, i.e., the integration of necessary conditions and the separation of sufficient conditions, represent pure causal modes, in which system complexity is modeled as the outcome of either necessary or sufficient conditions. However, between these two extremes lie many intermediate forms, where partially necessary or partially sufficient conditions lead to the emergence of system complexity.

Such cases arise when multiple conditions occur within a vast range of possibilities. For example, collective behavior may not emerge solely due to the entire composition of individuals within a system (integration), nor from the isolated actions of single individuals (separation), but rather due to the influence of certain core organizations within the system.

8.2 Causal Inference Framework of Computational Experiments

The nonlinear emergent characteristics of complex systems present significant challenges to causal inference. A core issue in computational experimental approaches is understanding in what sense and under what conditions ABMs can

Table 8.2 Sequential Logic Concepts of System Complexity.

Concept Name	Concept Explanation	Remarks
Sequential Causal Logic of System Complexity	System complexity C may emerge at any future time as a possible outcome within the sample space Ω of a branching process that originates from the beginning. Event nodes along this process lead to the eventual emergence of system complexity C. $C = X_{t-1} \wedge X_{t-2} \wedge \ldots \wedge X_{t-n+1} \wedge X_{t-n}$. The time index t denotes the moments preceding the occurrence of C. In general, all events prior to C collectively determine the possibility of its occurrence—that is, the "realizability" of C.	Each event exists independently of the preceding events and possesses a sequential probability based on conditional probability. $Pr(C) = P_{-n}.P_{-n+1}.P_{-n+2} \cdots P_{-1} = \prod_{t=0}^{n-1} P_i$ For the first (initial) event $P_{-n} = Pt(X_{-n})$; The second event $P_{-n+1} = Pt(X_{-n+1} \mid X_{-n})$; The final event $P_{-1} = Pt(X_{-1} \mid C)$
Low Sequential Probability of System Complexity	When the preceding events of outcome C (system complexity) have not yet occurred, the finite probability of C is always smaller than any individual probability of the preceding events and smaller than the minimum probability among them.	$Pr(C) < min(P_{-n}, P_{-n+1}, \ldots, P_{-1}) < P$
Dependency of Sequential Probability on the Occurrence Probability of Prior Events	From the perspective of sequential logic, the sequential probability of system complexity must account for the different impacts caused by variations in the occurrence probability of preceding events (P_i).	$\frac{\partial Pr(C)}{\partial P} = \wedge P^{\wedge-1}$ It is always positive, so $Pr(C)$ exhibits concavity in the direction of P.
Dependency of Sequential Probability on Process Length Variation	From the perspective of sequential logic, the sequential probability of system complexity must account for the different impacts caused by variations in the length ($\wedge$) of the branching process.	$\frac{\Delta Pr(C)}{\Delta \wedge} = P^{\wedge-1} - P^{\wedge}$ It is always negative, so $Pr(C)$ exhibits convexity in that direction.

Table 8.3 The Reverse Logical Concepts of System Complexity.

Concept Name	Concept Explanation	Remarks
Conditional Causal Logic of System Complexity	System complexity C arises from the simultaneous occurrence of necessary conditions.	$C_x = X_1 \wedge X_2 \wedge \ldots \wedge X_N$
	System complexity C occurs upon the occurrence of any one sufficient condition.	$C_z = Z_1 \vee Z_2 \vee \ldots \wedge Z_M$
Principle of Integration in System Complexity	The probability of system complexity C depends on the joint probability of the necessary events. The sequential logic mode is a special form of conditional logic.	$Pr(C_x) = \prod_{i=1}^{n} Pr(X_i) = P_1 * P_2 \ldots P_n = P^\theta$ θ denotes the number of necessary events leading to the occurrence of C $(2 < \theta < n)$, and P represents the probability of these events occurring.
Parallel Complexity	$Pr(C_Z) = 1 - (1 - Q)^\Gamma$, where Γ denotes the number of causal events leading to the composite event C (with $2 < \Gamma < m$), and Q represents the occurrence probability of these events.	It is easy to derive that the probability of system complexity not occurring is: $1 - Pr(C_Z) = (1 - Q)^\Gamma$
Hyper probability Principle	When multiple causal events lead to the emergence of system complexity C, the occurrence probability of C is greater than the occurrence probability Q of any individual causal event; it is also greater than the maximum occurrence probability among those causal events.	$Pr(C) > Max\{q_1, q_2, \ldots, q_m\} > Q$ It emphasizes that system complexity is characterized by the separation and integration of causal structures (where q_i denotes the occurrence probability of each individual causal event).
Occurrence Probability of Dependency-Redundant Events	The probability of the occurrence of system complexity event C varies with the change in the occurrence probability Q of the corresponding causal events.	$\frac{\partial Pr(C)}{\partial Q} = \Gamma(1 - Q)^{\Gamma - 1}$ Since this value is always positive, $Pr(C)$ exhibits inward concavity in the Q direction.
Occurrence Count of Dependency-Redundant Events	The occurrence probability of system complexity event C varies with the number of corresponding causal events Γ.	$\frac{\Delta Pr(C)}{\Delta \Gamma} = Q(1 - Q)^\Gamma$ Since this value is always positive, $Pr(C)$ exhibits inward concavity in the Γ direction.

effectively support causal reasoning. ABMs are capable of simulating sequences of events generated by low-level or small-scale interactions, and of gradually generating the outcomes associated with those sequences. In this way, they provide a means to explicitly model the causal mechanisms that may give rise to observed dependencies over time. However, despite their potential, ABMs often fall short in terms of generating empirical evidence or compelling arguments that convince audiences that the modeled mechanisms truly reflect those found in the real world. This section, therefore, reviews the intersection of ABMs and causal inference, aiming to clarify the nature of the relationship between ABMs and causal inference.

8.2.1 Horizontal Causation and Vertical Causation

In the study of causality within complex systems, Hoel et al. 's framework of causal emergence [8] highlights that micro-level dynamics are often noisy, leading to weaker causal relationships at that scale, whereas macro-level causal relationships are generally stronger and more discernible. The micro level refers to individual agents interacting according to specific rules, while the macro level encompasses the collective behaviors or emergent properties arising from those interactions. These emergent macro-level properties cannot be directly inferred from any single micro-level component, as they are the result of intricate interaction processes. This suggests that a system's overall behavior can be more complex and less predictable than the mere sum of its parts. While many current studies [8–11] focus on macro-level analysis, examining how inputs affect outputs, they often fail to illuminate the underlying causal mechanisms, which are crucial to understanding emergent phenomena.

Casini and Manzo, drawing on Hall's classification framework, divide causal theories into two categories: horizontal causation (causal inference) and vertical causation (causal explanation) [12]. From the horizontal perspective, causal mechanisms are understood as networks of variables in which a cause variable influences an outcome variable through a chain of intervening variables. In contrast, the vertical perspective conceptualizes causal mechanisms as "complex system" [13], where the behavior (or property change) of one individual can trigger behaviors or property changes in others, and interactions among micro-level individuals give rise to system-level phenomena or behaviors.

As shown in Figure 8.4, horizontal causation is primarily applied at the macro level and aligns with the "dependency view" of causation. It examines causality from a correlational standpoint, aiming to determine whether a "cause–effect" relationship exists between two (or two types of) events, phenomena, or variables and to evaluate the causal effect of the cause on the outcome. Vertical causation, on the other hand, belongs to the "generative view" of causation [14], which focuses

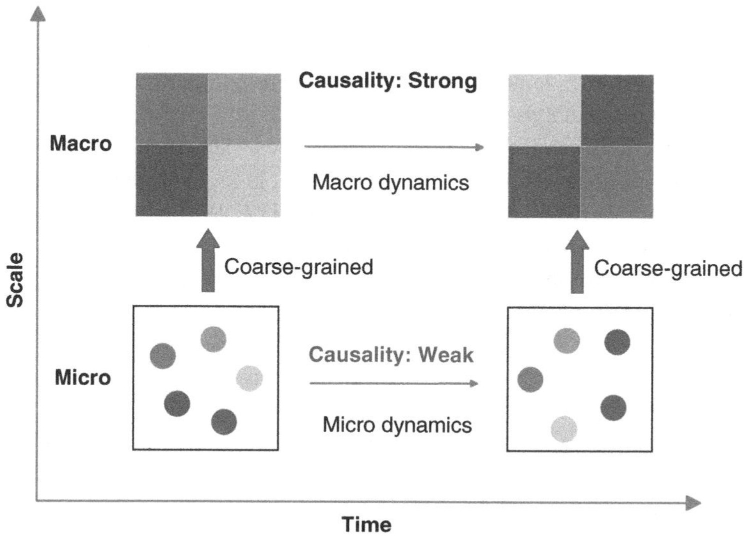

Figure 8.4 Abstract framework of causal emergence.

on analyzing how the cause leads to the effect, through what pathways, under what conditions, and via what processes. It emphasizes uncovering the complex pathways and processes that link causes to outcomes. In the context of complex systems, it is often necessary to reconstruct the process behind emergent phenomena in order to truly grasp the underlying causal mechanisms.

Different perspectives on causality often correspond to different ways of "opening the black box." Horizontal causality holds that opening the black box means revealing the intermediate variables between treatments and outcomes. From this perspective, people should rely on quantitative tools to prioritize the identification of nonspurious relationships, establish counterfactual claims, and estimate unbiased quantitative parameters of these relationships when possible—all to infer from samples or test populations to unobserved target populations. Structural equation models (SEMs) and causal graphs are typical representatives of horizontal causality [15]. From this viewpoint, the quality of data and how to describe such data creatively are of great importance.

In contrast, vertical causality argues that opening the black box involves decomposing the system into its individual components and generating the macroscopic phenomena under study by reconstructing the dynamic interactions between these components. Based on this perspective, priority should be given to methods that provide credible narratives to explain observed patterns. This view maintains that dependencies do not constitute causal mechanisms themselves, but rather are manifestations of such mechanisms. In response to this viewpoint, simulation methods become powerful tools for investigating the complex dynamic details between events. In particular, ABM is regarded as a key tool for establishing causal mechanisms, as it can demonstrate that the dependencies under study can be derived from hypothetical dynamic narratives [16].

Lucas's critique of causal reasoning in macroeconomics—specifically regarding the claim that inflation causes employment—provides a classic example for understanding the differences between these two perspectives [17]. Within the "horizontal causation" camp, data-driven approaches have become the trend, establishing the central role of interventional methods. This has led to the sophisticated application of statistical techniques, the proliferation of time-series econometric models, and the development of vector autoregression (VAR) methods [18].

In contrast, the "vertical causation" camp has embraced theory-driven approaches as its mainstream, requiring macroeconomic models to be integrated with "microfoundations." Initially, this led to the widespread adoption of rational choice models, which calculate economic aggregates based on individual preferences and expectations [19, 20]. Subsequently, ABM was introduced to address issues related to agent heterogeneity and the aggregation of social networks [21–23]. Sociology—analytical sociology in particular—has followed a similar path: shifting from regression-based statistical methods for survey data to ABM, in order to tackle micro-to-macro problems when interdependent structures exist [24].

Scholars who endorse the perspective of horizontal causation tend to be skeptical of the uniqueness of agents as a modeling tool for social mechanisms. For instance, Morgan argues, "The originality based on mechanisms has been overstated.... The claim that mechanisms are connections of nonlinear functions does not serve as a valid argument for rejecting the use of variables, since the basic elements of input-output functions can be redefined as variables" [25]. This statement deserves special attention: It can be used to argue that, given that the theoretical formulation of mechanisms requires variables and functions, a set of structured intervention/mediator variables (i.e., mechanisms in the horizontal sense) can be regarded as "mechanism sketches," and multivariate statistics can serve as a direct tool for testing mechanism-based explanations.

From the perspective of vertical causation, however, this implication is incorrect, because the role of variables and functions in statistical models differs from their role in ABM. In the former case, variables and functions are used to detect patterns of average effects—patterns that may reflect the overall statistical characteristics of the hypothesized underlying mechanisms (composed of entities, attributes, and interactions). In contrast, in the latter case, variables and functions are employed to precisely represent the attributes, activities, and interactions of relevant micro-level entities. Furthermore, functions are used to initiate (i.e., simulate) the dynamics associated with the hypothesized mechanisms, with the goal of assessing under what conditions such dynamics give rise to the relevant macro-level phenomena.

Both horizontal causation and vertical causation possess irreplaceable advantages and inevitable shortcomings. Horizontal causation enables counterfactual verification through data analysis, yet it fails to illustrate the process by which causes lead to effects. In contrast, vertical causation can demonstrate the process of causes resulting in effects, but it struggles to obtain verification from real-world data. Horizontal causation can provide directional guidance for the experimental design of vertical causation. Such experimental design not only allows generative Ness to be reflected in correlational causal relationships but also ensures the stability and verifiability of causal mechanisms. Meanwhile, vertical causation can offer mechanistic explanations for horizontal causation. Specifically, conclusions drawn from horizontal causal analysis can be deduced through experiments, capturing temporal dependencies and nonlinearities within causal structures—thereby providing a potential explanation for emergent phenomena at the system level.

8.2.2 Forward Causation and Reverse Causation

The concepts of "forward causation" and "reverse causation" proposed by Gelman [26] distinguish between two fundamental types of causal inquiry. When one pursues a forward causal question, the aim is to understand and quantify "what will happen if we do X"; in contrast, interest in reverse causation involves asking "what caused Y." In the former case, researchers focus on a particular phenomenon (e.g., education) and seek to identify the consequences of its presence, absence, or variation (e.g., its impact on fertility). In the latter case, attention is drawn to an observed outcome, and the goal is to retrospectively trace the various phenomena that may have led to this outcome.

The potential outcome framework is widely used by statisticians and economists to address questions of forward causation or causal effects, with experiments being regarded as the canonical method for tackling this type of inference [27]. Once the effects of a given cause are established, a further question arises: Why did the intervention produce the observed outcome? This leads to what is known as causal explanation [28]. To answer this, one must return to reverse causation, designing a set of possible causes that could have produced a specific effect. This form of causal explanation attempts to open the black box of first-order causal dependencies through an interventionist approach. When the objective is to transform intervention effects into practical measures, it becomes essential to understand why an intervention works, whether the intervention's effectiveness is context-dependent, and how it may vary over time. Gelman and Imbens argue that asking "why" questions can provide explanatory hypotheses for understanding causal effects and help identify flaws in the assumed causal models [29].

Goldthorpe further clarified that causal relationships can be understood in at least three distinct ways: robust dependence, manipulability of outcomes, and generative processes [30]. In the first view, a causal claim depends on whether X continues to affect Y after introducing a set of other variables (denoted as Z) into the analysis. This perspective sees causality as contingent on controlling for confounding variables. Methods aligned with this view include time-series analysis, early causal path analysis, SEM, and, more broadly, regression-based multivariate analytical techniques used in survey data analysis. In the second view, causality is interpreted as follows: If a causal factor X is manipulated, then under appropriate controls of confounding variables, a systematic effect on the response variable Y will be observed. This manipulative view of causality is aligned with the methodology of randomized experiments. Lastly, when causation is understood as a generative process, the focus shifts to explaining the nature and validity of the relational processes on which causality depends. In this view, simulation methods are considered promising tools for explaining underlying mechanisms.

The first scenario corresponds to a forward causal research paradigm, where researchers use structured or unstructured observational data—obtained from surveys, censuses, or online crawling data—to estimate counterfactual associations between independent variable X and dependent variable Y. Causality is defined as a robust, counterfactual association between observed variables. Researchers typically use statistical models to test and estimate these relationships. Common tools include propensity score methods (matching, stratification, weighting), instrumental variable techniques, path analysis, SEM, and directed acyclic graphs (DAGs) linking causes, mediators, and outcomes. In recent years, machine learning methods have also gained traction, including causal trees, causal forests, Bayesian additive regression trees, genetic matching, double robust methods, and generalized boosted models.

The second scenario also falls under the forward causal paradigm but integrates both longitudinal and cross-sectional perspectives. In longitudinal designs, a treatment group undergoes an intervention while a control group does not, allowing researchers to observe how Y changes in response to a change in X, relative to when X is unchanged. This thus helps infer causality between X and Y. Since complex systems often involve phenomena at the system-level rather than individual-level, such interventions must be repeated across multiple units to test generalizability. From the cross-sectional perspective, researchers collect pre- and posttreatment data for both groups and apply statistical models to test causal hypotheses and quantify effects—for instance, using difference-in-differences (Did) or regression discontinuity designs to evaluate policy effects. However, due to practical constraints, field experiments in social science contexts are relatively rare. With the rise of the Internet, online experiments have emerged as a viable alternative.

The third scenario belongs to the reverse causal research paradigm, characterized by a longitudinal, generative perspective, and often operationalized through

ABM. In this framework, Y denotes the social phenomenon to be explained, X refers to a set of agents with diverse heterogeneous attributes, and rules specify agent behaviors and interaction mechanisms. Simulations illustrate how micro-level interactions give rise to specific macro-level phenomena. The actions and interactions of agents are the causes of emergent macro-patterns. Tracing how agent-level behaviors and relationships "step-by-step" lead to these patterns provides a mechanistic explanation of the observed outcomes.

However, since ABM parameters are entirely defined by the researcher, questions often arise about external validity, i.e., the extent to which the model reflects real-world mechanisms. Moreover, simulation research faces a tension between model simplicity and realism; that is, overly simple models risk being dismissed as thought experiments, while overly complex models may better mirror reality but impede abstraction and causal generalization.

8.2.3 Analytical Framework of Computational Experiments

Understanding causal mechanisms in complex systems requires not only identifying causal effects themselves but also explaining the underlying causal processes. Computational experiments serve as a bridge connecting these two types of causal analysis. By simulating the operation of an "artificial society," computational experiments can not only predict causal effects between system variables (i.e., forward causation) but also explore potential system causal mechanisms (i.e., reverse causation) by tracing simulated trajectories and counterfactual paths. This dual capability makes computational experimentation uniquely valuable in causal research: on the one hand, it provides algorithmic estimations of causal effects; on the other hand, it offers operational modeling tools for causal explanation.

This analytical power relies on the construction and manipulation of a World Model [31]—an abstract representation of the structural and evolutionary rules governing real-world systems. Such models fall into two categories: the first is an external, objective world model (e.g., the Sora model), which learns from data, video, or other information to simulate environmental state transitions with high precision; the second is an internal, agent-centered belief model, which encodes how agents understand and expect the world to behave. By embedding both types of models, computational experiments can simulate not only the dynamic evolution of the external environment but also agents' decision-making processes under different cognitive assumptions.

Within this computational experimental framework, forward causal analysis is typically implemented through intervention simulations—by altering the value of a system variable and observing changes in system outputs, one can uncover robust causal networks that describe causal interdependencies within the system. In contrast, reverse causal analysis relies on counterfactual trajectory generation,

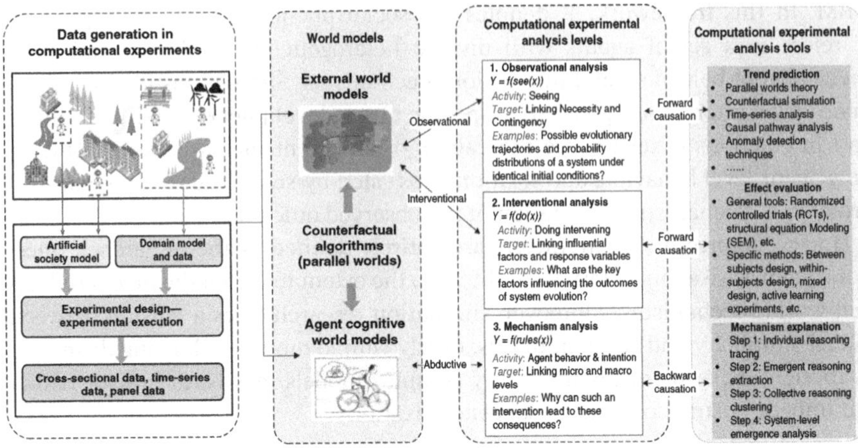

Figure 8.5 Causal inference framework of computational experiments. The left side of the diagram illustrates the artificial society model operating in parallel with the real world, along with the ongoing data generation process. The middle section presents a three-layer analytical framework for computational experiments. The right side lists the tools and methods corresponding to each analytical layer.

constructing alternative pathways that are similar to the actual pathway but differ in key variable values, thereby allowing for the inference of latent causal mechanisms. These two forms of causal analysis complement each other in simulation processes, thus jointly revealing the operating logic of world models and the dynamics of behavioral evolution.

Based on this theoretical foundation, Figure 8.5 presents a computational experiment analysis framework that explains how such simulations can be used to analyze world models within artificial societies. The framework consists of three layers of experimental analysis:

1) **The association layer (what is?):** Given a variable, what is the probability of an outcome occurring?
2) **The intervention layer (what if?):** What would happen if we intervened or changed a condition?
3) **The mechanism layer (how is?):** If a cause leads to an effect, what is the process through which it happens?

The following sections explore the specific techniques and methods used at each layer of this framework.

1) **Observational analysis (linking necessity and contingency):** The first layer corresponds to a process based on an objective world model and aligns with the forward causation research paradigm. The core idea of observational

analysis acknowledges the high uncertainty surrounding future outcomes and realization paths of events—manifested in the fact that the evolution of social systems involves multiple possibilities and is influenced by initial conditions and relevant external factors. Observational analysis focuses on modeling the possibility space of system evolution, identifying all potential states the social system might experience, along with their probability distributions. This process emphasizes the uncertainty in event evolution and uses scenario-based simulation data to reveal the randomness and unpredictability in system dynamics. Specifically, when facing multiple future possibilities, observational analysis seeks to quantify each trajectory and assess each trajectory's likelihood within a probabilistic framework. This approach not only effectively highlights the role of contingency in system evolution but also provides theoretical support for preparing responses for various possible futures.

2) **Interventional analysis (linking cause and effect):** The second layer also builds upon an objective world model and falls under the forward causation paradigm. Interventional analysis employs randomized controlled experiments to compare system trajectories across different hypothetical scenarios, observing how changes in intervention variables induce changes in outcome variables. This enables the identification of robust structural causal models (SCMs). Key components of interventional analysis include:

   Treatment/control groups: One scenario adopts an intervention A, while the control scenario does not; the differences in outcomes are then examined to determine whether intervention A caused changes in outcome B.

   Sample consistency: Experimental and control scenarios are configured identically, ensuring that differences in outcomes are due to the intervention rather than other confounding factors.

   Random sampling/assignment: Initial parameters of both experimental and control scenarios are randomly sampled within specified ranges. By running parallel experiments for control and treatment groups, the average treatment effect (ATE) can be used to quantify the intervention's impact on the system.

3) **Mechanism analysis (linking micro and macro):** The third layer represents the reverse causation research paradigm. This level involves not only analyzing the objective world model but also focusing on the agents' subjective world models. Here, Y denotes the social phenomenon to be explained; X represents a set of agents with heterogeneous attributes; and Rule refers to the agent's world model, revealing the internal cognitive structures and reasoning pathways that guide decision-making or actions. Mechanism analysis posits that agent actions and interactions are the root causes of macro-level phenomena and that mechanistic explanations clarify how these micro-level dynamics lead to emergent patterns. The internal world model of an agent functions like

"turning the black box into a white box": by analyzing behavioral trajectories and cognitive chains, one can understand how individual micro-level behaviors give rise to macro-level emergent outcomes. In this context, large language models (LLMs) are powerful tools that can assist in uncovering the reasoning logic and behavioral intentions of agents.

8.3 Causal Inference Layers in Computational Experiments

8.3.1 First Level: Observational Analysis

In complex social systems, there exists a strong nonlinear coupling and dynamic feedback between individual behaviors and environmental states. The evolution of the system is typically the result of interactions and strategic games among multiple agents, displaying high degrees of path dependence and multi-trajectory possibilities. Any observed event trajectory represents just one of many possible paths within a broader latent space, with its evolution jointly shaped by initial conditions, background variables, and stochastic disturbances.

The core task of observational analysis is to construct the possibility structure of system state evolution—namely, to identify all possible combinations of states, event sequences, and their probability distributions that could emerge under natural system dynamics. This analysis is based on an objective world model, typically developed through learning and modeling from historical observational data. The goal is to abstract the system's state transition mechanisms, variable causal dependency structures, and evolutionary patterns, thereby providing a foundational framework for understanding natural system evolution.

In this analytical process, the operational trajectory of the artificial society can be formalized as a time series of triplets consisting of state (S_t), event (E_t), and action (A_t): $\pi = \{(S_t, A_t, E_t)\}_{t=1}^{T}$. Based on this structure, causal relationships within a trajectory can be further defined. For instance, if event C causes event E in a realized trajectory π, there may exist a counterfactual trajectory π'—sufficiently similar to π—in which the state related to event C remains the same, but the state related to event E differs from that in π. This indicates that the same initial configuration of the artificial society may give rise to a wide variety of counterfactual trajectories, representing diverse traversals through the model's state space.

These simulated trajectories can be highly complex, and various forms of causal relationships may arise depending on which events occur or are omitted in trajectory π. These include:

1) **Basic causal relationship:** Event A causes event B (A → B), meaning A occurs in π and is both necessary and sufficient for B.

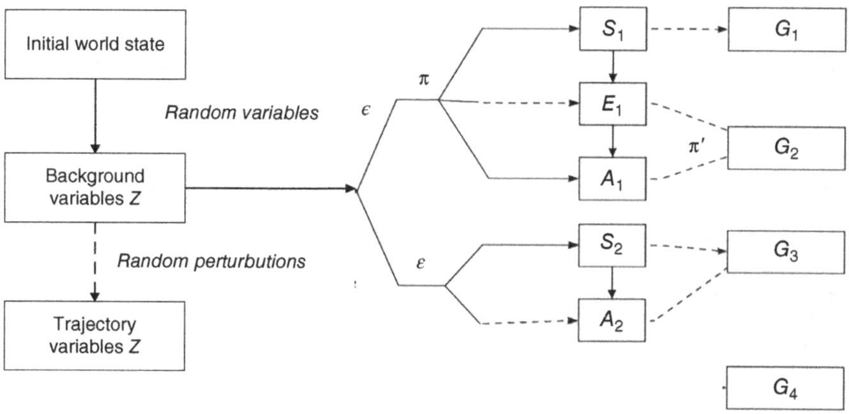

Figure 8.6 Illustration of multiple possible evolutionary trajectories in a complex system. The figure highlights the interactions among initial conditions (initial world state), background variables (Z), stochastic disturbances (ε), and nonlinear feedback. The actual system trajectory and counterfactual trajectory (π') are represented as sequences of transitions among states (S), events (E), and actions (A), ultimately leading to different system outcomes (G).

2) **Prevention:** Event A prevents event B from occurring. A's occurrence is both necessary and sufficient for the nonoccurrence of B.
3) **Transitivity:** If A causes B and B causes C, then A causes C.
4) **Absence-as-cause (counterfactual dependence on nonoccurrence):** If the nonoccurrence of A is both necessary and sufficient for the occurrence of B—and the absence of a condition is accepted as a cause—then A's absence can be considered the cause of B.
5) **Preventive transmission:** Based on prior explicit causal claims, if A causes B and B prevents C, then A prevents C (Figure 8.6).

The above trajectory structure not only facilitates the identification of causal paths among system variables but also enables the reconstruction of the system's state evolution network and underlying mechanisms based solely on natural observational data, without requiring any interventions. To further formalize the strength of dependencies among variables, we introduce conditional mutual information (CMI) to quantify the statistical coupling within variable triplets:

$$I(X;Y \mid Z) = \sum_{x,y,z} P(x,y,z) \log \frac{P(x,y \mid z)}{P(x \mid z)P(y \mid z)} \tag{8.1}$$

This metric is used to quantify the amount of shared information between variables X and Y, conditional on a third variable Z. If the CMI is significantly nonzero, it indicates that X and Y remain statistically dependent even after controlling

Table 8.4 Observational Analysis.

Input: A set of experimental datasets $\{T_1, T_2, \ldots, T_n\}$, domain knowledge factors $\mathbb{F}_{domain}$, and initialization conditions C_i for each experiment

Output: Set of key influencing factors $\mathscr{K}$, and analysis results of stochasticity vs. determinism

1. Initialize the simulation environment using domain knowledge $\mathbb{F}_{domain}$ and initial conditions C_i.
2. For each round of parallel simulation, perform the following steps:
   For each dataset T_i:
   - Collect simulation results.
   - Compare with baseline or alternative runs to identify emergent phenomena.
3. Aggregate results from all simulation rounds for comparative analysis.
4. Apply k-means clustering to identify diverse evolutionary outcomes of the system and estimate their respective probabilities.
5. Use unsupervised anomaly detection algorithms, combined with dynamic thresholds, to detect stochastic disturbances during simulation.
6. Return the final set of emergent patterns and key influencing factors $\mathscr{K}$.

for Z. In the construction of objective world models, CMI serves as a key tool for edge selection, structure learning, and latent causal graph construction, making it essential for observational analysis.

To gain a deeper understanding of deviations in system evolution, anomaly detection methods are typically employed to capture stochastic events that cause such deviations. These methods include (1) global anomaly detection, which identifies major temporal outliers across the entire dataset and (2) local anomaly detection, which targets more subtle deviations within smaller time windows.

ML techniques are often applied to uncover complex deviations from expected behavior, while rule-based approaches focus on identifying violations of predefined norms. In dynamic and high-dimensional environments, dimensionality reduction techniques such as principal component analysis (PCA) [32] and online learning algorithms are required to enable adaptive modeling.

By integrating causal discovery with anomaly detection, this observational analysis framework not only reveals the system's latent evolutionary pathways but also pinpoints critical potential deviation points in system evolution. Table 8.4 illustrates the procedural flow of conducting observational analysis using computational experiments.

8.3.2 Second Level: Interventional Analysis

While observational analysis outlines multiple possible evolutionary trajectories and their probability distributions for social systems, it is limited in its ability to identify clear causal relationships between those paths. To further uncover the underlying causal mechanisms, we introduce interventional analysis. This approach artificially manipulates system variables and constructs a comparison framework between treatment and control groups to allow for the inference of causal effects in a simulated environment.

In traditional social science research, conducting randomized controlled trials (RCTs) is often constrained by ethical concerns, high costs, and methodological reproducibility challenges. In contrast, computational experiments offer a natural counterfactual platform for causal inference. Through the construction and manipulation of virtual worlds, researchers can create two parallel worlds within a unified system model, one subjected to an intervention and the other allowed to evolve naturally, thus enabling a direct comparison of treatment effects under conditions free from confounding bias. The identification of treatment effects in computational experiments is further ensured by Pearl's back-door adjustment formula.

At its core, interventional analysis involves simulating and comparing system evolution under intervention and nonintervention scenarios. Using a unified world model, one can construct trajectories for both a treatment group (with intervention) and a control group (without intervention), thereby obtaining a counterfactual estimate of the causal effect. The expected difference under intervention is expressed in the following form:

$$Effect(x) = E[Y \mid do(X = x)] - E[Y \mid do(X = x')] \tag{8.2}$$

where $do(X = x)$ denotes an exogenous setting of the variable X, effectively "cutting" all causal paths leading into X from the system's mechanisms, while retaining only the direct or indirect causal effects of X on Y; and x' denotes the baseline value of X before intervention.

Within the computational experiment framework, to ensure identification of the estimated effect, Pearl's back-door criterion must be satisfied. Specifically, there must exist a set of variables Z that blocks all noncausal (spurious) paths between X and Y. In simulated systems, the set Z is typically selected using a combination of domain knowledge, sensitivity analysis, or structure learning algorithms.

The interventional effect is then computed using the following formula:

$$P(Y \mid do(X)) = \sum_{z} P(Y \mid X, Z = z)P(Z = z) \tag{8.3}$$

In computational experiments, the candidate set of confounding variables Z is identified using domain knowledge or sensitivity analysis. Repeated simulations are then used to obtain empirical estimates of both $P(Y|X, Z)$ and $P(Z)$, enabling effective identification of the true impact of the intervention variable on system evolution.

Interventional analysis is typically applied in computational experiments by modifying input variables and environmental parameters, observing output variations across simulation runs, and then hypothesizing causal relationships. Interventions are implemented by altering the assigned values of certain variables within the model. Common types of interventions include:

1) **Hard intervention:** Directly setting the value of a variable.
2) **Soft intervention:** Modifying the conditional distribution of a variable without fixing its value.
3) **Uncertain intervention:** The target or mechanism of the intervention is partially unknown; only the outcome is observable.
4) **No intervention:** No actual manipulation is performed; observational data are extracted from temporally naturally evolving trajectories.

For example, in epidemic modeling, interventional analysis can be used to trace transmission chains, estimate the blocking effect of intervention measures, or even locate "patient zero." In practice, analysis typically begins with a nonintervention counterfactual simulation, where only the initial infection status is set and the system evolves naturally. This is followed by a series of active intervention experiments, such as modifying transmission rates, incubation periods, or network topologies. In each experiment, one variable is exogenously fixed in the form $X = x$, and the resulting outcome distribution is obtained through simulation. The causal graph structure is then updated iteratively until it converges to a stable configuration or a predefined termination condition is met.

This process, as illustrated in Figure 8.7, constitutes a learning mechanism that combines experimental design with causal discovery. It begins with active intervention on variables to obtain causal feedback information, followed by the use of active Bayesian algorithms, such as strategies that maximize information gain, to plan the next experimental step. Through iterative cycles of this kind, the process ultimately yields a robust causal model with structured explanatory power. Table 8.5 presents the procedural flow for conducting interventional analysis in computational experiments.

8.3.3 Third Level: Mechanism Analysis

In the current field of social simulation, it is generally assumed that if the micro-level behavioral rules of agents possess a certain degree of realism, and the simulator can reproduce macro-level patterns observed in the real world, then the model is considered to have explanatory adequacy. However, this evaluation criterion has clear limitations: the fact that a certain mechanism can produce a certain phenomenon does not necessarily imply that this mechanism is the true cause of the phenomenon, especially when multiple agent behavior models can generate similar macro-level outcomes. The task of mechanism analysis is thus to further discriminate between them: Which micro-level mechanism is the genuine causal source behind the system's emergent behavior?

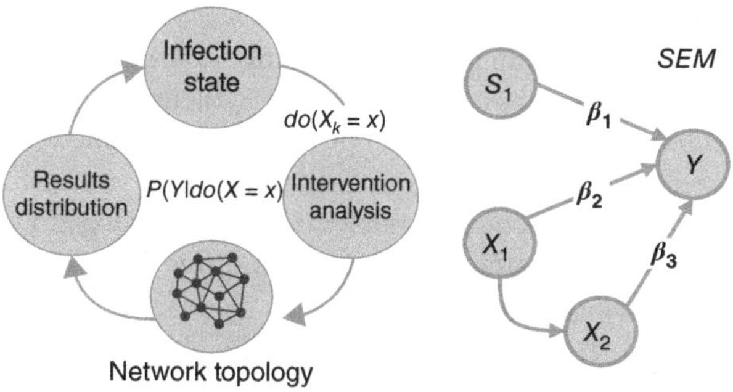

Figure 8.7 Intervention-based causal modeling framework (left) and its corresponding structural equation model (SEM) representation (right). The left panel illustrates an iterative interventional analysis loop in epidemic modeling, which includes initializing infection states, implementing intervention measures ($do(X = x)$), simulating outcome distributions ($P(Y \mid do(X = x))$), and dynamically updating the underlying network topology. The right panel presents the corresponding SEM diagram, depicting the structural relationships between exogenous variables (X) and the endogenous outcome variable (Y). Path coefficients ($\beta_1, \beta_2, \beta_3$) are used to quantify the strength of causal effects.

Table 8.5 Interventional Analysis.

Input: Set of influencing factors F

Output: System performance under intervention ($P_{treatment}$), system performance without intervention ($P_{control}$), causal effect estimate (ΔP)

1. Initialize prior beliefs over the causal graph G; initialize performance log $\Delta P \leftarrow \emptyset$.
2. Repeat until a termination condition is met:
   a) Select target intervention variable:

   $$X^* \leftarrow \text{argmax}_{X \in F} E[InfoGain(X)]$$

   b) Construct two simulation groups:
      - Treatment group T: perform intervention $do(X^* = x)$
      - Control group C: no intervention applied

   c) Simulate both groups to obtain system-level performance:
      - $P_{treatment} \leftarrow$ Simulate (T).
      - $P_{control} \leftarrow$ Simulate (C).

   d) Compute causal effect difference:

   $$\Delta P \leftarrow P_{treatment} - P_{control}$$

   e) Update the posterior distribution over the causal graph G based on the observed outcomes.
3. Return ΔP, $P_{treatment}$, $P_{control}$.

Mechanism analysis focuses on modeling the internal world model and decision-making mechanisms of individual agents, aiming to explain how system-level changes emerge from agents' interactions based on their micro-level cognitive structures (e.g., beliefs, intentions). We adopt Pearl's SCM framework to formally model such mechanisms. The SCM is represented as a triplet: $M = <U, V, F>$, where U denotes exogenous variables, V denotes endogenous variables, and F denotes the set of structural functions. Within this framework, counterfactual queries can be formally expressed as:

$$P(Y_x = y \mid X = x', Y = y') = \sum_x P(U = u \mid X = x', Y = y') \cdot 1[f_Y(x, u) = y] \quad (8.4)$$

where $X = x'$ and $Y = y'$ are observed facts, and Y_x represents the counterfactual outcome under intervention $X \leftarrow x$. The function $f_Y(x, u)$ is the structural function that describes how Y is generated from x and u, and $1[\cdot]$ is an indicator function selecting the path leading to the specified counterfactual result. This formula provides a three-stage pathway for counterfactual simulation: abduction, intervention, and prediction. It is used to trace the generative path of mechanisms and quantify potential system differences had a certain action not been taken. Computational experiments simulate the trajectory of events, enabling the estimation of the aforementioned mathematical quantities.

1) **Causal attribution stage:** Repeatedly run computational experiments with the initial observed conditions set as $X = x'$, $Y = y'$. Record the distribution of exogenous variables (such as random disturbances u) that lead to this result, approximating $P(U = u \mid X = x', Y = y')$, which corresponds to observational analysis.
2) **Intervention stage:** Through computational experiments, artificially intervene in the variable X in fact, that is, impose $do(X = x)$. And in the causal attribution process, run the system under each group of u values, and record the corresponding output $Y = f_Y(x, u)$.
3) **Prediction stage:** Construct an empirical distribution based on the intervention results and then estimate the counterfactual probability:

$$P(Y_x = y \mid X = x', Y = y') \approx \frac{1}{N}\sum_{i=1}^{N} 1[f_Y(x, u^{(i)}) = y] \quad (8.5)$$

where $u^{(i)} \sim (U \mid x', y')$ is from the attribution stage and N is the number of simulated samples. This process realizes the transformation of the counterfactual query formula under the SCM framework from abstract mathematical reasoning to a computational experimental method with the capabilities of observability, simulatability, and computability. This method reveals how the rules of the agents are manifested in the simulation trajectories and gradually accumulate into macroscopic phenomena.

In social simulation, the decision-making behavior of an agent is driven by its beliefs and intentions, and the changes of these beliefs often produce macroscopic effects through interactions [33]. Through observational and intervention analysis, combined with the belief update process of the agent, the causal relationship between micro-behaviors and macroscopic emergence can be clearly depicted. Mechanism analysis strengthens the dual modeling path of the "world model." (1) On the one hand, it relies on the "objective world model" to describe the evolution law of macroscopic phenomena. (2) On the other hand, starting from the "internal world model" of the agent, it generates micro-behavior trajectories and identifies system emergence through cluster analysis.

Based on the conceptual design shown on the left side of Figure 8.8, the right side of Figure 8.8 presents the "Mechanism Analysis Framework Based on Multi-agent Intention Emergence." This approach primarily achieves two objectives. (1) Intention-to-behavior analysis: This stage aims to establish a mapping mechanism between individual intentions and behaviors. By computing the linear correlation strength between changes in intention states and behavioral changes, the model identifies significant behavioral driving factors, thereby uncovering the pathways through which intentions influence individual decision-making behaviors. (2) Micro-to-macro analysis: This stage further tracks the collective aggregation effects of individual behaviors and identifies system-level emergent patterns driven by intentions. The specific details are as follows:

1) **Individual cognitive tracking**

$$
\begin{aligned}
I_s(i,t) &= LLM_{\phi_s}(M_{il}(t), M_{is}(t), O_i(t)) \\
I_r(i,t) &= LLM_{\phi_r}(M_{il}(t), M_{is}(t), O_i(t))
\end{aligned}
\tag{8.6}
$$

where $I_s(i, t)$ represents the bounded rational intention; $I_r(i, t)$ represents the perfectly rational intention; $M_{il}(t)$ denotes the long-term memory of A_i, which includes the integration of past experience and environmental interaction information; $M_{is}(t)$ stands for the short-term memory of A_i; $O_i(t)$ indicates the observation value of A_i, i.e., the decision-making context; LLM_{ϕ_s} represents the process by which the Observer extracts the bounded rational intention; and LLM_{ϕ_r} represents the process by which the Observer extracts the perfectly rational intention. This process is realized by invoking the thinking and reasoning ability of LLM.

2) **Emergent intention extraction**

To extract key emergent intentions, each agent is equipped with a dedicated Analyst, which analyzes the agent's intention process and extracts new emergent intentions from it. The emergence here refers to novel intentions and behaviors that arise in group interactions and have not appeared before; it reflects the new cognitions and new goals formed by agents in a dynamic environment. This process is divided into two parts: intention emergence and behavior emergence.

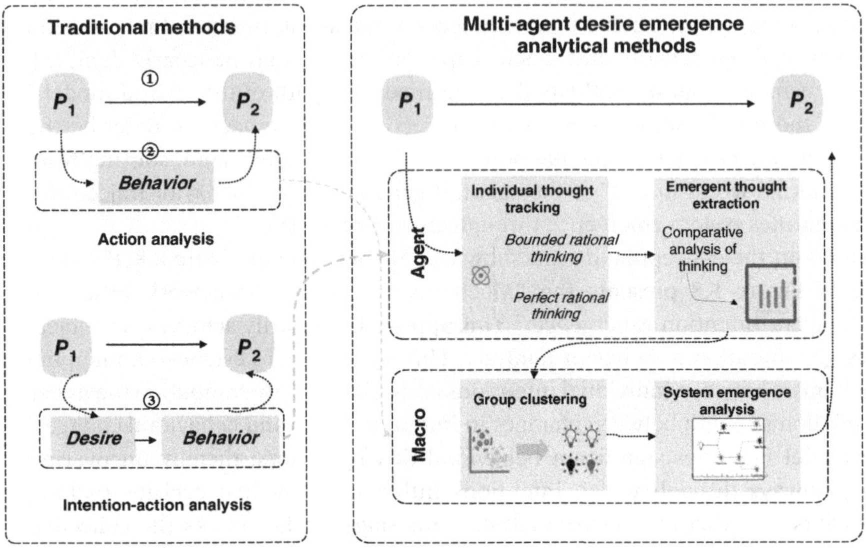

Figure 8.8 Mechanism analysis framework based on multiagent intention emergence.

In the process of intention emergence analysis, the set of intentions of all agents at time t is I_t. The Analyst can use clustering algorithms such as k-means to merge similar intentions in I_t, obtaining intention clusters C_t. By comparing C_t with the intention clusters C_{t-1} at the previous moment, the Analyst identifies new intention clusters C_{new}; then, the new intention clusters C_{new} are incorporated into the set of all emergent intentions $E_I(t)$ stored at time t, realizing the update and storage of intentions. The formula for emergent intention detection is as follows:

$$C_{new} = C_t - C_{t-1}$$
$$E_I(t) \leftarrow E_I(t-1) \cup C_{new} \tag{8.7}$$

In the process of behavior emergence analysis, the set of behaviors of all agents at time t is B_t. The Analyst identifies the new behavior set B_{new} by comparing B_t with the behavior set B_{t-1} at the previous moment; then, B_{new} is incorporated into the set of all emergent behaviors $E_B(t)$ stored at time t, realizing the update and storage of behaviors. The formula for emergent behavior detection is as follows:

$$B_{new} = B_t - B_{t-1}$$

$$E_B(t) \leftarrow E_B(t-1) \cup B_{new} \tag{8.8}$$

3) **Calculation of correlation between intentions and behaviors**
After identifying the emergent intentions and emergent behaviors of the group, it is necessary to calculate the temporal correlation between them. First, convert the intention emergence array and behavior emergence array into binary vectors. Among them, the intention vector E_{I-vec} indicates that if a new intention is generated, the corresponding position is 1, otherwise it is 0; the behavior vector E_{B-vec} indicates that if a new behavior is generated, the corresponding position is 1, otherwise it is 0. On this basis, the system uses the Pearson's correlation coefficient to measure the linear relationship between intentions and behaviors at different times, and its calculation formula is:

$$R \leftarrow Pearson(E_{I-vec}[: t-1], E_{B-vec}[: t]) \tag{8.9}$$

where $E_{I-vec}[:t-1]$ represents the binary sequence of intention emergence from the initial moment to t–1, and $E_{B-vec}[:t]$ represents the binary sequence of behavior emergence from the initial moment to t. Through analysis, it can reveal the lagged impact between group intentions and behaviors and provide a quantitative basis for the explanation of abnormal emergence mechanisms in complex systems. The detailed algorithm description is shown in Table 8.6.

8.4 Case Study: Algorithmic Behavior on Internet Platforms

With the development of mobile Internet and the sharing economy, O2O (online-to-offline) service ecosystems have emerged as a new model for service element integration. These ecosystems enable interaction and integration among various components, including online platforms (such as Maiduan, Didi, and Hello Bike), offline entities, online payment systems, service providers (such as couriers, drivers, and delivery riders), and consumers. By 2023, the scale of employment on Internet-based O2O platforms had reached an unprecedented level. Taking Maiduan as an example, the number of registered delivery riders had surpassed 10 million. Internet platforms analyze data from workers (e.g., Maiduan riders, Didi drivers) and feed the results back to these workers, making "Labor Order under Digital Control" possible. While this approach greatly improves labor efficiency, it has also given rise to a series of social issues, such as rider involution, algorithmic price discrimination ("data-enabled consumer exploitation"), and platform monopolies. This project adopts a "governance strategy + computational experiment" approach to study the O2O service ecosystem and uncover the underlying mechanisms of the platform economy. It also aims to explore how to design effective governance strategies that balance efficiency and fairness.

Table 8.6 Generation of the Intention Temporal Emergence Graph.

Input: Set of agents $A = \{A_1, A_2, \ldots, A_k\}$; time step T; where the memory of each agent A_i is $M_i = (M_{il}, M_{is})$, M_{il} denotes the long-term memory of the agent, M_{is} denotes the short-term memory of the agent; and O_i denotes the observation value of the agent.

Output: Intention emergence $\boldsymbol{E_I}$; behavior emergence $\boldsymbol{E_B}$; intention-behavior correlation R: i.e., $\boldsymbol{E_I}(t)$ vs. $\boldsymbol{E_B}(t+1)$.

---

1. Initialization: $\boldsymbol{E_I} = [\,]$, $\boldsymbol{E_B} = [\,]$, $R = [\,]$

2. For each time step $t = 1, \ldots, T$:

Current intention set $\boldsymbol{I_t} = [\,]$

Current behavior set $\boldsymbol{B_t} = [\,]$

For each agent $\boldsymbol{A_i}$ at time t:

$\boldsymbol{I_s(i,t)} = \boldsymbol{LLM_{\Phi_s}(M_{il}}(t), \boldsymbol{M_{is}}(t), \boldsymbol{O_i}(t))$ // Obtain bounded rationality intention

$\boldsymbol{I_r(i,t)} = \boldsymbol{LLM_{\Phi_r}(M_{il}}(t), \boldsymbol{M_{si}}(t), \boldsymbol{O_i}(t))$ // Obtain perfect rationality intention

$\boldsymbol{I_t(i)} = \{\boldsymbol{I_s(i,t)}, \boldsymbol{I_r(i,t)}\}$ // Get current intention of the Agent

$\boldsymbol{B_t(i)} = \boldsymbol{pi(O_i}(t), \boldsymbol{M_{il}}(t), \boldsymbol{M_{is}}(t))$ // Policy function *Pi* decides agent's behavior

$\boldsymbol{I_t} \leftarrow \boldsymbol{I_t(i)}$ // Add agent's intention to the intention set

$B_t \leftarrow \boldsymbol{B_t(i)}$ // Add agent's behavior to the behavior set

$\boldsymbol{M_{il}}(t) \leftarrow \textbf{Update}(\boldsymbol{M_{il}}(t), \boldsymbol{I_t(i)}, \boldsymbol{O_i}(t))$ // Update agent's long-term memory

$\boldsymbol{M_{is}}(t) \leftarrow \textbf{Update}(\boldsymbol{M_{is}}(t), \boldsymbol{B_t(i)}, \boldsymbol{O_i}(t))$ // Update agent's short-term memory

// Identify emergent intentions of the agent set

$\boldsymbol{C_t} = k\text{-}means(\boldsymbol{I_t})$ // Merge similar intentions to obtain intention clusters

$\boldsymbol{C_{new}} = \boldsymbol{C_t} - \boldsymbol{C_{t-1}}$ // Obtain new intentions generated by the agent group

$\boldsymbol{E_I}(t) \leftarrow \boldsymbol{E_I}(t-1) \cup \boldsymbol{C_{new}}$ // Store newly emergent intentions in the intention array

$\boldsymbol{E_I} \leftarrow \boldsymbol{E_I}(t)$ // Store intentions generated at time t

// Identify emergent behaviors of the agent set

$\boldsymbol{B_{new}} = \boldsymbol{B_t} - \boldsymbol{B_{t-1}}$ // Obtain new behaviors generated by the agent group

$\boldsymbol{E_B}(t) \leftarrow \boldsymbol{E_B}(t-1) \cup \boldsymbol{B_{new}}$ // Store newly emergent behaviors

$\boldsymbol{E_B} \leftarrow \boldsymbol{E_B}(t)$ // Store behaviors generated at time t

3. Calculate lagged correlation between intentions and behaviors

$\boldsymbol{E_{I-vec}} = [1 \text{ if len}(\boldsymbol{E_I(t)}) > 0 \text{ else } 0 \text{ for } t = 0 \ldots T-1]$

$\boldsymbol{E_{B-vec}} = [1 \text{ if len}(\boldsymbol{E_B(t)}) > 0 \text{ else } 0 \text{ for } t = 0 \ldots T-1]$

$R \leftarrow Pearson(\boldsymbol{E_{I-vec}}[:\boldsymbol{t-1}], \boldsymbol{E_{B-vec}}[:\boldsymbol{t}])$ // Calculate correlation

4. Return $\boldsymbol{E_I}$, $\boldsymbol{E_B}$, $\boldsymbol{R}$

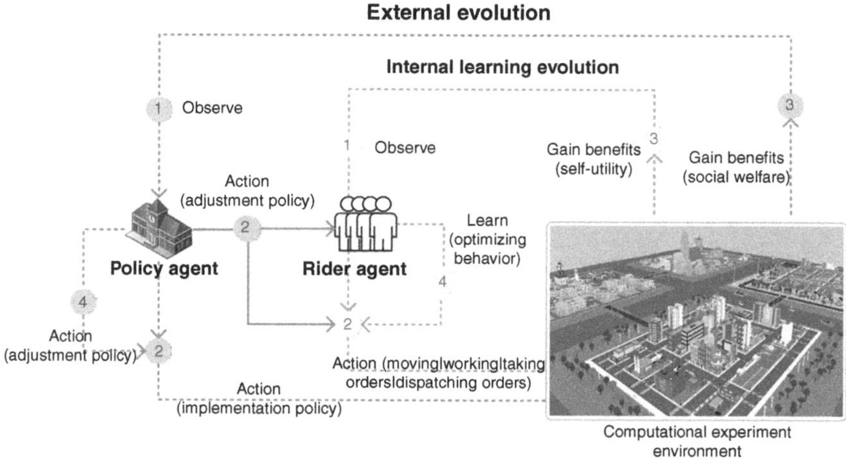

Figure 8.9 Governance strategy optimization framework based on two-level learning and evolution.

8.4.1 Algorithmic Mechanisms of O2O Platforms

The term "involution" was introduced by sociologist Clifford Geertz to describe a state of stagnation or regression in socioeconomic development. Within social ecosystems, involution refers to excessive competition among individuals or groups in a resource-constrained environment, leading to decreased efficiency of input–output ratios and a decline in overall welfare. In this study, rider involution manifests as workers extending their working hours and increasing labor intensity in order to compete for a limited number of orders. However, their income does not increase significantly; instead, this behavior leads to deteriorating health conditions and growing income disparity. This research employs an LLM to construct a realistic multiagent O2O platform simulation, aiming to analyze the causes and emergent dynamics of involution among delivery riders (Figure 8.9).

We propose an experimental system based on a two-level learning-evolution framework, consisting of an inner and an outer loop. In this system, the Policy Agent represents the platform, dynamically adjusting governance strategies to balance individual welfare and overall system stability. The Rider Agent, representing riders, adapts its behavior through learning to maximize individual utility. A bidirectional feedback loop connects the platform algorithm and the agent system: Riders' learning behaviors affect system performance, which in turn drives the platform to adjust its policies. Through repeated iteration, the system evolves toward a dynamic equilibrium. **Table A.8.1** presents the mapping between real-world delivery scenarios and the corresponding computational experiment system.

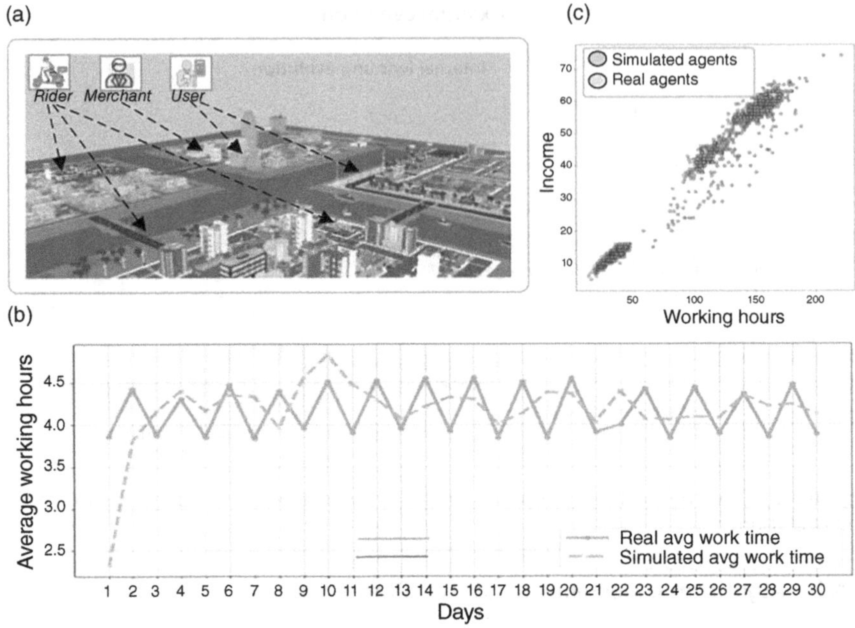

Figure 8.10 Comparison between the computational experiment system and real-world datasets. (a) O2O takeaway simulation system. (b) Comparison of average working hours. (c) Comparison between riders' working hours and income.

8.4.2 Experimental Objectives and Setup

This study aims to investigate the phenomenon of involution among rider agents within an O2O experimental platform. The experimental objectives are threefold:

1) **RQ1:** Can emergent phenomena be observed within the system?
2) **RQ2:** Can we intervene to identify the key influencing factors of the system?
3) **RQ3:** Can we uncover the underlying mechanisms behind the emergence to explain social dynamics through simulation?

All experiments were conducted using a locally deployed DeepSeek-R1-Distill-Qwen-32B-FP8-Dynamic LLM, with the temperature parameter set to 0 to ensure repeatability. The experimental environment consists of a multiagent system comprising five agent types: merchants, delivery riders, customers, government, and platform. For this study, we instantiated 100 rider agents.

As shown in Figure 8.10(a) , the physical environment is a grid-based virtual map simulating a city with 10 commercial zones. Each rider agent's reward function is designed to balance multiple objectives: minimizing working time, maximizing completed orders, and reducing labor costs. These objectives are weighted differently among riders to reflect varying decision-making priorities.

Rider agents are empowered by LLM-based intelligence, enabling autonomous learning and decision-making. Their decisions include selecting commercial zones, picking up orders, and executing random walk strategies. Throughout the 3600-step simulation—representing one month of activity—these rider agents continuously interact with the environment, leading to the dynamic evolution of the system.

To ensure the realism and validity of the experiments, the system was benchmarked using real-world data from the Zomato platform, which includes food delivery order records across multiple cities. The benchmark focused on simulating cross-city delivery scenarios, with key comparison metrics such as average effective working hours per rider per day and the relationship between working hours and order volume.

Figure 8.10(b) compares the average effective working hours per day, defined as the time spent actively delivering orders (excluding waiting and idle time), between the real-world dataset and the simulation results. The simulation yielded an average working time of 4.05 hours, which closely matches the real-world observation of 4.02 hours, with a mean absolute error (MAE) of 0.05 hours and a root mean square error (RMSE) of 0.06 hours, indicating minimal deviation. Furthermore, the Pearson's correlation coefficient of $r = 0.98$ demonstrates a strong linear correlation between simulated and real data.

These results collectively validate the accuracy and reliability of the proposed model in reproducing the distribution of real-world working hours.

Figure 8.10(c) illustrates the relationship between riders' working hours and the total number of orders accepted. In general, longer working hours correlate with more orders completed. Light Gray dots represent data from 1320 real riders, while medium gray dots denote data from the 100 simulated rider agents. The strong similarity between simulated and real data confirms the accuracy and robustness of the experimental setup.

8.4.3 Experimental Analysis and Conclusions

The experimental analysis consists of three levels. (1) Observational analysis: This phase identifies potential involution phenomena based on an involution index and analyzes the behavioral trajectories of rider agents over a 30-day period. (2) Intervention analysis: This phase tests the effects of three platform-level regulatory strategies: adjusting agent intelligence levels, optimizing interaction modes, and controlling order volume. (3) Mechanism analysis: This phase investigates the underlying connections between agent intentions and behaviors, providing in-depth explanations of the emergent phenomena observed within the system.

1) **Observational analysis (RQ1)**

This study adopts a repeated experimental approach to eliminate the influence of random factors. Under identical parameter settings, 10 independent runs were

Table 8.7 Comparison Between the Simulated O2O Service System and Real-World Dataset.

Metric	Value (h)
Actual average working time T_{real}	4.02
Simulated average working time T_{sim}	4.05
Mean absolute error (MAE)	0.05
Root mean square error (RMSE)	0.06
Pearson's correlation coefficient r	0.98

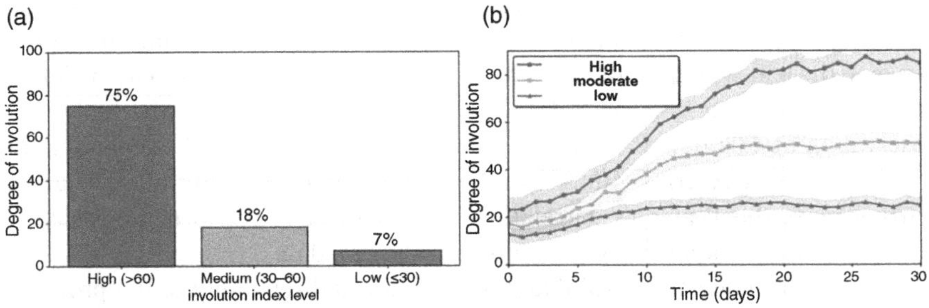

Figure 8.11 Analysis of involution levels in the experiment. (a) Experimental distribution of involution levels. (b) Temporal variation of involution degree in each experiment.

conducted. Following industry benchmarks, the involution index (as defined in **Table** 8.7) was used to classify the level of involution into three categories: low involution (index ≤ 30); moderate involution (30 < index ≤ 60); and high involution (index > 60).

As illustrated in Figure 8.11(a), the experimental distribution of involution levels shows that 75% of the simulations exhibited a high level of involution under highly competitive conditions of the O2O platform. This indicates a strong consistency in the emergence of involution effects across repeated trials. Figure 8.11(b) further shows that the degree of involution tends to increase over time in all experiment types, eventually stabilizing at a relatively high level.

One outlier experiment exhibited a low level of involution, which significantly deviated from the overall trend. This anomaly can be attributed to three stochastic disturbances: a temporarily optimal initial spatial distribution of rider agents, pseudo-random variations in the path-planning algorithm resulting in temporarily optimal routes, and a concentration of order assignments in low-density areas, which reduced competitive pressure.

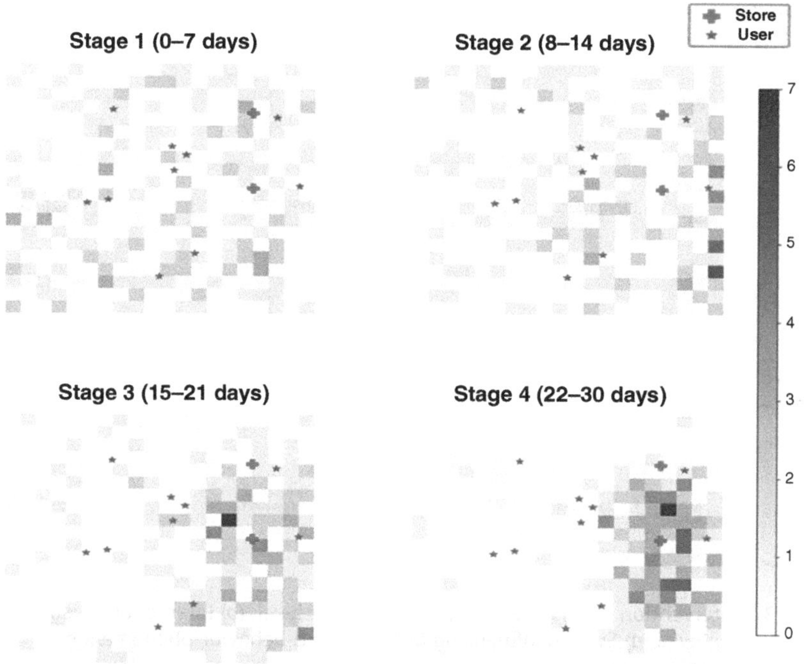

Figure 8.12 Heatmap of rider activity range.

As shown in Figure 8.12, we analyzed the temporal evolution of rider agents' activity range in experiments characterized by a high degree of involution. In the heatmap, color intensity represents the density of rider distribution. Over time, the initially random spatial distribution of rider agents gradually evolves into concentrated clusters around locations, such as merchants and users. This aggregation process visually reflects the emergence of involution behavior among rider agents.

2) **Intervention analysis (RQ2)**

To investigate the underlying mechanisms of involution, this phase of the study designed and conducted three targeted intervention experiments, each focusing on a specific influencing factor. The experiments explored the impact pathways of three core variables on involution dynamics: intelligence level adjustment (factor A), interaction mode transformation (factor B), and order volume fluctuation (factor C). By applying multidimensional data analysis methods, the study aimed to clarify how these variables collectively influence individual and group-level performance within work scenarios.

As shown in the boxplot visualization in Figure 8.13(a), the "intelligence-level adjustment" intervention revealed no statistically significant differences in

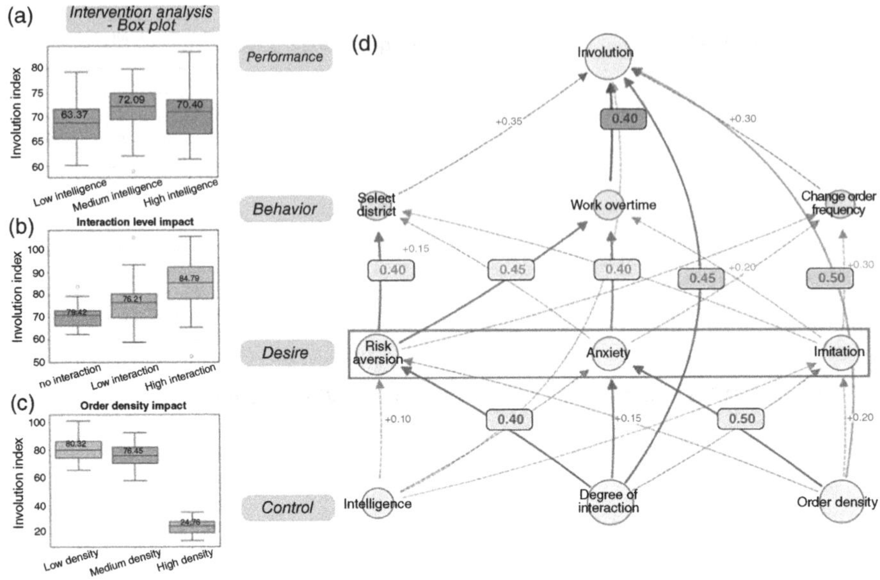

Figure 8.13 Intervention analysis results. (a) Intelligence level. (b) Interaction level. (c) Order density level. (d) System influencing factors-desire-behavior-phenomenon structural equation diagram.

involution levels among the high-, medium-, and low-intelligence agent groups. All groups fell within the high involution range, with involution indices spanning from 68.87 to 72.09. This result indicates that under the current experimental setup, enhancing agent intelligence has limited effectiveness in mitigating systemic involution.

In contrast, Figure 8.13(b) highlights the pronounced heterogeneity in system behavior resulting from changes in interaction mode (factor B). The introduction of a new interaction mechanism led to a significant increase in involution, with the involution index rising from 70.42 to 84.79 as interaction intensity increased. This suggests that higher levels of agent interaction may, in certain conditions, exacerbate involution rather than alleviate it.

Lastly, Figure 8.13(c) demonstrates that order volume (factor C) exerts a strong influence on the degree of involution. As the number of available orders increased, the involution index decreased steadily from 80.32 to 24.76, indicating a shift from a high-involution to a low-involution state. This finding underscores the crucial role of resource abundance—in this case, sufficient order volume—in promoting stable system operations for riders and alleviating competitive pressure within the platform.

Further analysis of the relationship between influencing factors and the degree of involution, as shown in Figure 8.13(d), was conducted using an SEM based on linear correlation calculations. The results reveal that interaction mode transformation and order volume are the two most critical factors affecting the experimental outcomes. Their respective standardized path coefficients are 0.45 and 0.50, indicating notable direct impacts on the observed level of involution within the system.

3) **Mechanism analysis (RQ3)**

Through clustering analysis of rider agents' intention sequences (as shown in Figure 8.14(a)–(c)), the experiment identified several representative types of intentions. Over time, the clustering structure evolved from an initially homogeneous form into a more diverse, multi-cluster distribution. This temporal shift reflects the increasing differentiation of agent decision-making as the simulation progresses.

Further correlation analysis conducted during the system's temporal evolution reveals a notable relationship between agent intentions and behaviors in shaping involution dynamics. As shown in Figure 8.14(d)–(f), factors such as business zone switching and order acceptance exhibit stronger correlations with intentions related to risk avoidance and anxiety, and these correlations grow more pronounced over time.

Notably, as highlighted in the medium-dark gray -marked region of Figure 8.13(d), these two types of intentions are highly associated with typical involution behaviors such as business zone selection and extension of working hours. Specifically, the risk avoidance intention demonstrates a correlation coefficient of 0.40 with regional selection and 0.45 with work hour extension. These findings underscore the critical role of individual intentions in driving and sustaining involution behaviors within the system.

As shown in Figure 8.14(g), a Sankey-based flow visualization method was used to trace the fine-grained transitions in agent behavior over a 30-day observation period. The decision-making intentions of rider agents underwent three distinct stages: an initial rule-following phase, followed around day 15 by a rise in anxiety-driven behaviors (increasing to 24%), and eventually transitioning to a dominant risk-avoidant phase, reaching 64% by the end of the 30-day observation period. Correspondingly, behavior patterns evolved from rest-focused, to aggressive order grabbing, and finally to risk-avoidant strategies.

To uncover the mechanisms driving these patterns, three intervention scenarios were evaluated. In the "enhanced interaction" setting (Figure 8.14(h)), agents overwhelmingly adopted risk-avoidant behaviors, indicating that deeper interaction intensity increases the tendency toward conservative decision-making. In the "enhanced intelligence" scenario (Figure 8.14(i)), the share of risk-averse intentions rose even further, suggesting that stronger cognitive capacity reinforces

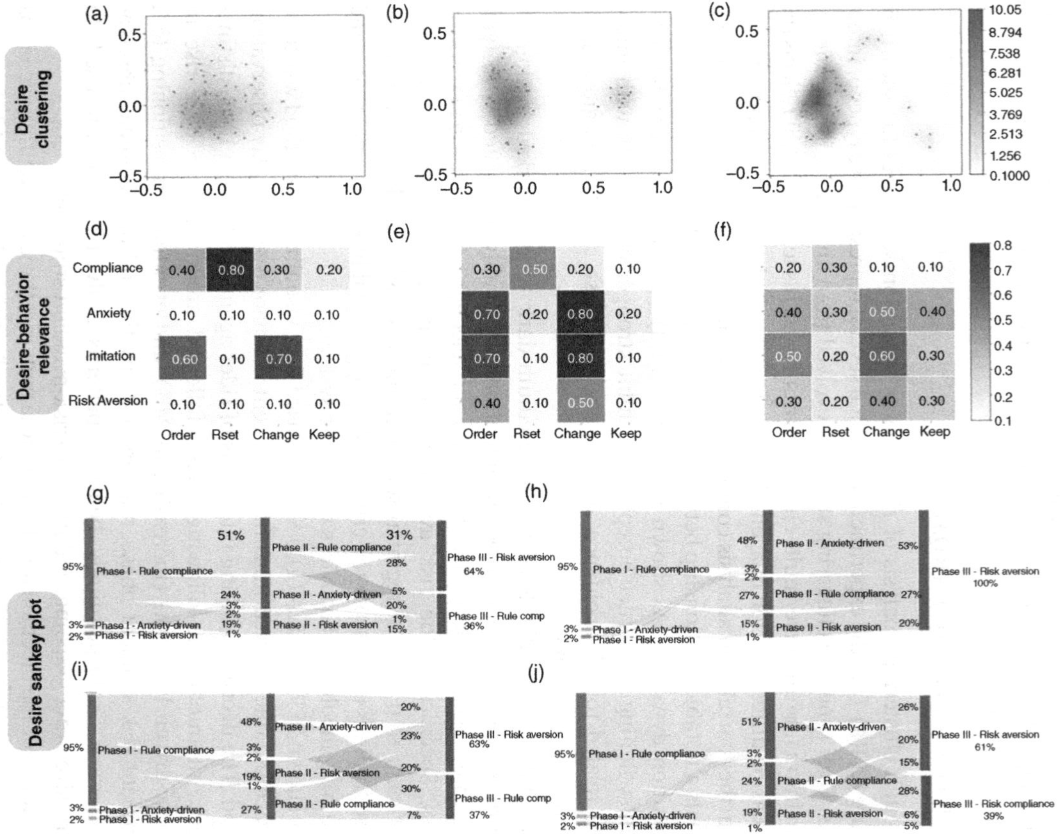

Figure 8.14 Temporal emergence of intentions. (a) Phase 1 0–10 days. (b) Phase 2 11–20 days. (c) Phase 3 21–30 days. (d) Phase 1 0–10 days. (e) Phase 2 11–20 days. (f) Phase 3 21–30 days. (g) Initial configuration. (h) Enhanced interaction. (i) Increased intelligence. (j) Higher order density.

caution. Conversely, in the "increased order density" scenario (Figure 8.14(j)), the proportion of risk-aware agents dropped to 61%, implying that higher order density-induced workloads reduce risk sensitivity and lead to more risk-taking behavior.

These findings indicate that interaction dynamics and environmental pressure are the two key drivers of involution. Enhanced interaction depth significantly amplifies agents' inclination toward risk avoidance, showing that changes in interaction modes directly affect decision-making styles. Meanwhile, higher order density—representing greater task loads—can suppress risk awareness and push agents toward competitive and aggressive strategies.

In summary, the emergence of involution is not driven by a single factor but is instead the result of the combined effects of interaction mechanisms, environmental stress, and cognitive ability. This conclusion aligns with the results of the observational and intervention analyses, highlighting the complexity of behavioral emergence in agent-based O2O systems. Summary of findings across the three analytical layers:

1) **Observational analysis:** Agents in the O2O system increasingly exhibit involution over time, with their behaviors converging and becoming more homogeneous.
2) **Intervention analysis:** Timely introduction of governance strategies—such as order-balancing mechanisms—during the anxiety-dominated phase can effectively mitigate over-competition and enhance system performance and stability.
3) **Mechanism analysis:** Order scarcity and anxiety diffusion trigger risk-avoidant intentions, which in turn drive cross-zone order grabbing and extended working hours, forming the core behavioral patterns that underlie rider involution.

8.5 Chapter Summary

Although computational experimentation has made significant progress, its development still faces numerous challenges. One of the key issues lies in bridging the gap between descriptive modeling (data analysis) and predictive modeling (meta-modeling). While scattered efforts have been made in this direction, there is still a lack of a systematic and mature framework for causal inference through computational experiments. To address this gap, this chapter proposes a multilayer integrated framework that combines computational experiments with causal reasoning. The framework is designed to tackle the challenges of identifying and analyzing causal relationships in complex social systems, which is mainly composed of three levels. (1) Observational analysis:

Based on the data from computational experiments, explores causal relationships between influencing factors and response variables, and uses anomaly analysis methods to capture system anomalies. (2) Intervention analysis: Implements various interventions to observe changes in the system, further evaluating causal effects. (3) Mechanism analysis: By correlating micro-level behaviors with macro phenomena, explains how "interventions" induce "outcomes" through what pathways, under what regulatory conditions, and via what processes. Finally, the phenomenon of rider involution on food delivery platforms is used as a case study to effectively demonstrate the validity, generality, and efficiency of the proposed causal inference framework.

References

1 Wang, F.Y. (2010). The emergence of intelligent enterprises: from CPS to CPSS. *IEEE Intelligent Systems* 25 (4): 85–88.

2 Yilma, B.A., Panetto, H., and Naudet, Y. (2021). Systemic formalisation of cyber-physical-social system (CPSS): a systematic literature review. *Computers in Industry* 129: 103458.

3 Xiao, X., Xiangning, Y., Deyu, Z. et al. (2023). The origin, current status, and prospects of computational experimentation methods. *Acta Automatica Sinica* 49 (2): 246–271.

4 Cioffi-Revilla, C. (2017). *Introduction to Computational Social Science: Principles and Applications*. Springer.

5 Fromm, J. (2004). *The Emergence of Complexity*. Kassel: Kassel University Press.

6 Newell, A. and Simon, H.A. (1972). *Human Problem Solving*. Upper Saddle River, NJ: Prentice-Hall.

7 Hamilton, J. (1994). *Time Series Analysis*, 1e. Princeton University Press.

8 Hoel, E.P., Albantakis, L., and Tononi, G. (2013). Quantifying causal emergence shows that macro can beat micro. *Proceedings of the National Academy of Sciences* 110 (49): 19790–19795.

9 Klein, B. and Hoel, E. (2020). The Emergence of informative higher scales in complex networks. *Complexity* 1–12.

10 Shalizi, C.R. and Moore, C.(2003). What is a macrostate? Subjective observations and objective dynamics. ArXiv:Cond-Mat/0303625.

11 Barnett, L. and Seth, A.K. (2023). Dynamical independence: discovering emergent macroscopic processes in complex dynamical systems. *Physical Review E* 108 (1): 014304.

12 Casini, L. and Manzo, G. (2016). *Agent - Based Models and Causality: A Methodological Appraisal*. Linköping University Electronic Press. The IAS Working Paper Series.

13 Glennan, S. (2002). Rethinking mechanistic explanation. *Philosophy of Science* 69 (3): S342–S353.

14 Psillos, S. (2007). What is causation? In: *The Episteme Reviews: Research Trends in Science, Technology and Mathematics Education* (ed. B. Choksi and C. Natarajan), 11–34. Bangalore: Macmillan India.

15 Pearl, J. (2009). *Causality: Models, Reasoning, and Inference*. Cambridge: Cambridge University Press.

16 Anzola, D. (2020). Causation in agent-based computational social science (Ch. 5). In: *Advances in Social Simulation: Looking in the Mirror, Springer Proceedings in Complexity* (ed. H. Verhagen, M. Borit, G. Bravo, and N. Wijermans), 47–62. Cham: Springer International Publishing.

17 Lucas, R. (1976). Econometric policy evaluation: a critique. In: *The Phillips Curve and Labor Markets, Volume 1 of Carnegie-Rochester Conference. Series on Public Policy* (ed. K. Brunner and A. Meltzer), 19–46. Amsterdam: North-Holland.

18 Sims, C. (1980). Macroeconomics and reality. *Econometrica* 48 (1): 1–48.

19 Kirman, A. (1992). Whom or what does the representative individual represent? *Journal of Economic Perspectives* 6 (2): 117–136.

20 Hoover, K. (2008). Does macroeconomics need microfoundations? In: *Philosophy of Economics* (ed. D. Hausman), 315–333. Cambridge: Cambridge University Press.

21 Tesfatsion, L. (2002). Agent-based computational economics: growing economies from the bottom up. *Artificial Life* 8 (1): 55–82.

22 Tesfatsion, L. (2006). Agent-based computational economics: a constructive approach to economic theory. In: *Handbook of Computational Economics. Agent-Based Computational Economics*, vol. 2 (ed. L. Tesfatsion and K.L. Judd), 831–880. Amsterdam: North-Holland Elsevier.

23 Arthur, W.B. (2021). Foundations of complexity economics. *Nature Reviews Physics* 3: 136–145.

24 Manzo, G. (2020). Agent-based models and methodological individualism: are they fundamentally linked? *L'année Sociologique* 70 (1): 197–229.

25 Morgan, S.L. (2005). *On the Edge of Commitment: Educational Attainment and Race in the United States*. Stanford, CA: Stanford University Press.

26 Gelman, A. (2011). Causality and statistical learning. *American Journal of Sociology* 117 (3): 955–966.

27 Dawid, A.P., Faigman, D.L., and Fienberg, S.E. (2014). Fitting science into legal contexts: assessing effects of causes or causes of effects? *Sociological Methods & Research* 43 (3): 359–390.

28 Sampson, R.J., Winship, C., and Knight, C. (2013). Translating causal claims. Principles and strategies for policy-relevant criminology. *Criminology & Public Policy* 12 (4): 587–616.

29 Gelman, A. and Imbens, G. (2013). Ask why? Forward causal inference and reverse causal questions. National Bureau of Economic Research Working Paper Series, No. 19614(10.3386/w19614).

30 Goldthorpe, J.H. (2001). Causation, statistics and sociology. *European Sociological Review* 17 (1): 1–20.

31 Zholus, A., Ivchenkov, Y., and Panov, A. (2022). Factorized world models for learning causal relationships. *ICLR2022 Workshop on the Elements of Reasoning: Objects, Structure and Causality.*

32 Shyu, M.L., Chen, S.C., Sarinnapakorn, K., and Chang, L. (2003). A novel anomaly detection scheme based on principal component classifier. In: *Proceedings of the IEEE Foundations and New Directions of Data Mining Workshop*, 172–179. Piscataway, NJ: IEEE.

33 Rao, A.S. and Georgeff, M.P. (1995). BDI agents: from theory to practice. *Icmas* 95: 312–319.

9

The Fifth Step – Maturity Evaluation of Computational Experiments

In the computational experiment method, although experimental verification is the last step, it is undoubtedly the most important component. From the perspective of causal dependence, computational experiment design and analysis can answer specific "hypothesis" questions, namely, "why the cause of interest leads to a certain specific result." However, the flexibility of computational experiments also leads to the arbitrary nature of their construction process, which, in turn, affects the reliability of the results. In order to establish a counterfactual dependency between the computational experiments and the real world to be explained, verification and evaluation of computational models are particularly important. To address this challenge, this chapter first gives the theoretical basis for verification of computational experiments, that is, the simultaneous existence of the "reality" of "theory," "input" and "output" can maximize the potential of "artificial society" as an "imitation" device. Then, using the epidemic model as the research object, a framework for evaluating the capability maturity of computational experiments is proposed, which mainly includes two aspects. (1) Model self-evaluation: By comparing the implemented outcomes with the expected results, we can determine whether the model meets its intended objectives. (2) Comparison between different models: By comparing the total evaluation scores of the implementation of various models, models with relatively mature comprehensive capabilities can be identified. Finally, this evaluation framework is used to analyze and assess several representative COVID-19 computational models, helping decision-makers identify more mature and valuable models and guiding modelers in determining the directions for future model improvements (Figure 9.1).

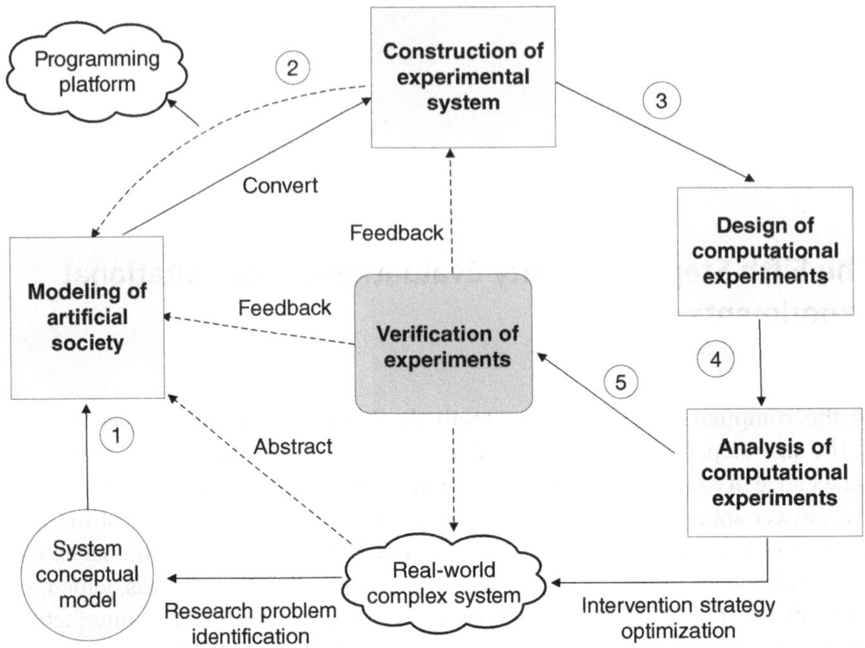

Figure 9.1 The status of experimental system construction in computational experiment methods.

9.1 Theoretical Foundation of Computational Experiment Validation

As shown in Figure 9.2, computational experiments typically treat the relationship "model assumptions + model parameters = model behavior" as a priori research premises. Model construction essentially involves formulating a hypothesis and assuming that it reflects the governing rules of the real-world system. By executing computational experiments and observing the resulting model behaviors, researchers attempt to provide causal explanations for real-world phenomena. The quality of input data and domain knowledge exerts a direct influence on the final outcomes of computational experiments; poor input may result in model behaviors that deviate significantly from real-world observations. To build an effective experimental system, it is essential to ensure the availability and reliability of data and knowledge sources at the input stage. The notion of "realism" refers to the extent to which a model can capture the operational mechanisms of the real world, enabling the model to be considered credible. According to Sugden's theory, "if we are to make inductive inferences from the model world to the real world, we must recognize some salient similarities between the two worlds" [1].

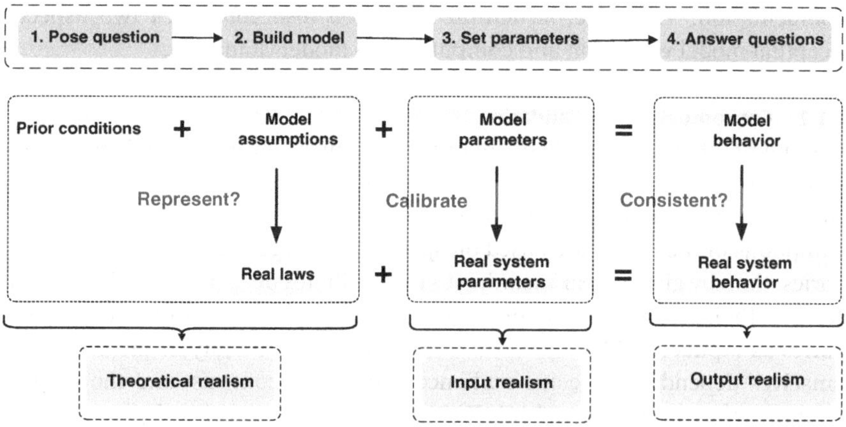

Figure 9.2 Classification of experimental validation.

In order to enhance the reliability of computational experiment results, we enhance the "parallelism" between agent-based model (ABM) and the real-world mechanisms it intend s to describe from three levels: "theory," "input," and "output." (1) Theoretical realism: The extent to which the assumptions used in model design are explicitly based on sociological and/or psychological theories(supported by empirical evidence), thereby justifying the rationality of ABM's micro-level assumptions. (2) Input realism: The degree to which empirical and/or experimental data are employed to calibrate the initial values, distributions, and functional forms in the model design, thus ensuring that the initially assumed model mechanisms, along with their specific parameters and functions, provide the most reasonable account. (3) Output realism: The extent to which the model's output results are consistent with empirical and/or experimental data, serving to demonstrate how effectively the computational model can generate emergent phenomena observed in the real world.

9.1.1 Theoretical Realism and Model Analysis

A conceptual model is an abstraction of a real system, which, when implemented via programming, becomes an executable computational model. Structural validation is the process of verifying model construction to ensure that the model can reflect the internal structural characteristics of the real system and that the intermediate processes through which assumptions generate model behavior are correct. That is, to ensure that the hypothetical model is a sufficient condition for producing behavioral outcomes. The process of agent-based social modeling includes two steps: conceptual model design and computational model implementation. Therefore, structural validation also comprises two aspects: conceptual model validation and computational model validation.

9.1.1.1 Conceptual Model Validation

The trend of "theoretical realism" requires that micro-level assumptions be supported by explicitly articulated sociological or psychological theories, which may be supported by empirical or experimental evidence. In this context, the behavioral assumptions of the model are based on specific sociological and/or psychological theories, thereby giving clear meaning to the attributes of agents, their behavioral rules, and their responses to other agents. However, the benefits of using specific theories as inputs for ABMs are still limited, as the credibility of these theories themselves depends on specific audiences, historical contexts, and the empirical support they have received [2]. Furthermore, sociological and psychological theories themselves have limitations and often portray agents in fundamentally different ways. In this sense, having only micro-level theories provides limited help in making ABMs more realistic.

At this stage, the main focus is to examine the consistency of the conceptual model with the research objectives, the given assumptions, existing theories, and evidence, ensuring that the model's simplifications do not seriously affect its credibility or the understanding of researchers regarding key features of the real system. At the same time, the conceptual model should meet the requirements of internal completeness, consistency, and correctness. The rationality of the model is mainly evaluated through three methods: expert judgment, theoretical comparison, and empirical data fitting. These methods are often qualitative and lack strict verification. By comparing and learning from existing models, researchers can greatly enhance the degree of validation of the conceptual model by (1) clarifying the scope of application and limitations of the model; (2) showing that mainstream theories adopt the same assumptions or that similar studies by other researchers adopt the same assumptions; (3) developing and analyzing a set of models with the same core assumptions but different additional assumptions; and (4) establishing equation-based models that exhibit similar behavioral characteristics to the original ABM.

When designing a conceptual model, a prominent issue is the lack of a unified and effective method to balance the complexity of the real system and the analyzability of the model results. This often leads to arbitrariness in the choice of model structure, resulting in excessive emphasis on the "realism" of the model while neglecting its overall applicability. Pattern-oriented modeling (POM) has emerged in recent years within the field of systems modeling as a promising solution to this problem [3]. POM holds that patterns are a characteristic of systems, implicitly containing and conveying information about the system's underlying operating principles. For example: "What kinds of patterns characterize the internal mechanisms of the system?" and "What variables and processes must the model include in order for these patterns to emerge?" By extracting information from multiple patterns to build an effective model, researchers can achieve a balance between model complexity and analyzability, thereby enhancing the verifiability of the conceptual model.

9.1.1.2 Computational Model Validation

Social scientists often criticize the black-box nature of ABMs [4, 5]. Some researchers even regard "epistemological opacity" as a defining feature of ABMs [6]. The ability of ABMs to generate data relevant to causal inference depends on systematic reliability checks, and such checks are undoubtedly necessary to determine whether the mechanisms designed in the simulation model can produce the observed outcomes. Model analysis refers to the strategies employed during the execution of computer programs to understand and describe a series of events, behaviors, and feedback—in other words, it (model analysis) involves a thorough examination of the internal workings of an ABM to understand the processes triggered by its encoded mechanisms. Revealing the operational mechanism of an ABM is crucial for assessing its value in causal reasoning: It is important to know not only whether an ABM can generate data relevant to causal inference but also how it generates such data. This allows for the evaluation of the credibility of the explanatory mechanisms in the ABM and ultimately helps determine whether the model describes processes that could occur in the real world.

At this stage, the main focus is to check whether the algorithms, code, and operating conditions are consistent with the conceptual model. The conceptual model is transformed through programming into a computational model that can run on a specific computer system—a process that involves numerous factors and is prone to subtle errors. Since ABMs are discrete models, and there are significant differences among various programming languages and computer systems, programming details can have a major impact on model results. Currently, the main methods used for computational model validation include (1) employing mathematical techniques based on Markov process theory, which can precisely describe the sequence of states in the evolution of an ABM [7, 8]. (2) Introducing the main mechanisms of the model sequentially, making it easier to assess their roles in the model dynamics and the relative weights of these mechanisms in the outcomes. (3) In addition, using ABM as an iterative simulation tool, data can be collected at different levels of analysis and specific measurements can be calculated to analyze various aspects of model dynamics [9].

To simplify and accelerate the development of computational models, researchers have developed many modeling platforms (including model libraries and analysis functions) for model development and result analysis [10–13].

However, there are significant differences in coding styles among different tools, making these tools incompatible with one another. To facilitate computational model validation, it is necessary to establish a unified computational model development platform and adopt a separation of model development from execution. This approach enables the same model code to be executed on different platforms, thus producing identical experimental results. The benefit of this approach is that one only needs to verify whether the model design is effective and credible (herein, the model design is simply a set of specifications), without needing to check the reliability of the program implementation of the model.

9.1.2 Input Realism and Parameter Calibration

The advantage of ABM lies in its ability to handle heterogeneity, complex interactions among agent behaviors and networks, and feedback across multiple levels of analysis, thereby simulating various aspects of social phenomena—data related to which are often difficult to obtain in reality. This is also the main reason why ABMs have become widely adopted across disciplines [14, 15]. When ABM is used for causal inference, empirical calibration becomes a reasonable requirement. The trend of "input realism"—that is, using parameters and functions in ABMs that are estimated from empirical or experimental data, also known as "empirical calibration"—increases the likelihood of reproducing real-world patterns or trends within the virtual world.

In computational experiments, "model assumptions + model parameters = model behavior" is usually regarded as a priori condition. When building a model, one is essentially proposing a hypothesis and assuming that this hypothesis reflects the laws of the real system. By running computational experiments and observing the resulting model behaviors, researchers can make causal explanations for the real system. The quality of input data will directly affect the final results of computational experiments and may even lead to outcomes that deviate significantly from reality. To build an effective model, it is essential to ensure the availability and reliability of data from the source. Many researchers [16, 17] have noticed the issue of data availability. It is generally believed that a correct, usable, and large dataset needs to meet the properties of consistency, accuracy, completeness, timeliness, and entity identity. For computational experiments, the following factors need to be given special consideration.

9.1.2.1 Credibility of Data

The process of data collection is, in fact, an artificial process carried out according to specific requirements, and thus, this process is inherently biased from the outset. It is also impossible to guarantee complete authenticity during the collection process. Since there is no absolute "truth," data collected in strict accordance with scientific principles and involving less subjectivity are closer to being "real" and are considered more credible. Generally, data from government agencies and authoritative institutions are more reliable. In most cases, however, the quality of data, especially its authenticity, is difficult to guarantee. Although, compared to simulation data, data from the real world are indeed a strength of statistics research, the authenticity of data from social media is often questioned, primarily in three aspects:

1) **Representativeness of data:** The population using the Internet and various social platforms is not randomly distributed within the total population—for example, elderly individuals use the Internet less frequently. When representativeness cannot be ensured, increasing the sample size does not necessarily improve the reliability of research conclusions.
2) **Authenticity of data:** The individual or group behavior patterns obtained from social media data may not be consistent with those in real life. For instance, some people are very active online but are quiet and unsociable in their real life. Moreover, big data only records certain aspects of people's lives, not the whole picture.
3) **Reliability of data:** In many cases, data are records left behind when users use certain products (such as search engines or recommendation algorithms). It should be noted that our behavior patterns may be influenced or altered by these algorithms, causing the corresponding data to be "distorted." For example, a significant portion of social media users are bots, which can manipulate online topics and influence people's perceptions of certain events. Conclusions drawn from such data, which are mixed with bot behaviors, may deviate significantly from reality.

9.1.2.2 Availability of Data

From the perspective of causal dependence, when ABM is well empirically calibrated, elements such as the distribution of agent attributes, agent behaviors, and environments have corresponding counterparts in the real world. This means that it is reasonable to regard the macro-level emergence in ABM as a counterfactual reference for the real world. However, the availability of data raises an issue for data calibration: the question of when can simulated data and empirical data be considered sufficiently close. This is crucial not only for determining the generative capacity of micro-level agent behaviors but also for assessing to what extent specific interventions might cause us to deviate from observed outcomes—and this assessment is the key to counterfactual reproduction.

In statistics, sample size is directly related to the accuracy of inferential estimates, and there are various standard methods for evaluating the degree of fit between two datasets [18]. Given a fixed population, the larger the sample size, the smaller the representational bias of the statistical estimator; conversely, the smaller the sample size, the greater the estimation error. When estimating computational model parameters, for the same population (such as the same region), having more data can reduce the probability of random errors and problems of insufficient representativeness. In many cases, existing data resources do not contain the data we intend to study, or the data directly related to the research question are very scarce. As a result, researchers look for proxy variables for their studies, and the degree of correlation between the proxy variables and the direct variables will affect the experimental results.

9.1.2.3 Processuality of Data

As Macy and Flache pointed out, a simple input–output mapping does not demonstrate whether the internal functions of an ABM reflect real-world mechanisms [19]. After all, the generative capacity of an ABM ultimately depends on the execution process of the computer program describing the hypothetical mechanisms. Therefore, it is necessary to compare the simulation process with real-world processes to effectively validate an ABM. In other words, "output" realism should be complemented by "process" realism. Therefore, data calibration requires abundant longitudinal data. As Hansen and Heckman noted when discussing simulation models, using only cross-sectional data at the beginning of a simulated dynamic process is of limited value [20]. Thus, the processuality of data is an important factor affecting experimental results, especially for projects with strong timeliness requirements. However, an unavoidable issue is that data collection takes time, and data acquisition is subject to delays—this is particularly prominent in the early stages of sudden emergency events, during which reporting delays are significant. For computational models, the longer the historical time series data used and the closer they are to the current time point, the better the experimental results.

9.1.2.4 Granularity of Data

In fact, when it comes to mechanism design, the flexibility of ABMs comes at a cost: finer granularity requires more detailed information in order to construct model components at different levels of analysis (e.g., actors, networks, groups, structures). "Parameter calibration" can involve agent attributes, the functional forms of behavioral rules, as well as the networks and geographic locations into which agents should be embedded. All of these elements can be directly derived from surveys, censuses, or digital data. Therefore, calibration using the best available data helps to further reduce uncertainty. Generally, micro-level characteristics of the model are easier to calibrate: (1) the quantity and types of agent heterogeneity are mostly derived from empirical data; (2) the specific behavioral rules agents should adopt are often based on survey data in the form of functions (similar to regression), but these functions cannot finely describe the mechanisms behind agents' decision-making; and (3) the correct representation of the environment in which agents are situated is relatively easy, but embedding ABMs into network structures is challenging [21, 22].

The strength of ABMs lies in simulating aspects of social mechanisms that existing data (or models) cannot address, but parameter calibration requires that this method be constrained by those data. Ultimately, for ABMs to truly be effective, a balance must be struck between designing fine-grained mechanisms (linking scarce input data to puzzling real-world patterns) and data-driven mechanisms (using detailed input data). Since data are in fact limited, the more we want to constrain simulations empirically, the more we must reduce the granularity of hypothesized mechanisms. As Fagiolo et al. pointed out, the pursuit of empirically calibrated and validated ABMs contains an implicit conservative stance: It may lead us to design types of mechanisms that fit the available data, which limits the full potential of ABMs in mechanism design [23]. The more an "ABM" emphasizes empirical calibration and validation, the less it can be used to gain insights into phenomena for which data are lacking.

9.1.3 Output Realism and Result Validation

The trend of "output realism," also known as "result validation," is mainly used to evaluate whether the results of computational experiments are consistent with the behavior of real systems. Real data can be collected from the real world, while the model's output data are generated by its operational mechanisms. By comparing the data produced by the model with real data, researchers can infer whether the model "appropriately" reflects reality, thus achieving the purpose of result validation. In addition, "result validation" imposes constraints on both the meaning of model inputs and the form of model outputs, in order to reduce inconsistencies between the causal relationships described by the simulation and the causal relationships in the real world. On the one hand, the more specific the real-world target that an ABM aims to reproduce, the more necessary it is to define a precise set of initial assumptions to generate this target; on the other hand, to assess whether specific interventions applied within the ABM will affect the probability of observed outcomes, it is essential to clearly specify the output results.

It should be noted that, regardless of how far "output realism" is pushed through "result validation," it is always possible to find at least one other ABM that is based on a different set of micro-level assumptions and can also fit real-world phenomena. Therefore, in this sense, even empirically validated ABMs cannot provide sufficiently convincing data and arguments. The standard practice of mechanism analysis is to observe possible causal associations and hypothesize mechanisms that might link these causal associations. However, the possible correspondence between causal associations and hypothesized mechanisms does not imply causality, unless it can be shown that the association can only be caused by the hypothesized mechanisms.

Therefore, in addition to result validation within the limits of data availability, researchers also systematically conduct various reliability checks on ABMs. These reliability checks can be used to assess the variability of specific simulation results in certain respects, including (1) sensitivity analysis, which evaluates how changes in model parameter values on changes impact ABM simulation results; (2) robustness analysis, which assesses how changes in internal model details impact ABM simulation results; and (3) stochastic analysis, which involves repeated simulations under the same set of parameter values to evaluate the variability of ABM simulation outputs. These tools can be regarded as supplements to result validation, assessing in different ways the dependence of ABM's macro-level simulation results on its model assumptions, thereby ultimately reducing the uncertainty caused by data limitations.

The "model assumptions" include all aspects of a given ABM, since there is often insufficient data to determine what these aspects should look like in the real world. "Model assumptions" generally include "formal assumptions" and "substantive assumptions." The former refers to general properties of the model, such as the probability distribution types of specific model parameters, the functional forms of certain agent behavioral rules, and the network topology of agent interactions. In contrast, substantive assumptions involve the core explanatory mechanisms of agents, such as the logic of agent actions and the influence of local interactions on agent behavior—these substantive assumptions usually come from domain-specific theories and knowledge. Both formal and substantive assumptions contribute to enabling causal inference. In fact, when ABM assumptions (i.e., those parts not supported by data) are changed, the stability of ABM behavior—if maintained—increases the observer's confidence that the macro-level simulation results can be regarded as a reference for the real world, rather than merely a result of model assumptions. This is why it is necessary to systematically assess the dependence of macro-level simulation results on model assumptions using reliability tools.

9.1.3.1 Sensitivity Analysis

Sensitivity analysis can be conducted in various ways [24, 25]. "Global" sensitivity analysis primarily provides a comprehensive description of model behavior across the entire parameter space. First, once a complete mapping of simulation inputs and outputs is generated, it becomes easier to identify errors and/or logical inconsistencies within the model. Systematic numerical exploration of ABM elements lacking data and/or with insufficient data can help avoid empirical calibration of inconsistent and/or irrelevant aspects of the model. Second, since empirical or experimental estimates used as model inputs are themselves subject to uncertainty, understanding agents' behavior across a broader parameter space is crucial for properly interpreting the empirical data employed. When properly conducted, sensitivity analysis helps to counter criticisms that ABM results are unreliable due to their dependence on specific numerical inputs. However, global sensitivity analysis can be challenging, especially as the dimensionality of the parameter space continues to increase. In such cases, heuristic evolutionary algorithms provide a possible solution [26]. Additionally, sampling-based "local" sensitivity analysis methods can also be utilized.

9.1.3.2 Robustness Analysis

If sensitivity analysis focuses on the values of agents' parameters, then robustness analysis targets the "gears and wheels" inside the ABM, aiming to assess the impact of changes to these internal components on the outcomes of an ABM [9]. Key components of an ABM include (1) the probability distributions of agent attribute values; (2) the functional forms linking agents' attributes to their behaviors; (3) the spatial/relational structure of agent interactions; and (4) the execution order of agents' rules and the timing of their actions. There is often a lack of empirical data to comprehensively calibrate all these aspects, which has led critics to argue that ABM results are unreliable because ABM results depend on too many freely variable internal simulation details. Robustness analysis is used to evaluate the extent to which the macro-level results of an ABM depend on its internal components. The less an ABM's outputs are influenced by internal mechanisms, the more confident we can be in using it to infer the functioning of the real world. Bruch and Mare provided a robustness analysis method focused on the behavioral level of ABMs and applied it to the Schelling model, finding that the degree of spatial segregation actually depends on the functional form of agents' residential preferences [27]. Similarly, Manzo et al. studied various imitation heuristics of agents and found that the so-called causal effect of certain network features on the speed of innovation diffusion may actually be reversed, depending on the role played by the most central nodes in the network during the diffusion process [28].

9.1.3.3 Stochasticity Analysis

ABMs typically include various stochastic components. For example, agents' attributes are often initialized according to probability distributions, agents' behavioral tendencies are probabilistic, and the order in which agents are activated is random. As a result, a single simulation run can be regarded as a specific realization of an unknown stochastic process. For a given set of initial conditions, the outcome of each ABM simulation may differ. Stochasticity analysis refers to the process of running multiple simulations under the same set of parameter values to quantify the variability in ABM simulation results.

Therefore, stochasticity analysis is distinct from "sensitivity analysis," which evaluates the impact of potential variability in the given parameter values on ABM outcomes. In this sense, stochasticity analysis can be considered a special form of sensitivity analysis [9]. When global sensitivity analysis is combined with repeated model simulations, researchers generate a multidimensional space in which each combination of parameter values produces a distribution of outcomes. Appropriate measures of dispersion can then be used to describe each distribution and assess the degree of overlap between distributions, thereby evaluating the likelihood of insignificance between two interventions. In this way, ABMs can generate probabilistic predictions of outcomes. This procedure is particularly important for ABMs that lack sufficient empirical data. If changes in parameter values (or subsets of these parameter values) do not lead to significant differences in simulation results, our confidence in the reliability of ABM outcomes is strengthened.

9.2 Model Evaluation Framework in Computational Epidemiology

9.2.1 Classification of Computational Epidemiological Models

Epidemiology is an important application of computational experiments [29, 30], where researchers use computational models to help leaders grasp the overall situation and contain the escalation of emergencies (Figure 9.3). In the field of epidemiology, computational experiments serve three main functions.

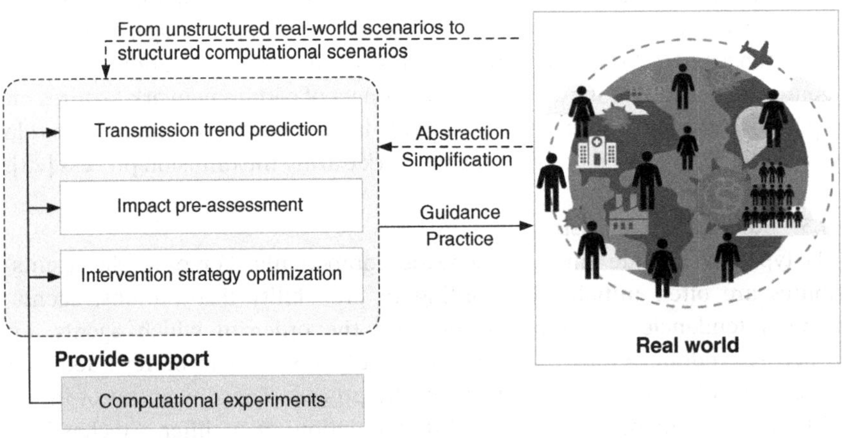

Figure 9.3 Schematic diagram of computational experiments in the field of epidemiology.

(1) **Prediction of transmission trends:** Accurate prediction of transmission trends is a crucial prerequisite for implementing effective prevention and control measures and serves as an important references for matters such as medical resource allocation. (2) **Impact preassessment:** The spread of an epidemic not only poses direct threats to public health but also has indirect impacts on social interactions and economic development. Therefore, quantitative analysis of these impacts is an important basis for emergency preparedness. (3) **Intervention optimization:** Dynamic intervention measures based on the development of the epidemic are a key objective of emergency decision-making. Intervention measures should keep the epidemic within the carrying capacity of the healthcare system and minimize harm to people and society.

At the end of 2019, the outbreak caused by the novel coronavirus (SARS-CoV-2 [31]) developed into a pandemic in less than two months. The World Health Organization named the disease caused by this viral infection "COVID-19" [32]. The emergence and continued spread of the epidemic placed higher demands on personnel in all fields, requiring leadership to organize decision-making, mobilize, and coordinate various resources. Since the outbreak of COVID-19, experts and scholars have been conducting research on the epidemic, and a wide variety of models have emerged [33–36], but the results vary greatly. For example, the estimated values of the basic reproduction number (R0) vary significantly. R0 is a quantitative parameter used in epidemiology to measure the transmission potential of infectious diseases and it is an important research focus in the field. Zhao et al. from the Chinese University of Hong Kong [37] estimated the R0 range to be 2.24–3.58. Jung et al. from Hokkaido University in Sapporo, Japan [38], predicted an R0 value of 1.6–4.2. Wu et al. from the University of Hong Kong [39] estimated R0 to be 2.47–2.8.

Modeling approaches for epidemic models can be broadly divided into two categories: data-driven statistical models and simulation models. Data-driven statistical models mainly include growth models, regression models, and probabilistic statistical methods. Simulation models primarily include system dynamics models, ABMs, and stochastic simulation methods.

Data-driven statistical models: Growth models such as the Logistic model and the Richards model are commonly used to study S-shaped cumulative growth processes, such as sustained population growth in resource-limited environments. Therefore, they have been used to fit epidemic curves [40, 41], just as they were once used to predict the spread of SARS [42], and have achieved good results. Regression models accurately measure the degree of correlation between the target variable and influencing factors as well as the fit of the model by studying the specific dependencies between them, making them highly applicable in forecasting. Probabilistic statistical methods are based on statistical inference using probability theory, with Bayesian thinking serving as a theoretical foundation for these approaches.

Simulation models: The Susceptible-Infectious-Recovered (SIR) and Susceptible-Exposed-Infectious-Recovered (SEIR) models are classical infectious disease models that adopt a system dynamics approach [43]. These models partition the population into different compartments and simulate disease transmission based on the feedback mechanisms of system components. ABMs [44] capture complex epidemic dynamics by explicitly modeling individual-level attributes, behaviors, and the micro-level transmission processes of disease. When analytical solutions are intractable to obtain, stochastic simulation methods can be employed to obtain probabilistic approximations. These modeling approaches play a crucial role in the description, explanation, and prediction of epidemic outbreaks.

9.2.2 Computational Experiments and Model Evaluation

For the computational experiment method to become a powerful tool for studying complex systems, the issue of validation—namely, model evaluation—must be effectively resolved. Existing model validation methods can be roughly divided into two categories: process validation and result validation. Process validation includes conceptual model validation and computational model validation. Conceptual model validation mainly uses three methods: expert judgment, theoretical comparison, and empirical data fitting. Computational model validation usually involves code-level checks, and these checks are very challenging in terms of code availability and the cost of code review. Result validation often directly relies on established statistical methods, such as regression analysis, confidence interval (CI), and hypothesis testing. However, these methods also have some limitations. First, it is essential to ensure the availability of real-world data. Second, the results from the real world and the experimental system must be obtained under the same initial conditions, environment, and parameters. Currently, Sargent's validation framework based on the model development process has a significant influence, including conceptual model validation, computational model validation, and operational validity [45]. However, researchers still mainly focus on result validation, while process validation is primarily of concern to software engineering researchers.

Different epidemiological models use different methods and data, so even when the research subjects and objectives are the same, the conclusions they reach may differ. The wide variation in R0 estimates is undoubtedly a concern for the public and policymakers, and this wide variation directly leads to difficulties in the application of computational models. Although the flexibility of computational models makes this approach applicable to various fields, it also results in considerable arbitrariness in model structure and in the modeling process. Therefore, for computational experiment methods centered on computational models to be widely adopted, this issue must be addressed. In this context, model evaluation and comparison can help identify more reliable models.

Most existing methods for evaluating epidemiological models use statistical techniques to calculate estimation errors and assess model performance based on the model results. De Angelis et al. [46] argue that Bayesian methods can tackle the significant challenges posed by multiple data sources in epidemiological model evaluation. Tabataba et al. [47] proposed a framework that combines different features of forecasts, error measurements, and ranking schemes to evaluate epidemic forecasts and demonstrated the practicality of this framework by assessing six influenza forecasting methods in the United States. Gibson et al. [48] believe that the best approach should be determined by the actual modeling problem and that multiple methods should be used to assess the suitability of epidemiological models. In terms of modeling approaches, evaluation is usually limited to one or a few types of models, which often have poor generalizability and fail to enable comparison between different types of models. Moreover, evaluating only the results does not help modelers trace the sources of bias, offering little assistance for future model improvement.

Therefore, we propose a general evaluation framework based on the entire modeling process, aiming to enhance researchers' understanding of the sources of bias by exploring factors that affect model capability maturity. As shown in Figure 9.4, the proposed framework enables two types of evaluation: self-assessment of individual models and comparison between different models. The evaluation results can not only help us identify models with better overall performance but also facilitate taking corrective measures in future modeling efforts, thereby improving

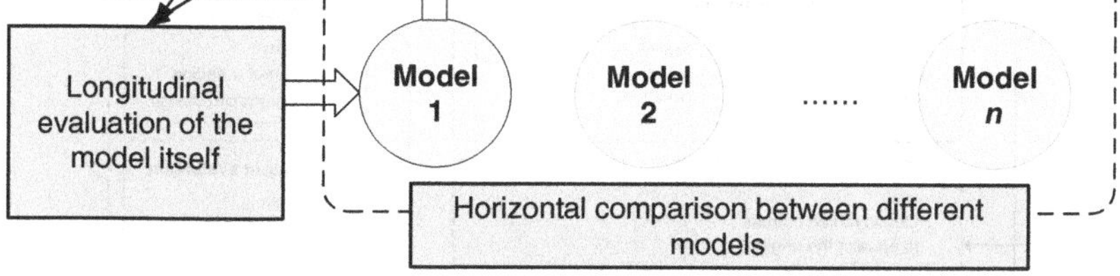

Figure 9.4 Structure of the comprehensive evaluation framework.

the practical applicability of models. To validate the effectiveness of the proposed evaluation framework, we conducted a case study using several representative COVID-19 models. These models were evaluated and compared to assess the framework's capability in benchmarking and model performance analysis.

9.2.3 Capability Maturity Evaluation Framework

As shown in Figure 9.5, the complete modeling process is divided into four steps. (1) **Problem formulation.** The first step in software development is requirements analysis. Similarly, before formal modeling, we need to clearly define the problem to be solved, thus providing clear direction for the entire research process. (2) **Research object identification.** After clarifying the problem, we need to determine the research object, that is, which variables will be used to measure whether the target outcome is achieved and which factors influence the target outcome. (3) **Model construction.** Based on the research question and objective, we use appropriate methods to build a computational model based on epidemiological characteristics. (4) **Answering the question.** The outputs of the model are obtained and analyzed to answer the question posed in the first step.

Based on steps 1 and 2, we can conduct a preliminary expectation evaluation of the model; based on steps 3 and 4, we can conduct an implementation evaluation

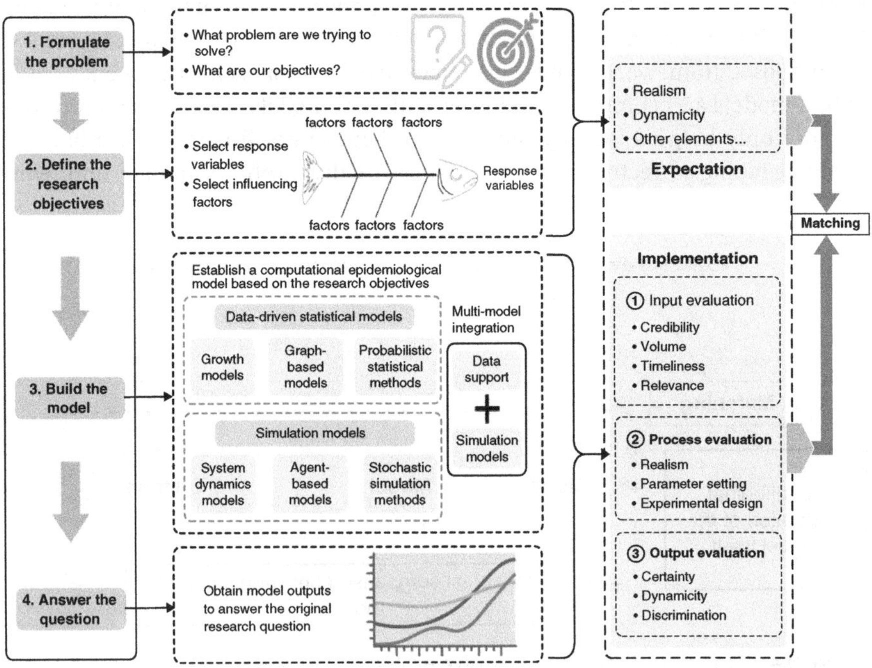

Figure 9.5 comprehensive evaluation framework for the capability maturity of computational epidemiological models.

of the model. The results of implementation evaluation tell us how the model performs, and the implementation evaluation results of multiple models can help us identify models with higher capability maturity. The elements in expectation evaluation are the same as those in implementation evaluation. By matching the results of expectation evaluation with those of implementation evaluation, we can determine whether the model itself meets expectations and addresses the proposed problem.

Whether it is a data-driven model or a simulation model, the implementation process includes input, modeling process, and output, and the quality of these three aspects affects the overall performance of the model. We cannot arbitrarily judge the merits of a particular modeling approach; instead, we should assess it based on the entire implementation process. Therefore, model implementation evaluation includes input evaluation, process evaluation, and output evaluation. Input evaluation assesses the reliability and effectiveness of input data sources (including data credibility, volume, timeliness, and relevance). Process evaluation evaluates the modeling process to ensure that the model reflects the internal structural characteristics of the real system, thereby ensuring that the intermediate processes by which assumptions generate model behavior are correct. This is a prerequisite for the model results to be consistent with the assumptions. Output evaluation assesses the usability of the model results. According to qualitative evaluation criteria, the evaluation results for each element are generally graded into 0.5–1 (low), 1.5–2 (medium), and 2.5–3 (high) [49].

9.3 Capability Maturity of Computational Epidemiological Models

9.3.1 Expectation Evaluation

A model is a simplification of certain aspects or levels of the real world. The choice of these aspects and levels depends on the modeler's research objectives. Therefore, an effective evaluation framework should consider the extent to which the model achieves its intended purpose. To this end, we introduce problem identification and objective prioritization techniques to evaluate the model's expectations based on the research question and objectives. The elements of expectation evaluation include credibility, data volume, timeliness, relevance, realism, parameter setting, experimental design, determinacy, dynamics, and discriminative power. The specific evaluation methods are as follows:

1) **Realism:** Model design aims to strike a balance between concreteness and abstraction at a specific level, so the realism of the model's description of the real world should be adjusted based on the research problem and objectives. If the research goal is for the results to be directly applicable to the real world, more real-world factors need to be considered. For example, studies on the

global spread of an epidemic under different conditions are primarily focused on theoretical exploration and are highly abstract, allowing many factors to be ignored; whereas studies on the optimal intervention measures for a specific region are often intended to directly inform government decision-making and thus require incorporating more real-world elements. The relationship between realism and the corresponding model is shown in Table 9.1.

2) **Dynamics:** The process of epidemic transmission is a time-varying dynamic process, but different research questions focus on different aspects, and the expectations for model dynamics also vary. The requirements for dynamics in models that predict the epidemic situation at a specific point in time are clearly different from those in real-time prediction models. Models that evaluate the impact of interventions need to capture and compare changes before and after interventions and, therefore, have higher requirements for dynamics. The relationship between dynamics ratings and corresponding models is shown in Table 9.2.

For the credibility, data volume, timeliness, and relevance of inputs, parameter settings, and experimental design during implementation, as well as the

Table 9.1 Realism Ratings and Their Corresponding Model Relationships.

Realism	**Corresponding Model**
0.5	Model for studying global epidemics
1	Model for studying epidemics in a specific country
1.5	Model for studying epidemics in a specific city/region
2	Model for evaluating the impact of interventions (global/national level)
2.5	Model for evaluating the impact of interventions (regional/city level)
3	Model for estimating the medical resource demand for epidemics

Table 9.2 Dynamics Ratings and Their Corresponding Model Relationships.

Dynamicity	**Corresponding Model**
0.5	Model for estimating the epidemic situation at a specific point in time
1	Model for estimating the epidemic situation after a specific number of days
1.5	Model for studying epidemic transmission risk or growth patterns
2	Model for estimating disease transmissibility (e.g., R0)
2.5	Model for evaluating the impact of interventions
3	Model for real-time monitoring of the epidemic development

determinacy and discriminative power of outputs, any model is expected to achieve optimal performance. Therefore, the evaluation results for all these elements are set to 3 (the highest level).

9.3.2 Input Evaluation

Valid model inputs are crucial for ensuring reliable model results. Both data-driven statistical models and simulation models require the guaranteed reliability of input data. Typically, epidemiological models use confirmed cases and death counts to define an epidemic, while other factors influencing the results (such as population mobility data and social data) vary with the research subject. The reliability of these input data will affect the model's results. In our evaluation framework, input evaluation includes the following factors as represented by Eq. (9.1), where C_I denotes credibility, V_I denotes data volume, T_I denotes timeliness, and R_I denotes relevance. If there are multiple inputs, the result is the average of the scores for the different input data.

$$\text{Input} = < C_I, V_I, T_I, R_I > \tag{9.1}$$

1) **Credibility (C_I):** The process of data collection is essentially an artificial process carried out according to specific requirements, and this process introduces bias from the outset. Furthermore, it is impossible to guarantee complete authenticity during the collection process. In this context, large-scale real-time data collected by third parties can be considered more credible, while data collected by researchers themselves for specific research questions may be less credible. For case data, credibility is also affected by factors such as testing capacity and coverage. Retrospective studies generally have higher credibility due to having more time for data verification, more complete theoretical results, and more mature techniques. The relationship between credibility scores and corresponding data type is shown in Table 9.3.
2) **Data volume (V_I):** In statistics, the size of the sample is directly related to the reliability of the results—that is, given a fixed population, the larger the sample size, the smaller the representational bias of the results. For the same population, more data can reduce the probability of random errors and problems of insufficient representativeness. For different types of data, a unified "quantity" cannot be simply used to measure data volume; the continuity of the data is a more important indicator. Compared to data from a single day, data collected over a continuous period is less likely to be affected by random errors and representativeness issues. The relationship between data volume scores and corresponding data continuity is shown in Table 9.4.
3) **Timeliness (T_I):** Disease transmission is a dynamic and uncertain process, influenced by numerous factors and prone to sudden events. As intervention

Table 9.3 Credibility Ratings and Their Corresponding Data Type.

C_I	Corresponding Data Type
0.5	Reported data before January 20, 2020 (daily reports by China's National Health Commission began on January 21)
1	Reported data from January 21 to January 31, 2020 (various nucleic acid testing products were approved by the end of January)
1.5	Reported data after February 2020
2	Data collected by researchers themselves for specific studies
2.5	Data obtained from retrospective studies; integrated data from multiple sources
3	Data collected and organized by third parties (e.g., telecommunications big data, transportation big data)

Table 9.4 Data Volume Ratings and Their Corresponding Data Continuity.

V_I	Corresponding Data Continuity
0.5	Data from a single day
1	Data from 2 to 5 consecutive days
1.5	Data from 6 to 10 consecutive days
2	Data from 11 to 15 consecutive days
2.5	Data from 16 to 30 consecutive days
3	Data from more than 30 consecutive days

measures improve and public awareness of epidemic prevention increase, the transmission situation changes rapidly, thus making the timeliness of input data a critical factor affecting model results. For epidemiological models, the more recent the data used, the more reliable the model. The relationship between timeliness ratings and corresponding data cutoff standards is shown in Table 9.5.

4) **Relevance (R_I):** Sometimes, the data we wish to study are not available in existing big data resources, or the direct data relevant to the study are incomplete. As a result, many researchers use indirect variables for their studies, such as population mobility data [50] and social media data [51]. However, the use of indirect variables affects the relevance of the data to the research objectives, thereby impacting the reliability of the model. We consider direct data to have higher relevance, while proxy variable data have lower relevance. The relationship between relevance ratings and corresponding data source type is shown in Table 9.6.

Table 9.5 Timeliness Ratings and Their Corresponding Data Cutoff Standards.

T_I	Corresponding Data Type and Cutoff Time
0.5	Case and social data (data cutoff more than 30 days before the end of the study); population mobility data and similar types (data cutoff more than 3 years before the study)
1	Case and social data (data cutoff 20–30 days before the end of the study); population mobility data and similar types (data cutoff 1–3 years before the study)
1.5	Case and social data (data cutoff 15–20 days before the end of the study); population mobility data and similar types (data cutoff more than 30 days but less than 1 year before the study)
2	Case and social data (data cutoff 10–15 days before the end of the study); population mobility data and similar types (data cutoff 20–30 days before the study)
2.5	Case and social data (data cutoff 5–10 days before the end of the study); population mobility data and similar types (data cutoff 10–20 days before the study)
3	Case and social data (data cutoff within 5 days before the end of the study); population mobility data and similar types (data cutoff within 10 days before the study)

Table 9.6 Relevance Ratings and Their Corresponding Data Source Type.

R_I	Corresponding Data Source Type
0.5	Nonepidemic type data from nontarget regions
1	Nonepidemic type data from a subset of the target region
1.5	Nonepidemic type data from the target region
2	Epidemic type data from nontarget regions
2.5	Epidemic type data from a subset of the target region
3	Epidemic type data from the target region

9.3.3 Process Evaluation

All models are the result of certain assumptions and reasoning. In the modeling process, differences in model assumptions, parameter settings, and methods can lead to significantly different results. When a new epidemic breaks out, theoretical research findings on virus transmission become an important basis for model construction, providing references for how well the model reflects the real world, as well as for parameter settings and experimental design. The following section provides a detailed introduction to process evaluation for computational epidemiological models, as shown in Eq. (9.2), where R_P denotes the realism of the description of the real world, P_P represents parameter settings, and E_P represents experimental design.

$$\text{Process} = < R_P, P_P, E_P > \tag{9.2}$$

1) **Realism (R_P):** The process of modeling disease transmission is essentially a process of abstracting and simplifying the real world, and this process should not compromise the credibility of the model—i.e., the simplification should retain key real-world features. For example, droplet and contact transmission are the main modes of COVID-19 spread, which means short-range transmission is the primary driver for the rapid spread of the epidemic. In this context, the frequency of human contact, or the density of social network connections, is a crucial factor influencing epidemic transmission. Key factors such as medical resource availability and intervention measures are also important for epidemic development. Therefore, in epidemic transmission systems, the virus, population, epidemic-related environment, and interventions are all essential components and should be the focus of abstraction and simplification. Regardless of the modeling approach used, accurately describing these elements is key to the model's capability. The relationship between realism ratings and corresponding models is shown in Table 9.7.
2) **Parameter setting (P_P):** Model results and their stability are highly dependent on parameter settings. Generally, parameters can be set manually by the researcher based on experience or obtained through ML. When set manually,

Table 9.7 Realism Ratings and Their Corresponding Model Relationships.

R_P	Corresponding Model
0.5	Virus (uncertainty ignored), population (assumed homogeneous), environment (not considered), interventions (not considered)
1	Virus (uncertainty ignored), population (assumed homogeneous), environment (not considered), interventions (simplistically considered)
1.5	Virus (uncertainty ignored), population (simple classified), environment (not considered), interventions (simplistically considered)
2	Virus (uncertainty ignored), population (simple classified), environment (simply abstracted), interventions (simplistically considered)
2.5	Virus (uncertainty considered), population (assumed heterogeneous), environment (simplistically abstracted), interventions (dynamically changing)
3	Virus (uncertainty considered), population (assumed dynamically heterogeneous), environment (concrete and realistic), interventions (dynamically changing)

Table 9.8 Parameter Setting Ratings and Their Corresponding Parameter Source Standards.

P_P	Corresponding Parameter Source Standards
0.5	Epidemiological parameters (referencing data from other coronaviruses), other parameters (data published by official organizations or institutions more than 3 years prior)
1	Epidemiological parameters (referencing data from other coronaviruses), other parameters (data published by official organizations or institutions 1–3 years prior)
1.5	Epidemiological parameters (referencing early research reports), other parameters (data published by official organizations or institutions 6+ months to 1 year prior)
2	Epidemiological parameters (referencing early research reports), other parameters (data published by official organizations or institutions 1+ months to 6 months prior)
2.5	Epidemiological parameters (obtained through real-time big data machine learning), other parameters (data published by official organizations or institutions 10+ days to 1 month prior)
3	Epidemiological parameters (referencing the latest laboratory research findings), other parameters (data published by official organizations or institutions within 10 days prior)

modelers usually rely on past experience, existing theoretical results, reports, and other relevant sources; in principle, the latest published theories or research findings can be considered closest to objective reality. Parameters obtained in real time through ML using big data are also highly reliable. The relationship between parameter setting ratings and corresponding parameter source standards is shown in Table 9.8. If there are multiple parameters, the result is the average of the scores for the different parameters.

3) **Experimental design (E_P):** In most cases, many factors can influence experimental results, and model outputs alone cannot identify the key factors driving those experimental results. Causal theory is an important tool for helping us uncover the objective laws of complex systems. To conduct causal inference, it is necessary to statistically adjust nonexperimental data to eliminate the effects of preexisting differences between intervened and nonintervened groups, or to use active interventions to isolate the target variable from other variables. Therefore, experimental design is particularly important. We argue that models with standardized experimental designs are more reliable. In this context, a standard experimental design should meet the following criteria: randomization, controlled trials, replication, and factorial design. The relationship between experimental design ratings and corresponding design standards is shown in Table 9.9.

Table 9.9 Experimental Design Ratings and Their Corresponding Design Standards.

E_P	Corresponding Experimental Design Standards
0.5	No experimental design implemented
1	Simple control design implemented
1.5	Achieves one of the following: control, replication, or randomization
2	Achieves two of the following: control, replication, or randomization
2.5	Implements control + replication + randomization
3	Implements randomization + control + replication + factorial design

9.3.4 Output Evaluation

The practical needs for epidemic prediction models mainly include assessing epidemic situations and optimizing intervention strategies. Models for assessing the epidemic situation typically use the number of infections, deaths, and the basic reproduction number (R0) as outputs. Models for optimizing intervention strategies compare scenarios with different control measures to identify the most effective intervention strategy or combination of strategies. Traditional output validation methods include classical hypothesis tests such as the U-test and t-test, which are typically conducted by the modeler. This paper's output evaluation focuses on the determinacy of the output results, the feasibility of dynamic monitoring, and the ability to distinguish different entities in the system. This can be represented by Eq. (9.3), where C_O denotes determinacy, D_O denotes dynamics, and D'_O denotes discriminative power. If there are multiple outputs, the result is the average of the scores for the different outputs.

$$\text{Output} = < C_O, D_O, D'_O > \tag{9.3}$$

1) **Determinacy:** One way to improve model reliability is to make the model more realistic; however, this often leads to an increase in the number of model parameters, and the uncertainty in these parameters—stemming from cognitive uncertainty—leads to greater uncertainty in the output. High uncertainty means that more caution is needed when applying the model's conclusions to practice. Incorporating uncertainty analysis is a core component of computational model maturity, and modelers should employ various methods to reduce this uncertainty, such as sensitivity analysis and Monte Carlo simulation. Typically, a 95% CI is provided for output results Eq. (9.4), and the width of this interval can be used to measure the uncertainty of the output; Eq. (9.5) shows the calculation method for uncertainty, where w_i is

Table 9.10 Determinacy Rating Scale.

C_O	Uncertainty	R^2
3	(0, 0.1)	(0.95, 1)
2.5	(0.1, 0.3)	(0.9, 0.95)
2	(0.3, 0.5)	(0.85, 0.9)
1.5	(0.5, 0.8)	(0.8, 0.85)
1	(0.8, 1.5)	(0.7, 0.8)
0.5	(1.5, 3)	(0.6, 0.7)
0	(3, +∞)	(0, 0.6)

the result of the i-th output variable, Pr denotes probability, and μ is the mean of the output results. In addition, the coefficient of determination (R^2), a statistic that measures goodness of fit, can be used to assess the determinacy of regression models. Specific rating criteria are provided in Table 9.10.

$$Pr(c_{i1} \leq w_i \leq c_{i2}) = 0.95 \tag{9.4}$$

$$\text{Uncertainty} = \frac{1}{n}\sum_{i=1}^{n}\frac{c_{i2} - c_{i1}}{\mu} \tag{9.5}$$

2) **Dynamics (D_O):** The significance of epidemiological models lies in providing early warnings, assisting relevant authorities in responding in a timely manner, and providing references for the allocation of medical and health resources. However, disease transmission itself is an uncertain and dynamic process; thus, a mature epidemiological model should generate dynamic outputs. A model with dynamic outputs is more instructive for real-life applications and holds certain value for future applications. The relationship between dynamics ratings and corresponding output dynamic characteristics is shown in Table 9.11.
3) **Discriminative power (D_O'):** For disease risk prediction models, discriminative power is an important dimension for evaluating their maturity. In this context, "discriminative power" refers to the ability to correctly distinguish between individuals at high risk and those at low risk for a given outcome. Therefore, a mature predictive model should be able to differentiate the risk levels of individuals from different regions or with different characteristics for a specific outcome. The relationship between discrimination power ratings and corresponding distinction capabilities is shown in Table 9.12.

Table 9.11 Dynamics Ratings and Their Corresponding Output Dynamic Characteristics.

D_O	Corresponding Output Dynamic Characteristics
0.5	Outputs only the situation at a specific point in time
1	Outputs the situation at several discrete time points
1.5	Outputs an epidemic curve
2	Outputs parameters representing transmission dynamics
2.5	Outputs the situation under different intervention measures
3	Provides daily updated results

Table 9.12 Discriminative Power Ratings and Their Corresponding Distinction Capabilities.

D_O'	Corresponding Distinction Capabilities
0.5	Population (no distinction between different groups in results), region (no consideration of subregions)
1	Population (no distinction between different groups in results), region (only a few subregions considered)
1.5	Population (only one characteristic considered), region (only a few subregions considered)
2	Population (only one characteristic considered), region (several subregions considered)
2.5	Population (two characteristics considered), region (several subregions considered)
3	Population (characteristics such as gender, age, medical history considered), region (all subregions considered)

After obtaining the evaluation results for all elements in the implementation module, their sum gives the total implementation evaluation score. The higher the total implementation evaluation score, the higher the capability maturity level of the model, i.e., the better its overall performance. In addition, by matching the results of each element in the implementation evaluation with the results of the corresponding element in the expectation evaluation, we can determine how effectively the model achieves its intended objectives. If the implementation evaluation results meet or approach the expectation evaluation results, it indicates that the model has satisfactorily achieved its intended goals and addressed the originally proposed problem.

9.4 Case Study of Computational Epidemiological Models

Since the outbreak of COVID-19, numerous research institutions and scientists around the world have devoted themselves to researching the pandemic, hoping to provide support for decision-makers. As a result, the number of developed models and published papers has increased exponentially, which provides a large number of computational epidemiological models as research subjects for this study. Therefore, this chapter selects several COVID-19 models for case analysis to verify the effectiveness of the comprehensive evaluation framework.

The epidemic process of infectious diseases can generally be divided into the introduction period, diffusion period, outbreak period, decline period, and termination period. Although COVID-19 may coexist with humans for a long time, the role of computational models in predicting transmission trends, assessing epidemic impact, and optimizing intervention strategies is particularly prominent during the early ascending phase of the outbreak. The COVID-19 models selected in this study are all from this stage, and Table 9.13 shows the characteristic attributes of these models in chronological order. The "time" attribute in the table refers to the submission date of the paper or the date it was made publicly available online; if there is a revised version, the revision date is used.

The case studies selected in this chapter include the model developed by the Imperial College London team (hereafter referred to as model 1), the model developed by the Chinese University of Hong Kong team (hereafter referred to as model 2), the model developed by Professor Joseph T Wu's team at the University of Hong Kong Faculty of Medicine (hereafter referred to as model 3), the model developed by the Georgia State University team (hereafter referred to as model 4), the model developed by Dr. Jayson Jia's team at the University of Hong Kong (hereafter referred to as model 5), the model developed by Academician Zhong Nanshan's team at the National Clinical Research Center for Respiratory Diseases (hereafter referred to as model 6), the model developed by Dr. Adam J Kucharski's team at the London School of Hygiene & Tropical Medicine (hereafter referred to as model 7), the model developed by epidemiologist Ashleigh Tuite's team at the University of Toronto (hereafter referred to as model 8), and the model developed by Professor Pinar Keskinocak's team at the Georgia Institute of Technology (hereafter referred to as model 9). These models include those developed by both Chinese and international researchers and institutions, and the research subjects cover both domestic and international regions. In terms of modeling approaches, these models encompass both data-driven statistical models and simulation models, thus ensuring the comprehensiveness of model selection.

The nine selected COVID-19 models were scored using the automated computational epidemiological model evaluation system developed in this study. Figure 9.6

Table 9.13 Characteristic Attributes of the Model.

Model ID	Time	Research Question	Research Subject and Method	Input Data Source	System Element Description	Parameter Setting	Experimental Design	Result
1. Imai et al. [52]	2020/1/17	Estimate the number of clinical cases that may have occurred in Wuhan by January 2020.	Estimate the scale of similar cases in Wuhan based on the number of cases detected outside China.	The number of confirmed cases detected outside China as of January 16, 2020, was referenced from reports.	Virus (uncertainty not considered); population (homogeneous); environment (not considered); interventions (not considered)	The population of Wuhan was referenced from the WHO report on January 5, 2020; epidemiological parameters were based on early reports and studies; the daily number of international outbound passengers from Wuhan International Airport was adjusted according to 2018 IATA data.	The baseline scenario was compared with four alternative scenarios.	As of January 12, 2020, there were 1723 (95% CI: 427–4471) symptomatic 2019-nCoV cases in Wuhan.
2. Zhao et al. [37]	2020/1/27	Estimate the basic reproduction number of the novel coronavirus.	Model the epidemic curve of cases in mainland China using an exponential growth model.	The time-series data of cases in mainland China from January 10 to 24, 2020, were obtained from the Wuhan Municipal Health Commission.	Virus (uncertainty not considered); population (homogeneous); environment (not considered); interventions (simply considered)	Epidemiological parameters were obtained by fitting case numbers and referencing parameters from SARS and MERS.	Six different reporting rates and different serial interval (SI) scenarios were compared.	After the diagnostic protocol was released on January 17, 2020, and many medical supplies arrived in Wuhan, if reporting increased by 2–8 times, the average R0 value of 2019-nCoV ranged from 2.24 (95% CI: 1.96–2.55) to 3.58 (95% CI: 2.89–4.39).

3. Wu et al. [39]	2020/2/4	Estimate the epidemic scale in Wuhan from December 1, 2019, to January 25, 2020, and predict the public health risks of the epidemic in China and globally.	Infer the number of infections in Wuhan based on the number of cases exported from Wuhan to cities outside mainland China, and estimate the number of cases exported from Wuhan to other cities in mainland China; develop a composite SEIR meta-population model to simulate the epidemic in all major cities in China since December 2019.	As of January 28, 2020, data on cases exported from Wuhan to regions outside mainland China were obtained from reports published by the Chinese Center for Disease Control and Prevention.	Virus (uncertainty not considered); population (simple classification); environment (not considered); interventions (simply considered)	Population mobility data into and out of Wuhan were referenced from official airline guides, the Tencent database, and 2019 data from the Wuhan Traffic Management Bureau; epidemiological parameters were based on previous studies of SARS-CoV and MERS-CoV.	Randomization was implemented, and different scenarios were set for the infectivity of zoonotic diseases.	Under the baseline scenario, the R0 of 2019-nCoV was 2.68 (95% CI: 2.47–2.86). As of January 25, 2020, 75,815 people (95% CI: 37,304–130,330) were infected in Wuhan. Chongqing, Beijing, Shanghai, Guangzhou, and Shenzhen imported 461 (95% CI: 227–805), 113 (57–193), 98 (49–168), 111 (56–191), and 80 (40–139) cases from Wuhan, respectively.

(Continued)

Table 9.13 (Continued)

Model ID	Time	Research Question	Research Subject and Method	Input Data Source	System Element Description	Parameter Setting	Experimental Design	Result
4. Roosa et al. [53]	2020/2/11	Real-time prediction of the epidemic in China.	Calibrate the generalized logistic growth model, Richards model, and subepidemic wave model using the daily cumulative reported cases from Hubei Province and other provinces.	Cumulative daily confirmed case reports for each province in China from January 22 to February 9, 2020, were collected from the National Health Commission of China.	Virus (uncertainty not considered); population (homogeneous); environment (not considered); interventions (not considered)	Model parameters were determined by fitting the model to daily confirmed case data.	Not reported	In the following 5 days, Hubei would have an average of 7409–7496 new confirmed cases, and other provinces would have an average of 1128–1929 new cases. As of February 24, the average cumulative number of cases in Hubei was 37,415–38,028, and in other provinces, 1588–13,499. As the epidemic continued, daily online forecasts were published 10 days in advance.
5. Jia et al. [50]	2020/2/18	Predict the transmission risk and growth patterns in different regions of China.	Explore the spread of COVID-19 using population outflow data from Wuhan to different regions of China (determined by mobile phone records).	Population mobility data from January 1 to 24, 2020, were provided by a major Chinese telecom operator; daily infection data from January 26 to February 19 were obtained from the government.	Virus (uncertainty not considered); population (homogeneous); environment (spatiotemporal dynamic changes); interventions (not considered)	Parameters were determined by fitting case data from government reports up to February 19, 2020, using nonlinear least squares (Levenberg–Marquardt algorithm).	Population mobility factors were compared with other factors, and other models were also compared.	The epidemic curve and risk index were obtained. Compared with other indicators, population geographic mobility was more effective in estimating the epidemic situation in other regions of mainland China. The goodness of fit (R^2) was 0.772 on January 24 and 0.946 on February 19.

6. Yang et al. [54]	2020/2/27	Forecast the epidemic trend of COVID-19 in China before and after January 23, 2020, under public health interventions.	Improve the SEIR model by considering population dynamics, combine domestic population migration data with the latest COVID-19 epidemiological data to predict the epidemic, and then use 2003 SARS data for artificial intelligence training.	Epidemiological data before and after January 23, 2020 (from January 16 to 25), were sourced from reports by the National Health Commission of China; migration data were obtained from an online program; SARS data were sourced from an archived news website.	Virus (uncertainty not considered); population (dynamic changes, simple classification); environment (not considered); interventions (simply considered)	The incubation period was obtained from reports; the SEIR model was fitted to epidemic data from Hubei Province to obtain infection and recovery/mortality rates for Hubei; the number of susceptible individuals was set as the resident population; population migration data were from Baidu Migration's 2020 Spring Festival data.	Randomization was implemented; the LSTM model was optimized using the Adam optimizer and run for 500 iterations.	Based on current interventions, Hubei, Guangdong, and Zhejiang were expected to reach epidemic peaks on February 20, and mainland China on February 28, with a peak of 59,764 cases (95% CI: 51,979–70,172) and a total epidemic size of 122,122 cases (95% CI: 89,741–156,794). The trend would gradually decline by the end of April. If interventions were delayed by 5 days, the epidemic size would triple.
7. Kuchar-Ski et al. [55]	2020/3/11	Estimate changes in epidemic transmission in Wuhan from December 2019 to February 2020 (before and after Wuhan's lockdown).	Fit a stochastic transmission dynamics model (an improved SEIR model) with case data from Wuhan and international export case data from Wuhan.	Case data and international export case data for Wuhan from the outbreak to the first few days of February 2020 were collected from multiple datasets.	Virus (uncertainty considered); population (simple classification); environment (not considered); interventions (simply considered)	Model parameters were determined by fitting the model to COVID-19 case data.	Randomization was implemented, and different scenarios under various assumptions were compared.	The median daily reproduction number in Wuhan dropped from 2.35 (95% CI: 1.15–4.77) in the week before travel restrictions on January 23, 2020, to 1.05 (0.41–2.39) in the week after.

(Continued)

Table 9.13 (Continued)

Model ID	Time	Research Question	Research Subject and Method	Input Data Source	System Element Description	Parameter Setting	Experimental Design	Result
8. Tuite et al. [56]	2020/4/2	Evaluate how different intervention measures can be used to control the pandemic and reduce the burden on the healthcare system.	Establish an age-structured SEIR model to describe the spread of COVID-19 in the population of Ontario, Canada.	The initial 750 cases were based on 150 cases reported in Ontario on March 19, 2020, assuming a 20% reporting rate.	Virus (uncertainty considered); population (different age structures and health conditions considered); environment (not considered); interventions (dynamically changing)	Contact patterns within and between age groups were based on the POLYMOD study; comorbidity estimates for each age group were obtained from the Canadian Community Health Survey (CCHS) in Ontario; epidemiological parameters were sourced from published studies.	Randomization was implemented; the baseline scenario was compared with various combinations, with each scenario repeated 100 times.	With limited testing, isolation, and quarantine, it was estimated that 56% (95% CI: 42–63%) of individuals in Ontario would be infected. At the epidemic peak, the number of hospital (non-ICU) and ICU cases would be 107,000 (95% CI: 60,760–149,000) and 55,500 (95% CI: 32,700–75,200), respectively. All interventions would delay and reduce the peak, with restrictive physical distancing being the most effective.

9. Keskin-Ocak et al. [33]	2020/4/29	Predict the number and situation of infections in Georgia under continued social distancing, as well as the demand for local hospital medical resources.	Develop an agent-based simulation model to predict the spread of infections.	County-level confirmed COVID-19 cases and deaths were collected from The New York Times; county-level hospitalization data were obtained from the Georgia Department of Public Health, covering data through the end of April.	Virus (uncertainty considered); population (uncertainty and heterogeneity considered); environment (more realistic); interventions (dynamically changing)	Population-related parameters were obtained from Georgia; COVID-19 parameters were obtained from the literature.	Randomization was implemented; different scenarios under various combinations were compared, and the average was taken from 30 runs.	Shelter-in-place and voluntary quarantine would reduce the infection peak to below 53,000 and delay the peak from April to July or later. Increasing shelter-in-place from 4 to 5 weeks would reduce cumulative infections and deaths by 2–9% and 3–11%, respectively. Under different scenarios, cumulative deaths ranged from 6660 to 19,430. The situation and medical resource needs for different counties were distinguished.

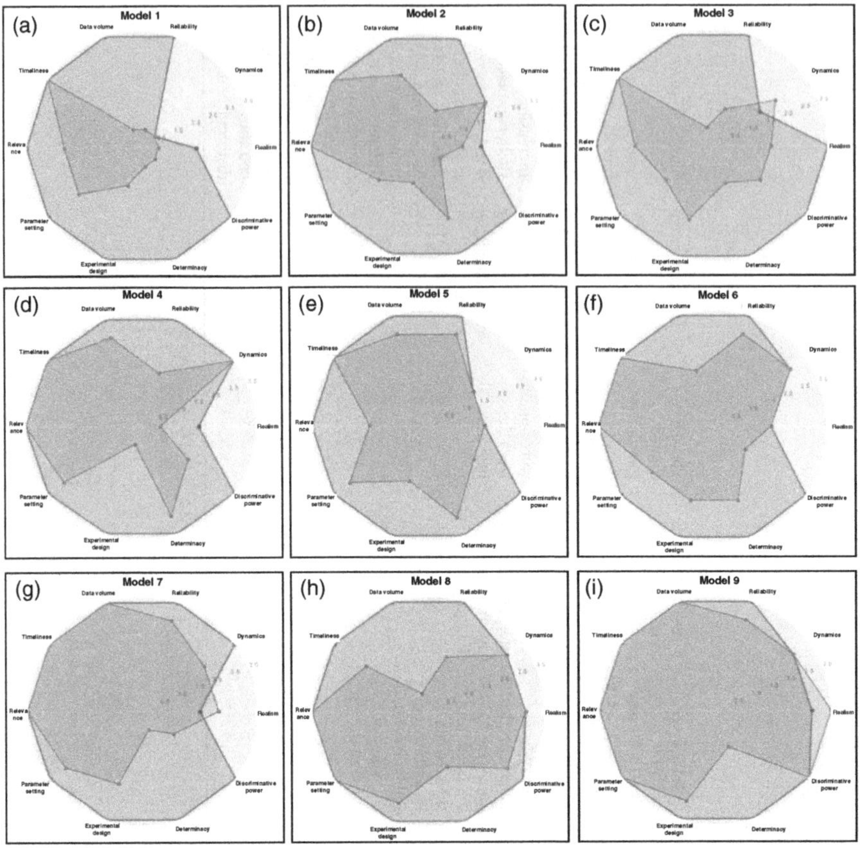

Figure 9.6 System demonstration diagram of model evaluation results. (a) Evaluation results of model 1. (b) Evaluation results of model 2. (c) Evaluation results of model 3. (d) Evaluation results of model 4. (e) Evaluation results of model 5. (f) Evaluation results of model 6. (g) Evaluation results of model 7. (h) Evaluation results of model 8. (i) Evaluation results of model 9.

presents a system demonstration of the evaluation results for each model. The main part of the figure shows the matching between the expectation and implementation evaluation results: the shape formed by Red – Medium gray line segments connects red dots representing the evaluation results of each element in the expectation evaluation, while the shape formed by blue line segments connects blue dots representing the evaluation results of each element in the implementation evaluation. If the shape formed by the blue line segments approaches or matches that of the red line segments, the model can be considered to have met its expectations; if the blue dot for a particular element is far from the red dot, it indicates that this element is an area where the model needs improvement.

Table 9.14 Implementation Evaluation Results of Each Computational Epidemiological Model.

Model ID	C_I	V_I	T_I	R_I	R_P	P_P	E_P	C_O	D_O	D'_O	Total Evaluation Score
1.	0.5	0.5	3	2	0.5	2	1	0.5	0.5	0.5	11
2.	1	2	3	3	1	1.5	1	2	2	0.5	17
3.	1	0.5	3	2	1.5	1.5	2	1	2	1.5	16
4.	1.5	2.5	3	3	0.5	2.5	0.5	2.5	3	1.5	20.5
5.	2.5	2.5	3	1.5	1.5	2.5	1.5	2.5	1.5	1.5	20.5
6.	2.5	1.5	3	3	1.5	2	2	2	2.5	1	21
7.	2.5	3	3	3	2	2.5	2	0.5	2	1	21.5
8.	1.5	0.5	2	3	2.5	3	2.5	1.5	2.5	2.5	21.5
9.	2.5	3	3	3	3	3	2.5	1	2.5	3	26.5

As can be seen from the figure, among all nine models, only one element in model 9 falls short of expectations, while the other elements almost all meet expectations effectively; in contrast, several elements in the other models show gaps between implementation evaluation results and expectations. The implementation evaluation results for models 1 through 9 are summarized in Table 9.14, which also lists the total implementation evaluation score for each model. Figure 9.7

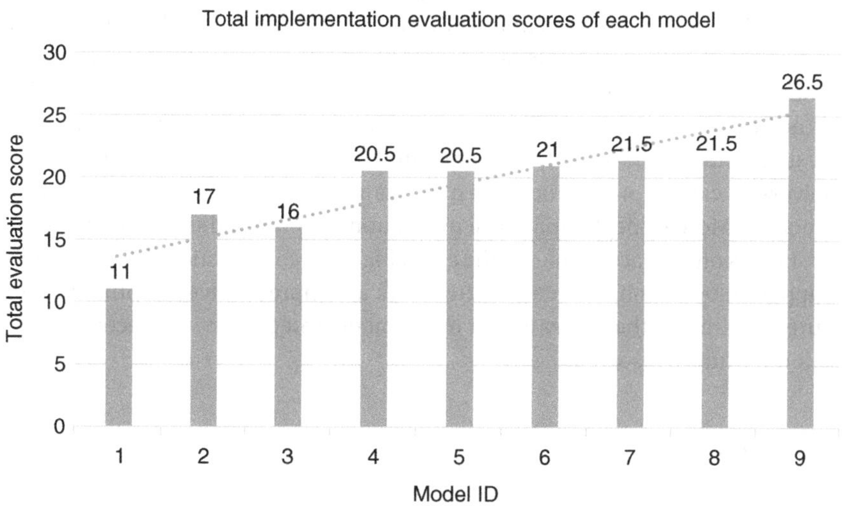

Figure 9.7 The trend of total implementation evaluation scores for models 1 through 9.

shows the trend of total implementation evaluation scores for models 1 through 9. It is evident that the capability maturity of epidemiological models exhibits an overall upward trend over time. This is consistent with reality: as the epidemic progressed, theoretical research on the virus and disease became more mature, more knowledge and data were accumulated, and the models became more consistent with the real world.

9.5 Chapter Summary

Due to their repeatability, computational experiments have been widely adopted in applied research where direct experimentation is prohibitively risky or costly. With their increasingly widespread application, computational experiments face several challenges. The issue of arbitrariness in the construction of computational models has not yet been resolved. While mathematical models can overcome this problem by strictly constraining the initial form of the model, the flexibility and creativity of computational models often lead modelers to continually add new features, making model validation more difficult. There is still no universally accepted standard to ensure model validity. Therefore, we have proposed a capability maturity evaluation framework for computational epidemiological models, and we have also conducted case studies on several typical COVID-19 models.

Although the capability maturity evaluation framework proposed in this chapter is designed for computational epidemiological models, the evaluation methodology is based on a general modeling workflow, which allows it to be extended to other domains for the evaluation and comparison of various computational models. The evaluation results go beyond single-point assessments by identifying the "weaknesses" of models and offering guidance for the refinement of models in future iterations. This work still has some limitations. Due to the generality of the framework, it cannot provide detailed quantitative analysis when evaluating specific models. More in-depth analytical evaluations require additional model-level data, which is often inaccessible. In the era of big data, computational experiments are supported by abundant data resources and a comprehensive scientific infrastructure. As a result, the evaluation of multi-model integration will become a key direction for future research.

References

1 Sugden, R. (2000). Credible worlds: the status of theoretical models in economics. *Journal of Economic Methodology* 7: 1–31.

2 Kalter, F. and Kroneberg, C. (2014). Between mechanism talk and mechanism cult: new emphases in explanatory sociology and empirical research. *Kolner Zeitschrift FurSoziologie Und Sozialpsychologie* 66: S91–S115.

3 Grimm, V., Revilla, E., Berger, U. et al. (2005). Pattern-oriented modeling of agent-based complex systems: lessons from ecology. *Science* 310 (5750): 987–991.

4 Morgan, S.L. and Winship, C. (2015). *Counterfactuals and Causal Inference: Methods and Principles for Social Research*, 2nde. Cambridge: Cambridge University Press.

5 Leon-Medina, F.J. (2017). Analytical sociology and agent-based modeling: is generative sufficiency sufficient? *Sociological Theory* 35 (3): 157–178.

6 Humphreys, P. (2009). The philosophical novelty of computer simulation methods. *Synthese* 169 (3): 615–626.

7 Izquierdo, L.R., Izquierdo, S.S., Galan, J.M., and Santos, J.I. (2009). Techniques to understand computer simulations: Markov Chain analysis. *Journal of Artificial Societies and Social Simulation* 12 (1): 6.

8 Gintis, H. (2013). Markov models of social dynamics: theory and applications. *ACM Transactions on Intelligent Systems and Technology* 4 (3): 53.

9 Railsback, S.F. and Grimm, V. (2019). *Agent-Based and Individual-Based Modeling: A Practical Introduction*, 2nde. Princeton: Princeton University Press.

10 http://www.swarm.org.

11 https://repast.github.io/index.html.

12 Cioffi-Revilla, C. and Mark, R. (2010). Mason rebeland: an agent-based model of politics, environment, and insurgency. *International Studies Review* 12 (1): 31–52.

13 Tisue, S. and Wilensky, U. (2004). Netlogo: a simple environment for modeling complexity. *International Conference on Complex Systems* 21: 16–21.

14 Auchincloss, A.H. and Roux, A.V.D. (2008). A new tool for epidemiology: the usefulness of dynamic-agent models in understanding place effects on health. *American Journal of Epidemiology* 168 (1): 1–8.

15 Smith, E.R. and Conrey, F.R. (2007). Agent-based modeling: a new approach for theory building in social psychology. *Personality and Social Psychology Review* 11 (1): 87–104.

16 Gao, J., Zhang, Y.-C., and Zhou, T. (2019). Computational socioeconomics. *Physics Reports* 817: 1–104.

17 Salganik, M.J. (2017). *Bit by Bit: Social Research in the Digital Age*. Princeton University Press.

18 O'Sullivan, D. and Perry, G.L.W. (2013). *Spatial Simulation: Exploring Pattern and Process*. Chichester: Wiley.

19 Macy, M. and Flache, A. (2009). Social dynamics from the bottom-up: agent-based models of social interaction. In: *The Oxford Handbook of Analytical Sociology*, CCCCh Ch. 11 (ed. P. Hedstrom and P. Bearman). Oxford: Oxford University Press.

20 Hansen, L.P. and Heckman, J.J. (1996). The empirical foundations of calibration. *Journal of Economic Perspectives* 10 (1): 87–104.

21 Rolfe, M. (2014). Social networks and agent-based models. In: *Analytical Sociology: Actions and Networks* (ed. G. Manzo), 237–260. Chichester: Wiley.

22 Smith, J.A. and Burow, J. (2020). Using ego network data to inform agent-based models of diffusion. *Sociological Methods & Research* 49 (4): 1018–1063.

23 Fagiolo, G., Moneta, A., and Windrum, P. (2007). A critical guide to empirical validation of agent-based models in economics: methodologies, procedures, and open problems. *Computational Economics* 30: 195–226.

24 Saltelli, A. (2000). What is sensitivity analysis? In: *Sensitivity Analysis* (ed. A. Saltelli, K. Chan, and E. Scott), 3–13. Chichester: Wiley.

25 Delli, G.D., Fagiolo, G., Gallegati, M. et al. (2018). *Agent-Based Models in Economics. A Toolkit*. Cambridge: Cambridge University Press.

26 Stonedahl, F. and Wilensky, U. (2010). Finding forms of flocking: evolutionary search in ABM parameter-spaces. *Proceedings of the MABS workshop at the Ninth International Conference on Autonomous Agents and Multi-Agent Systems.* Toronto, Canada.

27 Bruch, E. and Atwell, J. (2015). Agent-based models in empirical social research. *Sociological Methods & Research* 44 (2): 186–221.

28 Manzo, G., Gabbriellini, S., Roux, V., and M'Mbogori, F.N. (2018). Complex contagions and the diffusion of innovations: evidence from a small-N study. *Journal of Archaeological Method and Theory* 25 (4): 1109–1154.

29 Carley, K.M., Fridsma, D.B., Casman, E. et al. (2006). BioWar: scalable agent-based model of bioattacks. *IEEE Transactions on Systems, Man, and Cybernetics-Part A: Systems and Humans* 36 (2): 252–265.

30 Stroud, P., Valle, S.D., Sydoriak, S. et al. (2007). Spatial dynamics of pandemic influenza in a massive artificial society. *Journal of Artificial Societies and Social Simulation* 10 (4): 9.

31 Gorbalenya, A.E., Baker, S.C., Baric, R.S. et al. (2020). The species severe acute respiratory syndromerelated coronavirus: classifying 2019-nCoV and naming it SARS-CoV- 2. *Nature Microbiology* 5 (4): 536–544.

32 WHO (2020). Novel Coronavirus (2019-nCoV) Situation Report 22, Data as Reported by 11 February 2020. [Online]. Available: https://www.who.int/emergencies/diseases/novel-coronavirus-2019/situation-reports/.

33 Keskinocak, P., Oruc, B.E., Baxter, A. et al. (2020). The impact of social distancing on COVID19 spread: state of Georgia case study. *PLoS One* 15 (10): e0239798.

34 Li, Q., Guan, X., Wu, P. et al. (2020). Early transmission dynamics in Wuhan, China, of novel coronavirus-infected pneumonia. *The New England Journal of Medicine* 382: 1199–1207.

35 Xie, G. (2020). A novel Monte Carlo simulation procedure for modelling COVID-19 spread over time. *Scientific Reports* 10 (1): 13120.

36 Zhao, S., Musa, S.S., Lin, Q. et al. (2020). Estimating the unreported number of novel coronavirus (2019-nCoV) cases in China in the first half of January 2020: a data-driven modelling analysis of the early outbreak. *Journal of Clinical Medicine* 9 (2): 388.

37 Zhao, S., Lin, Q., Ran, J. et al. (2020). Preliminary estimation of the basic reproduction number of novel coronavirus (2019-nCoV) in China, from 2019 to 2020: a data-driven analysis in the early phase of the outbreak. *International Journal of Infectious Diseases* 92: 214–217.

38 Jung, S.-M., Akhmetzhanov, A.R., Hayashi, K. et al. (2020). Real-time estimation of the risk of death from novel coronavirus (COVID-19) infection: inference using exported cases. *Journal of Clinical Medicine* 9 (2): 523.

39 Wu, J.T., Leung, K., and Leung, G.M. (2020). Nowcasting and forecasting the potential domestic and international spread of the 2019-nCoV outbreak originating in Wuhan, China: a modelling study. *Lancet* 395 (10225): 689–669.

40 Hsieh, Y.-H. and Cheng, Y.-S. (2006). Real-time forecast of multiphase outbreak. *Emerging Infectious Diseases* 12 (1): 122.

41 Rozema, E. (2007). Epidemic models for SARS and measles. *The College Mathematics Journal* 38 (4): 246–259.

42 Hsieh, Y.-H., Lee, J.-Y., and Chang, H.-L. (2004). SARS epidemiology modeling. *Emerging Infectious Diseases* 10 (6): 1165–1168.

43 Kermack, W.O. and McKendrick, A.G. (1927). A contribution to the mathematical theory of epidemics. *Proceedings of the Royal Society of London. Series A, Containing Papers of a Mathematical and Physical Character* 115 (772): 700–721.

44 Epstein, J.M. (2009). Modelling to contain pandemics. *Nature* 460 (7256): 687.

45 Sargent, R.G. (2015). An introductory tutorial on verification and validation of simulation models. *Proceedings of the 2015 Winter Simulation Conference (WSC)*, Huntington Beach, CA, USA, pp. 1729–1740.

46 De Angelis, D., Presanis, A.M., Birrell, P.J. et al. (2015). Four key challenges in infectious disease modelling using data from multiple sources. *Epidemics* 10: 83–87.

47 Tabataba, F.S., Chakraborty, P., Ramakrishnan, N. et al. (2017). A framework for evaluating epidemic forecasts. *BMC Infectious Diseases* 17 (1): 345.

48 Gibson, G.J., Streftaris, G., and Thong, D. (2018). Comparison and assessment of epidemic models. *Statistical Science* 33 (1): 19–33.

49 Hills, R.G., David, C.M., and Jonathan, W.N. (2015). "V&V framework," Sandia Rep., Albuquerque, NM, USA, Tech. Rep. SAND2015-7455, Sep. 2015, ch. 3, pp. 23–31.

50 Jia, J.S., Lu, X., Yuan, Y. et al. (2020). Population flow drives spatio-temporal distribution of COVID-19 in China. *Nature* 582: 389–394.

51 Li, L., Zhang, Q., Wang, X. et al. (2020). Characterizing the propagation of situational information in social media during COVID-19 epidemic: a case study on Weibo. *IEEE Transactions on Computational Social Systems* 7 (2): 556–562.

52 Imai, N., Dorigatti, I., Cori, A. et al. (2020). *Estimating the Potential Total Number of Novel Coronavirus Cases in Wuhan City, China*. London, UK: Imperial College London. https://doi.org/10.25561/77149.

53 Roosa, K., Lee, Y., Luo, R. et al. (2020). Real-time forecasts of the COVID-19 epidemic in China from February 5th to February 24th, 2020. *Infectious Disease Modelling* 5: 256–263.

54 Yang, Z., Zeng, Z., Wang, K. et al. (2020). Modified SEIR and AI prediction of the epidemics trend of COVID-19 in China under public health interventions. *Journal of Thoracic Disease* 12 (3): 165–174.

55 Kucharski, A.J., Russell, T.W., Diamond, C. et al. (2020). Early dynamics of transmission and control of COVID-19: a mathematical modelling study. *Lancet Infectious Diseases* 20: 553–558.

56 Tuite, A.R., Fisman, D.N., and Greer, A.L. (2020). Mathematical modelling of COVID-19 transmission and mitigation strategies in the population of Ontario, Canada. *Canadian Medical Association Journal* 192 (19): E497–E505.

10

LLM-Based Agents and Social Simulation

In the agent-based model (ABM) era, the question of how to characterize the anthropomorphic features of agents has long challenged researchers. For over four decades, researchers have endeavored to create agents capable of exhibiting believable human-like behaviors. These computationally driven agents act consistently with past experiences and generate credible environmental responses. However, the spectrum of human behavior is vast and complex, rendering traditional agent models insufficient for representing the true decision-making processes of social entities. Large language models (LLMs), trained on large-scale datasets, demonstrate unprecedented capabilities in natural language and multimodal understanding, reasoning, and generation. Against this backdrop, researchers aim to employ LLMs as core controllers for agents, thereby constructing key anthropomorphic characteristics such as memory, learning, reasoning, natural language interaction, and so on. This chapter first elaborates on the cognitive architecture of LLM agents for social simulation, including decision-making processes, memory hubs, and emotional cognition, which are crucial for agents' role-playing in virtual worlds. Next, using *Stanford Smallville* as an example, it illustrates how LLM technology enables a sandbox environment akin to *The Sims*, where each agent possesses distinct profiles, communicates with others, and generates rich dialogue and behavioral data. It then summarizes key technical implementations for generative agents' characteristic features, guided by the hierarchical modeling framework for artificial societies presented in Chapter 4. Finally, the chapter presents representative cases of LLM agents applied in social simulation, spanning fields such as social sciences, psychology, and economics.

10.1 The Architecture of LLM-Based Agent

As illustrated in Figure 10.1, the literature [1] proposes a motivation-driven autonomous agent behavior generation framework (desire-driven autonomous agent, D2A) for simulating the complexity of human behavior. This approach establishes a dynamic value system that integrates Maslow's hierarchy of needs, which translates intrinsic human motivations (e.g., socialization, health, fulfillment) into specific activity decisions. The D2A model automatically generates and selects activities based on individual need priorities, ensuring behaviors align with intrinsic motivations rather than relying solely on external task instructions. The D2A framework demonstrates that generating more human-like, reasonable, and diverse daily activity sequences requires consideration of both rational components (perception, decision-making, action) and affective components (beliefs, emotions, psychological states, and so forth) of agents, thereby bridging the gap between intention and behavior. The following analysis proceeds along three dimensions: decision-making processes, memory hubs, and emotional cognition.

10.1.1 Decision Workflow

The core components of an LLM-based agent encompass Perception, Cognition, Decision-Making, and Action. While each module is often discussed individually, the primary strength of LLM-based agents lies in their ability to seamlessly integrate these components: continuously receiving sensory input, making situational assessments, and leveraging this information to drive action selection. The iterative feedback loops between these modules enable the agent to engage in highly adaptive, goal-directed behavior in response to environmental changes. Specific details are as follows:

10.1.1.1 Agent's Observation

Perception refers to the agent's ability to sense and interpret its environment through various modalities (e.g., vision, touch, hearing). The agent observes and interacts with this environment, collecting diverse state information (such as character positions and object quantities) to inform decision-making and planning. We formalize this process using a simplified mathematical model: Assume the true state of the world is unknowable, represented by a latent variable s^* (denoting the world's governing principles or fundamental truths). An individual agent exists within this world, subject to its dynamics. This agent cannot directly observe s^* but perceives the external world through observations o. Critically, while s^* remains unobservable, o is generated by s^*, enabling the agent to reason about s^* to some extent.

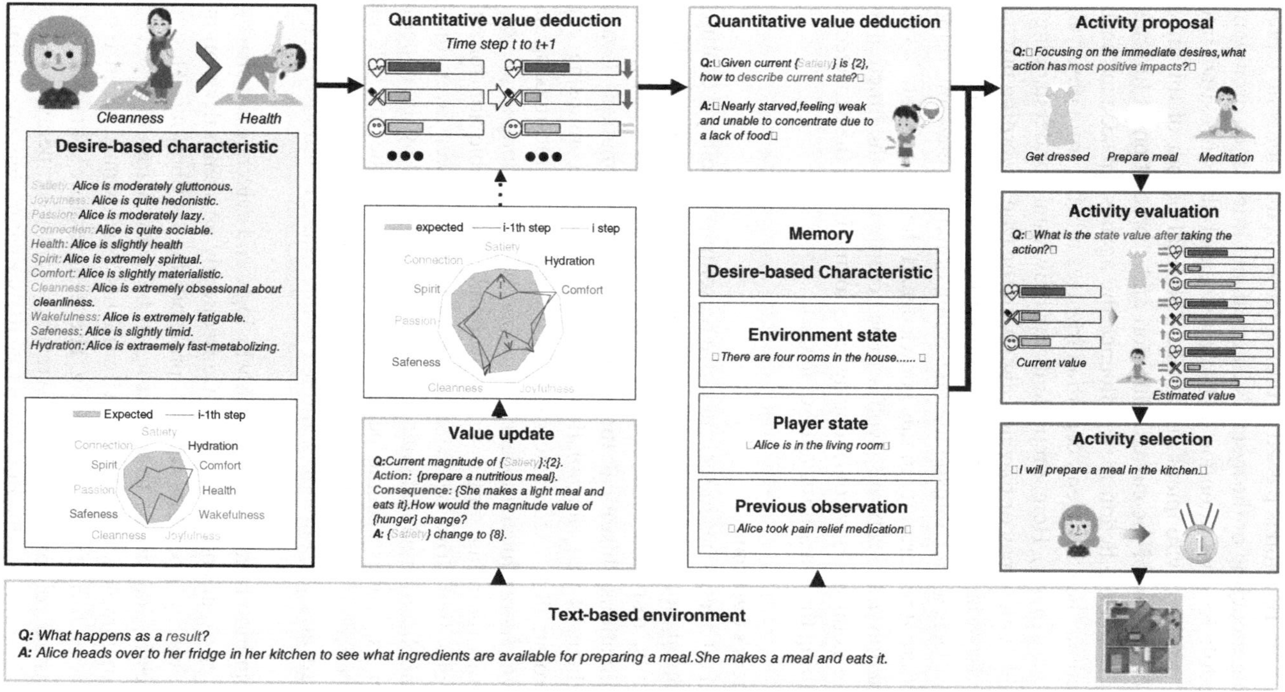

Figure 10.1 The desire-driven autonomy framework.

10.1.1.2 Agent's Orientation

The perceived information is passed to the cognitive module for state representation, including modeling of the system's external and internal environment. The cognitive module requires attention and selective capabilities to extract information relevant to the environment and goals. The final output of this modeling process is a definition of the current situation and a prediction of future states. This definition and prediction then influence the agent's subsequent decisions, plans, and actions. For the agent, cognition involves moving from observation to pattern discovery ($o \rightarrow s$). Here, s represents the individual's understanding of the external world, which may or may not align with the true governing principle s^*. According to Bayesian decision theory, posterior knowledge $p(s|o)$ obtained after observing o reflects a probabilistic understanding of natural laws and principles, indicating inherent uncertainty in this knowledge. If the natural state is s and the agent takes action u, the return (payoff) can be described by the function $R(s, u)$. The agent's decision-making continuously seeks to maximize the expected return, i.e., it is pursuing the maximum value of $R(s, u)$.

10.1.1.3 Agent's Decision

The power of AI agents lies in their iterative reasoning capability—continuously "thinking" throughout the problem-solving process. LLMs serve as the "brain core" of agents, responsible for task planning and decision-making based on current observations, historical memory, and goal information. Goal management influences plan formulation, which involves decomposing complex problems into smaller, executable steps. The agent's decision-making module comprises two hierarchical models: the upper layer is a long-sequence strategy model incorporating "intrinsic motivation," responsible for formulating macro-strategies that drive behavioral decisions in the lower layer; the lower layer is a symbolically modeled multitask model that, guided by the upper layer's macro-strategies, makes specific behavioral decisions at the micro-level. The interaction between these two layers relies on the agent's "memory," which integrates past and present into a unified whole and continuously shapes the future.

Human cognition is influenced not only by logic and evidence but also by emotions, intuition, experience, and values. Consequently, an agent's decision-making exhibits uncertainty, operating through probabilistic distributions. This "uncertainty" makes it challenging to predict agent behavior across diverse contexts, particularly in complex application environments. Agents must balance reliability and creativity during decision-making. The "temperature" parameter in LLMs also plays a critical role in reasoning and creativity. Lower temperature values yield more precise and predictable outputs, while higher values increase output diversity and randomness.

10.1.1.4 Agent's Action

Agent behavior is goal-oriented, reflecting its intention to alter the external world in pursuit of objectives. LLMs provide the foundation for the agent's "brain," while the behavior system determines its ultimate potential for achieving complex goals. The behavior system of an agent typically comprises three components: (1) all types of actions executable by the agent in simulation scenarios or downstream tasks, which may vary significantly based on different agent configurations; (2) the agent's learned actions within dynamic environments; and (3) tools, interfaces, or middleware accessible to the agent. Collectively, these elements define the scope and characteristics of the AI agent's action system. Agents autonomously generate and execute goal-oriented actions—a process formalized as trajectories $<o_t, s_t, a_t>$, where a_t is selected from action space A to transition the current state s_t to the next state based on observation o_t. By executing sequences of $<o_t, s_t, a_t>$, the agent is guided toward achieving its ultimate objectives.

10.1.2 Memory Core

While LLMs can simulate credible human behavior at individual time points, ensuring long-term consistency requires agents to possess an architecture managing ever-expanding memories—including the agent's current state, environmental feedback, and reward signals after action execution. Achieving credible human-like behavior becomes attainable if agents can store and retrieve relevant events and interactions over extended durations. This requires a "Summarize-and-Forget" mechanism that continuously aggregates, evaluates, and reflects on memory to derive higher-level inferences. By applying such reasoning, agents can create plans and responses that are meaningful to their immediate and long-term existence. Thus, the core lies in constructing a micro-world within the agent's brain: the memory stream. Building adaptive memory for LLM agents necessitates addressing three fundamental questions: First, how should memories be represented to capture diverse information types? Second, how do agent memories evolve—absorbing new experiences, adapting to changing contexts, and maintaining consistency? Finally, how can stored memories effectively enhance reasoning, decision-making, and overall agent performance?

As illustrated in Figure 10.2, agent memory mechanisms can be categorized into procedural memory (instructions), semantic memory (facts), and episodic memory (experiences). The agent's procedural memory combines LLM weights and agent code, analogous to the brain's core instruction set that fundamentally determines the agent's operational patterns and is typically nonmodifiable. The agent's semantic memory constitutes an "encyclopedia" of world facts, comprising information such as facts, concepts, and their interrelationships. Its implementation

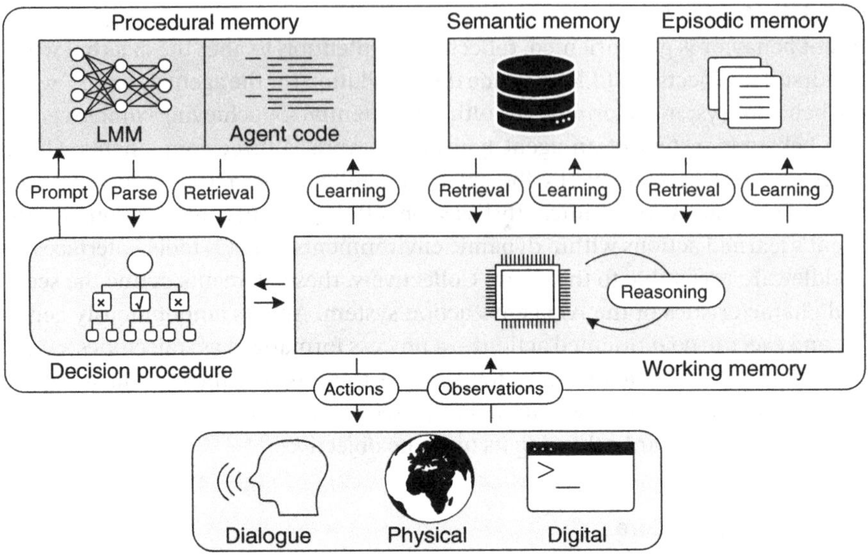

Figure 10.2 Memory mechanism of LLM agent. *Source:* [2] / LangChain, Inc.

primarily relies on vector database retrieval, integrated via retrieval-augmented generation (RAG) [3]. The agent's episodic memory includes internal state and memorized events (individual memory repository). "Internal state" manifests as numerical values (e.g., mood fluctuations) resulting from interactions between the agent's personality and long–/short-term goals. The "individual memory repository" stores past experiences—including linguistic descriptions, creation timestamps, and recent access times—capturing not only the agent's own actions but also observed behaviors of other agents or objects. Episodic memory functions like a "photo album," enabling diverse behavioral patterns based on unique past experiences.

From the perspective of memory utilization, existing research broadly falls into two categories. Write-and-Retrieve treats memory as an external knowledge base/memo, emphasizing rapid and precise information retrieval while maintaining contextual consistency. This read-write operation exerts only temporary influence, confined to the current task. Conversely, Learn-from-Memory treats memory as an intrinsic capability requiring analysis and comprehension of past experiences to extract knowledge and patterns, thereby incrementally adapting the agent's behavior for future tasks. This adaptive operation yields long-term influence generalizable across diverse tasks. Consequently, Learn-from-Memory approaches increasingly enable LLMs to derive cross-task insights, cultivating evolvable capabilities.

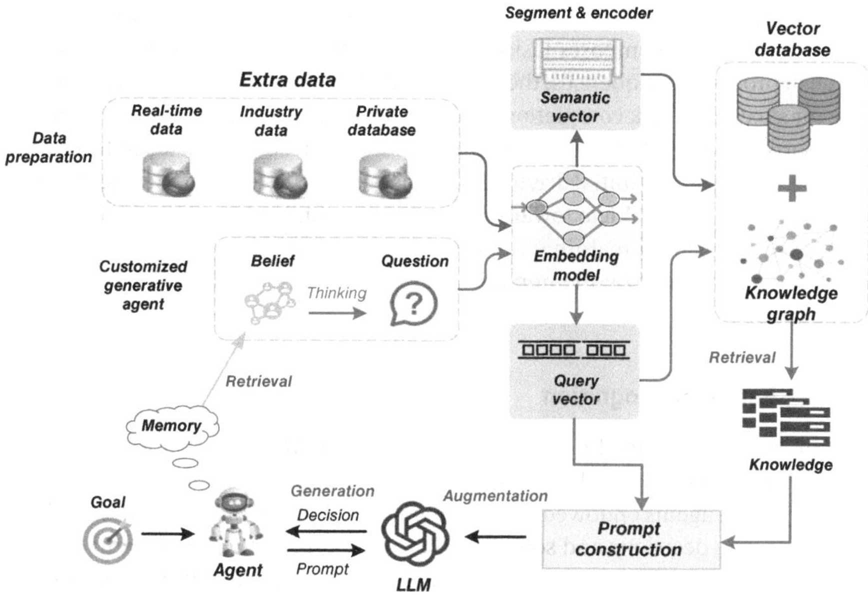

Figure 10.3 Workflow of RAG.

To achieve learning from memory, effective memory retrieval must first be implemented. Figure 10.3 illustrates the RAG architecture, where external knowledge bases are categorized into three types: real-time data, domain-specific data, and private data, respectively, providing connections to new knowledge, professional expertise, and private domain information [4]. After document chunking and embedding encoding, external data is structured into a vector database aligned with the LLM agent's role, serving as its long-term memory to aid decision-making. Driven by objectives, the agent performs reasoning and analysis on existing memory data to generate retrieval queries. These queries undergo similarity measurement between embeddings and vector database texts to retrieve contextually relevant information. After the retrieval is completed, the information obtained from the retrieval is structured into formatted context information (prompt). Together with the original query information, it is embedded into the prompt template and then fed into the LLM, thereby enhancing the adaptability and accuracy of the LLM agent when dealing with complex problems. Ultimately, the LLM agent generates more relevant and precise decisions, integrating both original context and retrieved external database knowledge.

A vector database specializes in storing, managing, querying, and retrieving vectors. Its operational principle resembles an intelligent librarian who knows the content and location of every book (data); when provided a topic (query vector), it rapidly identifies the most relevant books. When an agent needs role-aligned responses, it first decomposes tasks using Chain-of-Thought (CoT) techniques to establish processing sequences, then invokes RAG to locate specific data or knowledge for successful task completion. Analogously, an LLM agent can be compared to a sports car: the LLM acts as the driver's brain (storing knowledge from pretraining, equivalent to initial driving lessons); the prompt functions as the steering wheel, directing the vehicle toward its destination; and RAG operates like a navigation app, providing real-time route updates. Without the vector database, the driver must rely solely on memory, risking errors or detours if conditions change unexpectedly.

10.1.3 Emotional Cognition

Marvin Minsky described emotions in *The Emotion Machine* as a way to adjust our thinking processes, helping us solve problems more flexibly and creatively. Similarly, LLM agents endowed with analogous emotional traits can enhance their ability to make decisions and solve complex problems in more human-like ways. Formally, an agent's internal state can be defined as a tuple $I = (\theta, M)$, comprising the foundation model θ and the mental state M. The foundation model θ begins with pretraining, established by acquiring extensive world knowledge—similar to how a human infant absorbs environmental information during development. The mental state can be further decomposed into distinct constructs, as shown in the following formula (10.1) [5]:

$$M = \{M^{mem}, M^{wm}, M^{emo}, M^{goal}, M^{rew}\} \tag{10.1}$$

Where M^{mem} represents memory, M^{wm} represents world model, M^{emo} represents emotional state, M^{goal} represents goal (utility function), and M^{rew} represents reward. The agent's mental state influences the foundation model θ. Various learning and optimization mechanisms enable the agent to integrate new observations and experiences, maintain, continuously update, and autonomously optimize its rich internal mental states (including memory, goals, emotional states, reward mechanisms, and a comprehensive world model), thereby acquiring domain-specific knowledge and logical reasoning capabilities. Table 10.1 presents the major categories of emotion theories and corresponding explanatory notes.

Emotion constitutes a critical component of human intelligence and is likely to become an essential element or design consideration for LLM agents. However, integrating emotion into LLM agent modeling remains at an early stage. This promising yet challenging domain requires interdisciplinary collaboration across psychology, cognitive science, and AI ethics to systematically translate

Table 10.1 The Basic Theoretical Model of Emotional Psychology.

Name	Definition	Figure Illustration
Categorical Theory	Ekman's basic emotion theory identifies six fundamental emotions (anger, disgust, fear, happiness, sadness, and surprise), recognizable and expressible through specific facial expressions across cultures [6]. For LLM agents, categorical frameworks provide taxonomies to identify user emotions and generate appropriate responses. However, they face criticism for oversimplifying the human emotional experience and may fail to capture cultural variations in emotional expression.	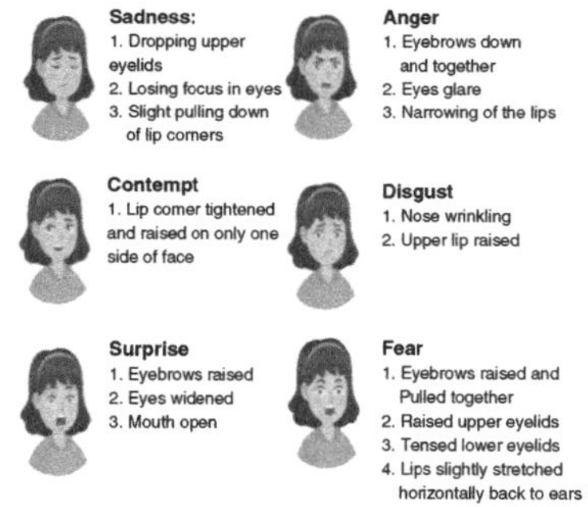
Dimensional Model	Russell's circumplex model maps emotions onto two primary axes: value (pleasant—unpleasant) and arousal (activation—deactivation) [7]. This framework allows granular control over generated content, enabling humans or agents to modulate tone along continuous scales rather than switching between discrete emotional states.	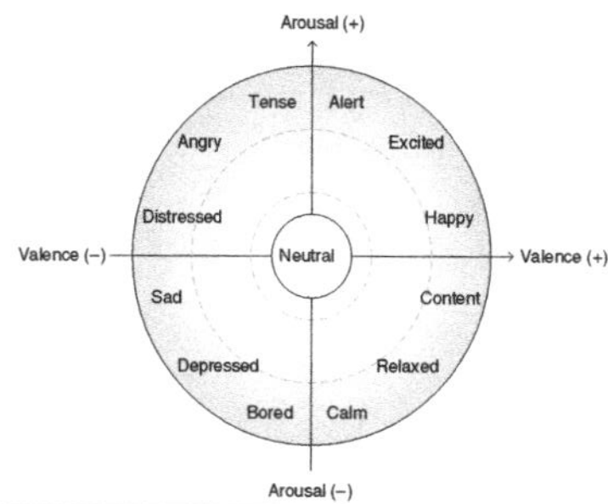

(Continued)

Table 10.1 (Continued)

Name	Definition	Figure Illustration
Mixed and Elemental Framework	Plutchik's Wheel of Emotions arranges eight basic emotions in a circular structure featuring intensity gradients and dimensional properties, enabling representation of complex emotional blends (e.g., love as a fusion of joy and trust) [8]. For LLM agents, these models provide computational frameworks to evaluate textual inputs and select contextually appropriate emotional responses, thereby enhancing behavioral coherence and perceived empathy.	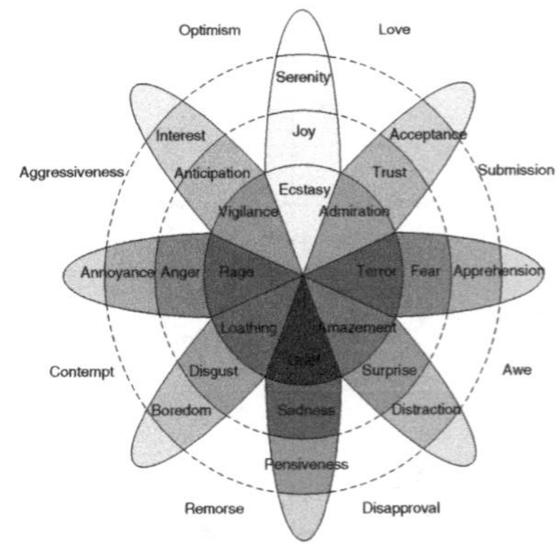
Neuro-cognitive Perspective	Damasio's Somatic Marker Hypothesis posits that emotions emerge through body–brain interactions, guiding decision-making by linking physiological states to anticipated outcomes. LeDoux's amygdala-centric model reveals dual-pathway processing of emotional stimuli [9]. LLM agents have begun implementing analogous multi-pathway information processing architectures, integrating "fast" emotion detection with "slow" deliberative reasoning to achieve rapid safety responses and deeper emotional comprehension [10, 11].	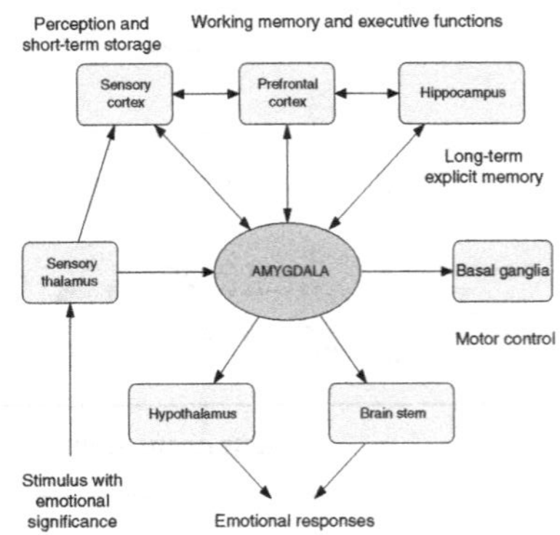

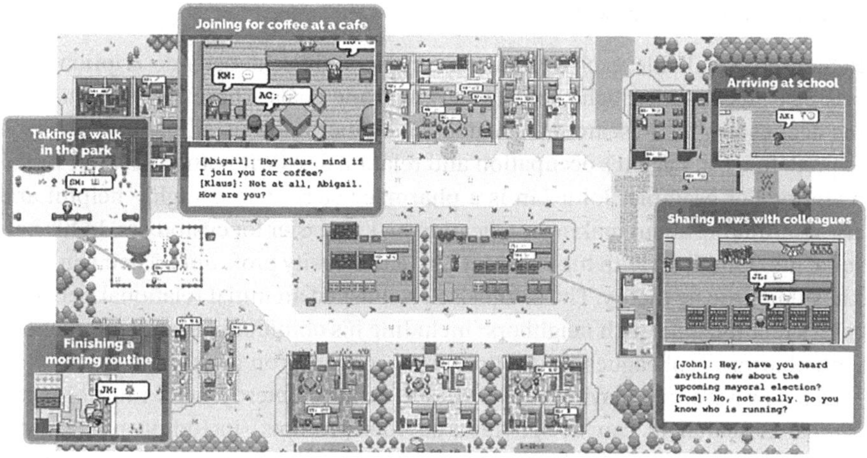

Figure 10.4 Panoramic view of Stanford Smallville. *Source:* [12] / ACM (The Association for Computing Machinery).

psychological and neuroscientific theories into LLM agent internal processes. For instance, dimensional models (valence/arousal/dominance) can serve as latent states influencing generation and shaping subsequent agent actions. Hybrid approaches offer balance: LLMs may first identify a discrete category (e.g., "fear"), then quantify its intensity and regulate dimensional parameters for nuanced dialogue. Such emotion-infused architectures may develop more consistent "emotion" over time, analogous to how humans sustain emotional states rather than resetting with each interaction. Explicit alignment with psychological theories also enhances interpretability: designers can debug or refine agent responses by comparing them against established emotion constructs (Figure 10.4).

10.2 Stanford Smallville

In 2023, researchers from Stanford and Google created 25 generative agents [12] using LLM technology within a sandbox environment resembling *The Sims*. Each agent had a distinct role and could communicate with others, generating rich dialogue and behavioral data. Traditional NPCs rely on prescripted dialogues and actions, in which specific lines are triggered at predetermined steps. With the advent of ChatGPT, these traditional game NPCs can now self-generate conversations using only key input information, breaking the constraint of rigid, prewritten scripts. The following section uses "Stanford Smallville" as a case study to examine the design details of generative agents.

10.2.1 Agent's Role Settings

In Stanford Smallville, each agent has a corresponding character description comprising traits, abilities, behaviors, constraints, and goals. These descriptions form the agent's profile. Researchers created 25 distinct roles, using natural language to predefine each agent's occupation and relationships with others as their initial memory. For example: John Lin is a pharmacy clerk who is highly helpful and always thinking about making medication pickup easier for customers. In his personal life, John lives with his wife Mei Lin (a university professor). Their son Eddy Lin studies music theory. The setup also established an initial relational network: John's relationships with neighbors (including his opinions of them), interactions with pharmacy colleagues, and John's usual conversation topics with them.

LLMs provide role-playing capabilities to achieve heterogeneity in LLM-based agents. Researchers can design tailored role prompts (including name, hobbies, daily routines, behavioral rules, and other elements) based on scenario requirements. In addition, they can provide these prompts to LLM-based agents to simulate designated roles. Basic information and settings serve as personalized parameters for the sandbox environment. Personality and lifestyle complement occupational identities and inter-agent relationships by including intrinsic characteristics, current status (researchers initialized two events: an election and a Valentine's Day party), and daily agendas (outlining routines to maintain temporal credibility). All of the abovementioned content is initialized into the agent's memory.

Each role begins with a predefined text segment sketching its framework to trigger subsequent interactions. The remaining development is handled autonomously by AI. Every role's actions are represented by emojis and summarized in one sentence within the system. Each agent perceives nearby characters, communicates in natural language, and autonomously decides whether to pass by or initiate conversations based on personality and relationships. For experimental purposes, researchers may intervene—not by traditional script input but by applying subtle guidance at key branching points. The AI autonomously handles the remainder (Figure 10.5).

10.2.2 Agent's Environment

A sandbox is a simulated or virtual environment constructed by humans, primarily used in software development and gaming. Within this environment, agents can interact more freely and experiment with various actions and strategies. "Smallville" uses a common small-town infrastructure as its base, including cafes, bars, parks, schools, dormitories, houses, and stores. It also defines objects within these spaces, such as kitchens in houses and stoves in kitchens. The agent–environment interface refers to how agents interact with and perceive their

Name: David Smith (age: 25), David Smith, a 25-year-old male photographer, is compassionate, caring, ambitious, and optimistic. He enjoys watching sci-fi and comedy movies and provides feedback and ratings to the recommendation system. He demands high standards for movies and the recommendation system and may criticize both. The observation about David watching "The Neon Bible" aligns with his interest in drama films and explores themes of faith, family, and coming-of-age. **Profile**

It is August 18, 2023, 12:00 a.m. **Context**

Most recent observations: David Smith enjoys and finds captivating films that have captivating plots, humorous elements, thought-provoking themes, delve into complexities of human nature and sexual desire, uplift viewers, and have vibrant and engaging performances by the cast.

Observation: David Smith has just finished watching Neon Bible, The (1995): "The Neon Bible" is a drama film set in the 1940s in a small southern town in the United States. It follows the story of a young boy named David who is struggling to understand the complexities of the world around him. David \'s mother is mentally unstable and his father is absent, leaving him to navigate the challenges of adolescence on his own. As he tries to make sense of his surroundings, he turns to religion and finds solace in the teachings of his local preacher. However, his faith is tested when he discovers the secrets and hypocrisies of those around him. The film explores themes of faith, family, and coming-of-age in a poignant and powerful way. **Memory**

All occurrences of movie names should be enclosed with <>. David Smith has not seen this movie before. Imagine you are David Smith, how will you feel about this movie just watched? Please share your personal feelings about the movie in one line. Please act as David Smith well. **Instruction**

Figure 10.5 Agent's feel about a movie motivated by prompts.

environment. Through this interface, agents can understand their surroundings, make decisions, execute actions to influence the environment, and acquire new information for subsequent thinking and planning. This cycle iterates continuously multiple times.

As shown in Figure 10.6, "Stanford Smallville" represents the sandbox environment (areas and objects) as a tree data structure, where edges denote containment relationships in the sandbox world. This tree is converted into natural language

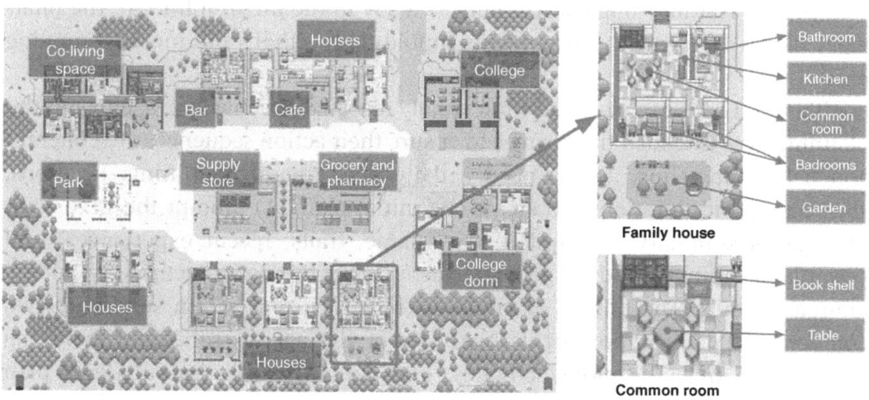

Figure 10.6 Sandbox world environment. Each area is labeled. The root node describes the entire world, branch nodes describe areas (such as houses, cafes, stores), and leaf nodes describe objects (such as tables, bookshelves). The agent remembers the world they see, along with the state when they observed it. *Source:* [12] / ACM (The Association for Computing Machinery).

for agents. For example, "stove" as a child of "kitchen" → rendered as "there is a stove in the kitchen." Before moving, the agent traverses from the root of the agent–environment Tree, enabling the LLM to locate the optimal area. For example: If Eddy plans to "take a short walk around his workspace" and needs to identify corresponding locations, the system queries the LLM for path solutions. Agent movement is guided by both the "LLM brain" and the sandbox game engine: When the LLM directs movement to a location, a walking path is computed in the Smallville environment, and the agent is moved along this path.

Users can enter this sandbox world through dialogue interactions with agents to influence object states. Two methods exist. (1) Users can manually modify environmental object states and inform the LLM. For example: When Isabella makes breakfast in the morning, the user can input "Change selected object" to the LLM, switching the state of the kitchen stove from "on" to "burning." Isabella would notice this the next moment, then turn off the stove and restart breakfast preparation. (2) Users can directly query object states via the LLM. For example: If Isabella outputs the action "make espresso for a customer," querying the object state through the LLM would change the Hobbs cafe coffee machine state from "off" to "brewing coffee."

10.2.3 Agent's Planning for a Day

In the town, agents' daily activities include (1) interacting with other agents and the world, (2) establishing memories and relationships, and (3) coordinating joint activities. Their behaviors evolve over time. Each day, every character follows their own preset "routine." Figure 10.7 depicts a day in the life of John Lin, the pharmacy clerk. As for what fresh events unfold daily—that's the most fascinating part.

Although LLMs can generate contextually reasonable behaviors, agents require planning over extended timeframes to ensure their action sequences are coherent and believable. Thus, agents need a daily plan roughly outlining: locations, start times, and durations. To create the initial plan: (1) prompt the LLM with the agent's summary description (e.g., name, traits, recent experiences) and yesterday's summary and (2) generate the full plan top-down recursively, adding granular details. The official example features a current action state with two components: daily requirement (overview of the agent's schedule, including approximate times/locations/activities) and daily schedule (minute-level plan entries formatted as: [Primary activity description (Specific action description), Estimated duration]) (see Figure 10.8).

When an agent has a plan, it must rely on the LLM to decide whether and how to act. The specific approach is illustrated in Figure 10.9. After generating the current action, the agent can then continue generating the next plan.

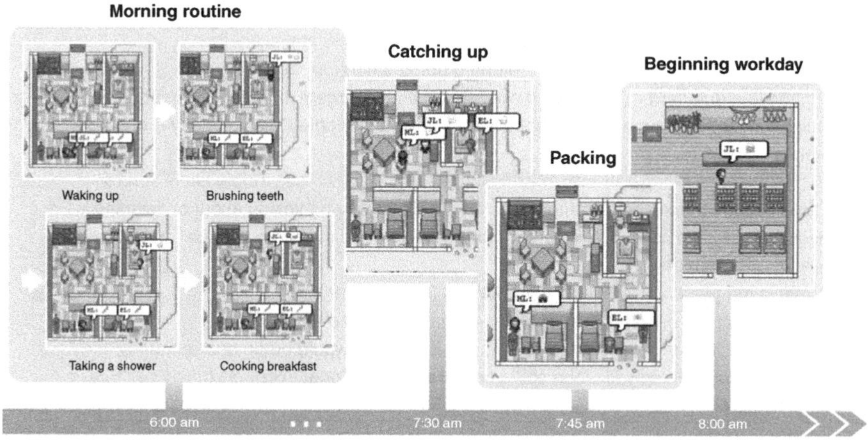

Figure 10.7 John Lin's morning routine: He wakes up around 6 a.m. and completes his daily routine, which includes brushing his teeth, showering, and eating breakfast. Before starting work, he briefly chats with his wife Mei and son Eddie. *Source:* [12] / ACM (The Association for Computing Machinery).

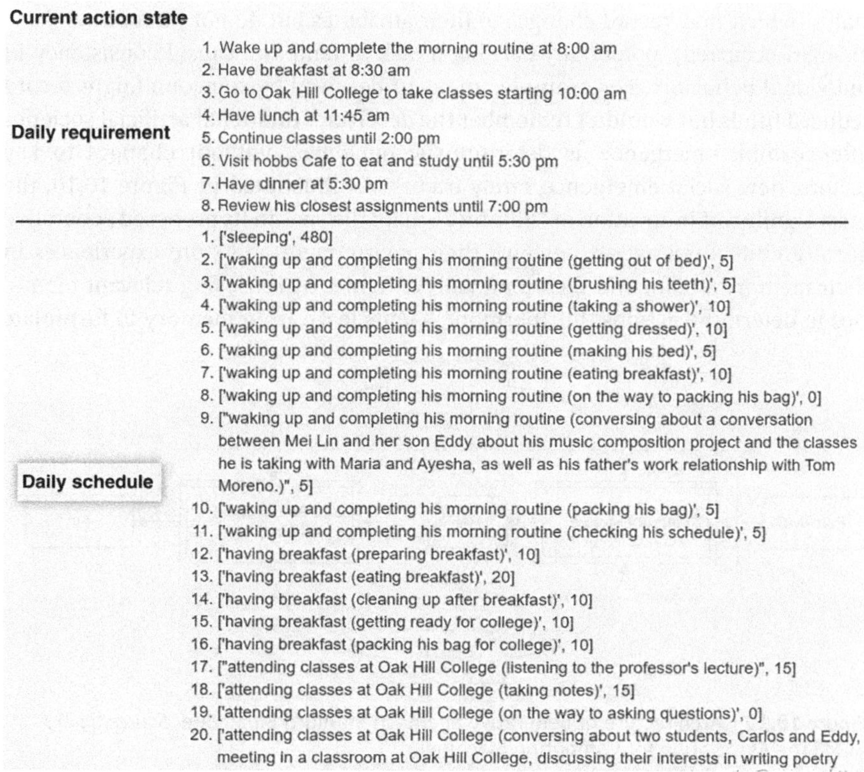

Figure 10.8 An example of current action state. *Source:* [12] / ACM (The Association for Computing Machinery).

The agent asks whether to act and what to do

Agent's Summary Description: It is February 12, 2023, 4:56 pm.

John Lin's status (last action): John returned home early from work.

Observation (the current time-step's observed memory stream in the environment, including the current memories of other agents): John saw Eddie walking briefly around his workplace.

Summary of relevant contextual information from John's memory: Eddie Lin is John's son. Eddie Lin has been composing music for his class. When thinking about or listening to music, Eddie Lin enjoys walking around the garden. (This summary was generated by answering two prompts:

"What is the [observer]'s relationship with the [observed entity]?"

and" [Observed entity] is [action status of the observed entity] ")

Should John respond to the observation, and if so, what would be an appropriate response? (The final reply is: John could consider asking Eddie about his music composition project)

Figure 10.9 The agent generates a sample of the current behavior.

10.2.4 Agent's Memory and Reflection

In ABM, agents typically make behavioral decisions based solely on current states (which may record changes to their attributes but do not retain how those changes occurred), potentially causing a lack of temporal causal consistency in individual behaviors. For example, an agent deceived by someone might record reduced funds but wouldn't remember the deceiver. Traditional artificial societies often exhibit emergence as deterministic outcomes: without changes to key factors, new social emergence rarely occurs. As illustrated in Figure 10.10, the most significant innovation of "Stanford Smallville" lies in its proposed generative agent architecture: agents perceive their environment and store experiences in their memory stream and then plan daily activities by retrieving relevant memories to determine actions. Furthermore, agents learn from memory to formulate

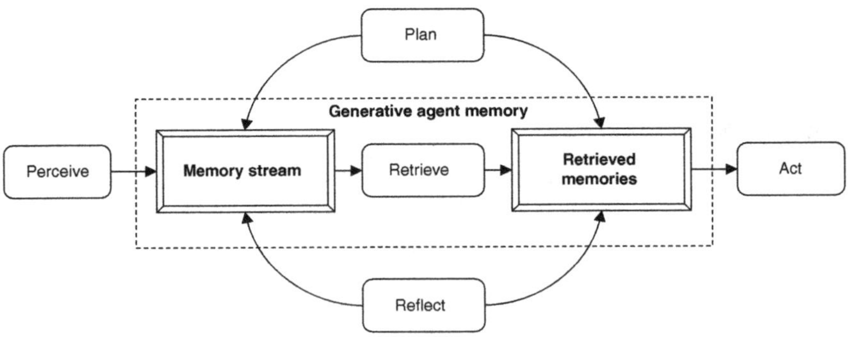

Figure 10.10 Architecture of generative agents in Stanford Smallville. *Source:* [12] / ACM (The Association for Computing Machinery).

longer-term plans and generate higher-level reflections. When executing actions, agents retrieve high-value, relevant memories—particularly those containing past successes with similar objectives—to refine current behaviors.

The core architectural challenge for LLM agents lies in ensuring retrieval and synthesis of the most relevant portions from their memory streams amid vast volumes of generated events and memories. As formalized in Eq. (10.2), retrieval scoring for all memory records incorporates three metrics: recency, importance, and relevance. Recency employs an hourly exponential decay function where more recent memories receive higher scores. Importance distinguishes ordinary memories from core memories, assigning elevated scores to significant memory objects (directly requiring the LLM to output an integer score). Relevance computes cosine similarity between the "current situation" and "memory situation," allocating higher scores to maximally relevant objects (Figure 10.11).

$$score = \alpha_{recency} * recency + \alpha_{importance} * importance + \alpha_{relevance} * relevance \tag{10.2}$$

When agents are equipped solely with raw observational memories—fragmented memory snippets—they struggle to generalize or draw inferences, resulting in suboptimal quality. Consequently, Stanford Smallville introduces a novel memory category: Reflection. Periodic reflections are triggered when the cumulative importance score of perceived events exceeds a predefined threshold. The reflection process begins with the agent determining reflection content by identifying

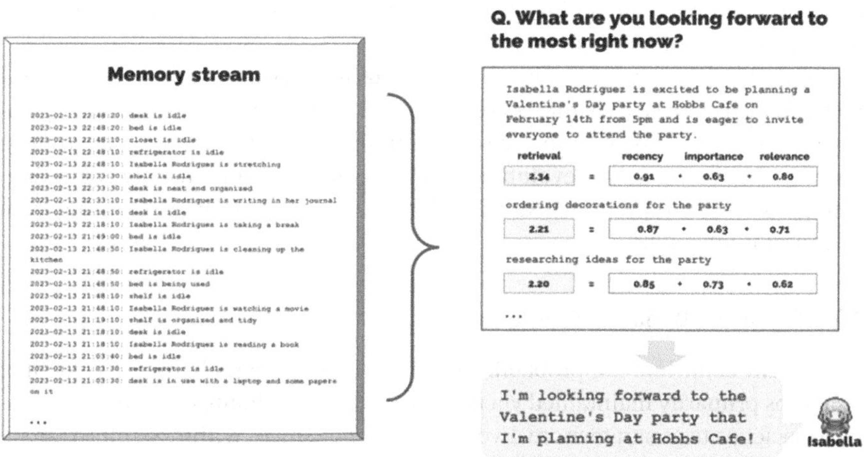

Figure 10.11 Memory stream + retrieval (when an agent wants to get the current action, it asks the agent: "What are you looking forward to most right now?" and gives the top memory experiences with high correlation scores). *Source:* [12] / ACM (The Association for Computing Machinery).

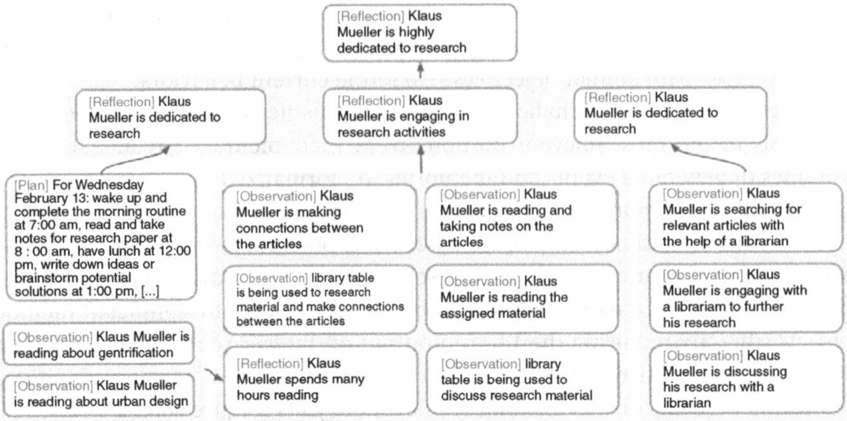

Figure 10.12 Klaus Mueller's tree of reflection. The agent's observations of the world (represented by leaf nodes) are recursively synthesized to derive Klaus's self-reflection—he is highly committed to his research. *Source:* [12] / ACM (The Association for Computing Machinery).

answerable questions based on recent experiences. Methodology: by using the 100 most recent records from the agent's memory stream, the system queries the LLM: "Given only the information above, what are the three most salient high-level questions we can answer about the subjects in the statements?" The model then generates three questions, which serve as retrieval queries to gather relevant memories (including prior reflections) for each question, and then integrates them into the memory stream.

Agents do not solely rely on historical records to determine subsequent actions; they can also achieve dynamic self-evolution (e.g., altering initial goals and planning strategies) based on feedback or communication logs. Furthermore, as illustrated in Figure 10.12, self-evolution enables agents to autonomously adapt their configurations or objectives, extending beyond the scope of mere learning from historical interactions.

10.2.5 Social Behavior Emergence

LLM agent behaviors are probabilistic, with causal linkages between sequential scenarios primarily maintained via memory. Such a mechanism also enables artificial societies to generate more diverse social emergence. Within Stanford Smallville, three emergent social behaviors have been observed: information diffusion, relationship memory, and coordinated cooperation.

1) **Information diffusion:** Historically, RL-based agents communicated implicitly through learned behaviors. In contrast, LLM-based agents communicate

via natural language processing (NLP), which offers humans a more transparent and explicit interaction paradigm. For example: At the grocery store, Sam informs Tom about his local election candidacy. The next day, Tom and John hear additional updates about Sam from other sources and begin discussing Sam's electoral prospects.

2) **Relationship memory:** Over time, Smallville agents form new relationships and remember interactions with others. For example: Initially unfamiliar with each other, Sam and Latoya Williams meet during a walk in Johnson Park. They introduce themselves, and Latoya mentions her photography project. During a later encounter, Sam asks: "Hello Latoya, how is your project going?"
3) **Coordinated cooperation:** Generative agents coordinate organically. For example: Isabella Rodriguez (Hobbs Cafe) plans a Valentine's Day party (5–7 p.m., February 14). When seeing friends and customers, she invites them proactively. On the afternoon of February 13, she starts to decorate the café. Her regular customer and close friend Maria Lopez arrives. Isabella requests Maria's help with decorations, and Maria agrees. Additionally, Maria's profile notes her secret crush on Klaus Mueller. That evening, Maria invites Klaus to the party—he accepts (Figure 10.13).

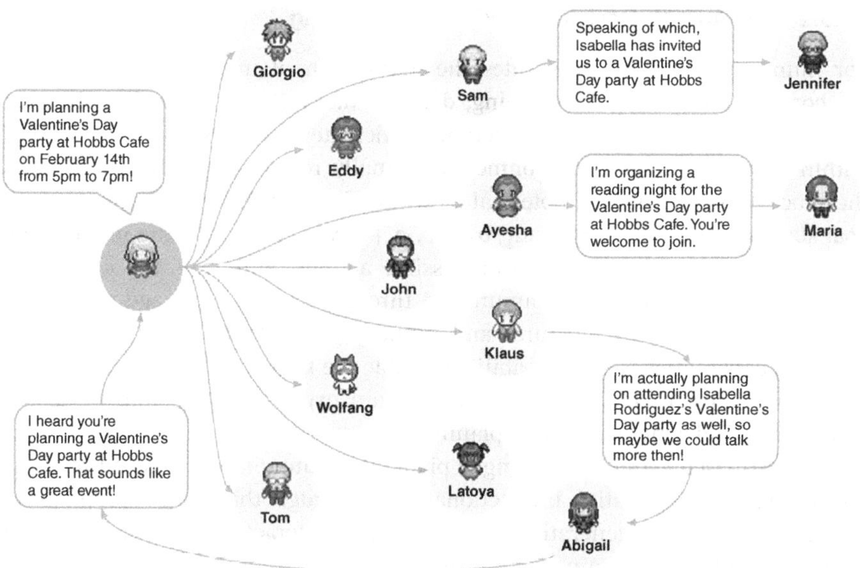

Figure 10.13 How the party was spontaneously guided by AI, the relationships between these AI entities, and the interactions triggered by the party. It involves memory retrieval, reflection through environmental observation, and decision-making. *Source:* [12] / ACM (The Association for Computing Machinery).

These scenarios may seem simple and playful at first, but their true power unfolds progressively. For instance, when applied to virtual reality (VR) worlds, generative agents serving as NPCs could make Westworld-like visions attainable. "In Westworld, every NPC possesses its own thoughts, personality, plans, and relationships—even more complex elements emerge autonomously." What researchers from Google and Stanford have created is essentially a stylized miniature version of *Westworld*.

10.3 LLM-Based Agent's Capability Pool

In Section 4.2, we presented a hierarchical capability framework for simulation agents to guide their behavior in open worlds. AI agents powered by LLMs "have transitioned from being passive input processors to becoming active agents capable of solving multi-step challenges through reasoning and interacting with environments in a manner similar to human-like problem-solving." This section elaborates on how to leverage LLMs to achieve simulation agents' autonomy, adaptability, sociality, and emergence.

10.3.1 L1: Agent's Decision-Making and Planning

For humans, reasoning constitutes the foundation of intelligence, serving as the cornerstone for problem-solving, decision-making, and critical analysis. It empowers individuals to organize actions, anticipate outcomes, and achieve goals within complex, dynamic environments. Formally, reasoning can be described as the process of constructing potential pathways from an initial state to a desired goal state, denoted as: $P: S_0 \rightarrow \{a_1, a_2, \ldots, a_n\} \rightarrow S_g$, where S_0 is the initial state, $\{a_1, a_2, \ldots, a_n\}$ is the sequence of possible actions, and S_g is the goal state. Although LLMs primarily communicate through natural language or specific texts, their internal architectures and training methodologies confer inherent planning capabilities. These capabilities enable the transformation of inputs into actionable steps, reducing reliance on domain expertise while improving the adaptability and generalization of planning systems.

For LLM-based agents, reasoning is pivotal for autonomous decision-making, converting raw information into actionable knowledge, thereby enabling logical inference, hypothesis generation, and purposeful interaction with the world. In human cognition, reasoning manifests through multiple strategies: deductive reasoning: applying universal rules to specific cases; inductive reasoning: generalizing from specific instances; and abductive reasoning: constructing plausible explanations from incomplete data. Within LLM agents, reasoning is formally defined as the action selection process based on mental states, serving as the

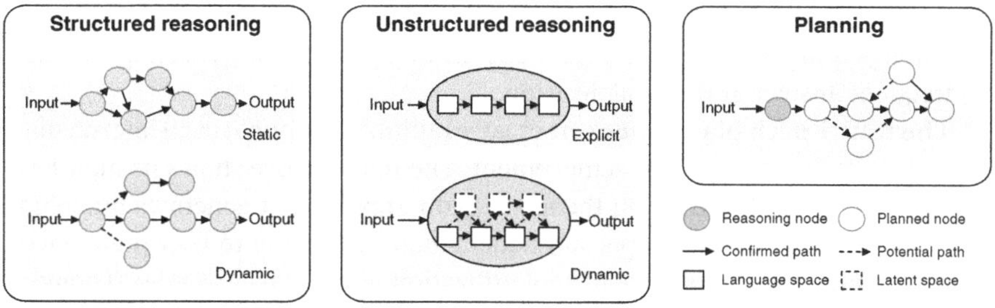

Figure 10.14 Comparison of reasoning paradigms in LLM agents.

critical bridge between perception and action. Precisely, given a mental state M_t at time t, reasoning is formalized as a function: $R(M_t) \rightarrow a_t$, where a_t denotes the selected action (Figure 10.14).

LLMs serve as the agent's "brain" responsible for thinking and planning, and well-engineered prompts constitute the most direct method to activate LLM agents' reasoning capabilities. For complex tasks, LLMs decompose such tasks into multiple relatively simple subtasks. Each subtask contains an independent subgoal, and the problem is solved step by step through a divide-and-conquer approach. Users may employ prompt engineering and related frameworks to guide agents' reasoning and planning, enabling more effective interaction with the environment and task completion. Implementation of reasoning actions follows two distinct approaches, namely, structured reasoning and unstructured reasoning.

10.3.1.1 Structured Reasoning

Structured reasoning represents a systematic approach to problem-solving that uses explicit organizational frameworks to guide the inference process. Unlike unstructured methods, it makes the composition of reasoning steps explicit, formally expressed as: $R_s = R_1 * R_2 * \ldots * R_n$, where each R_i denotes a discrete reasoning step with clear logical dependencies. In this formulation, each reasoning node is a well-defined computational unit, and connections between nodes represent deterministic information flow path. This methodology facilitates more systematic exploration of solution spaces, robust decision-making through deliberate, stepwise analysis, and high interpretability and traceability throughout the reasoning process.

1) **The single-path plan generator** generates linear sequences of plan steps to achieve user goals. The figure above illustrates its high-level graphical representation. After receiving and interpreting user goals, it coordinates the creation of intermediate steps for other agents/tools and determines task priorities for incremental goal completion. Critically, each step has exactly

one subsequent step (e.g., CoT). Additionally, self-consistency sampling is employed for querying the base model multiple times and selecting the most frequent answer as the final decision.

2) **The multi-path plan generator** creates multiple options at each intermediate step in the process of goal achievement. The figure above shows its high-level representation. Building on the single-path approach, it generates branching choices at every step, adapts subsequent steps according to user preferences, and dynamically adjusts agent/tool utilization. For instance, Tree-of-Thoughts (ToT) exemplifies this design pattern.

While planning paradigms help agents better solve complex tasks, they may yield less predictable outcomes than deterministic workflows. Currently, researchers are exploring more sophisticated planning architectures. Graph of Thoughts (GoT) extends tree-based methods to arbitrary graph structures, enabling more complex reasoning patterns to capture interdependencies between steps [13]. This approach permits connecting seemingly unrelated reasoning branches, facilitating finer-grained exploration of solution spaces. Furthermore, given the stochastic nature of LLMs, integrating parallel sampling with aggregated reasoning enhances inference performance [14]. Task decomposition constructs a space containing solution trajectories—encompassing multiple goal-oriented paths and their interrelations [15, 16]. This space enables sampling diverse potential solutions [17], with deep exploration through techniques like reflection, review, and knowledge-guided parallel sampling [18].

10.3.1.2 Unstructured Reasoning

Compared with structured reasoning (which clearly organizes reasoning steps), unstructured reasoning (Ru) takes the form of a whole $Ru = f\ (M_t)$, where the composition remains implicit and flexible. In this mode, the inference process is encapsulated in a single function mapping without explicitly defining intermediate steps or state transitions. This method uses language models' inherent ability to generate coherent reasoning without imposing strict structural constraints, with intermediate reasoning processes occurring explicitly in the language space or implicitly in the latent space. Notably, this dual framework reflects human cognition, where structured reasoning is similar to our explicit logical deduction process, while unstructured reasoning reflects our abilities in intuitive problem-solving and pattern recognition.

While recent research demonstrates the potential of LLMs in task planning, numerous research opportunities and directions remain unexplored. First, integrating multi-source environmental feedback poses challenges for LLMs, particularly regarding effective measurement and integration of holistic system data to enhance planning accuracy and adaptability. Future studies require planning systems to enable more precise environmental perception and more robust decision-making. Second, due to insufficient world knowledge, LLMs often lack deep understanding of world dynamics, relying on pattern recognition rather than genuine causal reasoning, which impedes their ability to manage subgoal interactions and environmental changes. Future work could explore restructuring task planning workflows by leveraging collaboration between large and small models, while integrating heuristic search and statistical methods to improve problem-solving speed and quality. Finally, given the diverse application scenarios of LLMs in task planning, a core issue is the absence of precise evaluation metrics for performance measurement. Future research should establish more nuanced evaluation frameworks to ensure both the viability and efficiency of LLM planning approaches.

10.3.2 L2: Agent's Learning Evolution

A key aspect of cognitive intelligence involves understanding how the world operates through direct interaction and accumulation of experience. This understanding encompasses how the environment responds to different actions and the consequences these actions entail. The self-enhancement and adaptive evolution mechanisms of LLM agents refer to their ability to establish and refine their memory, reward comprehension, and world models through continuous environmental interaction. By learning from successes and failures, they achieve comprehensive mastery of their operational domains. The feasibility of partial mental state learning is demonstrated through distinct approaches for different components: memory (M^{mem}), reward (M^{rew}), and world model (M^{wm}). We categorize mechanisms enabling individual adaptability into memory-anchored learning and parameter-based learning.

10.3.2.1 Learning Based on Memory (World Model Unchanged, Prompts Adjusted)

The reflective capability of LLM agents resembles memory replay, enabling the evaluation of prior decisions and subsequent environmental feedback, thereby significantly enhancing individual adaptability [19, 20]. The reflection paradigm constitutes a self-feedback mechanism where agents iteratively assess their output quality or decisions before final actions, thereby optimizing reasoning processes. This procedure assists agents in correcting errors and minimizing inaccuracies, continuously improving decision quality, and enhancing adaptability to better meet user requirements. As illustrated in Figure 10.15, several representative reflection paradigms include:

1) **Self-reflection:** Enables agents to generate feedback on planning/reasoning processes and provide self-improvement guidance. Figure 10.15 demonstrates

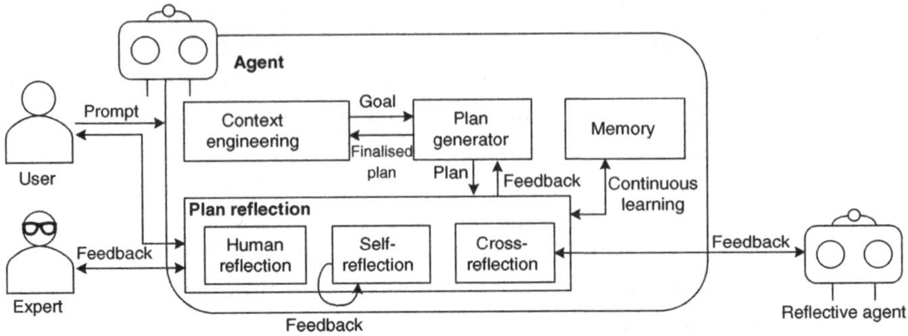

Figure 10.15 LLM agent's self-reflection.

an advanced graphical representation of self-reflection. Specifically, reflection constitutes an optimization process formalized as iteratively reviewing and refining agent-generated responses. Reflexion [21], a seminal work in this domain, conducts post-action reflection on the causes of success or failure. These insights are stored as experiential knowledge and incorporated into subsequent decision prompts. This approach, termed verbal RL, achieves RL-equivalent outcomes (success promotion and failure avoidance) through nonparametric means.

2) **Cross-reflection:** Employs distinct agents or foundation models to provide feedback for improving generated plans and associated reasoning processes. When an agent fails to produce accurate results or precise planning steps through self-reflection, users may instruct it to invoke a specialized reflection agent. The latter reviews and evaluates the original agent's outputs and reasoning steps, providing targeted improvement suggestions. This process iterates until the reflection agent confirms plan validity. Additionally, multiple agents can be queried for comprehensive response generation.
3) **Human-reflection:** Collects human feedback to refine plans and effectively align with human preferences. When users convert goals and constraints into prompts, agents first create plans that include sequences of intermediate steps. Subsequently, the constructed plan and its reasoning process undergo human review—either by the user or by domain experts—to verify feasibility and utility. Users and experts provide comments identifying steps that can be updated or replaced. Plans undergo iterative refinement until receiving user and expert approval.

10.3.2.2 Learning Based on Parameters and Interactions (Modifying the World Model)

World models constitute the core of agent learning systems, laying the foundation for understanding the dynamics of the environment and enabling more

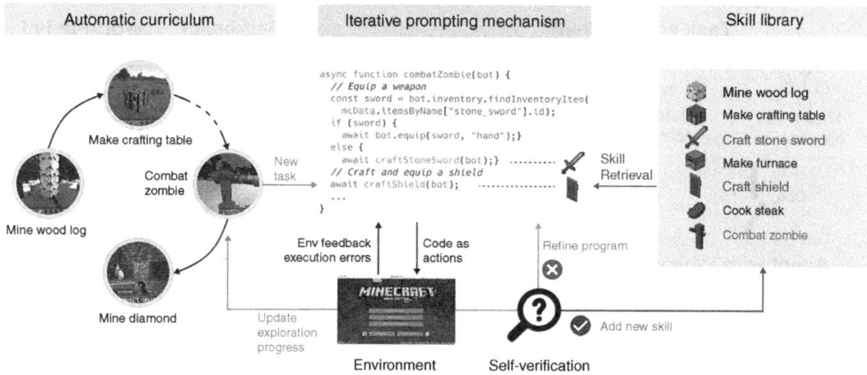

Figure 10.16 Three key components of VOYAGER agent. *Source:* [22] / Cornell University.

effective planning, reasoning, and decision-making in complex environments. Optimizing reward comprehension through environmental interactions is key to refining agents' internal world models. The Voyager agent [22], a paradigmatic example, possesses autonomous will and decision-making capabilities. Without human intervention, it freely explores Minecraft worlds, updates its skill library through environmental interactions (including autonomous discovery of new items, unlocking skill trees, traversing diverse terrains), and applies the learned skills to unseen tasks. As illustrated in Figure 10.16, Voyager agents achieve human-level performance via three key components. (1) Automatic curriculum: Agents autonomously explore worlds and discover new tasks. (2) Skill library: Agents store and retrieve skills to adapt to environmental demands. (3) Iterative prompting: Agents develop task-proposing capabilities through iterative refinement.

ARMAP refines environmental reward models by distilling data from agent action trajectories, establishing foundations for advanced learning [23]. Research [24] explicitly employs LLMs as world models to predict consequences of future actions, effectively modifying the agent's world understanding M^{wm}. ActRe [25] builds upon language models' inherent world comprehension, constructing tasks from trajectories to iteratively enhance agents' capabilities as both world models and reasoning engines. REMEMBERER [26] adopts a semiparametric strategy: it stores all observation-action pairs as memories, updating Q-values for each pair post-action based on rewards via Q-learning algorithms. During decision-making, the LLM retrieves relevant experiences from memory using similarity functions, incorporating actions with the highest/lowest Q-values into prompts as Encouraged and Discouraged directives.

Agent-Pro [27], another representative framework, requires LLMs to autonomously construct self and environmental beliefs in complex games

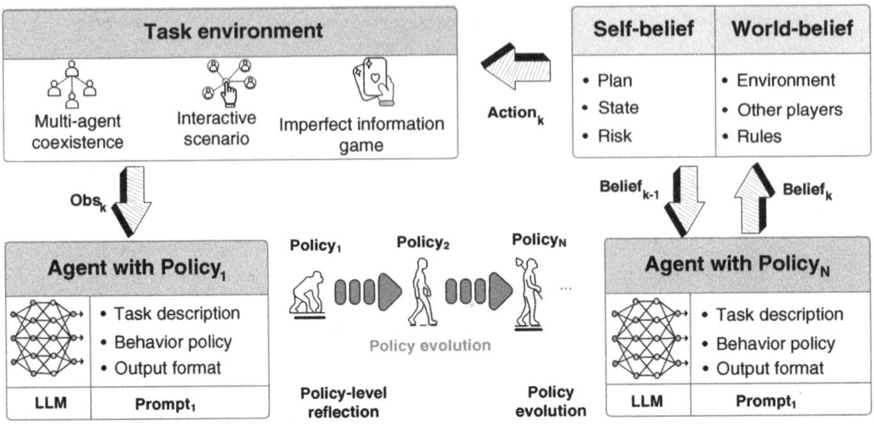

Figure 10.17 The architecture of Agent-Pro. *Source:* [28] / Cornell University.

like Blackjack and Texas Hold'em. Beliefs constitute high-level cognition encompassing the current game states, opponents' playing styles, game rules, and strategies. As illustrated in Figure 10.17, post-action, the LLM reflects not on actions (unlike prior approaches) but on beliefs. It prioritizes a rational and consistent understanding of the environment over action quality. Post-game, the LLM comprehensively rewrites its prompt—including task descriptions, action policies, and output examples—and based on memorized beliefs, achieving holistic evolutionary learning beyond mere action imitation.

However, this approach of adjusting local actions based on feedback often leads to suboptimal decisions—while offering short-term environmental adaptation, it may adversely affect long-term planning. To address challenges in long-term task planning (including issues such as low accuracy and executability), Wang et al. proposed the DEPS framework [29]. This framework enhances task planning success rates in open-world scenarios through an interactive process of Description, Explanation, Planning, and Selection (DEPS). Similarly, Sun et al.'s AdaPlanner [30] achieves adaptive plan adjustments by integrating intra-plan revision and extra-plan revision methodologies.

10.3.3 L3: Agent's Social Network

In the ABM era, designing interactions between agents poses significant barriers, requiring developers to manually define communication protocols, primitives, and interaction modes manually. Now, LLM agents interact seamlessly using natural language, substantially reducing development complexity for interactive features. Inspired by sociological theories of human interaction, LLM agent interaction networks are categorized into three types: consensus-oriented interaction, collaborative learning, and task-oriented coordination. These interactions form distinct paradigms, each presenting unique challenges related to cooperation, competition, coordination, and decision-making. Diverse agent interactions give rise to complex evolving networks over time. More details follow.

1) **Consensus-oriented interaction**
   This type of interaction primarily coordinates divergent agent goals and is crucial for integrating heterogeneous knowledge to establish unified understanding—significantly aiding collective decision-making. Consensus methods include discussions, debates, negotiations, reflections, and voting. Debates expose competing hypotheses, while negotiations resolve conflicting priorities and resource constraints. Agents iteratively share perspectives and assumptions to reach mutual understanding. GPTSwarm [31], a graph-based agent collaboration framework, structures agent collaboration via graph designs, where information flow and edge connections form discussion groups, excluding agents persistently providing erroneous views. RECONCILE [32] employs roundtable discussions with multi-cycle dialogues and confidence-based voting. Debates are vital for consensus building, hallucination reduction, and complex problem-solving. In Multi-agent Debate (MAD) [33], agents argue in tit-for-tat patterns under referee supervision to stimulate creative solutions.
2) **Collaborative learning**
   Collaborative learning typically occurs among similar agents. Although architecturally similar, these agents accumulate distinct memories and experiences due to their unique behaviors and environmental interactions with the environment. By collectively solving problems, these agents share experiences to enhance their problem-solving capabilities. Over time, each agent improves its skills through continuous interactions, leading to individual evolution. Notably, consensus-oriented interactions focus on aligning different perspectives to achieve knowledge integration and belief adjustment, whereas collaborative learning interactions emphasize peer knowledge construction and experience sharing, with a focus on mutual improvement and individual growth. Agents can continuously acquire team experiences through exchanging personal insights to achieve adaptive improvement. They can also guide their own strategies by observing others' behaviors and outcomes, a process termed observational learning.

   MAS-CTC [34] is a scalable multi-team framework that enables different teams to simultaneously propose various task-oriented decisions as insights, which are then exchanged at critical stages. Different agent teams utilize greedy pruning mechanisms and aggregation mechanisms to eliminate low-quality content, thereby improving software development performance.

AutoAgents [28] employs a knowledge-sharing mechanism where agents exchange execution results to enhance communication, acquiring long-term, short-term, and dynamic memories from others. In iAgents [35], human social networks are mapped onto agent networks, with agents continuously engaging in proactive interactions to acquire information necessary for task completion.

3) **Task-oriented interaction**
Agents achieve shared goals through coordinated task decomposition. Interactions focus on processing upstream outputs to generate inputs for downstream agents per task dependencies, rather than engaging in complex debates. These exchanges rely on explicit deliverables rather than direct negotiation. MACNET [36] structures agents as a DAG, a graph structure with no cycles. Supervisor agents issue instructions. Executor agents implement solutions, ensuring only optimized sequences are executed to completion. In a study [37], leader agents decompose complex goals into concrete tasks to manage workflows; executor agents perform actions (e.g., resource gathering).

10.3.4 L4: Agent's Social Emergence

In social simulation systems, agents can be categorized as homogeneous (identical traits) or heterogeneous (distinct traits) based on whether they have identical or distinct roles, capabilities, and action spaces. Homogeneous agents share identical functions, the same behavioral spaces, and identical observation spaces, whereas heterogeneous agents exhibit diverse perspectives and skills. Specifically, agent heterogeneity manifests across three dimensions: character-level heterogeneity, observation-space heterogeneity, and action-space heterogeneity.

1) **Character-level heterogeneity**
Agent role diversity influences how agents solve problems and interact. Most current LLM agents belong to this category [38, 39]. For example, in software development, agents may play roles such as programmers, product managers, or testers. In medical diagnosis, agents might represent cardiologists, oncologists, or pediatricians, each with distinct areas of expertise. Each role's unique perspective and expertise contribute to more robust collective decision-making. Although these heterogeneous agents may share identical action spaces, such as document authoring [40] (e.g., code, requirement reports, or test reports) or diagnostic suggestions provision [41], their roles influence the outcomes of these behaviors.

2) **Observation-space heterogeneity**
In social simulations, observation-space heterogeneity refers to differences in the range of information agents can observe or perceive within their environment. For example, in the *Werewolf* games, some agents (e.g., werewolves)

can see teammates' identities, seers can obtain designated players' identities, while others (e.g., villagers) cannot see any player's true identity. Similarly, in the *Avalon* game, different roles possess distinct observation spaces, thereby influencing players' strategies and communication. In these settings, each agent's perceptual capabilities or observation space directly correlate to their role within the system. In social simulations, differences in agents' observations typically affect their decision-making, communication, and coordination with other agents.

3) **Action-space heterogeneity**
Due to physical or functional constraints, agents exhibit differences in the set of executable actions. This is particularly important in virtual and physical environments, where agents may possess different capabilities based on their design or intended purpose. In virtual environments like the *Werewolf* and *Avalon* games, different roles have distinct abilities or skills. For example, in Werewolf, werewolves may have the ability to communicate secretly, while villagers may be limited to voting or observing. Such differences require agents to collaborate based on their unique abilities, facilitating strategic learning during interactions, such as teamwork, trust, and deception. This heterogeneity demands efficient collaboration and action coordination among agents, often involving task allocation according to their respective strengths.

Due to inherent randomness in LLMs and environments, these agents typically evolve along different trajectories. Even agents initially homogeneous in personality and behavioral space may develop heterogeneous behaviors through repeated simulations. For example, as shown in a study [42], agents initially sharing identical behavioral spaces and personalities developed different roles after multiple rounds of environmental interactions: some agents focused on food gathering, while others specialized in weapon crafting. Similarly, a study [43] observed initially homogeneous agents developing distinct language usage patterns, emotional expressions, and personalities after group interactions. These emergent behaviors demonstrate possible transitions from homogeneous to heterogeneous systems. Heterogeneity among individual agents strengthens interaction dynamics, which are essential for complex social network emergence and strategy refinement. It is crucial to emphasize that social emergence encompasses not only changes in individual capabilities but also behaviors that exceed the capacities of individual agents. Through sustained engagement in collective tasks and self-reflection on shared environments, agents continuously refine their reasoning and decision-making abilities. This implies new, complex behaviors can spontaneously arise from agent interactions—including actions beyond predefined protocols, such as trust, strategic deception, adaptive camouflage, and emergent cooperation—triggering deeper social dynamics [44, 45].

Through recursive interactions, agents inevitably form cooperative strategies, ultimately evolving into social contracts, organizational hierarchies, and labor division. Ultimately, this marks a transition from basic cooperative behaviors to complex social constructs, leading to cultural norms and conventions [46]. These norms form the foundation of social order, reducing conflict and improving coordination among agents, resulting in more stable and organized societies.

These findings highlight LLM agents' ability to develop complex behaviors through interaction and learning, demonstrating potential for exploring emergent behaviors in collective intelligence scenarios. Notably, such emergent behaviors depend on agents' memory and reflection mechanisms. In social simulations, shared context and environmental information significantly expand agents' accessible memories, thus enabling them to build upon past interactions, refine strategies, and adapt more effectively to dynamic environments [47]. The Sid project models large-scale agent societies, providing further evidence for the emergence of social norms and role specialization [42]. In this study, agents autonomously formed specialized social roles. These roles were not predefined but emerged naturally as agents interacted within environments and established collective rules. Simulations also highlighted the importance of collective democratic procedures in adhering to and modifying these collective rules. This process of norm-creation and role-specialization leads to better-organized and more adaptive governance structures within societies. In a study [48], the evolution of cultural and religious beliefs in agent societies was also observed, occurring through agent-driven idea selection—mirroring real-world societal change.

10.4 Applications of LLM-Based Agent Simulation

A primary application scenario for LLM agents is social simulation. Research in this domain is rapidly expanding, covering multiple fields such as social sciences, game development, psychology, economics, and policymaking. The key to using LLM-MA in world simulations lies in their exceptional role-playing capabilities, crucial for authentically depicting a wide range of roles and perspectives within simulated worlds. Environments for world simulation projects are typically designed to reflect specific simulated scenarios, with agents configured via diverse profiles to align with this context. Unlike problem-solving systems focused on agent cooperation, world simulation systems involve diverse agent management and communication methods, reflecting the complexity and diversity of real-world interactions. In the following sections, we introduce several simulation studies across different fields.

10.4.1 Economics and Finance

LLM agents are employed to simulate economic and financial transaction environments, primarily because they can serve as implicit computational models of human behavior. In these simulations, agents are endowed with specific resources and information, assigned predefined set of preferences, and permitted to explore their actions within economic and financial contexts. This is similar to economists' modeling of "Homo economicus," a concept that describe people as rational individuals pursuing their own interests in certain economic theories [49]. Given LLM agents' enhanced capabilities in textual comprehension and complex decision-making, they demonstrate significant potential for research in economics and finance. Explorations in related domains are presented in Table 10.2.

Table 10.2 LLM Agent in Economics.

Type	Agent Model	Applications
Market Simulation	LLM agents simulating behaviors of various market participants (supply-side and demand-side participants, competitors, and regulators).	Predict and simulate data (product prices, market share, market structure and transaction completion rate) applicable to scenarios such as retail purchases, bidding, bargaining, and collaborative bidding.
Financial Market Analysis	LLM agents simulating behaviors of market participants (investors, institutions, and regulators).	Provide insights into market volatility and risks; e.g., by modeling investor trading behaviors and information dissemination processes, LLM agents can predict fluctuations in stock prices, exchange rates, and interest rates.
Macroeconomic and Policy Simulation	LLM agents simulating fiscal/monetary policies implementation (covering governments, enterprises, and individuals).	Predict changes in macroeconomic indicators (GDP, inflation rate, unemployment rate).
Social and Economic Network Analysis	LLM agents modeling information dissemination, resource distribution, and trust establishment in social and economic networks.	Enhance understanding of network economics evolution; simulate behaviors of consumers, enterprises, and governments to explore network effects, information asymmetry, and market failure.

Traditional economics research relies on empirical data, making it difficult to delve into the micro-level behaviors; rule-based or ML ABMs also struggle to accurately simulate complex human economic behaviors. The rise of LLMs enables more authentic economic simulations. A study [50] employs LLM agents to simulate a virtual town featuring two restaurant agents (competitors agents executing operations such as menu modification, chef management, and advertising launch via APIs) and 50 customer agents (evaluator agents with differentiated attributes including income, taste preferences, and health conditions, categorized as individuals and groups). The simulation runs for 15 days. Restaurants competing for customers, who provide ratings and feedback after dining. If one restaurant exits, the simulation terminates early. Specifically, Figure 10.18 illustrates the daily operational workflow of restaurant agents.

Experimental results reveal that LLM agents' competitive behaviors align with classical socioeconomic theories. (1) Micro-level: Restaurant market strategies conform to classical theories, including differentiation (focusing on signature dishes), imitation (adopting competitors' strengths like local ingredients), and customer orientation (adjusting menus based on needs, e.g., offering sugar-free options for diabetics); customer decisions are influenced by multiple factors (demand, reputation, price, and other factors), with distinctions between individual and group decisions (in terms of decision-making, individuals prioritize reputation, while groups favor exploration). (2) Macro-level: Matthew effect (the "rich-get-richer" phenomenon) emerges, where popular restaurants attract more customers through positive reviews and high ratings, creating a "stronger-become–stronger" cycle; group customers are more willing to try new restaurants, providing opportunities for less popular, underperforming restaurants, and disrupting the Matthew effect's positive feedback loop, mitigating "winner-takes-all" outcomes; restaurants continuously optimize dishes to compete, with both restaurants' average dish ratings increasing by 0.26 and 0.22 on a five-point scale over the 15 days, respectively.

In economic simulations, LLM agents are employed to simulate human or economic agent decision-making, and the action space for agent interactions, as well as agent states, critically influence experimental outcomes. First, effectively representing agent states and interactive action spaces in order to accurately simulate economic agents' decision processes constitutes a significant challenge. Second, the anthropomorphic authenticity of LLMs presents another major challenge. Third, conducting large-scale macroeconomic analyses may require numerous LLM agents, posing difficulties regarding system performance and higher token consumption.

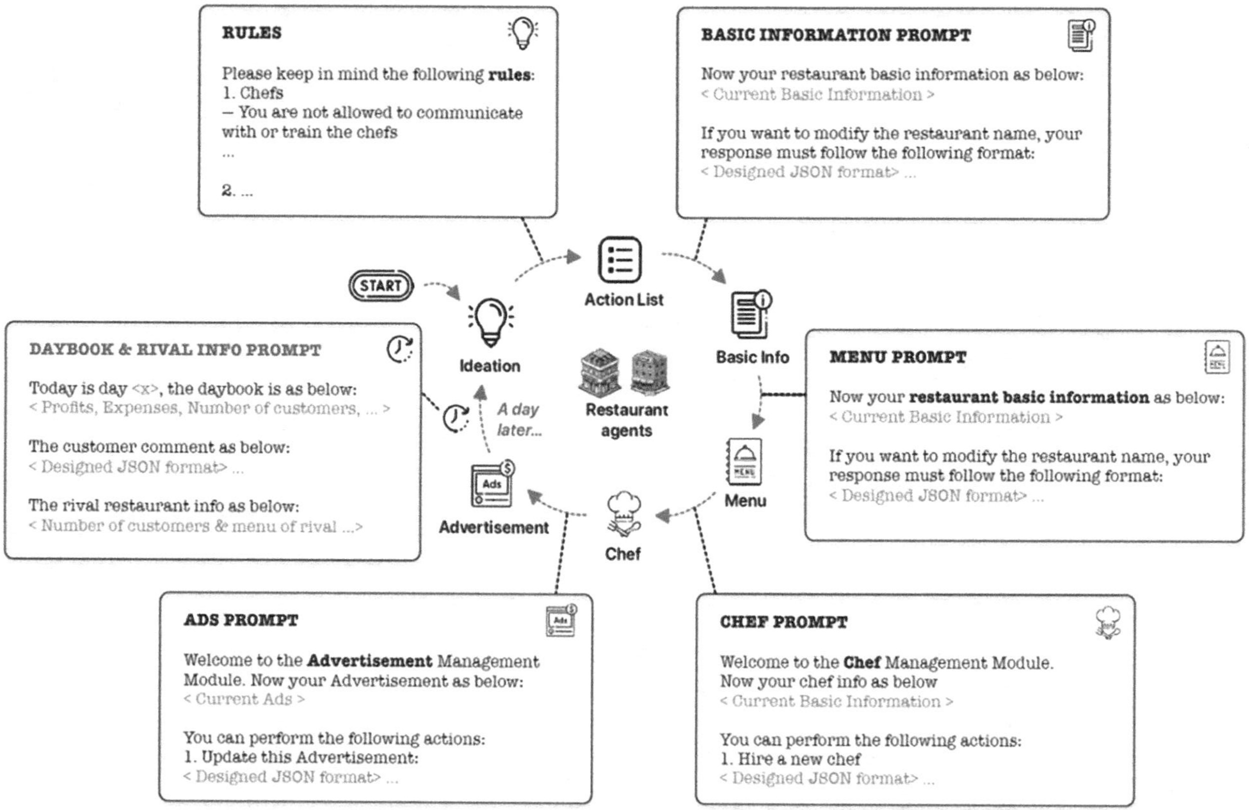

Figure 10.18 Overview of the restaurant operation process between competitors (i.e., two restaurants). Each day, the restaurant receives daily logs and information about competitors. Then, the agent manages the restaurant under the guidance of basic information prompts, menu prompts, chef prompts, and advertising prompts. *Source:* [50] / Cornell University / CC BY 4.0.

10.4.2 Political Science

Similar to simulations in game development and economic scenarios, policy formulation requires robust decision-making capabilities to address real-world and dynamically complex issues. LLM agents can simulate virtual government entities to study the impacts of diverse policies on different communities. These simulations provide policymakers with valuable insights, aiding in understanding and anticipating the consequences of their decisions, as noted in a study [51]. Key explorations of LLM agents in political science are summarized in Table 10.3.

A study [52] introduces the WarAgent system developed by Rutgers University and University of Michigan, leveraging the powerful language processing capabilities of LLMs combined with multiagent system interactions to simulate and analyze inter-nation decision-making processes and the evolution of international relations, including World War I (WWI), World War II (WWII), and China's ancient Warring States Period (WSP). As illustrated in Figure 10.19, within WarAgent, each agent represents a nation or organization capable of autonomous decision-making based on respective goals, resources, historical contexts, and current situations, while engaging in complex interactions with other agents. Notably, this approach enables WarAgent not only to provide deep insights into historical events but also to explore "counterfactual histories" under different strategic choices, thereby offering researchers an experimental platform

Table 10.3 LLM Agents in Political Science.

Type	Agent Model	Applications
Political Simulation and Prediction	LLM agents simulating behaviors/interactions of political process participants (political party competition, voter behavior, and policymaking).	Predict development trends of political events, election results, and policy effects.
Political Decision Analysis	LLM agents simulating behaviors/interactions in political decision-making processes	Evaluate pros/cons and impacts of policy options; simulate interactions between governments, political parties, and interest groups to provide insights into policy impacts.
International Relations Research	LLM agents modeling international politics, country interactions, and conflicts.	Enhance understanding of international politics complexity/risks; explore international trade, military conflicts, and diplomatic interactions.

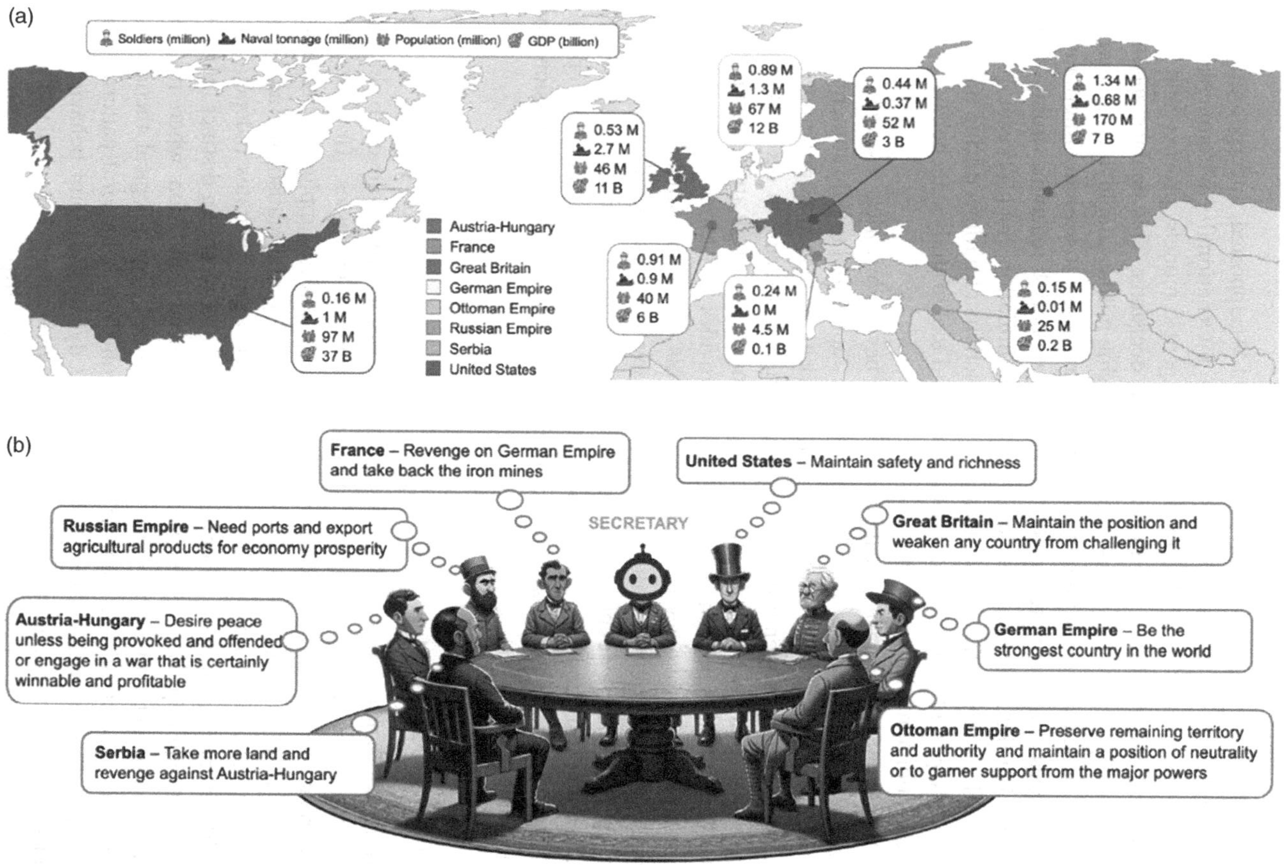

Figure 10.19 World War I simulation setting. *Source:* [52] / Cornell University / CC BY 4.0.

to investigate historical possibilities and informing modern conflict prevention and resolution strategies.

In WarAgent, agent decision-making is LLM-supported, incorporating not only historical data but also contemporary international environments and cross-nation inter-agent interactions. WarAgent's models process and analyze vast historical datasets to inform agent decisions. This enables agents to dynamically adjust strategies—such as modifying military deployments, altering foreign policies, or seeking new alliances—in response to evolving international landscapes. Furthermore, agents anticipate potential reactions from others, enhancing the realism and dynamism of the simulation. Such decision-process simulations mirror the actual reasoning and actions of nations during historical conflicts, providing researchers with a powerful tool to study state behaviors and international relations dynamics.

WarAgent's potential applications extend far beyond historical research. In education, WarAgent serves as an interactive educational tool to help students better understand international relations and historical events. By simulating diverse historical conflict scenarios, educators can engage students in decision-making processes, thereby deepening their comprehension of history and global politics. Additionally, for policymaking, WarAgent provides a risk-free environment to test diplomatic strategies and policy options. This allows policymakers to evaluate potential consequences of different approaches without real-world consequences, supporting more evidence-based decisions. Furthermore, for military strategists, WarAgent functions as a decision-support tool to explore strategic options under varied scenarios and assess their impact on conflict progression and outcomes.

10.4.3 Psychology

In psychological simulation research, multiple agents are employed to simulate humans with diverse characteristics and thought processes. However, one approach differs from social simulation in that it directly applies psychological experiments to these agents, focusing on observing and statistically analyzing their diverse behaviors. In this approach, each agent operates independently without interacting with other agents, essentially representing distinct individuals. Another approach more closely resembles social simulation, where multiple agents interact and communicate. In this context, psychological theories are used to understand and analyze emergent collective behavioral patterns. This method promotes the study of interpersonal dynamics, group behaviors, and group dynamics, providing insights into how individual psychological traits shape collective actions. Key applications of LLM agents in psychology are summarized in Table 10.4.

Table 10.4 LLM Agents in Psychology.

Type	Agent Model	Applications
Psychotherapy and Counseling	LLM agent simulating interactions and influences in psychotherapy/counseling process.	It helps to deeply understand the basic principles of psychotherapy and counseling psychology and supports patients receiving psychotherapy.
Cognitive Modeling	LLM agent simulating individual cognitive processes (perception, memory, thinking, and decision-making).	Reveal core principles of cognitive psychology; analyze cognitive biases and strategies.
Emotion and Motivation Modeling	LLM agent modeling emotional/motivational processes (assessing emotional responses, interests, drives).	Explore fundamental principles of emotion and motivation psychology.

Currently, LLM agents primarily focus on mental health support and psychological experiment simulation, enabling the development of NPCs with human psychological characteristics and autonomous cognition. For example, inspired by *The Sims*, a study [53] constructs a sandbox environment named "Quint's Cafe" based on a café prototype to verify the multifaceted influence of social networks and subjective consciousness on human preferences and personality formation. The café comprises four areas: public zone, dining area, reading zone, and viewing area, where three LLM agents (Anty, computer science student; Agnes, psychology student and Anty's partner; Qunit, coffee robot) perform activities, such as acting, communicating, and perceiving.

As illustrated in Figure 10.20, the core of the AFSPP framework includes two components. (1) Preference shaping: Using the frequency of agents' specific behaviors as metrics to investigate the influences of social networks (via attitudes from intimate agents—positive or negative, subjective or objective) and subjective consciousness (identity, sensory perception, prior knowledge, reflection, planning). (2) Personality shaping: Employing MBTI (Myers-Briggs Type Indicator) as a universal personality metric and SD3 (Short Dark Triad) for negative traits to examine impacts of social networks (attitudes from intimate agents—such as gentleness, arguments, breakups, and proposals) and subjective consciousness (identity). This framework lowers barriers for agent-based psychological experiments through low-complexity sandbox worlds, enabling the substitution of agent-based experiments for high-risk or hard-to-replicate human experiments (e.g., intimate relationship crises), and it successfully replicates human psychological findings

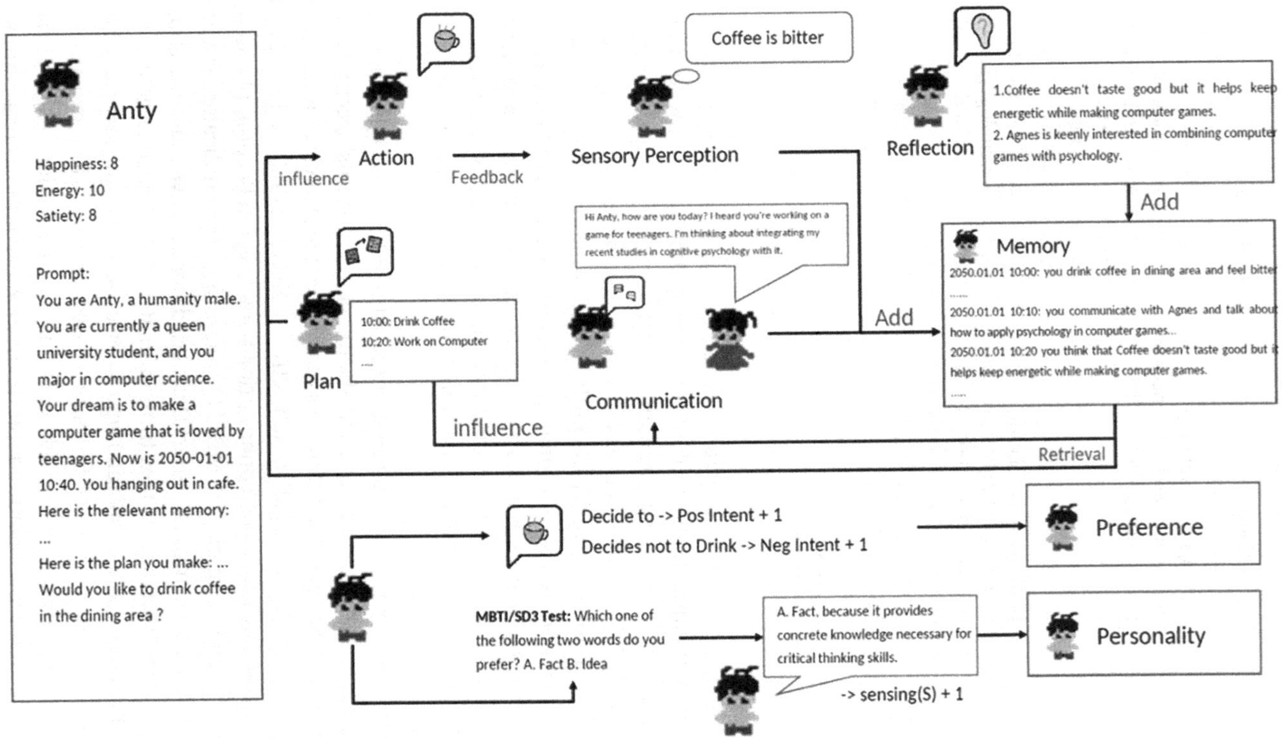

Figure 10.20 Overview of LLM agent behavior and components in Qunit Cafe (Mini sandbox world). *Source:* [53] / Cornell University.

(e.g., RIASEC-MBTI correlations between career interest types and personality types), validating its effectiveness in simulating human behaviors.

However, research in this domain faces skepticism. Human preferences and personalities are multifaceted and continuously evolve under environmental and subjective influences. However, current studies frequently employ preset prompts to initialize the preferences and personalities of LLM agents—an approach that is inconsistent with human attributes. Although some research explores LLMs' intrinsic traits, it often neglects their iterative evolution process in human-like societal contexts. Notably, future studies may address these challenges by developing more effective methodologies to enhance the performance of LLM agent in mental health support and psychological experiment simulations.

10.4.4 Recommendation System

Recommendation systems play a pivotal role in contemporary information dissemination, shaping users' personal preferences and cognitive processes. However, most such systems rely on traditional supervised learning methods, exhibiting significant discrepancies between offline metrics and online performance. This disconnect creates a schism between academia and industry, and it has become a bottleneck for future development. An experimental platform capable of providing human-like operations, analyzing human intent, and encoding human cognitive processes could test recommendation models' efficacy, which would thereby bridge the academic–industrial divide. The core challenge lies in enabling LLMs to precisely simulate user operations within recommendation systems.

The application of LLM agents in recommendation systems parallels their application in psychology, as both fields involve extrinsic and intrinsic human-related factors (e.g., cognitive processes and personality traits). To bridge the offline–online performance gap, agent4Rec [54] introduces an LLM agent-based simulation platform that models complex user interactions in recommendation environments. As illustrated in Figure 10.21, the simulator comprises two core aspects: LLM-driven generative agents (the red components—the right components) and the recommendation environment (the blue components—the left components). User and item profiles are initialized using real-world datasets (e.g., MovieLens-1M, Steam, and Amazon-Book). The recommendation algorithm serves as an adaptable system component, generating item recommendations presented to agents on a page-by-page basis. Agents equipped with specialized memory and action modules tailored for recommendation scenarios simulate diverse behaviors including viewing items, rating items, providing feedback, turning pages, and exiting.

Agent4Rec demonstrates that LLM agents can effectively simulate authentic user preferences and behaviors, provide insights into phenomena such as

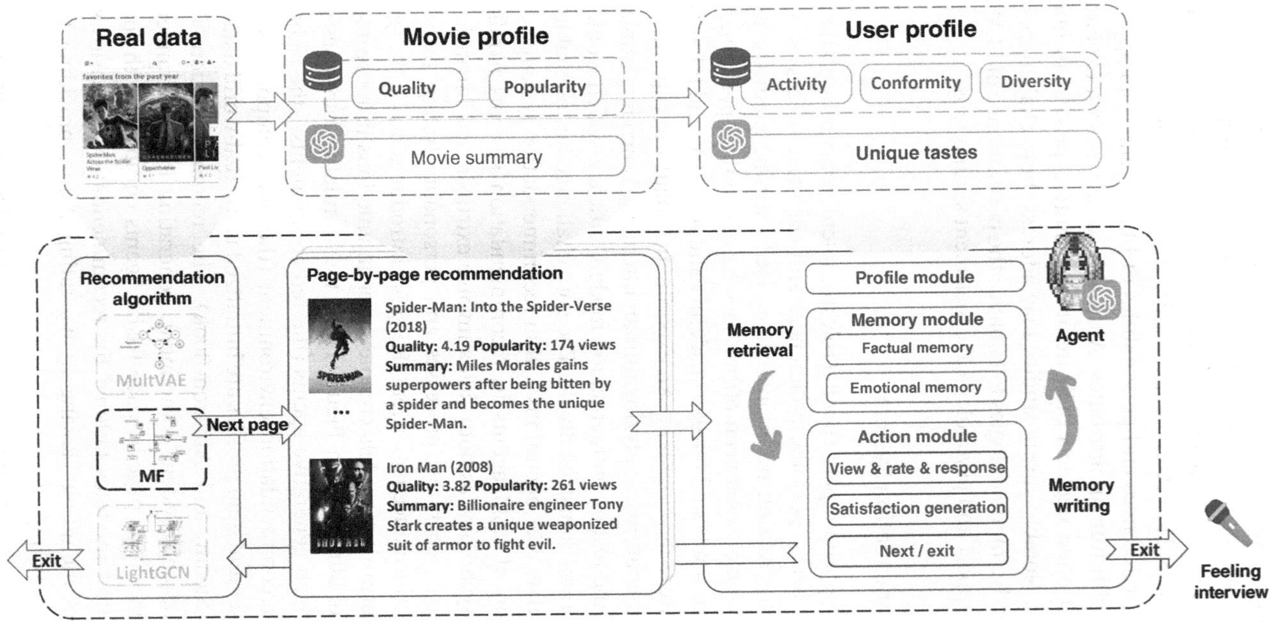

Figure 10.21 The overall framework of Agent4Rec. *Source:* [54] / ACM (The Association for Computing Machinery) / CC BY 4.0.

information cocoons, and help uncover key causal relationships in recommendation tasks. Notably, in this work, agents simulate users without inter-agent communication. Differing from the Agent4Rec framework, A study [55] treats both users and items as agents, emphasizing simulation of user–item interactions and propagating preferences among agents to capture the core essence of collaborative filtering.

Evaluating the simulator's effectiveness and limitations requires extensive experimentation. From the user perspective, focus lies on assessing the degree of user–agent alignment. Specifically, this involves evaluating to what extent agents ensure consistency with real users' social characteristics, personalities, and preferences necessitates various metrics and statistical tests. From the recommendation system perspective, different algorithms can be assessed using metrics spanning multiple dimensions, such as average number of recommended and viewed movies, average user ratings, user engagement duration, and overall user satisfaction. Agent feedback collected from the simulator serves as augmented data, enabling iterative training and refinement of recommendation strategies. Subsequently, feedback-driven recommendation enhancements are evaluated using standardized offline metrics. Combining simulated feedback with traditional offline evaluation enables a comprehensive assessment of recommendation algorithms, which helps further explore the simulator's potential applications in recommendation systems.

10.5 Chapter Summary

Recent years have witnessed the emergence of LLMs, offering a promising pathway to enhance the simulation capabilities of ABMs, with related works proliferating rapidly. This chapter focuses on exploring areas where large model agents can advance social simulation, alongside representative application scenarios. First, we propose a cognitive model architecture for LLM agents, which sorts out various mechanisms that influence decision-making, including memory, state, emotion, goals, learning, and other factors, hoping to facilitate the modeling of LLM agents; then, we elaborate on technical details of "Stanford Smallville" to provide readers with practical implementation insights; subsequently, guided by the agent capability hierarchy proposed in Chapter 4 of this book, we introduce methods to extend agents' anthropomorphic characteristics using large model techniques; finally, we summarize recent research achievements of LLM agents across social simulation domains—economics and finance, political science, psychology, and recommendation systems—and discuss the associated challenges.

References

1 Wang, Y., Chen, Y., Zhong, F. et al. (2024). Simulating human-like daily activities with desire-driven autonomy. arXiv preprint arXiv:2412.06435.

2 Cognitive Architectures for Language Agents. https://blog.langchain.dev/memory-for-agents/.

3 Arslan, M., Ghanem, H., Munawar, S. et al. (2024). A Survey on RAG with LLMs. *Procedia Computer Science* 246: 3781–3790.

4 Zhou, D., Xue, X., Ma, Q. et al. (2024). Federated experiments: generative causal inference powered by LLM-based agents simulation and RAG-based domain docking. *IEEE/CAA Journal of Automatica Sinica* 11 (11): 1–4.

5 Liu, B., Li, X., Zhang, J. et al. (2025). Advances and challenges in foundation agents: From brain-inspired intelligence to evolutionary, collaborative, and safe systems. arXiv preprint arXiv:2504.01990.

6 Ekman, P. (1992). An argument for basic emotions. *Cognition & Emotion* 6: 169–200.

7 Russell, J. (1980). A circumplex model of affect. *Journal of Personality and Social Psychology* 39: 161–178.

8 Plutchik, R. (1980). A general psychoevolutionary theory of emotion. *Theories of Emotion* 1: 3–31.

9 LeDoux, J.E. (1998). *The Emotional Brain: The Mysterious Underpinnings of Emotional Life*. Simon and Schuster.

10 Fei, H., Li, B., Liu, Q. et al. (2023). Reasoning implicit sentiment with chain-of-thought prompting. *Annual Meeting of the Association for Computational Linguistics*. https://doi.org/10.48550/arXiv.2305.11255.

11 Sun, X., Li, X., Zhang, S. et al. (2023). Sentiment analysis through LLM negotiations. arXiv preprint arXiv: 2311.01876.

12 Park, J.S., O'Brien, J., Cai, C.J. et al. (2023). Generative agents: interactive simulacra of human behavior. *Proceedings of the 36th Annual ACM Symposium on User Interface Software and Technology*, pp. 1–22.

13 Besta, M., Blach, N., Kubicek, A. et al. (2023). Graph of thoughts: solving elaborate problems with large language models. *AAAI Conference on Artificial Intelligence*.

14 Setlur, A., Rajaraman, N., Levine, S., and Kumar, A. (2025). Scaling test-time compute without verification or rl is suboptimal. arXiv preprint arXiv:2502.12118.

15 Yao, S., Yu, D., Zhao, J. et al. (2023). Tree of thoughts: deliberate problem solving with large language models. *Advances in Neural Information Processing Systems* 36: 11809–11822.

16 Hao, S., Gu, Y., Luo, H. et al. (2024). LLM reasoners: new evaluation, library, and analysis of step-by-step reasoning with large language models. arXiv preprint arXiv:2404.05221.

17 Zhang, J. and Liu, K. (2024). Thought space explorer: navigating and expanding thought space for large language model reasoning. *2024 IEEE International Conference on Big Data (BigData)*, pp. 8259–8251. IEEE.

18 Xiong, S., Payani, A., Kompella, R. et al. (2024). Large language models can learn temporal reasoning. arXiv preprint arXiv:2401.06853.

19 Anokhin, P., Semenov, N, Sorokin, A. et al. (2024). Arigraph: learning knowledge graph world models with episodic memory for LLM agents. arXiv preprint arXiv:2407.04363.

20 Xu, L., Hu, Z., Zhou, D. et al. (2023). Magic: investigation of large language model powered multi-agent in cognition, adaptability, rationality and collaboration. arXiv preprint arXiv:2311.08562.

21 OpenAI (2022). Introducing chatgpt. https://openai.com/blog/chatgpt/.

22 Wang, G., Xie, Y., Jiang, Y. et al. (2023). Voyager: an open-ended embodied agent with large language models. arXiv preprint arXiv:2305.16291.

23 Chen, Z., Chen, D., Sun, R. et al. (2025). Scaling autonomous agents via automatic reward modeling and planning.

24 Gu, Y., Zhang, K., Ning, Y. et al. (2024). Is your LLM secretly a world model of the internet? model-based planning for web agents. arXiv preprint arXiv:2411.06559.

25 Yang, Z., Li, P., Yan, M. et al. (2024). React meets actre: autonomous annotation of agent trajectories for contrastive self-training. *First Conference on Language Modeling*.

26 Zhang, D., Chen, L., Zhang, S. et al. (2023). Large language models are semi-parametric reinforcement learning agents. *Advances in Neural Information Processing Systems* 36: 78227–78239.

27 Zhang, W., Tang, K., Wu, H. et al. (2024). Agent-pro: learning to evolve via policy-level reflection and optimization. arXiv preprint arXiv:2402.17574.

28 Chen, G., Dong, S., Shu, Y. et al. (2023). Autoagents: a framework for automatic agent generation. arXiv preprint arXiv:2309.17288.

29 Wang, Z., Cai, S., Chen, G. et al. (2023). Describe, explain, plan and select: interactive planning with llms enables open-world multi-task agents. *Advances in Neural Information Processing Systems* 36: 34153–34189.

30 Sun, H., Zhuang, Y., Kong, L. et al. (2023). Adaplanner: adaptive planning from feedback with language models[J]. *Advances in Neural Information Processing Systems* 36: 58202–58245.

31 Zhuge, M., Wang, W., Kirsch, L. et al. (2024). Gptswarm: language agents as optimizable graphs. *Forty-first International Conference on Machine Learning*.

32 Chen, J.C.Y., Saha, S., and Bansal, M. (2023). Reconcile: round-table conference improves reasoning via consensus among diverse llms. arXiv preprint arXiv:2309.13007.

33 Liang, T., He, Z., Jiao, W. et al. (2023). Encouraging divergent thinking in large language models through multi-agent debate. arXiv preprint arXiv:2305.19118.

34 Qian, C., Dang, Y., Li, J. et al. (2023). Experiential co-learning of software-developing agents. arXiv preprint arXiv:2312.17025.

35 Kim, Y., Park, C., Jeong, H. et al. (2024). Mdagents: an adaptive collaboration of llms for medical decision-making. *Advances in Neural Information Processing Systems* 37: 79410–79452.

36 Qian, C., Xie, Z., Wang, Y. et al. (2024). Scaling large language model-based multi-agent collaboration. arXiv preprint arXiv:2406.07155.

37 Chen, J., Jiang, Y., Lu, J. et al. (2024). S-agents: self-organizing agents in open-ended environments. arXiv preprint arXiv:2402.04578.

38 Sun, J., Dai, C., Luo, Z. et al. (2024). LawLuo: a multi-agent collaborative framework for multi-round Chinese legal consultation. arXiv preprint arXiv:2407.16252.

39 Wu, D., Shi, H., Sun, Z. et al. (2023). Deciphering digital detectives: Understanding LLM behaviors and capabilities in multi-agent mystery games. arXiv preprint arXiv:2312.00746.

40 Hong, S., Zheng, X., Chen, J. et al. (2023). Metagpt: meta programming for multi-agent collaborative framework. arXiv preprint arXiv:2308.00352, 3(4): 6.

41 Tang, X., Zou, A., Zhang, Z. et al. (2023). Medagents: large language models as collaborators for zero-shot medical reasoning. arXiv preprint arXiv:2311.10537.

42 AL, A., Ahn, A., Becker, N. et al. (2024). Project sid: many-agent simulations toward AI civilization. arXiv preprint arXiv:2411.00114.

43 Takata, R., Masumori, A., and Ikegami, T. (2024). Spontaneous emergence of agent individuality through social interactions in LLM-based communities. arXiv preprint arXiv:2411.03252.

44 Du, S. and Zhang, X. (2024). Helmsman of the masses? evaluate the opinion leadership of large language models in the werewolf game. arXiv preprint arXiv:2404.01602.

45 Jin, X., Wang, Z., Du, Y. et al. (2024). Learning to discuss strategically: a case study on one night ultimate werewolf. *Advances in Neural Information Processing Systems* 37: 77060–77097.

46 Ren, S., Cui, Z., Song, R. et al. (2024). Emergence of social norms in generative agent societies: principles and architecture. arXiv preprint arXiv:2403.08251.

47 Han, S., Zhang, Q., Yao, Y. et al. (2024). LLM multi-agent systems: challenges and open problems. arXiv preprint arXiv:2402.03578.

48 Vallinder, A. and Hughes, E. (2024). Cultural evolution of cooperation among LLM agents. arXiv preprint arXiv:2412.10270.

49 Horton, J.J. (2023). Large language models as simulated economic agents: what can we learn from homo silicus? National Bureau of Economic Research.

50 Zhao, Q., Wang, J., Zhang, Y. et al. (2023). Competeai: understanding the competition dynamics in large language model-based agents. arXiv preprint arXiv:2310.17512.

51 Axtell, R.L. and Farmer, J.D. (2025). Agent-based modeling in economics and finance: past, present, and future. *Journal of Economic Literature* 63 (1): 197–287.

52 Hua, W., Fan, L., Li, L. et al. (2023). War and peace (waragent): large language model-based multi-agent simulation of world wars. arXiv preprint arXiv:2311.17227.

53 He, Z. and Zhang, C. (2024). Afspp: agent framework for shaping preference and personality with large language models. arXiv preprint arXiv:2401.02870.

54 Zhang, A., Chen, Y., Sheng, L. et al. (2024). On generative agents in recommendation. *Proceedings of the 47th International ACM SIGIR Conference on Research and Development in Information Retrieval*, pp. 1807–1817.

55 Zhang, J., Bao, K., Wang, W. et al. (2024). Prospect personalized recommendation on large language model-based agent platform. arXiv preprint arXiv:2402.18240.

11

Large Language Model Agents and Workflow

According to their purposes, large language model (LLM) agents can be divided into two categories. (1) Simulation agents, which strive to be more human-like and authentic, emphasizing the emotional and interactive characteristics of agents, and aim to generate scenarios and capabilities beyond designers' initial planning. The uncertainty and diversity generated by LLMs may become a unique advantage during the simulation process. (2) Autonomous agents, which leverage LLM as a "powerful general problem solver" to achieve more intelligent and automated problem-solving capabilities. When an agent is given a goal, it can create tasks, determine the priority of the task list, execute the tasks, and repeat this process until the goal is completed. Due to the high accuracy requirements, external tools are more frequently needed to help reduce the negative impact of LLMs' uncertainty. Table 11.1 presents a comparison between the two.

Nowadays, the market is flooded with various intelligent terminal products, ranging from smart speakers and smart watches to floor-cleaning robots, and from nutrition-tracking apps to children's storytelling robots. The huge potential demonstrated by LLMs in reasoning and planning capabilities precisely meets humans' expectations for autonomous agents. These agents can seamlessly integrate language understanding with actionable capabilities. By invoking external tools or systems, they can perform precise operations to achieve specific goals, thus becoming intelligent intermediaries between humans and machines. For example, LLM-driven agents can independently query databases, retrieve relevant information, and use this information to provide actionable insights, thereby seamlessly merging human intent with machine precision in increasingly complex workflows.

As application scenarios become increasingly complex, LLM-driven single agents are beginning to evolve toward multiagent collaboration. Compared with a single agent, a multiagent system (MAS) can effectively simulate the cooperation patterns of human groups in solving problems, emphasizing diverse agent configurations and collaborative workflows among agents. A MAS decomposes

Table 11.1 Differences Between Simulation Agent and Autonomous Agent.

Category	Simulation Agent	Autonomous Agent
Initial Motivation	Set internally by researchers (for simulation purposes)	Serving humans, with tool attributes
Accuracy Requirement	Generally, more emphasis on anthropomorphism	High
Memory	Short-term memory + long-term memory	Mainly relies on short-term memory in most cases
External Tools	Fewer used	More used
Current Applications	Stanford Smallville, Voyager, etc.	AutoGPT, MetaGPT, etc.

complex tasks into multiple subtasks and designs a dedicated agent for each subtask, allowing each agent to focus on a specific function. This makes the system structure clearer and easier to maintain and debug. At the same time, LLMs are creative systems with the characteristics of "hallucinations," and using multiple agents to complete tasks will achieve more stable and superior results than using a single agent [1]. In contrast, designing complex prompt engineering processes for a single agent can easily make the system difficult to understand and maintain.

Computational experiments involve many steps such as artificial society modeling, experimental design, experimental analysis, and experimental verification, which still present a certain threshold for nonprofessionals to utilize. To reduce the difficulty of use, the entire computational experiment process can be split into different tasks, and LLM agents playing different roles can assist in complete these tasks for users, similar to human division of labor and professional collaboration. In this context, how to use autonomous agents to improve the automation level of computational experiment platforms has become a key issue that urgently needs to be addressed. This chapter focuses on introducing the workflow, collaboration modes, and technical challenges of autonomous agents, thereby laying a foundation for promoting the automation of computational experiments.

11.1 Collaboration Framework of AI Agents

Traditional artificial intelligence (AI) systems require users to continuously provide clear instructions, which greatly increases the difficulty of use for nonprofessional users. As LLM technology becomes increasingly mature, how to enable AI

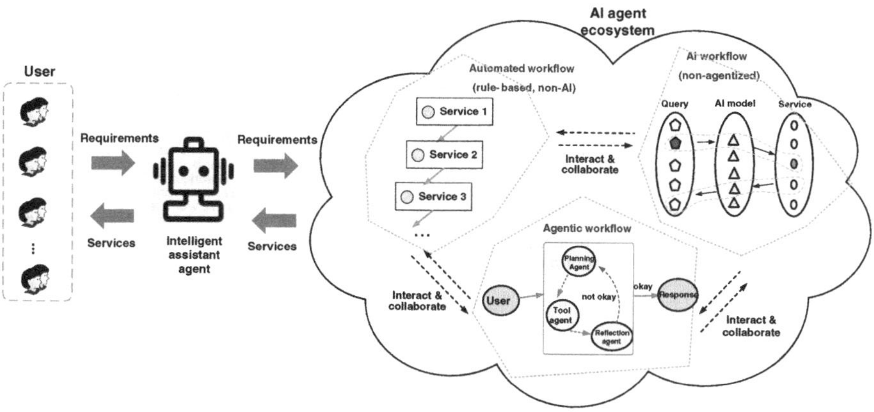

Figure 11.1 Schematic diagram of future AI system operation.

systems to understand user intentions, autonomously plan execution paths, and even adjust strategies in complex environments has become a key issue for their practical application. In an LLM agent system, each agent is equipped with different roles, initial knowledge, and its own goals. Agents interact with other agents to collect and process information; make independent decisions based on their goals, knowledge, and observations; and then participate in problem-solving. Multiagents perform complex tasks through collaboration, competition, or hierarchical organization, and effectively solve problems through diverse and goal-oriented interactions. Figure 11.1 illustrates the operational process of future AI systems [2]: (1) users interact with the intelligent assistant agent to express their needs; (2) the intelligent assistant agent parses the user's intentions and then searches for suitable service providers from the AI agent ecosystem; and (3) service agents can collaborate to complete the task, mainly involving three types of collaborative workflows: traditional non-AI workflows, which rely on fixed rules, execute according to preset steps, and cannot adapt to changes; general AI workflows, which use LLMs to perform tasks but lack dynamic decision-making or autonomy; and agentic workflows, which dynamically execute tasks through AI agents and are more responsive, adaptive, and self-evolving than traditional workflows. The following will introduce the details of collaboration between service agents from three aspects: value flow, collaboration mechanism, and interaction protocol.

11.1.1 The Value Flow of AI Agents

The value circulation mechanism of an MAS refers to the comprehensive performance of various agents in terms of service quality and efficiency during the value creation process. The effectiveness of value circulation not only affects customer

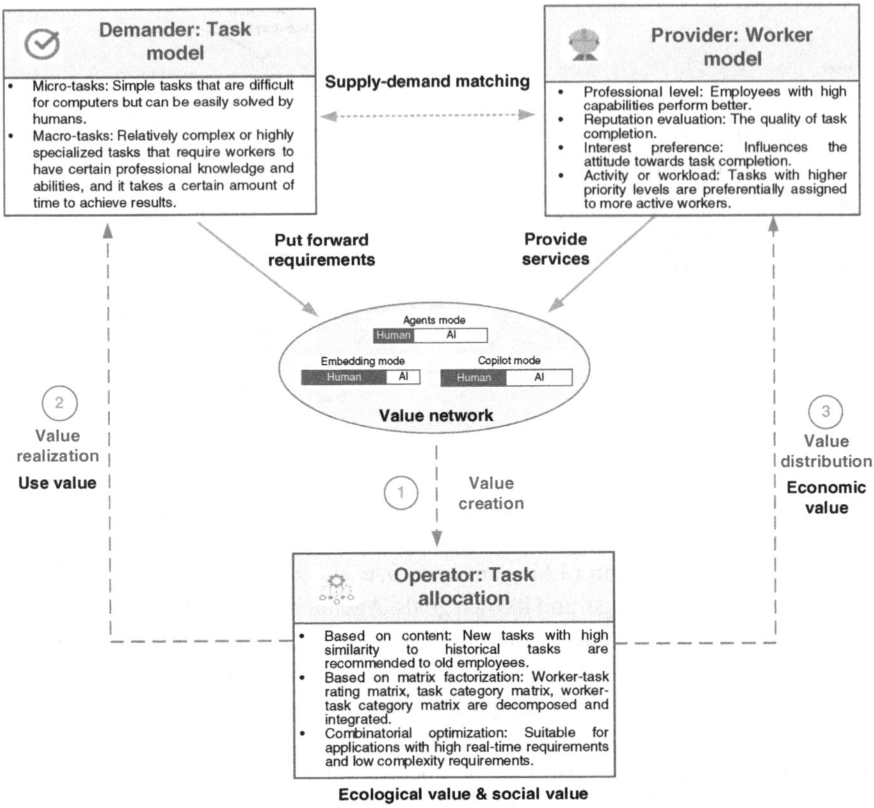

Figure 11.2 The value circulation mechanism of multiagent systems.

satisfaction and loyalty but also influences the costs and benefits of these agents. As shown in Figure 11.2, the value circulation mechanism provides an integrated mechanism for the comprehensive evaluation of supply agents, demand agents, and operation agents: service matching captures demand-side perceived value, while value creation reflects the value generated and circulated by operational agents, and value distribution represents the value obtained by supply agents. The specific details are as follows:

11.1.1.1 Service Matching—Collaborative Intent Understanding

The value-driven matching process connects providers and demanders: First, the task sequence of demanders serves as the driving force for the operation of MASs, typically described by a task model including micro-tasks and macro-tasks. Second, providers offer value-added services to demanders within a specific timeframe to meet their specific needs, which is generally modeled by MAS parameters such as professional proficiency, reputation evaluation, interest preferences, and activity load. Third, as a bridge between providers and demanders, operators need to construct diversified service chains to achieve the supply–demand matching. The entire process consists of a series of related behaviors and decisions, expressed by the following formula:

$$S_I \times S_C \times K_{match} = \Delta \tag{11.1}$$

where S_I represents the demander, which is generally measured by service quality, that is, whether it can provide reliable, accurate, timely, professional, and friendly services in accordance with the expectations and requirements of customers. Common measurement indicators include customer satisfaction, service personalization, etc. S_C represents the provider, which is generally measured by service efficiency, that is, whether it can complete the most service tasks with the least resources and time. Common measurement indicators include service productivity, service utilization rate, service response time, service processing time, service delivery speed, etc. K_{match} represents the service matching strategy, including content-based matching, matrix decomposition-based matching, and combinatorial optimization, etc., and needs to be continuously adjusted according to the current supply–demand matching status. Δ is the final service matching result, indicating whether the service chain completes the user's needs and whether the demander transfers profits to the provider.

11.1.1.2 Value Creation—Collaborative Task Decomposition

Service value is used to measure whether providers can create valuable service outcomes for customers, including meeting their functional, emotional, and social needs. Common measurement indicators of service value include service revenue, service cost, service profit rate, service return rate, etc. MASs generally decompose a large complex task into many micro-tasks and then assign complex workflows to numerous participating agents. The collaborative decision-making process of multiagents is crucial to ensuring the efficient operation of the system and the successful completion of tasks. According to the concept of value engineering, the value calculation formula for the multiagent collaboration process is expressed as follows:

$$Value =< Outcome, Cost, Model_{Eva} > \tag{11.2}$$

where *Outcome* represents the overall benefit of crowd-intelligence collaboration in the supply–demand matching process, including the profits obtained by the supplier and the user satisfaction of the demander. *Cost* denotes the total cost of providing services from the supplier to the demander, which may be influenced by the characteristics of the service network, such as network density, network scale, and clustering coefficient. $Model_{Eva}$ represents the value calculation model at the system level, which can be the difference or quotient between *Outcome* and *Cost*. Generally speaking, there are two main directions to achieve more value: increasing output (through service innovation or improving service efficiency) or reducing costs.

11.1.1.3 Value Allocation—Collaborative Decision-Making Consultation

Multiagent collaboration requires value allocation to attract and retain participating entities. This mechanism, often the focus of platform operators, serves as a bridge between service providers and demanders. It needs to maintain a balance between the coordinator's control and participants' autonomy. For a specific agent in the crowd-intelligence space, its value creation performance can be evaluated by individual effectiveness, calculated according to the following formula:

$$Utility(i) = Reward(i) - Cost(i) \tag{11.3}$$

where *Utility*(i) represents the effectiveness of the i-th service entity; *Reward*(i) denotes the revenue of the i-th service entity, and *Cost*(i) represents the consumption undertaken by the i-th service entity to obtain the revenue. Both *Reward* and *Cost* need to be interpreted in combination with specific scenarios, and the corresponding specific indicators of *Reward* and *Cost* vary across different scenarios. Differences in value allocation strategies will affect the behavioral initiative of service-providing entities, thereby influencing the sustainable development of MASs.

11.1.2 The Collaboration Mechanism of AI Agents

As is well known, the reasoning approach of LLMs is based on predicting the next token from the current context, which is a recursive decoding mechanism. Its advantage lies in ensuring that the model remains sufficiently focused when generating output. For information recipients, the model's output appears structured, logically coherent, and relevant to the topic. However, this also limits the potential of LLMs, as it constrains them to focus on expressing a single train of thought or a single role. Therefore, when using an LLM, constructing multiple agents with different roles allows the LLM to think divergently in different directions, which can benefit performance on certain tasks.

The advantages of a single-agent system lie in its relatively simple structure, making it easy to be implemented and controlled. However, its drawback is the limited capability to handle complex problems. For problems requiring multifaceted knowledge and skills, it may fail to provide effective solutions. MASs excel in their distributed and collaborative nature, enabling them to efficiently address complex, distributed problems. Each agent can focus on tasks within its specific domain, tackling system-wide issues through collaboration

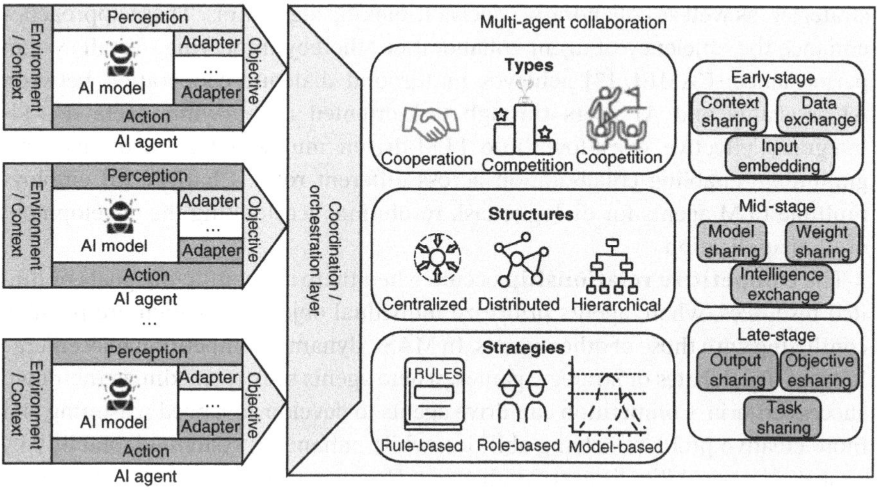

Figure 11.3 The collaboration mechanism of multiagent systems.

and communication. Moreover, MASs feature superior flexibility and scalability, allowing them to adapt to dynamically changing environments. Nevertheless, MASs face challenges such as coordinating behaviors among multiple agents, managing conflicts, and managing resource competition.

In the collaboration process of LLM agents, each agent consists of a language model m (serving as a neural processor), current goal o, environment e, input perception x, and corresponding output/action y. As shown in Figure 11.3, the core focus of the mechanism is on collaborative channels c, which facilitate coordination and orchestration among agents. These channels are defined by their participants (involved agents), type, structure, and strategy. Different agents may participate in different collaborative channels, each with distinct interaction types, and collectively achieve the overall system goals. The collaborative framework includes collaboration types, collaboration strategies, communication structures, and decision-making methods, with specific details as follows:

11.1.2.1 Collaboration Types

The **cooperative relationship** requires agents to align their individual goals with collective goals, jointly solve problems, make collective decisions, and complement each other's skills. Advantages of cooperation include the ability to assign subtasks to agents with specific strengths, its simple design and execution, and reliance on clear goals. However, cooperation also faces challenges such as potential inefficiencies from goal misalignment, exacerbation of a single agent's failure, and difficulties in coordinating actions within dynamic environments. In cooperative relationships, researchers primarily focus on role and task allocation strategies, as well as collaborative decision-making algorithms. These approaches enhance the efficiency of agent collaboration, thereby improving overall system performance. CAMEL [3] achieves multiround dialogue cooperation between AI assistants and AI users through task-oriented role-playing. MetaGPT [4] integrates effective workflows into LLM-driven multiagent collaborative programming, enabling collaboration across different roles. ChatDev [5] employs multiple LLM agents for dialogue task resolution, accelerating the development of LLM application.

The **competitive relationship** occurs when there are conflicting goals or limited resources, where agents prioritize individual objectives, which are possibly conflicting with those of other agents. In MASs, dynamic competition may emerge in tasks like debates or strategic games, where agents seek to maximize their own success criteria. Competition can drive agents to develop advanced reasoning and more creative problem-solving abilities, while enhancing system adaptability by pushing the capability limits of each agent. However, competition also introduces challenges, including the need for mechanisms to ensure that competition benefits overall system goals and prevents excessive competition in cooperative environments. In competitive relationships, considerations include designing effective competitive strategies, information hiding techniques, and adversarial training techniques. These techniques help agents gain advantages in competition to achieve their objectives. ChatEval [6] improves the efficiency and effectiveness of handling complex tasks by adopting competition among a group of LLM agents.

The **coopetition relationship** is a strategic combination of cooperation and competition, enabling agents to collaborate on certain tasks to achieve shared goals while competing in others. For example, in simulated negotiation scenarios, agents with different or even conflicting interests make trade-offs to reach mutually beneficial agreements. In such hybrid relationships, agents must balance cooperation and competition to achieve their objectives. Xu et al. [7] involved multiple LLM agents in the *Werewolf* game, where each agent cooperates or betrays others under asymmetric information to fulfill its role's goals. Similarly, Light et al. [8] engaged LLM agents in the *Avalon* game, where each agent must make decisions during dynamic game phases and negotiate with others through cooperation or deception to accomplish its assigned role's objectives. Inspired by human behavior, Corex [9] integrates diverse collaboration paradigms such as debate, review, and retrieval, which collectively enhance the authenticity, faithfulness, and reliability of the reasoning process.

11.1.2.2 Communication Structures

As shown in Figure 11.4, the communication topologies for multiagent collaboration can generally be divided into three main categories: centralized topology, decentralized topology, and hierarchical topology.a

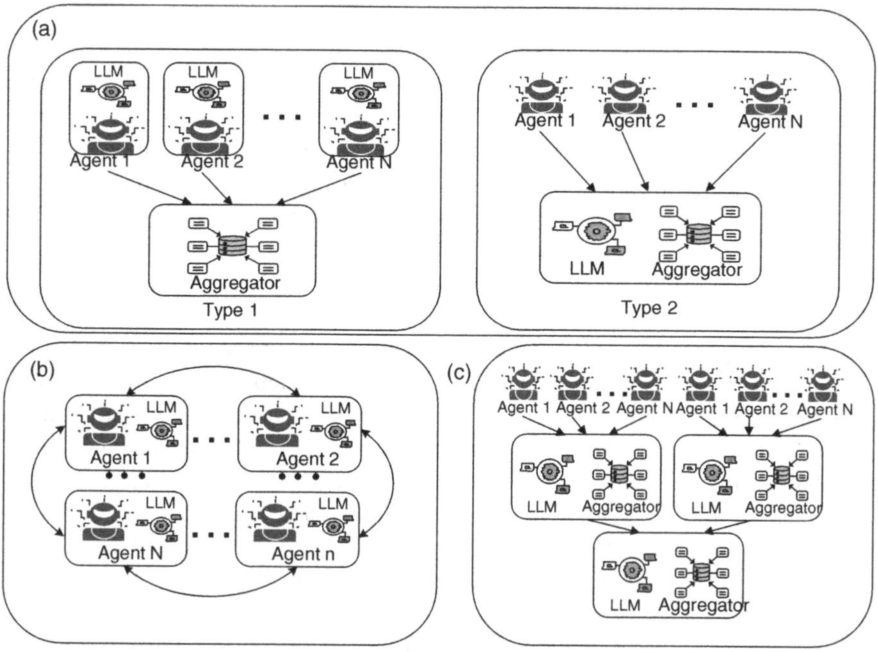

Figure 11.4 Communication structures of multiagent systems (centralized, decentralized, hierarchical). (a) Centralized. (b) Decentralized. (c) Hierarchical.

The **centralized structure** (also known as the star structure) is an implementation where each agent is connected to a central agent. In this structure, a specific agent serves as the central hub, with the goal of managing, controlling, and coordinating interactions among participants within the system. Other agents rely on the central agent for communication and task allocation. The LLM can be placed either in the central agent or in each peripheral agent.

The **decentralized structure** distributes control and decision-making among agents. In this architecture, all agents have equal status and jointly participate in the collective discussion of tasks. Each agent operates based on local information and possible communication with a limited number of other agents, requiring complex algorithms for interaction and decision-making. Information exchange between agents takes place on a common platform, ensuring that each agent can access all relevant information sent by other agents. This structure is typically used in fields such as robotics, network systems, and distributed AI.

The **hierarchical structure** is layered, where agents at each level have different functions and mainly interact within their own level or with agents at adjacent levels. In hierarchical relationships, researchers focus on developing efficient hierarchical control structures, information transmission mechanisms, and task decomposition methods. These technologies enable agents to collaborate effectively across different levels, thereby enhancing the overall performance of the system. Hierarchical relationships are usually manifested as a tree structure, in which parent node agents are responsible for task decomposition and task assignment to child node agents. Child node agents follow the arrangements of their corresponding parent nodes and return aggregated information. Currently, research on hierarchical relationships in LLM-based agent systems is still evolving, with only a limited number of levels having been explored.

11.1.2.3 Collaboration Strategies

Rule-based protocols: Interactions between agents are strictly controlled by predefined rules, ensuring that agents coordinate their actions in accordance with system-wide constraints. Agents act based on specific rule sets rather than probabilities or role-specific inputs. For example, agents can achieve efficient collaboration without deviating from predefined protocols by imitating human collaboration rules (such as debate rules and majority rules).

Role-based protocols: In an MAS, the system utilizes different predefined roles or task divisions, where each agent operates on segmented goals based on its domain knowledge to support the overall goals of the system. For example, the AgentVerse model simulates human collaboration by assigning specific responsibilities to each agent and enhances consistency through adherence to roles. This method, by distinguishing the roles of each agent, enables them to work proactively and avoid overlaps.

Model-based protocols: The uncertainty of input perception may affect the actions of agents. In such an environment, model-based protocols can provide flexibility in decision-making, where each agent can develop probability-based expectations for decisions based on input analysis, current environmental data, and shared collaborative goals. For instance, some studies have explored how to enable agents to make decisions through Theory of Mind (ToM) reasoning, by considering the possible mental states of their peers, which can thus improve task alignment even when agents face different goals.

11.1.2.4 Decision-Making Approaches

The differences in decision-making approaches directly determine the effectiveness of collaboration and the overall performance of the system. According to their architectural characteristics, decision-making approaches are divided into two main categories: authoritarian decision-making and collective decision-making.

Authoritarian decision-making refers to the approach where an authoritarian agent is responsible for aggregating the state information or local observations of all agents and then formulating clear decision-making guidelines. This method utilizes a global perspective to improve decision-making, thereby ensuring the reliability of system performance and the successful achievement of task objectives. The studies [10–12] demonstrated the single-agent decision-making process, where a single LLM synthesizes different viewpoints, resulting in more objective and comprehensive. Additionally, [13, 14] proposed a weighted integration method using ranking, scoring, or checklists, which enhances the robustness of the decision-making procedure.

Collective decision-making refers to the situation where multiple agents reach a decision without a central authority, primarily relying on interactive methods such as voting or negotiation. This approach distributes decision-making power among agents, enabling the system to adapt to changes while maintaining robustness and scalability. Compared with voting-based methods, debate-based decision-making focuses on structured interactions among agents to achieve optimal outcomes. In studies [15, 16], agents adopt a restrained stance, leveraging communication channels between themselves to build consensus through repeated discussions. To address the "cognitive silo" issue, some systems employ a universal retrieval knowledge base, allowing agents to access the same shared knowledge throughout the debate [17].

In future research, researchers expect to use game theory, auction mechanisms, and negotiation techniques to address the task allocation problem among cooperative agents. Additionally, the Distributed Constraint Optimization Problem (DCOP) provides an important framework for studying collaborative decision-making for cooperative agents. In other types of relationships, cooperative games and MORL have become key frameworks for exploring the delicate balance needed between cooperation and competition.

11.1.3 The Interaction Protocols of AI Agents

Based on the types of information exchanged, researchers can divide MAS into two categories. (1) **Systems with unstructured information:** Agents engage in activities such as discussion, negotiation, persuasion, or criticism through natural language to achieve their goals. Unstructured messages can express abstract ideas, emotional subtleties, or implicit contextual suggestions, making them highly suitable for creative and discovery-oriented problem spaces. (2) **Systems with structured information:** Compared to natural language, structured information features higher consistency, lower parsing complexity, and fewer ambiguities, making it more suitable for efficient and low-cost communication among agents. In some implementations, the information exchanged between agents is structured into different components, and this structure is designed to facilitate parsing and utilization by the receiving agent. For example, the exchanged information may include fields specifying the sender, recipient, message type, and instructions on how the recipient can parse or use the content.

In the interaction of LLM agents, unstructured and structured messages play complementary roles. While structured messages offer accuracy, consistency, and computational efficiency—making them suitable for operational and deterministic operations—unstructured messages can provide rich, contextual semantic representations, enabling agents to handle ambiguous, creative, and highly dynamic situations. These two modes together lay the foundation for effective and adaptive multiagent collaboration.

Developers typically design agent architectures and communication mechanisms for specific domains or tasks, including interactions between agents and the environment, between agents and humans, and interactions among agents. However, most existing systems lack a unified communication framework, leading to fragmented and isolated ecosystems. MASs, tools, environments, and data sources often operate independently, making it difficult for agents to interoperate or share functionalities. Additionally, the burden of learning and implementing specialized protocols falls on humans, and almost all current protocols are manually designed—a labor-intensive process that often lacks semantic flexibility and scalability. To address these issues, researchers have proposed several new agent communication protocols. Figure 11.5 presents the two most typical protocols: MCP and ANP.

The model context protocol (MCP) [18], developed by Anthropic, focuses on enabling LLM agents to access structured tools and data. Whether developers are building AI-driven Integrated Development Environment (IDEs), enhancing chat interfaces, or creating custom AI workflows, MCP provides a standardized way to connect LLMs with the required context. MCP follows a typical client–server architecture: the client uses the MCP protocol to connect to the MCP server, then accesses various information and tool capabilities of the server via the protocol.

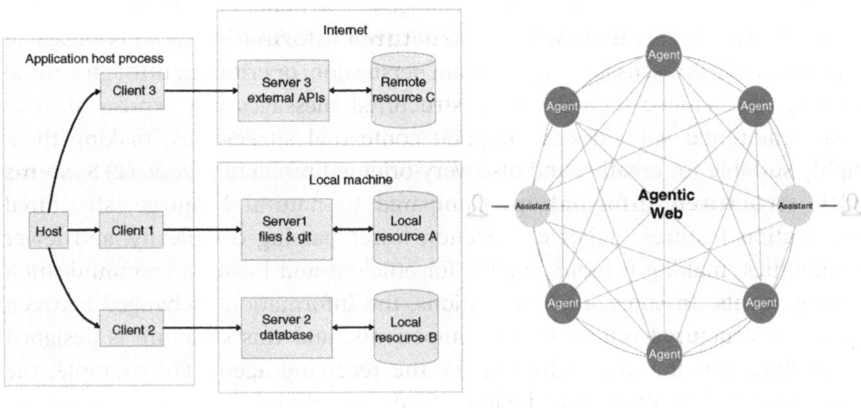

Figure 11.5 Schematic diagram of MCP protocol and ANP protocol.

It adopts a fully centralized approach based on OAuth authentication, with interactions limited to JSON-RPC 2.0 messages. Although it lacks a meta-protocol layer or semantic negotiation functions, its simple and streamlined architecture makes it a practical choice for tool use cases with clearly defined Application Programming Interface (APIs). However, MCP sacrifices flexibility and scalability, requiring the manual registration of supported functions.

The agent network protocol (ANP) [19] is an open protocol framework designed for the agentic Web. ANP implements decentralized identity authentication, enabling any two agents to connect. Meanwhile, it includes an agent description specification to facilitate more efficient data exchange and collaboration among agents. ANP adopts a typical peer-to-peer (P2P) architecture, meaning any agent can use ANP to connect to any other agent. The main reason for choosing the P2P architecture is to allow agents to both proactively initiate requests to other agents and receive requests from others at any time. The protocol includes a meta-protocol layer, enabling agents to negotiate the adoption of application-level protocols (e.g., HTTP, JSON-RPC, natural language), supporting semantic protocol selection based on agent capabilities. However, it does not explicitly supported the reuse of public protocols yet.

Table 11.2 Differences Between MCP and ANP.

	MCP	ANP
Worldview	Model-centric, where the entire Internet serves as its context and tools.	Agent-centric, where each agent holds equal status, forming a decentralized agent collaboration network.
Architecture Protocol	Typical client–server (CS) architecture protocol.	Typical peer-to-peer (P2P) architecture, allowing any agent to proactively establish connections with other agents.
Authentication Mechanism	MCP functions as a "USB interface" for models, based on OAuth standards, enabling clients to easily access current Internet resources.	ANP serves as the "Email" for agents, based on W3C DID standards, addressing interconnection between agents. Simply with the counterpart's ID, one can proactively send connection requests.
Organization Information	Based on JSON-RPC technology, essentially an API call.	Based on Linked-Data technology from the semantic web, aiming to build a data network that is AI-accessible and AI-understandable.

The strength of ANP, different from MCP, lies in the fact that with just a prompt and an HTTP function, agents can communicate and collaborate with any other agent. Compared with MCP, ANP has two major features that make it more suitable for agent communication: decentralized identity authentication and a P2P protocol architecture. Agents only need an anp_tool.py file (a prompt and an HTTP function) to achieve connection and communication with any agent. The simple access process stems from the concise design of the protocol. In the ANP, any two agents are completely decoupled, the protocol contains no complex concepts, and it supports stateless protocols. AI has changed not only the software development model but also the software operation and collaboration models. In the future, software may not require too much complex logic. Table 11.2 presents the specific differences between the two protocols.

11.2 The Orchestration Methods of Multiagents

Since each person's abilities and energy are limited, professional matters should be handled by professionals. People perform their own duties, with clear processes, rights, and responsibilities. This is the most suitable division of labor for the development of human society. With the advancement of technology, the limitations of a single agent have become increasingly prominent, especially when dealing with complex tasks that require the collaboration of multiple professional capabilities. This has driven the leap from "individual combat" to "army group collaboration," as evidenced by the rise of multiagent collaboration frameworks. This evolution is similar to the transition from single-core CPUs to distributed computing in computers. The core difference is that between agents; there is not just simple task allocation but also complex collaboration mechanisms such as dynamic role transformation, memory sharing, and conflict resolution. To achieve a higher level of complex coordination among multiple agents, it is necessary to study task orchestration methods among them. This section introduces three types of orchestration methods: process-based (task-oriented), role-based (collaborative learning), and interaction-based (consensus-reaching).

11.2.1 Process-Based Agent Collaboration

Task-oriented process collaboration requires effective agent coordination and task decomposition strategies to achieve shared goals. The interaction of agents mainly involves processing upstream outputs and generating results for downstream agents based on established task dependencies, rather than engaging in complex discussions or debates. The agent workflow K can be formally expressed as: $K = \{(N, E) \mid N \in \mathcal{N}, E \in \varepsilon\}$, where $\mathcal{N} = \{N(M, \tau, P, F)\}$ represents the set of

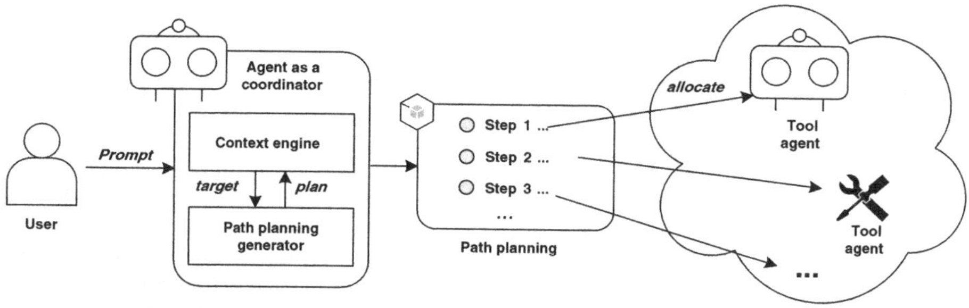

Figure 11.6 Process-based agent collaboration.

LLM agent nodes, with M, τ, P, and F denoting the available language model, temperature parameter, prompt space, and output format space, respectively, and E represents the edges between different agent nodes. This formulation encompasses the structural components and operational parameters that define the behavioral characteristics of agent workflows. Process-based agent collaboration mainly involves three aspects: agent discovery (identifying suitable agents), workflow construction, and workflow optimization (Figure 11.6).

11.2.1.1 Agent Discovery

The user proposes a goal to the interface agent, which then decomposes the goal into fine-grained tasks. Subsequently, the interface agent can query the tool/agent registry based on a series of metrics (such as capability, price, and context window) to select the most appropriate tool agent to fulfill the user's instruction. Based on the returned information, the interface agent can invoke the corresponding tool agent and allocate tasks to it. It should be noted that the registry can be implemented in various ways, such as a coordinating agent with a specific knowledge base, or blockchain-based smart contracts, and can also be extended into a dynamic market for tool/agent service transactions.

11.2.1.2 Workflow Construction

Most existing agent workflows are based on LLMs and agents within them interact in natural language. Users describe specific needs through natural language, and the agent system automatically designs and deploys a complete pipeline based on the user's description: The coordinator agent consists of various LLMs, the middle layer decomposes tasks into workflows, and the execution layer calls various tool agents to complete specific operations and then verify and summarize the results. Agent workflows can adopt different representation paradigms, which will significantly affect the dimensionality of the search space and the applicable optimization algorithms. Table 11.3 presents three typical representation paradigms.

Table 11.3 Representation Paradigms of Agent Workflows.

Name	Description	Example
Graph-Based Notation	This paradigm is particularly suitable for scenarios that require clear structural operations. It can visualize the topological structure of the workflow and represent the hierarchical, sequential, and parallel relationships between nodes.	LangChain is one of the earliest agent frameworks [20], and LangGraph is an extension of LangChain [21], which regards the graph architecture as the best way to define and orchestrate agent workflows. Each node represents a specific task or function, and edges indicate the transitions between these tasks. This graph-based approach provides fine-grained control over application processes and states and is particularly suitable for complex workflows that require advanced memory functions, error recovery, and human–machine collaborative interactions.
Neural Network Architecture	This paradigm excels at capturing nonlinear relationships between nodes.	This kind of workflow can exhibit adaptive behaviors through learnable parameters and is particularly effective in scenarios that require dynamic adjustments based on inputs and feedback.
Code-Based Notation	This paradigm provides the most comprehensive expressive capability.	Supports linear sequences, conditional logic, loops, and integration of graph and network structures. This approach enables precise control over workflow execution and effectively leverages the inherent code generation capabilities of LLMs.

11.2.1.3 Workflow Optimization

A workflow involves multiple agents collaborating to complete a task, and each agent carries the risk of "making mistakes," which will to some extent affect the final outcome. Error accumulation leads to the snowballing of minor errors in each step. For example, when AutoGPT is asked to arrange a business trip, it might first misremember the date (2% error), then choose the wrong city (3% error), then book the wrong hotel (5% error), and ultimately reserve a Valentine's Day couple suite for the boss. This is why the task success rate of the state-of-the-art model GPT-4 is only 14.9%. As the complexity of the workflow increases, challenges in state management, error handling, and context maintenance also intensify. Therefore, workflow optimization is essential, requiring not only the tracking of process progress and the understanding of dependencies between steps but also the ability to smoothly recover from interruptions or failures.

Given a task T and an evaluation metric L, the goal of workflow optimization is to discover the optimal workflow K^*. Here, $K^* = argmax_{k \in \mathscr{K}} L(K, T)$, where $\mathscr{K}$ represents the search space of workflows, and $L(K, T)$ typically evaluates multiple aspects, including task completion quality, computational efficiency, and execution latency. This optimization objective reflects the practical challenges faced when deploying agent workflows, necessitating a balance between task effectiveness and resource constraints. Optimization measures require implementing a robust state management system across the multistep workflow and incorporating clear validation checkpoints in the multistep workflow. It is also essential to establish comprehensive error-handling logic for each step and design failure response mechanisms to ensure appropriate handling in case of abnormal situations.

As shown in Figure 11.7, HuggingGPT is a typical process-based agent collaboration system, where LLMs act as key controllers and numerous expert models serve as key collaborative executors [22]. Its workflow is divided into four stages. (1) Task planning: LLMs such as ChatGPT first parse user requests, then conduct corresponding task decompositions, and plan task sequences and dependencies based on their knowledge, enabling the LLMs to know the scheduling order of each component in the workflow. (2) Model selection: After determining the workflow sequence, it is necessary to select the corresponding model. For example, if the user's requirement is to generate an anime-style image based on a text prompt, the LLM should learn to download the corresponding LoRA plug-in from Civitai and then load the model locally. (3) Task execution: Expert models execute the assigned tasks on inference nodes and record the execution information and inference results and send them to the LLM. (4) Response generation: The

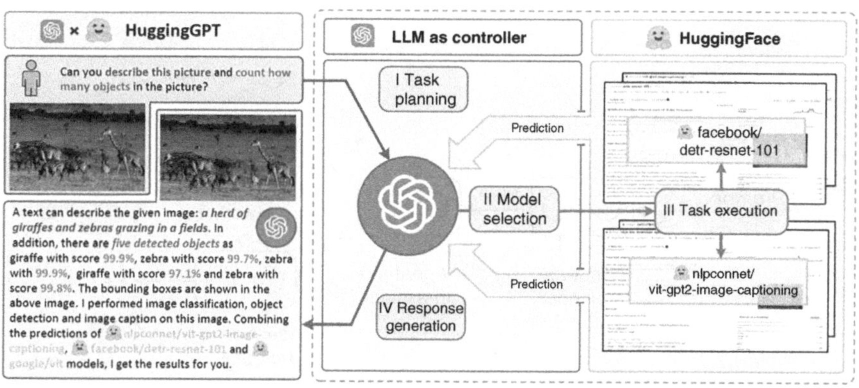

Figure 11.7 Workflow of HuggingGPT. *Source:* [22] / Cornell University.

LLM summarizes the execution process logs and inference results and returns the summary results to the user.

11.2.2 Role-Based Agent Collaboration

This type of collaboration assigns tasks based on the roles of agents and makes role-specific decisions. Figure 11.8 shows the role-based collaboration, where agents coordinate in a hierarchical manner. Specifically, the agent acting as a planner can generate a multistep plan by decomposing the user's goals into a chain of tasks. Subsequently, the agent functioning as an allocator can coordinate task distribution, meaning some tasks can be completed by the allocator itself, while others are delegated to specific agents based on their capabilities and expertise. Additionally, if no suitable agent is available, the agent serving as a creator can be invoked to establish a new agent with a specific role. This is achieved by the creator agent providing necessary resources, clear objectives, and initial guidance to ensure a seamless transition of tasks and responsibilities.

In the collaborative process, each agent accumulates different memories and experiences due to its unique behaviors and interactions with various environments. By solving problems together, these agents share experiences to enhance their strategic learning and task-solving abilities. When participating in collaborative learning interactions, agents will observe the behaviors of others to update their context and memory. For example, an agent can learn optimal strategies by observing discussions among other agents and adjust its own methods based on these observations, without necessarily reaching a consensus on a single "best" strategy. Despite the multiple benefits of collaborative learning interactions, there are some challenges. For instance, how to ensure fair knowledge exchange among agents with different capabilities and prevent the spread of errors and biases; how to promote learning while maintaining the diversity of agents; and how to

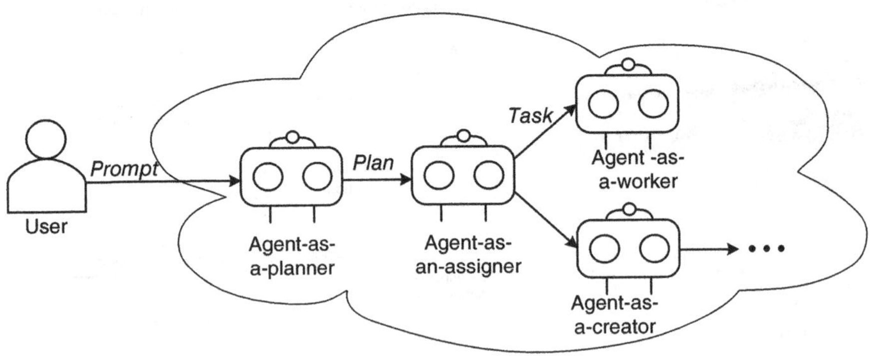

Figure 11.8 Role-based agent collaboration.

develop effective mechanisms that enable agents to selectively integrate others' knowledge.

Despite the difficulties posed by issues such as knowledge fairness, bias propagation, and scalability, MASs hold great potential for improvement in MASs, especially in dynamic and complex environments. Currently, there are several open-source multiagent collaboration frameworks, such as AutoAgents [23], AgentVerse [24], CAMEL [3], and AutoGen [25]. These frameworks all utilize LLMs for complex task solving, allowing the selection and customization of roles in multiagent configurations, and this capability accelerates the development of MASs. Several typical examples are introduced here.

11.2.2.1 The AutoAgents Framework

As shown in Figure 11.9, the AutoAgents framework can dynamically generate corresponding agents to handle a specific task (e.g., writing a novel about the awakening of AI). The process consists of two stages. (1) Draft stage: Through the cooperation of three predefined agents, namely, Planner, Agent Observer, and Plan Observer, it generates the agents and execution plans required to complete the task. (2) Execution stage: The generated agents are used to gradually execute the plan, and the execution process is monitored by a predefined agent, Action Observer.

During the draft stage, the planner is responsible for generating and refining agents and execution plans based on the task content. The agent observer is tasked

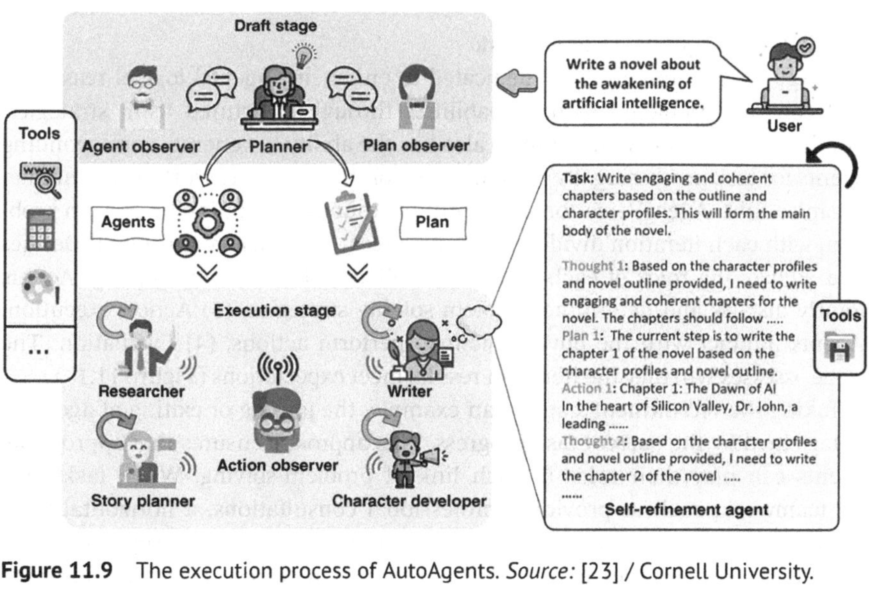

Figure 11.9 The execution process of AutoAgents. *Source:* [23] / Cornell University.

with analyzing the rationality of the generated agents and their alignment with the task requirements, and providing suggestions to the planner. The plan observer analyzes the rationality of the generated execution plans and their match with the task and also offers suggestions to the planner. The planner engages in multiple rounds of interaction with the agent observer and plan observer until no further feedback is given or the maximum number of interactions is reached, thereby deriving the final agents and execution plan. The execution plan comprises multiple steps, each completed by one or more agents. Each agent includes the following information: prompt (i.e., the agent's profile, objectives, and constraints), description, toolset, and suggestions.

During the execution stage, the execution plan is executed step-by-step, with each step completed by a single agent or multiple agents. The completion process of each step also undergoes multiple rounds of iteration until the current step is finished or the maximum number of iterations is reached. If a step's completion process involves multiple iterations by a single agent, the agent uses a CoT approach for gradual thinking, which the paper refers to as a "self-refinement agent." If a step's completion process involves multiple iterations by multiple agents, the agents use an interactive approach for gradual thinking, which is termed "collaborative refinement action" in the paper. Meanwhile, the paper introduces long- and short-term memory: The execution process of the plan employs long-term memory, while the iterative process of each step uses short-term memory. After each step is completed, the action observer analyzes the execution status of the plan up to the current step and adjusts the plan if necessary.

11.2.2.2 The AgentVerse Framework

The AgentVerse framework is dedicated to enhancing agents' logical reasoning and complex problem-solving capabilities through structured team strategies. Similar to AutoAgents, AgentVerse also has the ability to generate corresponding agents for task processing. By emulating the problem-solving workflow of human organizations, AgentVerse conducts multiple rounds of iteration for a given problem, with each iteration divided into four stages. (1) Expert recruitment: Defines and adjusts the roles of each agent. (2) Collaborative decision-making: Agents jointly discuss and determine problem-solving strategies. (3) Action execution: Agents interact with the environment to perform actions. (4) Evaluation: The stage assesses whether the iteration results meet expectations (Figure 11.10).

Taking the recruitment stage as an example, the joining or exiting of agents is determined by the current task progress. This approach ensures that appropriate agents can play their roles in each link of problem-solving. When tasks rely on teamwork, such as providing professional consultations, a horizontal team structure is more effective, where each agent proposes decisions separately, and then the decisions are integrated. In scenarios requiring clear responsibility

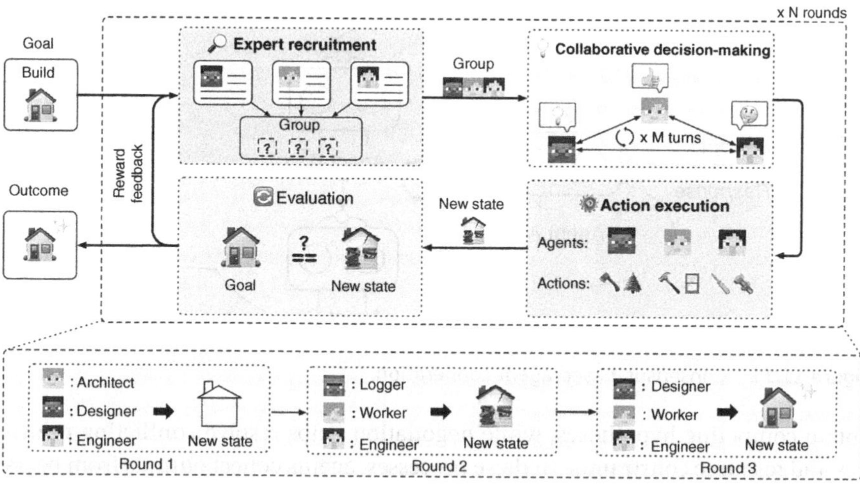

Figure 11.10 Illustration of AgentVerse. *Source:* [24] / Cornell University.

allocation and tool use, a vertical team structure is more suitable, where one agent proposes decisions, and other agents provide feedback, iterating multiple times until consensus is reached. This systematic methodology of AgentVerse enables it to precisely guide agent teams, enabling agent teams to flexibly respond to diverse task requirements, thereby enhancing the overall work efficiency of the team and improving the quality of problem-solving.

11.2.3 Consensus-Based Agent Collaboration

Although traditional collaboration methods have been widely explored and achieved good performance in multiagent task processing, in real life, the emergence of competition can prompt humans to exhibit more robust and effective behaviors. Therefore, agents can quickly adjust strategies through interaction, striving to choose the most advantageous or rational actions during competition to adapt to changes caused by other agents and the environment. This paradigm promotes the development of "task-agnostic" general methods. Through various collaboration modes such as debate, commentary, and retrieval, this paradigm collectively enhances the accuracy, credibility, and reliability of the reasoning process, enabling LLMs to "think outside the box," thereby overcoming hallucinations and delivering better solutions.

As shown in Figure 11.11, consensus-oriented collaboration primarily focuses on achieving knowledge integration and belief adjustment by synthesizing different perspectives. Methods for multiple agents to reach consensus typically include discussion, debate, negotiation, reflection, and voting. Debate enables agents to

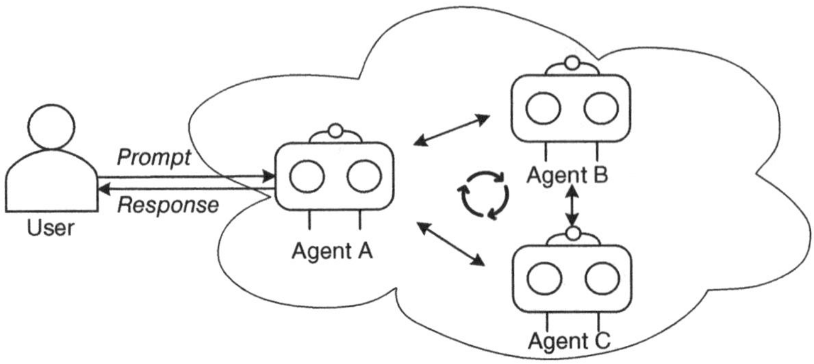

Figure 11.11 Consensus-based agent cooperation.

obtain competing hypotheses, while negotiation helps resolve conflicting priorities and resource constraints. In these processes, agents collect outputs from peers addressing the same problem and incorporate environmental feedback including numerical data and contextual details. These interactions allow agents to share views and hypotheses, gradually reaching a common understanding. In multiagent interactive games, designing fair rules and coordination mechanisms is one of the key challenges in achieving optimal decision-making, which involves issues such as task allocation, conflict avoidance, and resource sharing. The following introduces two of the most typical interactive collaboration methods:

Debate-based cooperation: Agents receive feedback from other agents and adjust their thoughts and behaviors through debates with other agents until consensus is reached. Users can send queries to an agent, which then shares the query (i.e., the problem) with other agents. Based on the shared problem, each agent generates its initial response, and then debates among agents are initiated. Agents will mutually share their initial responses with each other for validation, while providing guidance and potential planning directions. They aim to construct more comprehensive responses based on inclusive consideration and collective outcomes. Additionally, agents may use shared memory in some cases or allow mutual access to their respective memories to facilitate debates. This debating process can be iterative to enhance the performance of all participating agents. Debate-based cooperation can be concluded according to a predefined number of debate rounds, or agents will continue the process until consensus is achieved.

Voting-based cooperation: Agents can freely provide opinions and achieve a consensus through voting. Specifically, an agent can first generate a candidate response to the user's prompt, then hold a vote in which different suggestions are presented as options. Other agents are asked to submit their votes based on their capabilities and experience to select the most appropriate response option.

In this scenario, agents communicate in a centralized manner, with the original agent serving as a coordinator. The voting results will be formalized and sent back to the original agent, which can refine the candidate response as needed before answering the user.

In the MAD framework, multiple agents engage in "tit-for-tat" debate exchanges under the supervision of an LLM, guiding discussions toward a defined solution [10]. ChatEval [6] is a multiagent referee team that autonomously discusses and evaluates the quality of responses generated by different models, ultimately achieving a level comparable to that of human evaluators. CompeteAI simulates a virtual town with restaurant agents and customer agents to study competition between agents. The Formal Debate Framework (FORD) [26] enhances consistency among LLMs through organized debates, enabling more powerful models to guide consensus while weaker models adjust their perspectives. GPTSwarm [27] specifies collaborative relationships between agents through graphic design, with information flow and edge connections constructing basic group discussions. An agent will be excluded if it continuously provides incorrect opinions. RECONCILE [28] adopts a round-table discussion format, including multiple discussion cycles and a voting system based on confidence levels, integrating a reflection mechanism by learning from past discussions. Additionally, researchers have designed a communication game framework [7], using *Werewolf* game as a representative case to explore the framework's feasibility, as shown in Figure 11.12. During the game, emergent strategic behaviors and traits such as trust, confrontation, camouflage, and leadership can be observed.

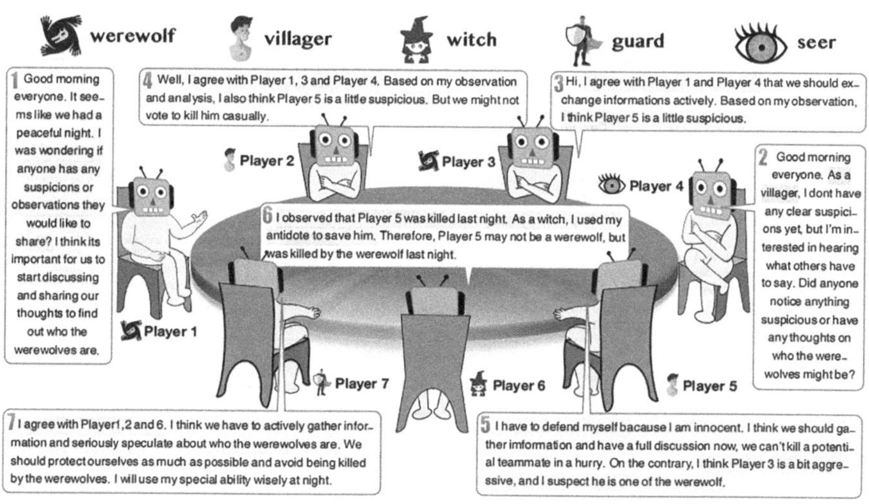

Figure 11.12 A snapshot of Werewolf game. *Source:* [7] / Cornell University / CC BY 4.0.

11.3 Automation of Computational Experiments

Computational experiments can serve as tools for AI applications in social sciences. The entire workflow typically involves generating hypotheses for target scientific problems, planning experimental scripts, conducting experiments and calculations, analyzing data, drawing conclusions, and revising hypotheses. However, the development of current computational experiments is a complex and labor-intensive process. Traditional methods of building such computational experiment systems from scratch require developers to deeply engage with complex technical details. In contrast, automating the entire computational experiment process using workflow agent technology can significantly reduce the need for manual intervention and reduce development costs.

Figure 11.13 shows a schematic diagram of the process automation of the computational experiment platform, which mainly includes three autonomous agents. (1) User-intent recognition agent: It analyzes the user intent (such as phenomenon explanation, trend prediction, strategy optimization) in accordance with user input and parses the experimental objectives in a certain format. (2) Decision theater: Based on the experimental purpose, experimental type, and existing knowledge in the knowledge base, it identifies the influencing factors and response variables that may be related to the experiment, determines

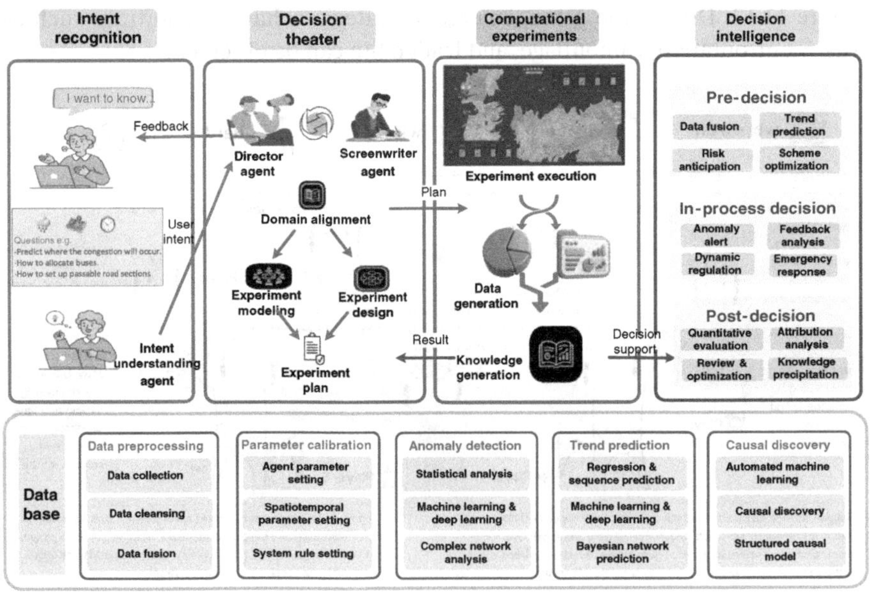

Figure 11.13 Schematic diagram of computational experiment automation.

the corresponding relationship between them, selects the most appropriate experimental method, designs the appropriate number of experimental groups, and generates the experimental scheme according to the factors determined in the previous steps. (3) Computational experiments and knowledge generation: It judges whether the generated experimental schema is reasonable, and evaluates whether the generated knowledge can verify the proposed hypothesis after the experiment is completed. If not, it feeds back to the system and prompt the system to reexecute and optimize the experimental design steps. The specific details are as follows:

11.3.1 User-Intent Recognition

Essentially, hypothesis generation is a process of discovering potential laws that govern data distribution and involve unobserved scientific phenomena. According to Karl Popper's view, scientific hypotheses must be falsifiable, and only those that remain valid after falsification are valid and credible. Typically, scientists verify or falsify hypotheses through experiments. A hypothesis is considered more valuable if it is comprehensive enough to explain various data and has a high degree of plausibility. To solve a scientific problem, a computational experiment platform first needs to formulate one or a few hypotheses based on high-value criteria, then design corresponding experimental schemes to conduct experiments to quickly minimizes the unknown space, use learning functions to process observations from hypothesis testing, and finally turn a proven valid hypothesis into guiding knowledge.

Accurately analyzing user needs is the primary step in computational experiments. Users can input natural language text, and the platform uses LLMs for parsing, while integrating external domain knowledge graphs and enhancing the reliability of LLM-generated results through RAG technology. Although most knowledge discovery processes rely on generating hypotheses and testing them in the real world, a significant portion of knowledge can be obtained purely through computational experiment deduction processes, such as forms of iterative reasoning and deep reasoning. Given all necessary prerequisites, deductive reasoning and inductive reasoning can be used to derive inferences from known prerequisites and data, thereby helping to prove or falsify hypotheses. According to the framework of computational sociology, user needs can be divided into three categories: description of social phenomena, explanation of social phenomena, and prediction of social phenomena. The specific details are as follows:

11.3.1.1 Describe: Inductive Reasoning

The description of social phenomena is mainly achieved through inductive reasoning, with specific techniques including pattern recognition and statistical learning.

Complex data analysis problems can be decomposed into different computational tasks, such as encoding, self-criticism, and regression analysis, to extract meaningful insights from complex datasets. Ideally, the extracted insights should be logically reasonable; otherwise, they must be discarded. If the fitting of analytical expressions to experimental data is successful, the results derived from the analytical expressions are regarded as equivalent to the results from experimental deduction. Conversely, if the fitting fails, it is necessary to increase the number of experimental points or narrow the fitting range until a satisfactory analytical expression is obtained. However, limitations in data coverage and limitations in the implementation of analytical algorithms may lead to the generation of illusory insights, which requires reliable data analyzers and reasoning tools to prevent overanalysis.

11.3.1.2 Explain: Deductive Reasoning

The explanation of social phenomena mainly relies on deductive reasoning. A fundamental question here is whether agents driven by LLMs can perform genuine reasoning. By definition, LLMs generate outputs by predicting the next token, which is a mechanism fundamentally different from human reasoning. From a result-driven perspective, the input–output behavior of agents phenomenologically demonstrates reasoning ability because the outputs they generate are meaningful when compared to a reference system that produces arbitrary responses. However, regardless of the perspective taken, this ability remains imperfect, especially when dealing with complex problems that are crucial for scientific knowledge discovery. This area still demands focused research attention.

11.3.1.3 Predict: Sample Optimization

The prediction of social phenomena requires optimization-seeking and demand-seeking in the experimental sample space. Optimization-seeking refers to finding a relatively optimized solution to a problem or selecting optimized parameter settings through computational experiment methods. Demand-seeking involves finding a feasible solution set that meets the demand goals through computational experiments. For some exploration problems with special properties, the optimization exploration method of meta-models can be adopted to improve exploration efficiency. This idea is based on generating a computational experimental system using as much real data as possible and then taking the experimental output results as the evaluation function. Experimental points are sampled in a certain interval, fit with a simple analytical formula (meta-model) based on the results, and then verified. If the verification is passed, this analytical formula can be used to find the optimal solution; if the verification fails, corresponding measures (such as increasing experimental points or narrowing the fitting interval) are taken, and the processes of iterative deduction, fitting, and optimization are repeated until a satisfactory optimization effect is achieved.

Although this is not perfect, if the quality of data and generation algorithms is high, it is sufficient to support the rationality of the model results to a certain extent.

11.3.2 Decision Theater

Computational experiments are essentially a form of logical model, and they examine the relationships between a few variables by controlling other variables. As more variables are considered, it becomes impossible to accurately predict the future behavior of the experimental system: on the one hand, the evolution of an event is often the result of the combined effect of multiple factors; on the other hand, due to the complexity and uncertainty of these factors, it is difficult to predict the evolutionary outcome of a system. In order to better cope with future uncertainties, the following steps need to be implemented. (1) Identify influencing factors: Determine which factors will affect the state and behavior of the system. (2) Analyze the paths and ways of influence: Determine how these factors affect the state and behavior of the system. (3) Explore deeper-level factors: Determine what is affecting these factors.

The experimental design scheme is a hierarchical regulation framework that combines game theory and cybernetics. The upper layer consists of macro-control strategy variables (which can be multivariate, including feedforward control strategies, feedback control strategies, correlation strategies, collaboration strategies, and indicator design), while the lower layer comprises multiple interconnected agent subsystems (including competition and cooperation) with different functions or goal orientations. Based on the domain knowledge graph, the experimental agent explores a feasible path between the concepts and parameters of interest, formulates relevant hypotheses, and expands them into a complete research plan that includes detailed hypothesis testing methods and standards. The entire scheme involves identifying influencing factors, designing response variables, and determining the approach to factorial experimental design. To enhance the reliability of the scheme, a specialized agent can be used to review, evaluate, comment on, and improve the proposed hypotheses, but this does not include the step of hypothesis testing through actual experiments. Figure 11.14 shows the specific execution process of the experimental design agent.

11.3.2.1 Feedback-Based Experimental Design

In the context of cybernetics, feedback mechanisms constitutes a closed-loop dynamic regulation system characterized by "circular causality," which can operate effectively in an open environment. A prominent advantage of the feedback-based design principle is that, when the system structure and environmental information are not fully known, subsequent experimental schemes can

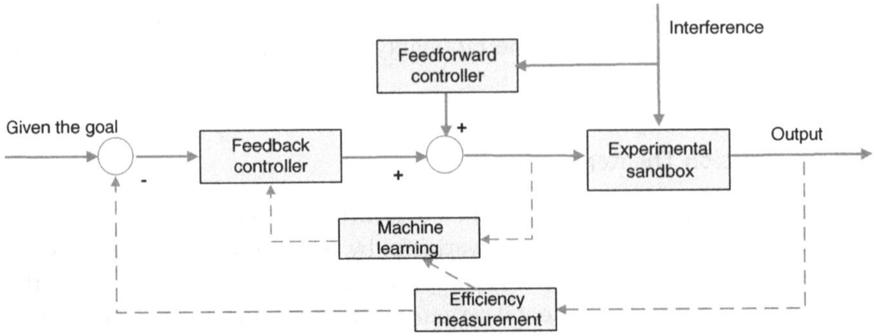

Figure 11.14 Specific execution process of experimental design agent.

be effectively adjusted solely based on the deviation data between the real-time output value of the system and the target value. Therefore, the feedback design strategy can, in principle, mitigate the impact of system structure uncertainty and external disturbances on the experimental scheme, and has strong robustness. The experimental design agent can conduct coverage analysis of the sample space based on the results of previous experiments and revise the experimental scheme according to the feedback.

11.3.2.2 Feedforward Experimental Design

The limitations of feedback-based experiments lie in the following: when the uncertainty of the simulation environment exceeds the capacity limit of the feedback mechanism, relying solely on the feedback mechanism may lead the experiment to encounter stability issues, and even lead to irreversible instability consequences. The principle of combining feedforward and feedback mechanisms is a basic experimental design strategy. The feedforward mechanism can be used to prevent or eliminate adverse output deviations that may be caused by uncertainty, and it is also an important mechanism to maintain experimental stability in an uncertain environment. However, the design and application of the feedforward mechanism often require necessary prior understanding or prediction of the evolutionary trend of the system and uncertain factors. In this regard, the assistance of domain knowledge helps to enhance the predictability of the system's evolutionary trend, and the experimental design agent can reduce the number of experiments and shorten the experimental time based on this. Considering the timeliness and adaptability of the experimental design strategy, feedforward and feedback mechanisms generally exist simultaneously and play their roles alternately.

11.3.2.3 Adaptive Experimental Design

Like feedback control, the learning mechanism of the experimental design agent is also a powerful means to address uncertainty, and they can complement each other effectively. The feedback mechanism can exert effective control over the simulation environment under the influence of large-scale uncertain changes and various strong disturbances. The learning mechanism, on the other hand, can extract useful processing patterns from computational experiment data, evaluate them based on basic facts, and continuously update the knowledge repository in the memory of the experimental agent, thereby providing a basis for the adjustment of feedback experiments. Therefore, in the computational experiment platform, a feedback loop can be formed between the experimental agent and the simulation environment, and ML can be used to continuously train and revise the experimental design strategy ("learning while regulating"), so as to identify possible adverse effects, contradictions, potential defects, and other issues and provide "reference," "estimation," and "guidance" for dealing with possible situations.

11.3.3 Computational Experiments and Knowledge Generation

The goal of computational experiments is to "explore as much unknown space as possible at the lowest possible cost" (referred to as "more, faster, better and more economical" [MFBM] for short). Therefore, the knowledge generation from computational experiments should be conducted from the following three aspects. (1) Micro-level knowledge: Starting with the intelligence level of individual agents to understand the coverage of individual diversity. (2) Meso-level knowledge: Using the value entropy model as a tool to combine individual behaviors with system performance, so as to understand the position of the current experiment in the sample space and point out the direction for the next experiment. (3) Macro-level knowledge: With the system effectiveness model as a reference, comprehensively considering the impact of linear characteristics and nonlinear emergence on the results, so as to understand the coverage of potential results (Figure 11.15).

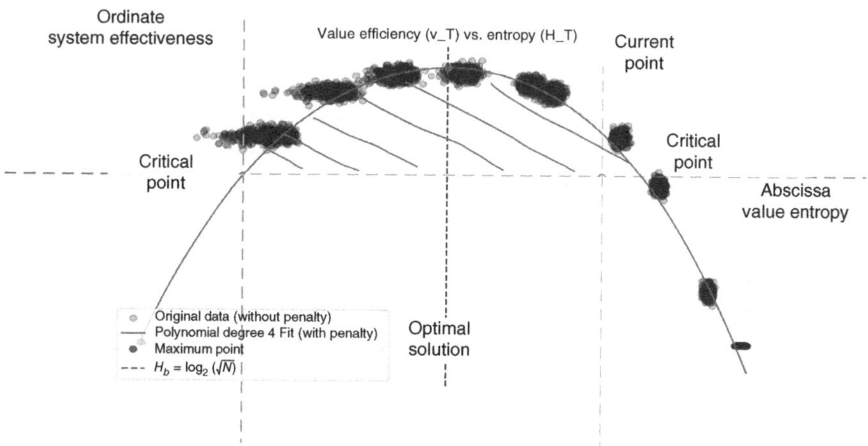

Figure 11.15 Performance measurement model for computational experiments.

11.3.3.1 Micro-level Knowledge: Agent Intelligence

In simulation deduction, an agent's intelligent behavior stems from accurate prediction of an uncertain world, involving the construction of predictive and adaptive models. From the perspective of knowledge discovery, the characteristics of world W are a dataset $X = \{x_1, x_2, \ldots, x_n\}$. During the agent's interaction with W, each dataset appears in experimental measurements or observations with probability $P_w(x)$. Here, we assume that data points may be correlated or not. In this context, the "world" is the set of all propositions. Given the parameter θ of the agent's world model and dataset x, the agent predicts a probability distribution $P_\theta(x)$. For scientific knowledge discovery, the agent's goal is to predict unexplored natural phenomena as accurately as possible. A smarter agent produces a better approximation of the real-world distribution $P_w(x)$. Thus, agent intelligence can be measured by the KL divergence (Kullback–Leibler divergence) or relative entropy between these two probability distributions [29]:

$$D_o(\theta) = \sum_{x \subseteq W} P_w(x) \log \frac{P_w(x)}{P_\theta(x)} \tag{11.4}$$

where $D_o(\theta)$ describes the difference between $P_w(x)$ and $P_\theta(x)$. More precisely, in the case of hypothesis testing, if we draw N samples from $P_w(x)$ and compare the results with the predictions of $P_\theta(x)$, an agent with a lower $D_o(\theta)$ will generate predictions that better align with reality. The intelligence level of an agent directly affects its individual effectiveness in simulation scenarios. At time t, the utility of an individual agent is a function u_i of its cumulative revenue $x^c_{i,t}$ and labor input $l_{i,t}$. Specifically, the utility function used here is a concave function where utility increases as income increases, while utility decreases linearly with cost:

$$u_i(x_{i,t}, l_{i,t}) = crra\left(x^c_{i,t}\right) - l_{i,t}, \quad crra(z) = \frac{z^{1-\eta} - 1}{1 - \eta}, \eta > 0 \tag{11.5}$$

where $x^c_{i,t}$ represents the cumulative revenue of the agent up to time t, and $l_{i,t}$ represents the cumulative cost of the agent up to time t. The *crra* model indicates that the utility u_i exhibits the characteristic of diminishing marginal revenue. The parameter η controls the degree of nonlinearity: the higher the value of η, the greater the deviation from linear behavior. Generally speaking, agents learn and evolve based on this utility function to adapt to changes in the external environment.

11.3.3.2 Meso-level Knowledge: Value Entropy Model

The macroscopic performance of social system deduction is not a linear accumulation of individual characteristics, and it is necessary to quantify the nonlinear effects generated by interactions among individuals. Such nonlinear effects are mainly reflected in two aspects. (1) The autonomous behavior of individuals may conflict with the interests of the entire system, exerting negative impacts on system performance. (2) Since individual behavioral decisions are based on local information, improvements in individual behavior cannot guarantee an enhancement in system performance. Entropy in physics is used to measure the degree of chaos among individuals from a systemic perspective. Inspired by this, the value entropy model is employed to quantitatively describe the nonlinearity of experimental scenarios and can be used to measure the coverage of real-world scenarios [30, 31]. The specific definition is shown in Eq. (11.6).

$$H_T = -\sum_{j=1}^{m} p_j log_2 p_j; \quad constraints: p_j = \frac{N_j}{N_{total}} \text{ and } \sum_{j=1}^{n} p_j = 1 \tag{11.6}$$

where p_j represents the distribution probability of agents in the j-th niche, N_j is the number of agents in the j-th niche, and N_{total} is the total number of agents in the system. The division of niches in the social system is based on the individual utility $Utility_{T,i}$ of agents. The calculation of value entropy (H_T) follows the following rules: First, according to the specified niche division criteria, use common sorting algorithms to sort the utilities of all individuals, assign individuals to appropriate niches, and obtain the total number of niches (denoted as m).

When the orderliness of the entire social system is higher, the speed of finding suitable nodes increases, and the efficiency of value generation is higher. The maintenance of the social system incurs costs, including heteronomous organization costs (i.e., costs for maintaining node order) and autonomous organization costs (i.e., costs for searching for nodes), from which the ecological niche with optimal entropy can be derived as $m = m_l = \sqrt{N_{total}}$. Taking the optimal entropy as a benchmark, the position of the experimental scenario in the value entropy space can be evaluated, including identifying extreme scenarios and critical scenarios.

11.3.3.3 Macro-level Knowledge: System Efficiency Model

In complexity science, researchers often use the product of system complexity and system scale to evaluate system competitiveness. Taking this as the theoretical basis, the macroscopic characteristics of social system deduction can be defined as the product of "the system's nonlinear characteristics (represented by value entropy)" and "the system's linear characteristics (represented by the summation of individual performances)." The specific formula is as follows:

$$\begin{aligned} System\ Utility_T &= Value\ entropy_T * Productivity_T \\ &= \left(e^{1-\left|\frac{H_T - H_{best}}{H_{best}}\right|}\right) * \left(\sum_{i=1}^{N_{total}} Individual\ Utility_{T,i}\right) \end{aligned} \tag{11.7}$$

where $Value\ entropy_T$ represents the nonlinear characteristics of the system, measured by $e^{1-\left|\frac{H_T - H_{best}}{H_{best}}\right|}$, and this metric can evaluate the relative gap between the system's current entropy (denoted as H_T) and the optimal entropy (denoted as H_{best})

of the system; *Productivity* represents the linear characteristics of the system, measured by $\sum_{i=1}^{N_{total}} Individual\ Utility_{T,i}$, that is, the summation of individual utilities of all agents in the system. When the system only has linear characteristics (indicating no interaction between agents), $Value\ entropy_T = 1$, and the system utility is reduced to the linear summation of individual utilities. However, it is worth noting that due to the nonmonotonic nature of value entropy, $Value\ entropy_T = 1$ is a necessary but not sufficient condition for a linear system. When $Value\ entropy_T > 1$, the system utility is greater than the sum of the individual utilities of agents. The system effectiveness metric can be used to determine the coverage of simulation results with respect to real-world scenarios.

11.4 The Challenges of Multiagent Collaboration

MASs are usually based on the same LLM. No matter how strong the capability of each agent is or how great the functional differences are, they can only be functional subsets of the LLM. There is no restriction that a specific task must be completed by a single agent. Therefore, the division of labor among agents will instead incur communication costs. When multiple LLM agents are used to solve the same task, they may initially have different solutions, but through continuous negotiation, they can eventually reach consensus. The problem of reaching consensus is prevalent in collective decision-making systems such as animal groups and human society. It is also a core research issue in multiagent collaboration. So far, there are still many unresolved issues regarding consensus-reaching in LLM-driven MASs. For example, can they eventually reach consensus? If so, how long will it take and which factors will affect the final consensus result? If not, what are the factors leading to this failure? The answers to these questions are crucial for the proper utilization of LLM agents.

11.4.1 Influencing Factors of Collaboration

A study [32] conducts a dedicated study on the "consensus-seeking" in LLM agent systems. The experimental design of this work is quite ingenious: the initial position of each agent is represented by a numerical value, and each agent's goal is to continuously adjust their positions. If all these agents eventually converge to the same position, consensus is considered achieved. Throughout the process, each agent can obtain the positions of other agents and develop adjustment strategies for its position based on this information. The paper discusses several factors that may affect agents' achievement of consensus, including agent personality traits, the number of agents, and communication topology. The specific details are shown in Table 11.4.

Table 11.4 Factors Affecting Multiagent "Consensus Reaching".

Type	Description	Legend
Influence of Personality Traits	When an agent has no clear personality traits, the most commonly used strategy is the average strategy, which takes the average of all agents' states as its own state at the next moment. As shown in the figure on the right, the three agents have different initial positions, but they all eventually converge to the average position.	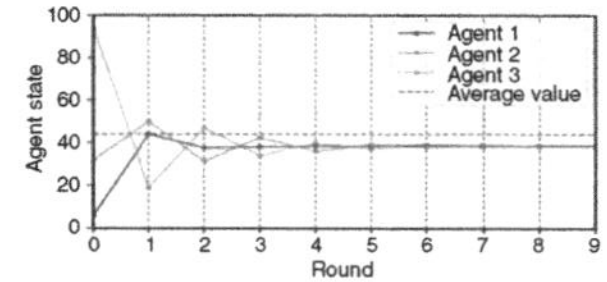
	The agent chooses the "compliance strategy," that is, the agent selects the positions of other agents as its target positions. As shown in the figure on the right, both agent 1 and agent 2 adopt the compliance strategy, taking each other's position as the target, so the states of the two agents will oscillate and fail to converge.	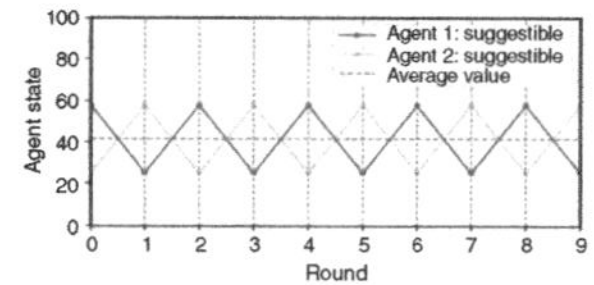
	If both agent 1 and agent 2 have stubborn personalities, they may each maintain their own states and fail to converge. As shown in the figure on the right, if both agent 1 and agent 2 adopt a stubborn strategy, their states will form parallel lines.	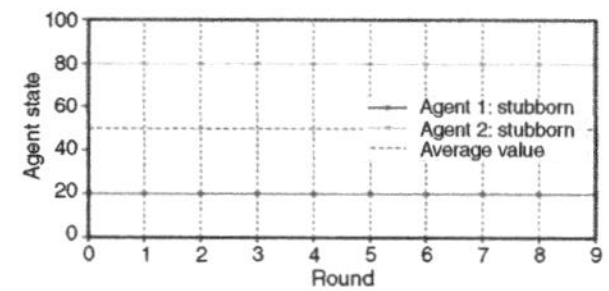
	An agent may also choose the "stubborn strategy," that is, the agent tends to expect other agents to move closer to itself while keeping its own original state unchanged. As shown in the figure on the right, agent 1 is stubborn and keeps its position unchanged; agent 2 and agent 3 choose to comply and move closer to it.	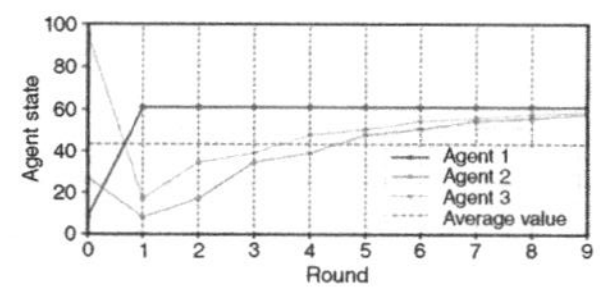

(Continued)

Table 11.4 (Continued)

Type	Description	Legend
	Sometimes agents may also experience hallucinations, leading to wrong choices. As shown in the figure on the right, agent 4 has obvious hallucinations, setting its target position at 81, which is far from the average position of all agents.	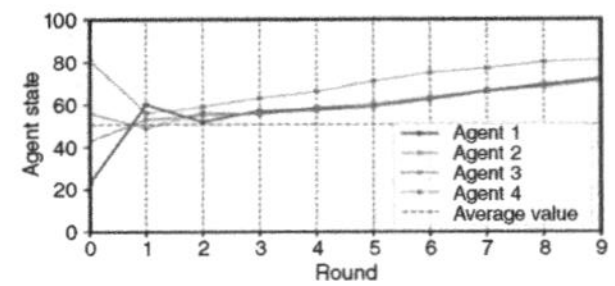
Influence of the Number of Agents	As the number of agents increases, the variance of the aggregation value decreases, and the mean gradually approaches the initial mean. This indicates to a certain extent that increasing the number of agents can alleviate randomness or hallucinations. The lower the model temperature parameter, the better the system stability; the higher the model temperature parameter, the greater the system dispersion.	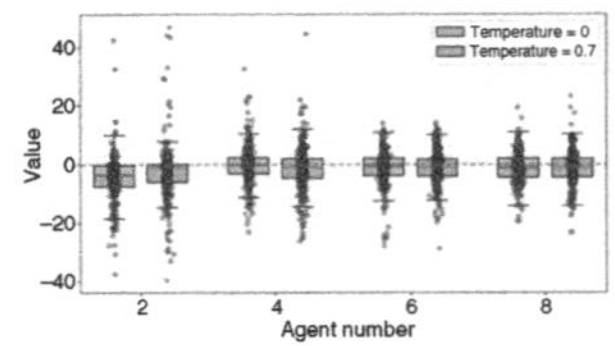
Influence of Communication Topology	Considering the topological connection of an undirected graph, the information flow is bidirectional at this time. If they are not fully connected, they can still converge, but the convergence speed will be slower.	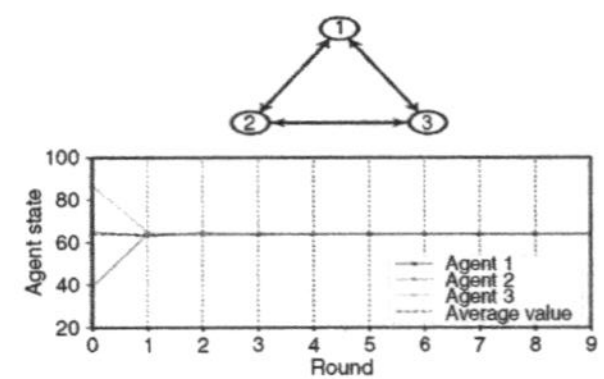

Considering the topological connection of an undirected graph, the information flow is bidirectional at this time. If they are not fully connected, they can still converge, but the convergence speed will be slower.

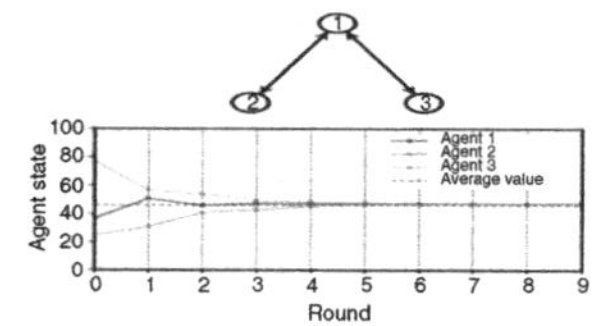

If the topological connection is directed, that is, the information flow is unidirectional at this time. Information flows from agent 1 to agent 2 and agent 3. Eventually, agent 2 and 3 will tend to the state of agent 1.

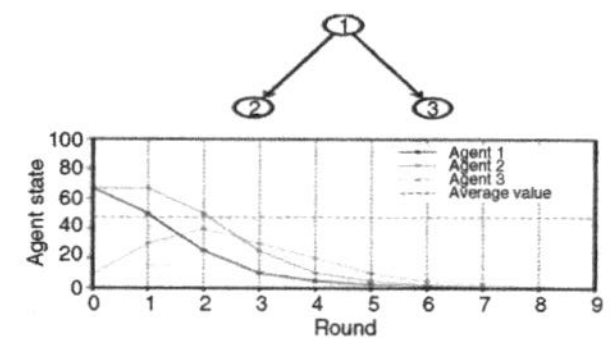

The topological structure adopts a chain topology, where the information of agent 1 flows to agent 3, and the information of agent 3 flows to agent 2. Eventually, the convergence value is also determined by agent 1.

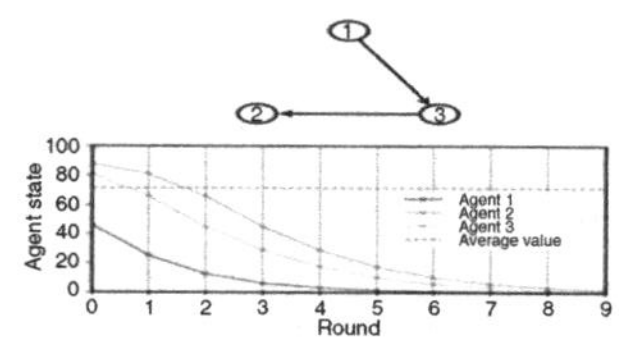

11.4.2 Causes of Collaboration Failure

Unlike deterministic workflows that follow fixed rules, agent workflows can dynamically respond to complexity, finely adjust their approaches based on feedback, and scale to handle more complex tasks. This adaptability makes agent workflows particularly important in scenarios requiring flexibility, learning, and decision-making capabilities. Despite their many advantages and innovative features, agent collaboration faces certain challenges and limitations. Due to the probabilistic nature of LLMs, LLM agents introduce inherent complexity to workflows. Table 11.5 outlines the advantages and disadvantages that LLMs bring to agent workflows.

Although the failure of multiagent collaboration can be partially attributed to the limitations of today's LLMs (such as hallucinations, misalignment), improving only the capabilities of the underlying models is insufficient to address all failures in MASs. Instead, effective MAS design requires in-depth organizational understanding: if there are flaws in the organizational structure, catastrophic failures can also result. Many failures in MASs stem from challenges in agents interaction, rather than the limitations of individual agents. According to the execution stages of MASs, Table 11.6 [33] presents the fine-grained failure modes that may occur in different stages.

In the final analysis, all types of failure may stem from a lack of proper verification or flaws in the verification process. If we assume that the verification agents can function perfectly, then all failures are detectable and avoidable. In many cases, we can view verification as the last line of defense against failures. Although many problems can indeed be attributed to insufficient verification, not all problems can be attributed solely to this factor. Other factors, such as poor specifications, flawed design, and inefficient communication, also contribute to failures. Therefore, a comprehensive understanding of and approach to addressing failures in MASs must take into account a broader range of factors.

11.4.3 The Development Direction of Collaboration

Current MAS architectures treat LLM agents the same as traditional programs, assuming LLM agents have high functional completeness by default while lacking redundancy. However, in reality, AI agents themselves often have design limitations, and external auxiliary tools are also insufficient. The contradiction between the two factors has created the current dilemma in LLM agent implementation. To make multiagent collaboration more robust against failures, general solution strategies are divided into two categories: (1) tactical methods, which involve direct modifications targeting specific failure modes, such as improving prompts, optimizing network topology, and enhancing dialogue management and (2) structural strategies, which are more comprehensive approaches with

Table 11.5 Advantages and Disadvantages of LLM Agent Workflows.

Category	Advantages	Disadvantages
Task Type	**Improved performance in complex tasks.** Agentic workflows (agent-driven workflows) decompose complex tasks into smaller manageable steps (through task decomposition and planning), offering flexibility to adjust and evolve according to task difficulty. By combining different modes, they can also be customized to provide modular design, enabling iterative upgrades as needs and complexity increase.	**Excessive complexity for simple tasks.** When used in simple workflows (such as form input or basic data extraction), AI agents may add extra overhead. Where deterministic, rule-based automation is sufficient, introducing agents may lead to inefficiency, additional costs, and even reduced performance.
Autonomous Capability	**Self-correction and continuous learning.** The reflection mode allows agentic workflows to evaluate their own behaviors, improve strategies, and enhance results over time. Using short- and long-term memory, they learn from past experiences, making each iteration more efficient and personalized. (RL requires manual design of reward functions, which is difficult to debug. In contrast, LLM can directly understand goals and feedback without predefined rewards.)	**Reduced reliability with increased autonomy.** As agents gain more decision-making power in workflows, their probabilistic nature may introduce unpredictability, making outputs less reliable and harder to control. It is crucial to implement and continuously maintain agent safeguards and regularly review their authorized permissions.
Comprehensive Consideration	**Operational efficiency and scalability.** Agentic workflows can automate repetitive tasks with high precision (if well-designed), reducing manual operations and operational costs in specific scenarios. They can also be easily scaled to handle larger workloads or complex systems.	**Ethical and practical challenges.** Not all decisions should be entrusted to AI systems. When using agents in high-risk or sensitive fields, careful supervision is needed to ensure responsible deployment and prevent unexpected consequences.

Table 11.6 Organizational Failure Modes of LLM Agent Workflows.

Category	Reasons	Performance
Preexecution: System Specification and Design Failures	In multiagent systems (MASs), task failures usually stem from incomplete or ambiguous instructions. However, even when clear specifications are provided, MAS may still exhibit inconsistent with user inputs or deviate from role specifications. This type of failure arises from deficiencies in both system design and task specifications.	(1) Unclear task descriptions. For example, the games generated by the ChatDev framework fail to meet the input format requirements. (2) Ambiguous role definitions. For instance, during the requirement analysis phase of ChatDev, the CPO (chief product officer) agent sometimes oversteps its authority, assumes the role of CEO, defines the product vision on its own, and makes final decisions.
During Execution: Agent Inconsistencies	MAS often suffers from low interaction efficiency, where ineffective communication occurs between agents, consuming computing resources without making substantial progress. Such failures stem from poor communication or behavioral conflicts between agents.	(1) Ineffective communication. For example, in a game development scenario, the programming agent engaged in seven rounds of interaction but failed to update the initial code. (2) Information concealment. For instance, the supervising agent instructed the telephone agent to retrieve information using an email as the ID, but after reading the document, the telephone agent found that a phone number should be used as the ID, yet still allowed the error to occur.
Postexecution Task Verification and Termination Failures	MAS may lack a dedicated verification step in their development, or may include verification agents that fail to perform their tasks effectively. Such failures stem from a lack of effective verification mechanisms or the premature termination of tasks.	For example, in the development scenario of a chess game, the verification agent only checks whether the code compiles, but does not run the program or ensure compliance with chess rules, leading to issues such as improper input handling.

system-wide impacts, including strong verification, enhanced communication protocols, uncertainty quantification, and memory and state management. These strategies require more in-depth research and more detailed implementation and are open research areas for future exploration. Table 11.7 presents several development directions that need attention in multiagent collaboration.

Table 11.7 Development Directions of Multiagent Collaboration.

Category	Method Introduction	Specific Technology
Tactical Methods	These tactical methods involve direct modifications targeting specific failure modes, such as optimizing prompts, adjusting the organizational structure of agents, and improving the interaction methods between agents. Although these methods may seem simple and effective, they may not cover all issues in practical applications or may lack practicality in certain situations.	**Optimizing prompts:** Clarify the instructions, agent roles, and task constraints in prompts. For example, prompts should clearly describe task objectives, define each agent's responsibilities, and encourage active dialogue between agents. In case of inconsistencies, agents should be able to reengage in discussions or retry the task. **Self-verification step:** After completing complex multistep tasks, add a self-verification step. By restating the solution, checking task constraints, and testing for errors, the agent can identify and address issues proactively. **Modular design:** Adopt a modular approach, using simple and well-defined agents instead of complex multi-task agents. This method not only improves performance but also simplifies the debugging process. **Clear roles and tasks:** Enhance the clarity of agents' roles and tasks by defining dialogue patterns and setting clear task termination conditions. For example, different agents can propose multiple solutions, and cross-validation can be used to ensure the accuracy of the results.
Structural Strategies	Structural strategies refer to methods that have a profound impact on the entire system structure, are crucial for improving the overall performance of MAS, and require more in-depth research and meticulous implementation.	**Strengthening the verification process:** Verification is a key link in ensuring the correctness of the system. Weak or insufficient verification mechanisms are one of the important causes of system failures. Even for experts, covering all edge cases is complex. Verification methods have different requirements in different fields (such as coding, quality assurance, and reasoning), and adapting verification across different fields remains an ongoing research challenge.

(Continued)

Table 11.7 (Continued)

Category	Method Introduction	Specific Technology
		Standardized communication protocols: Currently, LLM agents mainly communicate through unstructured text, which may lead to ambiguity. By clearly defining intentions and parameter formats, the alignment between agents can be improved, and formal coherence checks can be conducted during interactions. **Reinforcement learning training:** Train agents through reinforcement learning, enabling them to reward task-related behaviors and punish inefficient behaviors based on role-specific reward algorithms. **Probabilistic confidence:** Introducing probabilistic confidence in agent interactions can enhance the reliability of decision-making and communication. Agents can be designed to take action only when their confidence exceeds a predefined threshold. Conversely, when confidence is low, agents can pause to gather additional information. Furthermore, the system can benefit from adaptive threshold adjustment, where confidence thresholds are dynamically adjusted [34]. **Memory and state management:** Although generally regarded as attributes of a single agent, memory and state management are equally important in multiagent interactions. Effective management of shared memory and system state can improve context understanding and reduce communication ambiguity [35].

The combination of tactical approaches and structural strategies offers a comprehensive approach to address failure modes in MASs. However, the effectiveness and implementation details of these methods still need further research and refinement. Future studies should focus on how to select appropriate agent frameworks, considering factors such as task complexity, technical compatibility, performance requirements, development efficiency, and security compliance. This evaluation framework aims to help decision-makers choose the most suitable framework based on their business needs and technical environment. The architectural design of different frameworks determines their applicable scenarios and performance characteristics. LangGraph excels in managing communication via graph edges, allowing complex state transitions; AutoGen maintains good agent memory, making it suitable for dialogue-driven workflows; CrewAI handles basic state persistence through task outputs, simplifying process transitions through sequential and parallel processes.

11.5 Chapter Summary

The entire workflow of a computational experiment typically includes hypothesis generation for the target scientific problem, protocol planning, experimental implementation, data analysis, conclusion drawing, and hypothesis revision. However, the current construction of computational experiments is a complex and labor-intensive process that requires developers to be deeply involved in intricate technical details, which poses certain obstacles to the popularization and application of computational experiment methods. In contrast, automating the entire computational experiment workflow using multiagent collaboration technology can significantly reduce the need for manual intervention and lower development costs. To achieve full automation of the computational experiment workflow based on LLM agent collaboration, we must address the following questions: Can truly personalized needs be fully handled automatically by AI? Can it really understand our goals? Do the goals need to be optimized and adjusted during the process? Is manual intervention required in the intermediate process for confirmation? How to handle abnormalities if they occur? What if the intermediate process does not meet the user's requirements?

In traditional MASs, agents typically learn from offline training datasets using reinforcement learning. However, LLM agent systems mainly learn from immediate feedback, such as interactions with the environment or humans. Although these methods are effective for individual agents, they overlook the potential synergistic effects that may emerge in multiagent interactions and fail to fully leverage the inherent latent collective intelligence of the agent network. Therefore, collectively adjusting multiagents and achieving optimal collective intelligence remains a key challenge in LLM agent collaborative workflows. Furthermore, as the number of agents in the system increases, the complexity of ensuring effective coordination and communication rises significantly. For this reason, this chapter focuses on presenting collaboration frameworks and orchestration methods among multiagents, including optimizing agent workflows, coordinating task allocation among different agents, and reducing inter-agent communication redundancy. It then presents an automated solution for computational experiment processes, along with the potential challenges that may be encountered. In terms of specific implementation, it is advisable not to overly pursue general domains, as the goals and knowledge in general domains are difficult to align with users' needs. A more suitable direction is to focus on vertical domains, which enables setting clear constraints and making full use of industry expertise.

References

1 Rawte, V., Sheth, A., and Das, A. (2023). A survey of hallucination in large foundation models. arXiv preprint arXiv:2309.05922.

2 Xue, X., Li, G., Zhou, D. et al. (2022). The research roadmap of service ecosystem: a crowd intelligence perspective. *International Journal of Crowd Sciences* 6 (4): 195–222.

3 Li, G., Hammoud, H., Itani, H. et al. (2023). Camel: communicative agents for "mind" exploration of large language model society. *Advances in Neural Information Processing Systems* 36: 51991–52008.

4 Hong, S., Zheng, X., Chen, J. et al. (2023). Metagpt: meta programming for multi-agent collaborative framework. arXiv preprint arXiv:2308.00352, 3(4): 6.

5 Qian, C., Liu, W., Liu, H. et al. (2023). Chatdev: communicative agents for software development. arXiv preprint arXiv:2307.07924.

6 Chan, C.M., Chen, W., Su, Y. et al. (2023). Chateval: towards better LLM-based evaluators through multi-agent debate. arXiv preprint arXiv:2308.07201.

7 Xu, Z., Yu, C., Fang, F. et al. (2023). Language agents with reinforcement learning for strategic play in the werewolf game. arXiv preprint arXiv:2310.18940.

8 Light, J., Cai, M., Shen, S. et al. (2023). Avalonbench: evaluating LLMs playing the game of avalon. arXiv preprint arXiv:2310.05036.

9 Sun, Q., Yin, Z., Li, X. et al. (2023). Corex: pushing the boundaries of complex reasoning through multi-model collaboration. arXiv preprint arXiv:2310.00280.

10 Liang, T., He, Z., Jiao, W. et al. (2023). Encouraging divergent thinking in large language models through multi-agent debate. arXiv preprint arXiv:2305.19118.

11 Nair, V., Schumacher, E., Tso, G. et al. (2023). DERA: enhancing large language model completions with dialog-enabled resolving agents. arXiv preprint arXiv:2303.17071.

12 Kim, Y., Park, C., Jeong, H. et al. (2024). Mdagents: an adaptive collaboration of LLMs for medical decision-making. *Advances in Neural Information Processing Systems* 37: 79410–79452.

13 Si, C., Shi, W., Zhao, C. et al. (2023). Getting more out of mixture of language model reasoning experts. arXiv preprint arXiv:2305.14628.

14 Jiang, D., Ren, X., and Lin, B.Y. (2023). LLM-blender: ensembling large language models with pairwise ranking and generative fusion. arXiv preprint arXiv:2306.02561.

15 Xu, Z., Shi, S., Hu, B. et al. (2023). Towards reasoning in large language models via multi-agent peer review collaboration. arXiv preprint arXiv:2311.08152.

16 Liu, Z., Zhang, Y., Li, P. et al. (2024). A dynamic LLM-powered agent network for task-oriented agent collaboration. *First Conference on Language Modeling*.

17 Wang, H., Du, X., Yu, W. et al. (2023). Apollo's oracle: retrieval-augmented reasoning in multi-agent debates. CoRR.

18 Anthropic (2025). Model context protocol. https://www.anthropic.com/news/model-context-protocol (accessed 07 January 2025).

19 Yang, Y., Chai, H., Song, Y. et al. (2025). A survey of AI agent protocols. arXiv preprint arXiv:2504.16736.

20 Topsakal, O. and Akinci, T.C. (2023). Creating large language model applications utilizing langchain: a primer on developing LLM apps fast. *International Conference on Applied Engineering and Natural Sciences* 1 (1): 1050–1056.

21 Wang, J. and Duan, Z. (2024). Agent AI with langgraph: a modular framework for enhancing machine translation using large language models. arXiv preprint arXiv:2412.03801.

22 Shen, Y., Song, K., Tan, X. et al. (2023). Hugginggpt: solving AI tasks with chatgpt and its friends in hugging face. *Advances in Neural Information Processing Systems* 36: 38154–38180.

23 Chen, G., Dong, S., Shu, Y. et al. (2023). Autoagents: a framework for automatic agent generation. arXiv preprint arXiv:2309.17288.

24 Chen, W., Su, Y., Zuo, J. et al. (2023). Agentverse: facilitating multi-agent collaboration and exploring emergent behaviors in agents. arXiv preprint arXiv:2309.17288.

25 Wu, Q., Bansal, G., Zhang, J. et al. (2024). Autogen: enabling next-gen LLM applications via multi-agent conversations. *First Conference on Language Modeling*.

26 Xiong, K., Ding, X., Cao, Y. et al. (2023). Examining inter-consistency of large language models collaboration: an in-depth analysis via debate. arXiv preprint arXiv:2305.11595.

27 Zhuge, M., Wang, W., Kirsch, L. et al. (2024). Gptswarm: language agents as optimizable graphs. *Forty-first International Conference on Machine Learning*.

28 Chen, J.C.Y., Saha, S., and Bansal, M. (2023). Reconcile: round-table conference improves reasoning via consensus among diverse LLMs. arXiv preprint arXiv:2309.13007.

29 Liu, B., Li, X., Zhang, J. et al. (2025). Advances and challenges in foundation agents: from brain-inspired intelligence to evolutionary, collaborative, and safe systems. arXiv preprint arXiv:2504.01990.

30 Yu, X., Xue, X., Zhou, D. et al. (2025). Unlocking complexity: harnessing value entropy for advanced multidimensional utility evaluation in service ecosystems. *IEEE Transactions on Services Computing* 18 (4): 2456–2470.

31 Xue, X., Chen, Z., Wang, S. et al. (2020). Value entropy: a systematic evaluation model of service ecosystem evolution. *IEEE Transactions on Services Computing* 15 (4): 1760–1773.

32 Chen, H., Ji, W., Xu, L. et al. (2023). Multi-agent consensus seeking via large language models. arXiv preprint arXiv:2310.20151.

33 Cemri, M., Pan, M.Z., Yang, S. et al. (2025). Why do multi-agent LLM systems fail?

34 Horvitz, E. (1999). Uncertainty, action, and interaction: in pursuit of mixed-initiative computing. *IEEE Intelligent Systems* 14 (5): 17–20.

35 Packer, C., Fang, V., Patil, S.G. et al. (2023). MemGPT: towards LLMs as operating systems.

12

Roadmap of Computational Experiment Method

In an era where artificial intelligence (AI) is profoundly transforming scientific research paradigms, large language models (LLMs) are opening up entirely new avenues for social science research. This breakthrough extends beyond large-scale text processing to reshape traditional research paradigms, giving rise to the emerging field of AI-driven social science (AI4SS) [1]. Even more noteworthy is the integration of LLMs with computational experiment methods, enabling the creation of digital agents with complex cognitive abilities. This provides a novel experimental platform for studying the evolution of social systems—ranging from simulating micro-level individual behaviors to deducing macro-level social dynamics, and from generating theoretical hypotheses to predicting policy outcomes. LLMs are fundamentally transforming the social science research toolkit [2]. However, this technological innovation brings with it deep-seated challenges, characterized by "technological fragmentation and theoretical lag." Modeling approaches often rely on engineering-oriented parameter tuning rather than being guided by scientific principles; experimental workflows lack unified and traceable verification standards; and extrapolating conclusions from virtual to real-world scenarios remains constrained by local optima. In light of these challenges, this chapter discusses the future directions of computational experiment method.

12.1 Roadmap of Computational Experiment Method

As a core methodology connecting social science and intelligent science, computational experimentation plays a crucial role in reconciling the cognitive tensions between the two disciplines—social science emphasizes explaining processes, while intelligent science focuses more on predicting outcomes. This approach is driving paradigm shifts in fields such as economics, finance, military affairs, and political science. However, the study of complex social systems often involves high-dimensional nonlinearity, multiscale dynamic evolution, and uncertainty coupling mechanisms. Computational experiments powered by LLMs face three major bottlenecks: imprecise scientific abstraction, deficiencies in causal reasoning, and divergences between simulation outcomes and real-world observations. As shown in Figure 12.1, building a foundational scientific theory system for AI4SS computational experiments requires solving three core problems: real-world model mapping, experimental design and causal analysis, and extrapolation validation of virtual conclusions.

Question 1: At the modeling and experimental setup level, current research primarily focuses on using LLMs to achieve human-like agent modeling, and some scenarios have already achieved partial alignment with real-world patterns. However, there are still significant limitations in the scientific abstraction of complex systems by LLMs: excessive simplification may lead to the loss of essential system characteristics, while excessive anthropomorphization may result in computational infeasibility. For example, existing models struggle to balance the accuracy of individual behaviors with the ability to fit macro-level patterns.

Question 2: In the field of experimental design, the integration of causal inference with virtual experiments is beginning to emerge, and methods such as reinforcement learning and Bayesian networks have been used to optimize experimental pathways. However, the causal logic of virtual experiments still relies on manually preset intervention variables and lacks the ability to autonomously construct counterfactual reasoning frameworks, especially in highly coupled scenarios like social system governance and game-theoretic trend prediction. The interpretability and reproducibility of virtual experimental designs urgently require theoretical breakthroughs.

Question 3: Regarding the credibility of experimental results, current studies mostly use sensitivity analysis and uncertainty quantification for validation. However, inherent biases in LLMs—such as shifts in training data distribution and fixed inductive preferences—lead to systematic errors when extrapolating virtual conclusions to real-world applications. Recent research indicates that hybrid modeling based on symbolic regression can partially mitigate these biases, but it cannot fundamentally bridge the gap between the "perfect virtual experiment assumption" and the "incompleteness of real-world systems."

Overall, the current AI4SS theoretical framework has not yet established a mathematical foundation for mapping between virtual and real systems, nor has it formed a scientific methodology that integrates causal reasoning with virtual experiment design. It also lacks a systematic correction mechanism for LLM biases. Therefore, building a computational experimental scientific theory system for AI4SS requires breakthroughs at three levels: first, developing

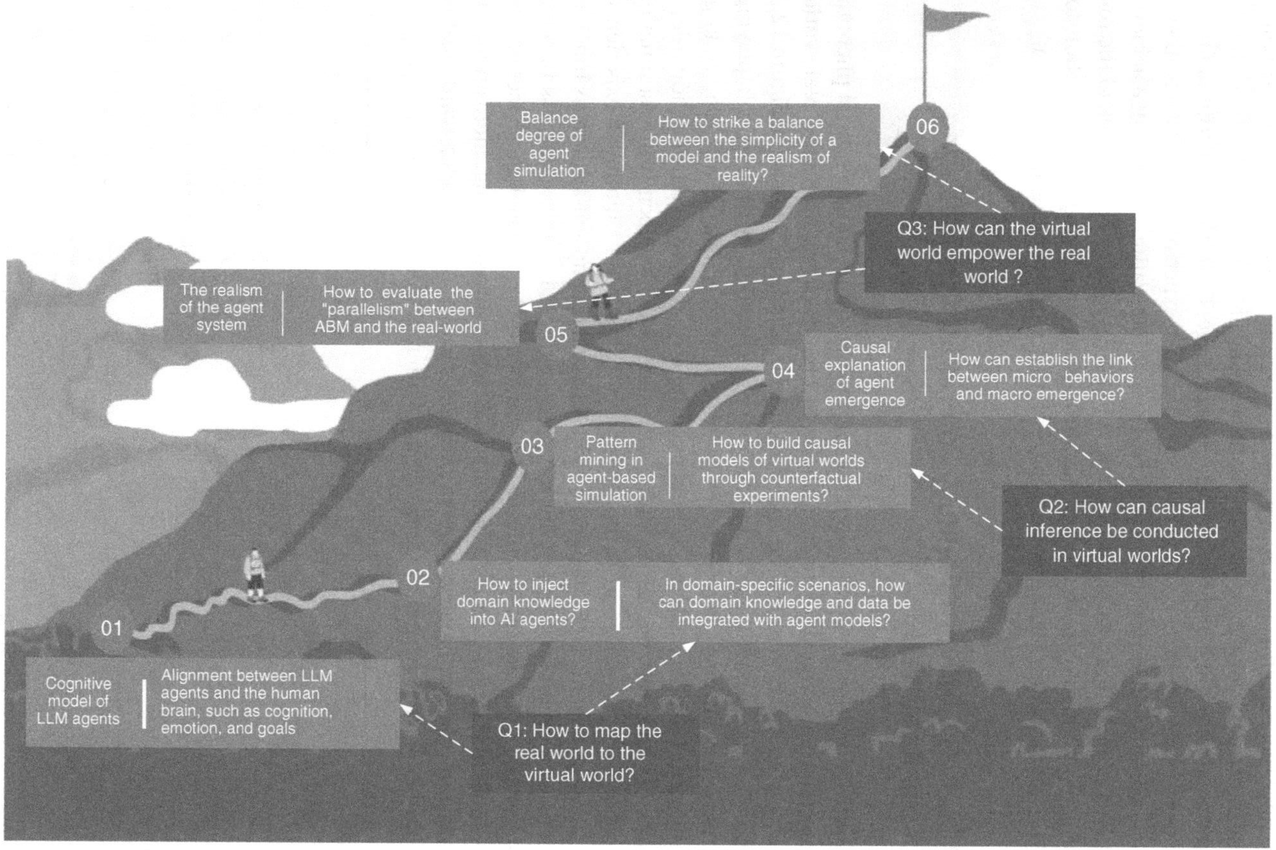

Figure 12.1 Roadmap for computational experiment method.

cross-scale modeling theories based on complexity science and intelligent science to solve the "scientific abstraction paradox"; second, establishing an interpretable causal-virtual hybrid experimental framework to address the logical discontinuity in virtual reasoning; and third, designing convergence verification criteria for virtual and real systems to create a pathway for virtual experiment results to empower real-world applications. This not only requires integrating cutting-edge achievements from AI and complexity science but also demands a redefinition of the role of computational experiments in scientific discovery, laying the foundation for AI4SS to move from being an auxiliary tool to driving theoretical disruption.

12.2 Q1: How to Conduct Computational Modeling of the Real World

Social simulation provides a window for understanding complex social phenomena by simulating human behaviors and characteristics from the real world, serving as a valuable tool for social science research. As shown in Figure 12.2, the powerful capabilities of LLMs—such as perception, reasoning, role-playing, and self-evolution—have brought new vitality to social simulation. A large number of related studies have emerged rapidly, covering both individual-level and societal-level simulations, along with many available open-source simulation tools [3–6]. However, in terms of consciousness and emotional experience, LLM-based agents lack the genuine subjective states and self-awareness inherent to human cognition. While fully replicating human-like consciousness in social simulation may be neither necessary nor desirable, psychological factors and emotional experiences play a profound role in human reasoning, motivation, and social interaction. Incorporating these aspects can guide research toward creating simulations that are more closely aligned with real-world dynamics.

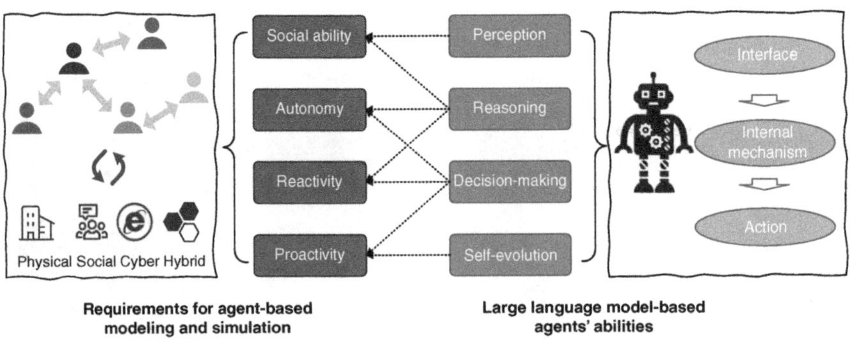

Figure 12.2 The correspondence between ABSS requirements and LLM agent capabilities.

All scientific experiments, whether based on real human subjects or LLM-generated agents, involve a sampling process. In past research, scholars have aimed to use comprehensive samples to ensure the reliability of results. In the context of social simulation with LLMs, the issue of sample validity shifts to the question of how closely an LLM should reflect real humans. This involves several trade-offs:

First, the trade-off between agent complexity and simulation accuracy. Providing more detailed personal experiences or interview data can enhance the LLMs' ability to understand and simulate individual behavior, but overly complex external knowledge introduces substantial workload and computational costs, while offering diminishing returns in simulation fidelity.

Second, the trade-off between diversity and stereotyping. Human datasets are always a sample of the broader population. How do we define when diversity is sufficient to cover marginalized groups?

Third, the trade-off between data quality and data security. Collecting more detailed, accurate, and richer real-world data increases the difficulty and cost of data acquisition, while also raising concerns about privacy and data security management. How can we ensure high-quality data collection without crossing ethical boundaries?

In response to these challenges, Gati Aher et al. proposed the Turing Experiment method in [7], selecting four classic sociology experiments and testing whether LLMs can generate representative simulated samples by varying names and genders across the entire human population dimension. Their findings show that LLM agents can reproduce social experiment results and capture gender biases in human behavior.

However, using LLMs for social simulation raises a series of new challenges. First, social biases embedded in LLM training data may be covertly encoded, leading to unfair or biased results. Second, LLMs tend to generate homogenized group representations, overlooking internal diversity within groups. Third, LLMs lack detailed representations of human inner psychological states and individual life experiences. Fourth, LLMs display significant inconsistency in simulation, often giving different answers to different formulations of the same question.

In this context, researchers need to establish a dual verification mechanism: testing whether the model accurately reflects group characteristics while also checking for systemic errors caused by data sampling bias.

A more difficult issue is the methodological paradox. When researchers attempt to remove gender or racial bias from a model through fine-tuning, they may inadvertently weaken the model's capacity to represent the true complexity of society. This realism vs. ethics tension is particularly acute in studies involving sensitive topics.

Currently, designing agents that can realistically simulate human behavior and social interaction remains an open problem. Although LLMs offer excellent language reasoning capabilities, most agent designs are still ad hoc, combining perception, memory, and planning modules in fragmented ways. They are far from achieving the kind of well-coordinated cognitive and functional integration found in biological systems such as the human brain.

Recent studies suggest that theories of human behavioral patterns and cognitive processes can improve the design of agent workflows, thereby enhancing LLMs' ability for social behavior simulation [8]. This is considered an important direction for future research. Figure 12.3 presents a brain-inspired cognitive architecture for LLM-based agents, which connects the rational components of an agent (such as perception, decision-making, and actions) with the emotional components (such as beliefs, emotions, and mental states), enabling the link from intention to behavior.

This cognitive model mainly consists of two parts. First, the left side represents the software modules that support agent decision-making. The agent's internal state and emotions are continuously updated through external inputs and internal loop processing, interacting with the memory system. The agent also possesses reflective learning capabilities, allowing it to adjust its goal orientation based on its current state. Second, the right side illustrates the agent's decision-making loop following the OODA cycle (Observe, Orient, Decide, Act). The agent processes inputs through the perception module, and the high-level cognitive controller receives the information. It then outputs an action option and a subgoal to guide the next layer of actions.

In addition, the evaluation of the realism of generated human behaviors still mainly relies on subjective human assessment, which poses challenges for scaling agent models. Developing reliable and scalable evaluation methods will be another key direction for future research.

12.3 Q2: How to Conduct Causal Reasoning in a Virtual World

In virtual environments, agents act in a goal-directed manner, reflecting their intent to change the external world in pursuit of objectives. The agent treats the LLM as its brain and interacts with the environment through an action system, which ultimately governs the agent's capacity to achieve complex objectives. This process can be mapped into a multistage workflow, involving the construction of the action space, learning to improve decision-making mechanisms, and integrating external states. These components form the basis of causal inference in virtual worlds, where the state and events at one time point influence the state and events at the next.

A central debate in cognitive science and neuroscience is whether the source of internal causal flow in agents originates from action or perception. The

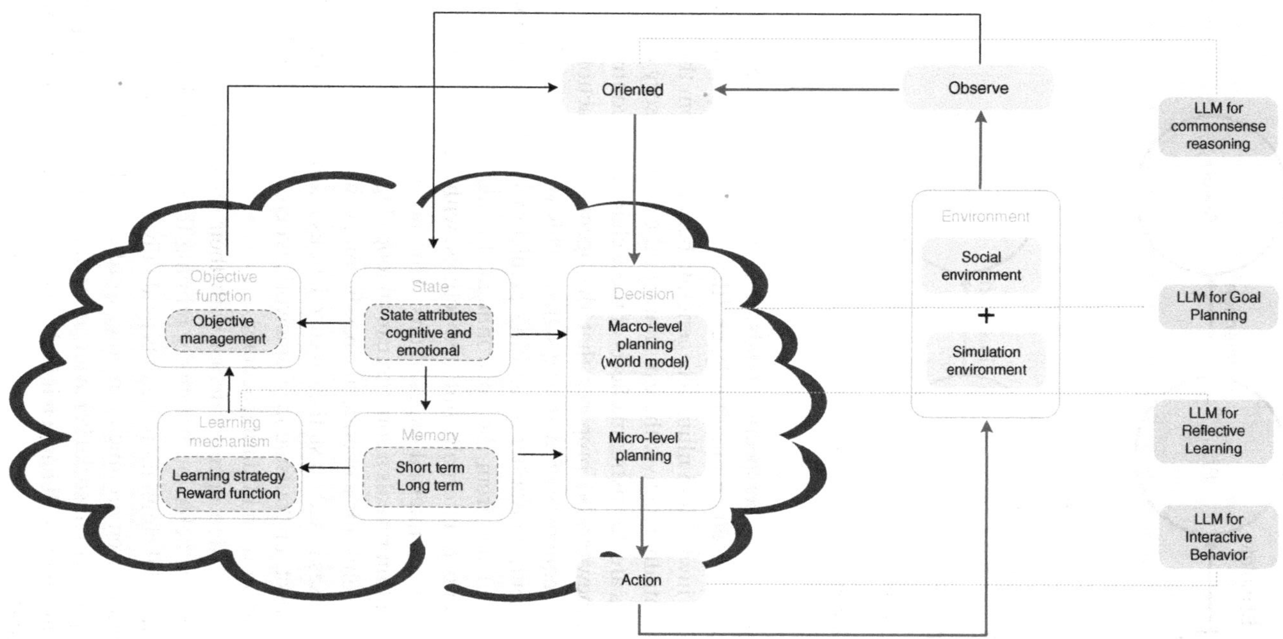

Figure 12.3 Brain-inspired cognitive model for LLM agents.

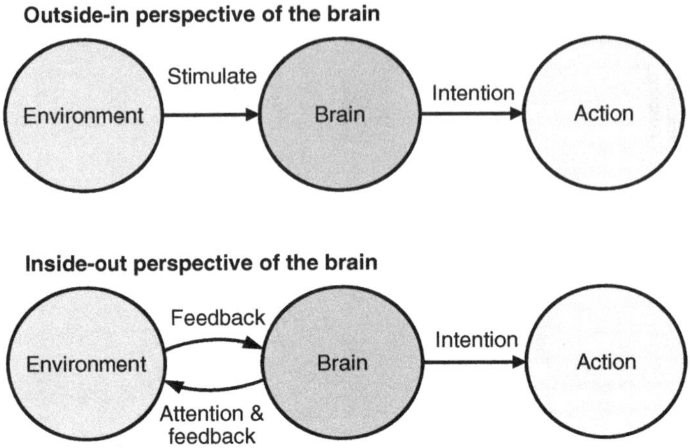

Figure 12.4 Comparison of the brain from "outside-in" and "inside-out."

traditional "outside-in" perspective holds that causal influence begins with external stimuli. The environment triggers the agent's perception, and these signals propagate inward and ultimately generate behavior. From this viewpoint, the agent is essentially reactive: the external world causes sensory changes, and the agent's actions are downstream effects of these changes. In contrast, Buzsáki's "inside-out" framework [9] proposes that the agent's own actions shape the meaning and consequences of incoming signals. This view implies the existence of a proactive agent, constantly generating predictions and motor commands, then receiving feedback from the environment. In this model, causality shifts from external events to internally initiated actions, with external stimuli serving to validate or correct internal expectations. This normal interaction between action and perception reveals a complex causal loop (Figure 12.4).

For workflow agents, most operate in a passive mode, waiting for user input and generating responses based on statistical correlations learned from large datasets. This passivity reflects an "outside-in" framework, where the agent's role is limited to responding rather than initiating actions.

In contrast, simulation agents interact with their environment through continuous spontaneous behavior, iteratively refining their internal states through causal loops. In social simulations, agents gradually develop implicit representations of the simulated environment. In other words, they build an internal "world model" that supports their social behavior generation, including helping the agent make predictions (what will happen if I take action A?), planning (what sequence of actions should I take to achieve goal B?), and counterfactual reasoning (what would have happened if I had made a different decision in the past?).

In computational experimental analysis, intention modeling of agents is the key link between micro-level individual behaviors and macro-level emergent phenomena. The agent's internal world model is critical for generating intentions and making decisions. If we can capture the evolution of an agent's world model, we can also grasp the causal process of the agent's evolution.

What is a world model? Here is a simple example. In your "memory," you know the weight of a cup of coffee. So when you intend to pick up the cup, your brain accurately "predicts" how much force to apply, and you lift the cup smoothly. But what if the cup happens to be empty? One might apply excessive force when lifting a cup that turns out to be empty. Your hand will immediately sense something is wrong. Then, your "memory" will update with a new piece of information: The cup might be empty. The next time you "predict," you won't make the same mistake.

The more you do, the more complex your brain's world model becomes, allowing you to make more accurate predictions about how the world will respond. This "first memorize, then predict" approach is how humans understand the world. This cognitive pattern is called a world model.

The core function of a world model is counterfactual reasoning—that is, even for decisions not yet made in reality, the model allows you to reason out the possible outcomes. If an accurate world model is available, you can repeatedly simulate trial and error within it to find the optimal decision for the real world.

As shown in Figure 12.5, LLM agents form implicit representations and basic cognition of the external world by storing observations gained through interactions with the environment, essentially creating an implicit world model [11]. The world model is fundamentally a highly structured and often predictive form of long-term memory. Memory provides the raw material for constructing the world model—knowledge and experience—while the world model, in turn, acts as an organizing framework that influences how new memories are encoded, consolidated, and retrieved. For example, a well-developed world model may prioritize storing surprising events because they highlight gaps in the agent's understanding. This cognition can be stored in textual form in a memory database, allowing the LLM agent to retrieve and access useful information and leverage experiential knowledge from past environmental interactions to make decisions.

The core of an LLM agent lies in its ability to construct and utilize its internal world model, which is a key step toward causal inference. A robust world model must handle multiple types of information, including physical laws, social norms, and behaviors of other agents, while continuously updating and refining itself based on new observations. In social simulation, agents need to constantly learn and update their world models, gradually forming a "cognitive map" of their surrounding environment, encompassing knowledge of the physical environment, local contexts, and human society.

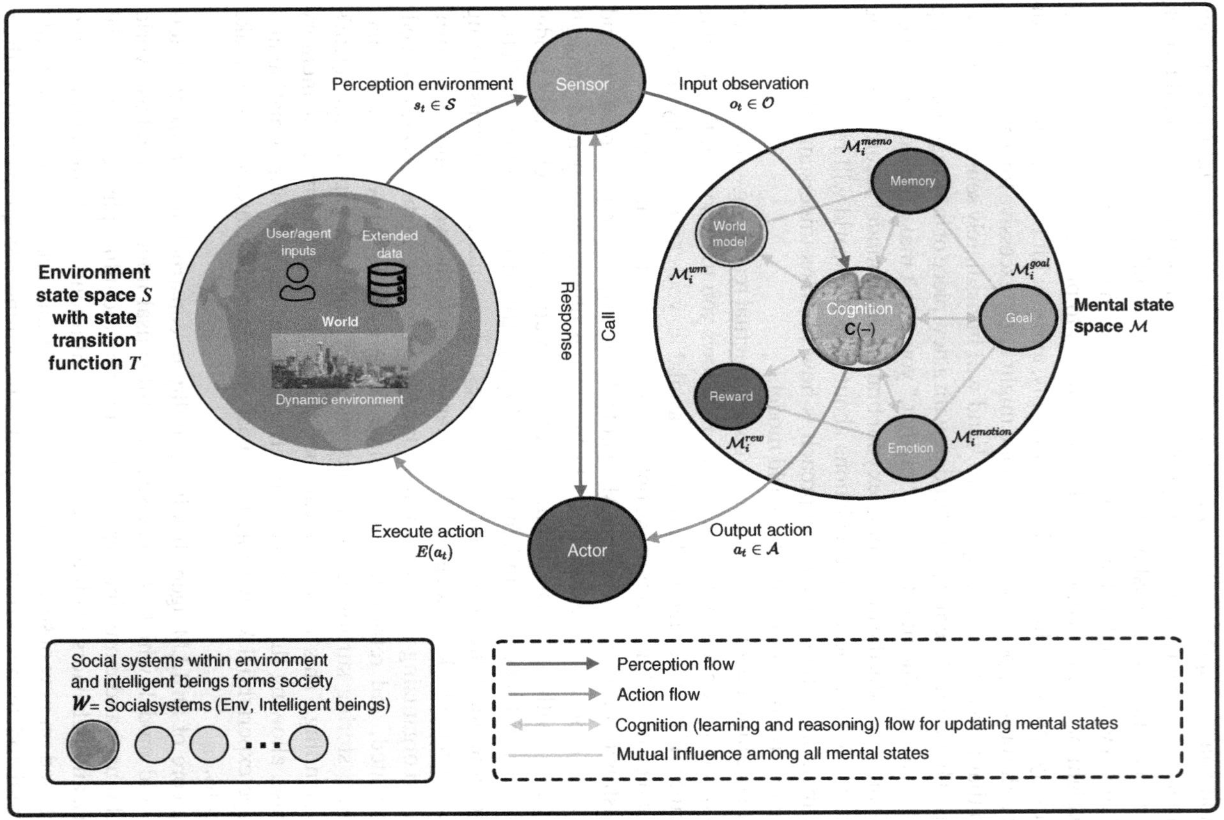

Figure 12.5 Overview of the general framework for the agent loop and agent society. *Source:* [10] / with permission of Oxford University Press.

However, developing effective world models faces significant challenges. These include how to appropriately abstract the complexity of the real world—balancing accuracy, complexity, and computational efficiency; how to efficiently learn, update, and ensure the generalization of these models; and how to implement efficient search algorithms to handle inherent uncertainties in prediction.

12.4 Q3: How to Ensure That Experimental Laws Hold in Reality?

Computational experimentation serves as a bridge between the virtual and real worlds, using counterfactual reasoning to eliminate infeasible strategies and transferring optimized strategies to the real world to guide actual system operations. In the virtual world, we can create extreme scenarios that are difficult to replicate in the current real world. By combining optimization theories from operations research under conditions of uncertainty and multiple objectives, intervention strategies can be continuously trained and refined to achieve the desired optimization effects. This approach intervenes in the current development trajectory and operational status from a future perspective, identifying potential adverse effects, conflicts, and risks of strategies, thereby providing "reference," "forecast," and "guidance" for possible contingencies. Therefore, the core goal of computational experimentation must be to empower the real world with reasoning results derived from the virtual world.

In recent years, LLM agents have frequently been used as tools for social simulation due to their human-like cognitive abilities, helping to understand and analyze complex human interactions and social dynamics. However, these studies are based on an unproven assumption that LLM agents behave like humans in simulations. To achieve the core goal of computational experimentation, we must confront a fundamental question: Can LLM agents truly simulate human behavior [12, 13]? Essentially, LLMs are trained machine learning models that generate text based on provided prompts. Their key characteristic is acquiring knowledge from training data, but it is difficult to determine which knowledge the model has retained or missed. As a result, LLMs cannot reliably judge the correctness of generated text and may produce incorrect, meaningless, or unrealistic outputs known as "hallucinations."

The modules defining roles, memory, planning, and execution within LLM agents face similar challenges, with no guaranteed accuracy or effectiveness. Hallucinations can take several forms: (1) logical fallacies, where the agent makes unreasonable decisions during reasoning; (2) fabricated facts, where the agent produces knowledge inconsistent with its role or historical context; and (3) ethical biases, where the agent's output skews in certain directions. Moreover, in multiagent interactions, hallucinations by one agent may cascade, amplifying erroneous information through the interaction network. Explainability is key to ensuring the credibility of social simulation results and their use in intelligent decision-making. Therefore, achieving alignment between LLM agents and real-world roles, i.e., ensuring AI agent behavior conforms to norms in social simulation scenarios, is a pressing challenge in this field. The following elaborates on this from three levels: factual alignment, cognitive alignment, and value alignment.

1) **Factual alignment**
   Factual alignment means the knowledge within the LLM agent's "brain" corresponds to objective facts, avoiding false, incorrect, or baseless reasoning, and accurately reflecting role performance in the real world. Achieving factual alignment typically involves integrating external knowledge bases with LLMs via RAG techniques. For example, when an agent plays a historical figure, it can access precise facts from knowledge graphs, including exact times, people, and events. A study [14] verified whether LLM agents can replicate human-like trust behaviors by using the Trust Game and its variants—effective tools in behavioral economics for quantifying trust-related decisions and actions. Multiple LLMs were tested, including closed-source models (e.g., GPT-4, GPT-3.5-turbo) and open-source models (e.g., Llama2, Vicuna variants). To simulate diverse human decision patterns, 53 roles representing different personalities or backgrounds were designed using the belief–desire–intention (BDI) framework as the basis for the language model's decision-making process. The study found notable consistency between LLM agents' and humans' trust behaviors, especially in GPT-4, providing empirical support for simulating human trust behavior.
2) **Cognitive alignment**
   Cognitive alignment means the LLM agent's thinking and reasoning processes align with human logic, enabling reasonable analysis, inference, and judgment similar to humans. It includes understanding causal relationships, logical sequences, and abstract concepts and solving problems and generating responses in ways that feel natural and consistent with human cognition. When LLM agents think similarly to humans, the parallelism between virtual and real world improves, making it easier to fit various social phenomena. Cognitive alignment is mainly achieved through (a) multi-turn reinforcement reasoning training—designing multiround experimental scenarios where agents interact with users or other agents, iteratively reasoning and providing feedback to simulate human thought progression; (b) chain-of-thought prompting—adding guiding information in inputs to encourage step-by-step reasoning (e.g., "first…, then…, finally…"); and (c) analogy and case-based

learning—providing many analogous examples and problem-solving strategies to teach agents human-like reasoning patterns for new problems.

3) **Value alignment**

Value alignment means the agent's behaviors, decisions, and outputs conform to the values and ethical standards of the roles they simulate. Human decision-making is not only based on rational principles but also influenced by emotions, psychology, and values. As social simulations gain widespread use across fields, their impact grows. If an agent's behaviors or outputs contradict role values, simulation results are unlikely to gain user acceptance. Given human society's diversity, roles played by LLM agents may embody different values. To meet diverse value alignment needs, a "decoupled value alignment" framework can be adopted [15]: (a) split fixed value alignment into immutable core values and customizable diverse values; (b) embed clear value rules for core values into LLM agents; and (c) provide users with convenient interfaces to customize diverse values so that roles from various cultural backgrounds can be accurately portrayed.

12.5 Traditional Simulation vs. Generative Simulation

Currently, computational models of complex social systems still adopt a "bottom-up" modeling approach, including traditional ABM and LLM-based agents. However, the entire process is extremely labor-intensive and unfriendly to users. Therefore, achieving "a complete and precise simulation world generated from a single sentence" has long been the goal pursued by simulation researchers. Ideally, we want to eliminate complex parameter configurations and automatically generate high-precision simulation results through natural language scene descriptions alone. This capability not only lowers the threshold for usage but also makes simulation unprecedentedly efficient and intelligent. Complex scenarios that would traditionally take hours or even days to simulate may, in the future, be completed within minutes—or even seconds—ushering in unprecedented opportunities for the development of intelligent societies.

On February 16, 2024, OpenAI released an innovative and revolutionary text-to-video LLM tool: Sora. As an LLM focused on video generation, Sora accepts text or image inputs and allows for video editing in multiple ways, including generation, concatenation, and extension. As a type of multimodal LLM, it has attracted significant attention from both academia and industry. Sora can understand the semantics of complex textual inputs and accurately represent real-world objects, actions, and backgrounds within intricate scenes. Its generative capacity lies in transforming textual descriptions into high-quality video content, seamlessly blending realism with imagined scenarios. This remarkable achievement opens new possibilities for video production, storytelling, and entertainment. Sora's capabilities mark a major milestone in the development of AGI, pushing the boundaries of what machines can accomplish in text-to-video synthesis.

Sora's training dataset consists of a large number of short videos. Each video is encoded into a high-dimensional feature representation, which is then compressed by the model into a more compact latent representation. In the latent space, these feature vectors are further divided into patches and organized temporally, forming spatiotemporal patches—known as time–space tokens. Each token records its frame sequence index (time) and row–column index within the frame (space). Furthermore, Sora utilizes a Transformer architecture to determine how to combine these time–space tokens based on text or image inputs in order to generate coherent video content. We selected three demo videos provided on Sora's official website (see Figure 12.6). For each video, we clipped a representative frame in playback order: "A woman walking through the streets of Tokyo," "A mammoth walking across a snowy field," and "A pirate ship inside a coffee cup."

In the abovementioned examples, we can observe significant differences between generative simulation (like Sora) and traditional simulation. Sora demonstrates a future development trend for world simulators: the ability to construct a virtual world that merges reality and imagination based on users' natural language descriptions, intuitively presenting the simulation and evolution of research subjects [16]. Technically, Sora follows a data-driven approach—"video data → model training → video generation"—while traditional artificial society modeling relies on a knowledge-driven approach—"domain knowledge → model

Figure 12.6 Screenshots of videos generated by Sora. *Source:* Generated with AI using Sora.

construction → simulation inference" (in knowledge-driven physics engines, efforts are made to make NPCs increasingly intelligent, until eventually replaced by real humans), which is fundamentally different. In traditional artificial society modeling, knowledge is embedded into models to ensure the simulation process conforms to physical laws. For world simulators like Sora, maintaining the validity of the simulation process also requires adherence to physical laws. In summary, as shown in Table 12.1, the differences are reflected in the following aspects.

Data-driven generative simulation technology holds tremendous potential in the field of computational experiments. It not only promotes advancements in robotics through efficient and intelligent simulation capabilities but also offers new solutions for virtual reality, game development, and related domains. Despite its certain limitations in terms of data dependency and interpretability, its flexibility and efficiency make it a strong complement to traditional simulations. With ongoing technological progress, generative simulation is expected to achieve breakthroughs in more fields, advancing scientific research and industrial practice to new heights.

Sora-generated videos can only be guided using vague prompts, making precise control difficult to achieve. Therefore, it serves more as a video generation tool rather than a counterfactual reasoning tool capable of accurately answering "what–if" questions. Since OpenAI has not disclosed the extent of differences between Sora's generated samples and its training data, it is currently difficult to comprehensively evaluate its generalization ability. Despite Sora's powerful capabilities, it still faces issues in simulating complex physical phenomena, understanding specific causal relationships, handling spatial details, and accurately describing events that change over time.

1) ***Correlation vs. causality. Statistical correlation derived from probability cannot precisely express the causality of physical laws, and the contextual relevance of natural language cannot achieve the precision of partial differential equations.***

   Sora cannot accurately express spatiotemporal causal relationships between different tokens. For instance, when a grandmother blows out birthday candles, the flames remain perfectly still. While the visuals are realistic and token transitions are smooth, the physical causality—such as the air blown affecting the flame—fails to manifest. This implies that the Transformer architecture can express statistical correlations between tokens but cannot accurately model physical causality. Although Transformers can to some extent understand and generate natural language, natural language itself is insufficient for precisely expressing physical laws, which currently can only be rigorously described by partial differential equations. This reflects an inherent limitation of probabilistic world models.

Table 12.1 Differences Between Generative Simulation and Traditional Simulation.

Dimension	Generative Simulation	Traditional Simulation
Core Principle	**Data-driven, based on generative models.** Relies on deep learning techniques to approximate the distribution of real data and generate samples accordingly. Suitable for complex, nonlinear systems, especially those that are difficult to model using traditional mechanistic approaches.	**Knowledge-driven, based on rule-based and physical models.** Builds executable models through in-depth understanding of mechanisms, often from physics or expert knowledge. Emphasizes explanatory logic and predictive accuracy, suitable for systems with clear structure and well-defined mechanisms.
Data Dependence	**Highly dependent on data quality and volume.** If the data is insufficient or of poor quality, performance may degrade significantly, or even produce results that deviate from reality. In addition, data diversity is critical for generalization.	**More reliant on theoretical assumptions and mechanistic rationality of systems, rather than direct dependence on data.** In some scenarios, models can be built without much empirical data, as long as theoretical mechanisms are sound.
Model Transparency	**Black-box and hard to interpret.** Often uses deep learning to automatically learn models from raw data, capturing statistical laws and internal distributions. The modeling process is usually automated and data-driven, rather than manually coded.	**White-box and interpretable.** Models are built and calibrated manually, requiring clear understanding of mechanisms. Calibration and interpretation of each model parameter is possible.
Applicable Scenarios	**Complex, nonlinear, hard-to-model systems.** For instance, through generated virtual traffic scenarios, it can help test robustness and safety of autonomous systems without real-world high-cost testing.	**Rule-based, structured systems.** For example, structural simulations in engineering or market simulations in economics can be implemented clearly and accurately.
Computational Complexity	**Expensive training, high cost.** Requires large-scale high-dimensional data (images, videos, multimodal). Usually consumes massive computing resources. Once trained, can quickly generate samples at low marginal cost.	**Computation based on explicit models.** In well-understood systems, may reduce complexity through simplification. However, high accuracy often increases computational demands.

Table 12.1 (Continued)

Dimension	Generative Simulation	Traditional Simulation
Result Accuracy	**Highly dependent on data quality, less robust to distribution shifts.** If training data is biased or unrepresentative, generated results may struggle with rare or edge cases, lacking social generalizability.	**Model-based inference, more reliable.** If the model is well-established, results can be quite precise. In extreme cases, limitations due to model assumptions must be carefully handled.
Flexibility	**High.** Learns from diverse data distributions and generalizes across variations. Can retrain with new data or fine-tune to adapt to novel inputs.	**Low.** Typically designed for a fixed system structure. When new requirements emerge, entire model may need to be rebuilt. Often performs poorly under new or rapidly changing situations.

2) ***Locally plausible but globally absurd. While Transformers can learn connection probabilities between adjacent spatiotemporal tokens, global coherence requires higher-level mathematical theories or deeper scientific and cultural knowledge—something current Transformers cannot truly comprehend.***
Sora can stitch together neighboring tokens plausibly, yet the overall video may violate common sense, lacking global consistency. For example: a chair floats mid-air. If the viewer focuses only on the central part of the screen, the video seems reasonable. Connections between token intervals are smooth and continuous. But globally, the chair defies gravity—something that contradicts everyday experience. This "locally coherent but globally absurd" outcome means that the Transformer has learned local token transition probabilities but lacks broader contextual awareness. In this case, the global notion is gravity—it's invisible in local frames, but omnipresent in the overall scene.
3) ***Absence of commonsense knowledge. Sora overlooks critical transitional states in physical processes. This is partly due to the scarcity of transitional samples and partly because diffusion models blur the boundaries of the steady-state data manifold, erasing the existence of critical states and resulting in abrupt transitions between different steady states.***
Current Transformers can learn local context but struggle with broader global context. This "global" context might be the gravitational field in physics, ergonomics in human factors, or species taxonomy in biology. Although every

training video implicitly contains global concepts, the tokenization process segments them, preserving only local token linkages. This results in outputs that are locally reasonable but globally flawed. For example, in Sora's generated "running treadmill in the wrong direction" video, each local segment appears reasonable, and token transitions seem natural, but the overall video is absurd—the treadmill and the runner move in opposite directions. This violates basic ergonomics, revealing a disconnect between local and global understanding.

12.6 Chapter Summary

In recent years, AI4SS has become a key driving force for breaking through traditional research paradigms. With the rapid rise of data-intensive computing capabilities and the growing demand for complex system modeling, computational experimentation—serving as the core method linking the real world with digital twins—is driving paradigm shifts in fields such as society, finance, economics, and military studies. However, complex social problems often involve high-dimensional nonlinear systems, multiscale dynamic evolution, and coupled mechanisms of uncertainty. Traditional modeling methods face inherent trade-offs between realism, interpretability, and computational efficiency. Meanwhile, LLM-powered computational experiments continue to face three core constraints: persistent gaps in scientific abstraction, discontinuities in causal reasoning, and limited generalizability from virtual to real-world settings. This chapter outlines a roadmap for the future development of computational experimentation in the AI4SS context. At the modeling level, it is necessary to address the scientific abstraction paradox in synthetic data generation by large-model agents, balancing modeling complexity with realism. At the experimental design level, the focus is on integrating causal inference frameworks with virtual experimental procedures to establish interpretable reasoning mechanisms. Finally, at the validation level, the challenge is to overcome the extrapolation barrier between virtual conclusions and real-world applications, particularly addressing how inherent LLM biases affect the credibility of experimental results.

References

1 Wang, X., Xue, X., Yan, R. et al. (2024). Sora for intelligent vehicles: a step from constraint-based simulation to artificiofactual experiments through dynamic visualization. *IEEE Transactions on Intelligent Vehicles* 9 (3): 4249–4253.

2 Xu, R., Sun, Y., Ren, M. et al. (2024). AI for social science and social science of AI: a survey. *Information Processing & Management* 61 (3): 103665.

3 Xue, X., Xiangning, Y., and Wang, F.-Y. (2023). ChatGPT chats on computational experiments: from interactive intelligence to imaginative intelligence for design of artificial societies and optimization of foundational models. *IEEE/CAA Journal of Automatica Sinica* 10 (6): 1357–1360.

4 Guo, T., Chen, X., Wang, Y. et al. (2024). Large language model based multi-agents: a survey of progress and challenges. arXiv preprint arXiv:2402.01680.

5 Wang, L., Ma, C., Feng, X. et al. (2024). A survey on large language model based autonomous agents. *Frontiers of Computer Science* 18 (6): 186345.

6 Ma, Q., Xue, X., Zhou, D. et al. (2024). Computational experiments meet large language model based agents: a survey and perspective. arXiv preprint arXiv:2402.00262.

7 Xi, Z., Chen, W., Guo, X. et al. (2025). The rise and potential of large language model based agents: a survey. arXiv 2023. arXiv preprint arXiv:2309.07864, 10.

8 Aher, G.V., Arriaga, R.I., and Kalai, A.T. (2023). Using large language models to simulate multiple humans and replicate human subject studies. *International Conference on Machine Learning*, pp. 337–371. PMLR.

9 Liu, B., Jiang, Y., Zhang, X. et al. (2023). Llm+ p: empowering large language models with optimal planning proficiency. arXiv preprint arXiv:2304.11477.

10 György Buzsáki, M.D. (2019). *The Brain From Inside Out*. Oxford University Press.

11 Liu, B., Li, X., Zhang, J. et al. (2025). Advances and challenges in foundation agents: from brain-inspired intelligence to evolutionary, collaborative, and safe systems. arXiv preprint arXiv:2504.01990.

12 Yan, M., Li, R., Zhang, H. et al. (2023). Larp: language-agent role play for open-world games. arXiv preprint arXiv:2312.17653.

13 Wang, Q., Tang, Z., and He, B. (2025). From ChatGPT to DeepSeek: can LLMs simulate humanity? arXiv preprint arXiv:2502.18210.

14 Xie, C., Chen, C., Jia, F. et al. (2024). Can large language model agents simulate human trust behavior? *Advances in Neural Information Processing Systems* 37: 15674–15729.

15 Huang, Y., Yuan, Z., Zhou, Y. et al. (2024). Social science meets llms: how reliable are large language models in social simulations? arXiv preprint arXiv:2410.23426.

16 Baolin, S. (2025). The ethical framework for value alignment in large-scale artificial intelligence models. *Studies in Ethics* 85 (3): 94.

A

Appendix

Chapter A.1

This book is the culmination of the efforts of many researchers in the field of computational experiments, and we would like to thank them for their significant contributions to the development of computational experiment method. To facilitate further exploration, Table A.1.1 lists key classical literature corresponding to each chapter, providing readers with direct references to delve into the technical details.

Table A.1.1 Classical Literature Related to Each Chapter.

Chapter	Related Classical Literature	Remark
1	• Wang, F.Y. (2004). Computational experiment methods and the analysis of complex system behavior and decision evaluation. *Journal of System Simulation* 05: 893–897. • Wang, F.Y. (2004). Computational theory and methods for complex system research. *China Basic Science* 5: 5–12. • Wang, F.Y. (2007). Toward a paradigm shift in social computing: the ACP approach. *IEEE Intelligent Systems* 22 (5): 65–67. • Wang, F.Y. (2010). Parallel control and management for intelligent transportation systems: concepts, architectures, and applications. *IEEE Transactions on Intelligent Transportation Systems* 11 (3): 630–638.	A pioneering work on computational experiments and ACP method.
	• Michel, M. (2017). *Complexity (First Impulse Series: Comprehensive Series)* (trans. L. Tang). Changsha: Hunan Science and Technology Press.	A book on complexity science, many of which contain amazing ideas.

(Continued)

Table A.1.1 (Continued)

Chapter	Related Classical Literature	Remark
	• Wang, F.Y. (2010). The emergence of intelligent enterprises: from CPS to CPSS. *IEEE Intelligent Systems* 25 (4): 85–88. • Xue, X., Li, G., Zhou, D. et al. (2022). The research roadmap of service ecosystem: a crowd intelligence perspective. *International Journal of Crowd Science* 6 (4): 195–222. • Lazer, D., Pentland, A., and Adamic, L. (2009). Social science. Computational social science. *Science* 323 (5915): 721–723. • Lazer, D.M.J., Pentland, A., Watts, D.J. et al. (2020). Computational social science: obstacles and opportunities. *Science* 369 (6507): 1060–1062. • Claudio, C.R. (2014). *Introduction to Computational Social Science-Principles and Applications*. London: Springer-Verlag.	A comprehensive discussion on the complexity of the intelligent society from the perspective of "cyber–physical–social" system (CPSS). The book systematically explains the causes and laws of complexity in the social field.
2	• Law, A.M., Kelton, W.D., and Kelton, W.D. (2007). *Simulation Modeling and Analysis*. New York: McGraw-Hill. • Gilbert, N. and Troitzsch, K.G. (2005). *Simulation for the Social Scientist*, 2e. London: Open University Press.	Computer simulation and complex system.
3	• Pearl, J. and Mackenzie, D. (2018). *Why: The New Science of Cause and Effect*. New York: Basic Books. • Wang, X., Xue, X., Yan, R. et al. (2024). Sora for intelligent vehicles: a step from constraint-based simulation to artificiofactual experiments through dynamic visualization. *IEEE Transactions on Intelligent Vehicles* 9 (3): 4249–4253.	A specific roadmap is provided for experiment method and causal inference, which is very valuable for beginners of the method.
4	• Xue, X., Yu, X., Zhou, D. et al. (2023). Computational experiments: past, present and perspective. *Acta Automatica Sinica* 49 (2): 246–271. • Wen, D., Yuan, Y., and Li, X.R. (2012). Artificial societies, computational experiments, and parallel systems: an investigation on a computational theory for complex socioeconomic systems. *IEEE Transactions on Services Computing* 6 (2): 177–185. • Xue, X., Shen, Y., Yu, X. et al. (2023). Computational experiments: a new analysis method for cyber-physical-social systems. *IEEE Transactions on Systems, Man, and Cybernetics: Systems* 54 (2): 813–826.	Representative papers that systematically discuss computational experiment method.

Table A.1.1 (Continued)

Chapter	Related Classical Literature	Remark
5	• Xue, X., Wang, S., Zhang, L. et al. (2019). Social learning evolution (SLE): computational experiment-based modeling framework of social manufacturing. *IEEE Transactions on Industrial Informatics* 15 (6): 3343–3355. • Zhou, D., Xue, X., and Zhou, Z. (2022). SLE2: the improved social learning evolution model of cloud manufacturing service ecosystem. *IEEE Transactions on Industrial Informatics* 18 (12): 9017–9026. • Zhou, D., Xue, X., Lu, X. et al. (2024). A hierarchical model for complex adaptive system: from generative agent to AI society. *ACM Transactions on Autonomous and Adaptive Systems* https://doi.org/10.1145/3686802.	The value of the SLE series of models lies in: it provides a hierarchical template for modeling social systems; on the other hand, its customized nature links computational experiments with various intelligent algorithms.
6	• Xue, X., Yu, X., Zhou, D. et al. (2024). Computational experiments for complex social systems, part III: the docking of domain models. *IEEE Transactions on Computational Social Systems* 11 (2): 1766–1780. • Xue, X., Yu, X., Zhou, D. et al. (2024). Computational experiments for complex social systems: integrated design of experiment system. *IEEE/CAA Journal of Automatica Sinica* 11 (5): 1175–1189.	Computational experiment step 2: Clarify how the artificial social model is connected to specific domain scenarios.
7	• Xue, X., Zhou, D., Yu, X. et al. (2024). Computational experiments for complex social systems: experiment design and generative explanation. *IEEE/CAA Journal of Automatica Sinica* 11 (4): 1022–1038.	Computational experiment step 3: Demonstrate how to reduce experimental costs through experimental design.
8	• Yu, X., Shen, Y., Xue, X. et al. (2025). Computational experiments for complex social systems: experiment analysis and causal inference. *IEEE/CAA Journal of Automatica Sinica* (Under Review). • Manzo, G. (2022). *Agent-based Models and Causal Inference*. John Wiley & Sons.	Computational experiment step 4: Demonstrate how to make causal inferences through computational experiments.
9	• Lu, M., Chen, S., Xue, X. et al. (2021). Computational experiments for complex social systems—part II: the evaluation of computational models. *IEEE Transactions on Computational Social Systems* 9 (4): 1224–1236.	Computational experiment step 5: Explain how to calibrate and verify computational experimental results.

(Continued)

Table A.1.1 (Continued)

Chapter	Related Classical Literature	Remark
10	• Park, J.S., O'Brien, J., Cai, C.J. et al. (2023). Generative agents: interactive simulacra of human behavior. *Proceedings of the 36th Annual ACM Symposium on User Interface Software and Technology*, pp. 1–22. • Hua, W., Fan, L., Li, L. et al. (2023). War and peace (waragent): large language model-based multi-agent simulation of world wars. arXiv preprint arXiv:2311.17227.	Representative work of LLM agent in social simulation.
11	• Chen, H., Ji, W., Xu, L. et al. (2023). Multi-agent consensus seeking via large language models. arXiv preprint arXiv:2310.20151.	Representative work of LLM agent in workflow applications.
12	• Lavin, A., Krakauer, D., Zenil, H. et al. (2021). Simulation intelligence: towards a new generation of scientific methods. arXiv preprint arXiv:2112.03235.	Some thoughts on the development roadmap of computational experiments.

Chapter A.2

Table A.2.1 Common Tools in Descriptive Statistics.

Category	Name	Explanation	Common Statistical Measures
Descriptive Statistics	Central tendency	Describes the tendency of data to cluster around a typical or representative value.	Mean: The total sum of all data divided by the number of data points. Median: The middle value when data are sorted. Mode: The value that occurs most frequently in the data.
	Variability	Measures the degree of dispersion in the data, i.e., how different the data points are.	Range: Difference between the maximum and minimum values. Variance: The average of the squared differences from the mean. Standard Deviation: The square root of the variance.
	Distribution	Describes how data are distributed across different values.	Histograms and box plots can help us visualize the distribution of the data.

Table A.2.2 Core Concepts in Statistics and Probability Theory.

Term	Explanation
Sample Space	Usually denoted by capital letter S, it includes all possible outcomes of an experiment. For example, when flipping a coin, the sample space can be {Heads, Tails}, indicating the two possible results.
Event	A subset of the sample space that includes one or more outcomes of interest. For example, in the context of flipping a coin, an event can include a single outcome (e.g., "Heads on first flip") or a set of outcomes (e.g., "Heads on first or second flip").
Random Variable	A random variable is a function defined on the sample space that maps outcomes to real numbers. It quantifies the results of a random phenomenon and allows us to perform mathematical analysis on them.
Mutually Exclusive and Independent Events	Events A and B are mutually exclusive if they cannot occur at the same time. Events A and B are independent if the occurrence of one does not affect the probability of the other.
Conditional Probability	Conditional probability expresses the probability of one event occurring given that another event has occurred. It is usually denoted as P(A|B), meaning the probability of event A occurring given that event B has occurred.

(Continued)

Table A.2.2 (Continued)

Term	Explanation
Bayes' Theorem	Bayes' theorem is a method of inference based on conditional probability. It allows us to update the probability of an event based on new information. Bayes' theorem is widely used in statistics and machine learning (see Ref. [10] in Chapter 2).
Expectation	Expectation is the average value of a random variable and represents the theoretical average of the outcomes in repeated independent trials. Denoted by $E(X)$, where X is the random variable.
Variance	Expectation is the average value of a random variable and represents the theoretical average of the outcomes in repeated independent trials. Denoted by $E(X)$, where X is the random variable.
Law of Large Numbers	As the number of observations increases, the sample statistics (such as mean) converge to the population statistics. This means that relative frequencies approach theoretical probabilities as the number of trials increases, thus validating probability models through empirical experiments.

Table A.2.3 Typical Probability Distribution Functions.

Category	Name	Explanation	Image
Discrete Probability Distribution	Bernoulli distribution	Describes the case where a random variable has only two possible values (usually success and failure). For the random variable X, the PMF is: $P(X = k) = kp + (1 - k)(1 - p)$ Here, p represents the probability of success, and $1 - p$ represents the probability of failure.	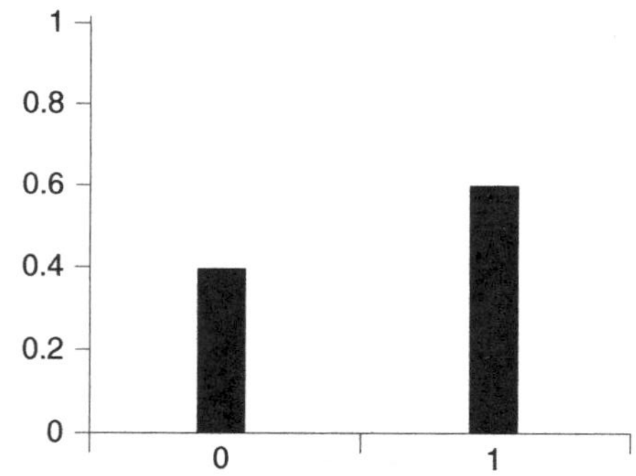
	Binomial distribution	Describes the number of successes in n independent Bernoulli trials, with the probability of success in each trial being p. For the random variable X, the PMF is: $P(X = k) = C_n^k * p^k * (1 - p)^{n-k}$ Here, k is the number of successes, n is the total number of trials, p is the probability of success in a single trial, and $C(n, k)$ denotes the binomial coefficient (n choose k).	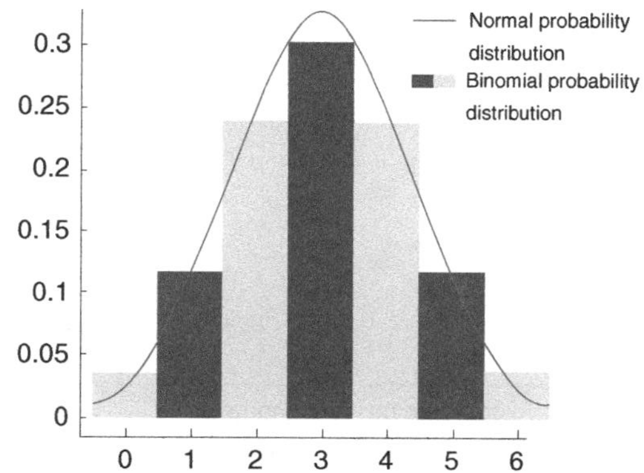

(Continued)

Table A.2.3 (Continued)

Category	Name	Explanation	Image
	Poisson distribution	Describes the number of events occurring in a unit time or unit space, where the events occur independently and at a constant average rate. For the random variable X, the PMF is: $P(X = k) = (\lambda^k * e^{-\lambda})/k!$ Here, k denotes the number of events, λ is the average number of events occurring in the given time or space unit, e is Euler's number (approximately 2.71828), and $k!$ is the factorial of k.	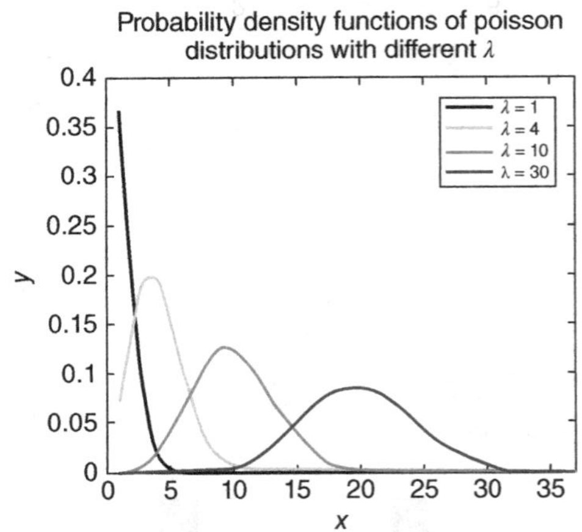
Continuous Probability Distribution	Normal distribution	The normal distribution, also known as the Gaussian distribution, is a statistical distribution that appears in many natural phenomena. For a random variable X, the PDF is as follows: $f(x) = \frac{1}{\sqrt{2\pi\sigma^2}} e^{-\frac{(x-\mu)^2}{2\sigma^2}}$ where x represents the value taken by the random variable, μ is the mean (center of the distribution), σ is the standard deviation (measure of spread), π is the circular constant, and e is the base of the natural logarithm.	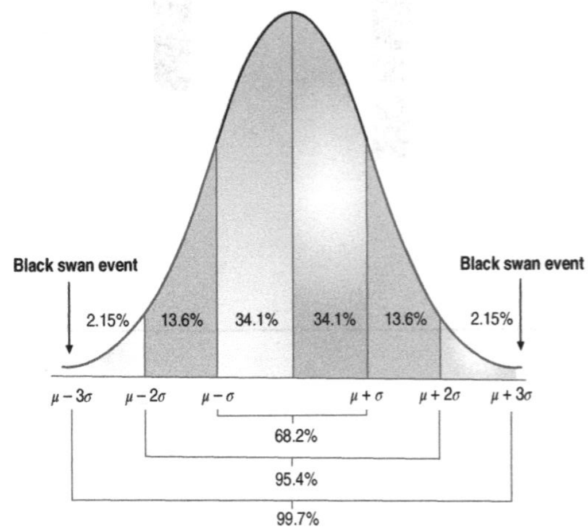

<table>
<tr>
<td>Exponential distribution</td>
<td>The exponential distribution describes the time intervals between independent random events, commonly used to model waiting times or lifespans. For a random variable X, the PDF is as follows:
$f(x) = \lambda * e^{-\lambda x}$
where x is the value taken by the random variable, and λ is the rate parameter, representing the reciprocal of the event occurrence rate.</td>
<td>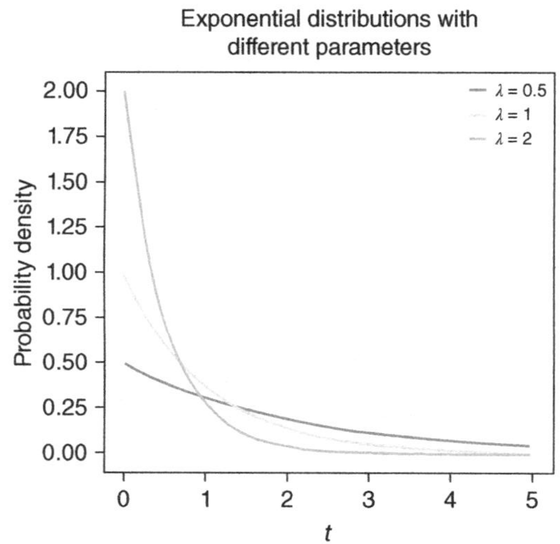
</td>
</tr>
<tr>
<td>Uniform distribution</td>
<td>Describes a random variable that can take any value equally likely within a specified range, with no specific center or bias. For a random variable X, the PDF is as follows:
$f(x) = \frac{1}{b-a}$, for $a \leq x \leq b$
where a and b are the lower and upper bounds of the interval, respectively. Within the interval a, b, the PDF is a constant, equal to $\frac{1}{b-a}$, meaning each value has equal probability of occurring.</td>
<td>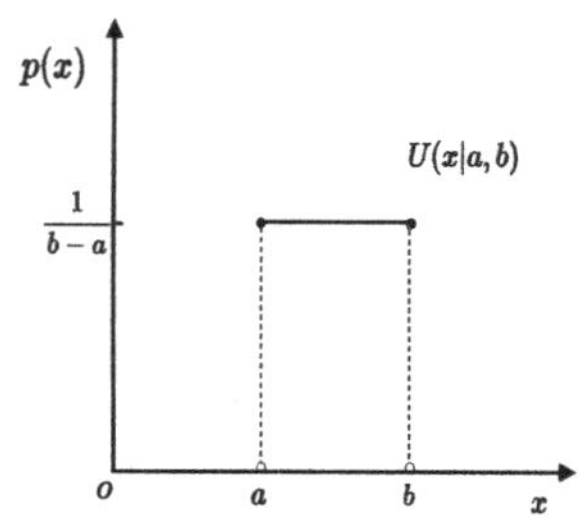
</td>
</tr>
</table>

Table A.2.4 Common Tools in Inferential Statistics.

Category	Name	Explanation	Common Statistics
Inferential Statistics	Sampling	Selecting a representative subset (sample) from the population, then analyzing the sample to draw conclusions about the population.	Mean, standard deviation, sample size
	Confidence interval	A statistical tool used to express the uncertainty of an estimated result.	Estimate ± confidence interval
	Hypothesis testing	A method to determine whether there are significant differences between two or more groups.	*t*-value, *p*-value, significance level α

Table A.2.5 Core Concepts of Inferential Statistics.

Term	Explanation	Example
Population	The entire set of objects under study. It is usually large and may be impossible to observe or measure in full.	If you want to understand the average tuition expenses of college students nationwide, then all college students nationwide constitute the population.
Sample	A subset of elements taken from the population, used to represent the whole in statistical analysis. The selection of a sample should be random to ensure representativeness.	If it's not possible to survey all college students nationwide, you can randomly select a portion of them to form a sample, and then conduct a survey on that sample.
Parameter	A fixed value that describes a characteristic of the population, such as the mean or standard deviation. It is constant for a given population.	If you want to know the average tuition expenses of all college students, then the "average tuition" is a parameter.
Statistic	A numerical measure that describes a characteristic of a sample. It is computed from sample data and used to estimate population parameters. Because it varies with different samples, different samples may yield different statistics.	If you calculate the average tuition from a sample of college students, that average is a statistic used to estimate the population's average tuition expense.

Table A.2.5 (Continued)

Term	Explanation	Example
Sampling Distribution	Describes the distribution of a statistic based on repeated random sampling. According to the Central Limit Theorem, when the sample size is large enough, the sampling distribution approximates a normal distribution. This is critical for statistical inference due to the useful properties of the normal distribution.	We are typically concerned with the sampling distributions of sample means, sample proportions, variances, and other statistics.
Standard Error	Measures the variability of a sample statistic from the population parameter. For sample means, the standard error is calculated using the formula: $SE = \sigma/\sqrt{n}$, where SE is the standard error of the sample mean, σ is the population standard deviation, and n is the sample size.	The standard error tells us how much the sample mean is likely to differ from the population mean.
Central Limit Theorem	Assume we have a population with a random sample, where the mean is μ, and the variance is σ^2. If we independently draw n samples from this population, as n increases, the distribution of the sample means will gradually approximate a normal distribution, with mean μ, and standard deviation $\sigma/\sqrt{n}$. One of the basic assumptions is that the samples must be independently and identically distributed (i.i.d.).	This theorem explains why the normal distribution appears so frequently in nature and the social sciences. Even if the original data are not normally distributed, the distribution of the sample means will tend to be normal, making the normal distribution a powerful and essential tool in statistical inference.

Table A.2.6 Common Resampling Methods Used in Statistics.

Name	Description	Example
Bootstrapping	A resampling method used to estimate the uncertainty of statistical distributions or parameters, especially suitable for small samples. The idea of "Bootstrap" is "pulling oneself up by one's bootstraps," which refers to repeatedly sampling from the original dataset with replacement to generate many pseudo-datasets. Typically, the steps include: first, specifying the number of samples; then drawing a bootstrap sample randomly with replacement from the original dataset; and finally, calculating the statistic (e.g., mean) for each bootstrap sample, thereby obtaining an empirical distribution of the statistic.	Suppose there is a small dataset, and bootstrapping is used to estimate the mean. 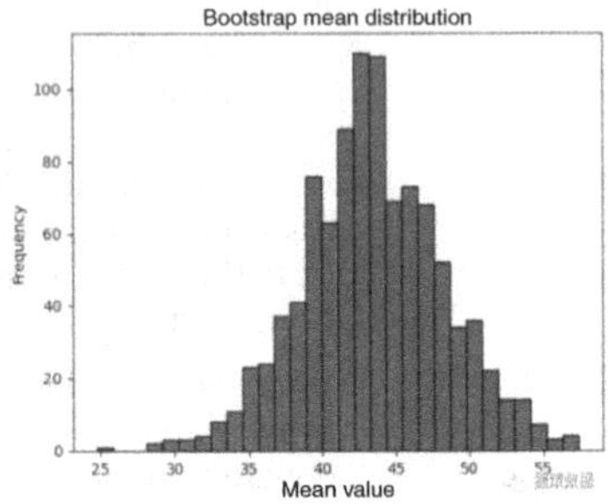
Cross-validation	A method used to evaluate model selection and model generalization ability. By splitting the dataset into different training and validation sets, the model is repeatedly trained and tested to evaluate its performance. This is especially useful when the sample size is limited. Common cross-validation methods include *k*-fold cross-validation and leave-one-out cross-validation (Leave-One-Out).	Python's scikit-learn can implement cross-validation. The cross_val_score function can be used to perform cross-validation and assess model performance.
Repeated Random Subsampling	This method randomly selects subsets from the original dataset multiple times for training and testing. Each time, a new training and testing set is randomly drawn to reduce the impact of randomness. It is commonly used for machine learning model evaluation, especially when the dataset is large or imbalanced.	Python's scikit-learn can implement repeated random subsampling. The cross_val_score function can also be used to conduct cross-validation and evaluate the model.

Table A.2.6 (Continued)

Name	Description	Example
Monte Carlo Simulation	Monte Carlo simulation involves using repeated random sampling to estimate statistical or computational quantities. It can be used to estimate probabilities or expected values when analytical solutions are difficult or impossible to obtain. For example, to estimate the probability of event A occurring, one can simulate the scenario many times, calculate the frequency of event A, and use it as an approximation of the probability of A. Monte Carlo simulation is a major class of computational experiments used for uncertainty quantification, probabilistic modeling, and algorithm design.	Randomly generate samples and test whether they fall within a circle. The proportion of samples that fall within the circle to the total number of samples can be used to estimate π. 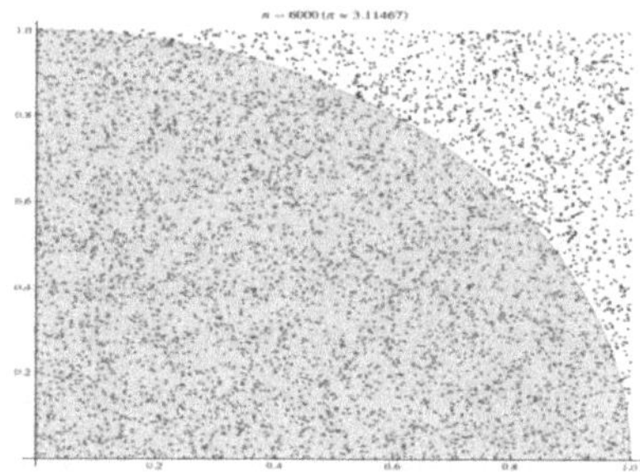

Table A.2.7 Classification of Parameter Estimation in Statistics.

Type	Subtype	Description
Point Estimation	Estimation of sample mean	If we want to estimate the population mean (μ), the most common method is to use the sample mean ($\bar{x}$) as the estimator. The formula is: $\hat{\mu} = \bar{x}$.
	Estimation of sample proportion	When we are concerned with the proportion of a certain attribute in the population (e.g., success rate, denoted as p), the proportion of successes in the sample can be used as the estimator. The formula is: $\hat{p} = \frac{\text{number of successes in the sample}}{\text{total sample size}}$
Interval Estimation	Confidence interval estimation	For the confidence interval of the population mean, we can calculate it using the sample mean, sample standard deviation, and either the t distribution or normal distribution. A common 95% confidence interval is expressed as: $\bar{x} \pm 1.96 \times \frac{s}{\sqrt{n}}$, where $\bar{x}$ is the sample mean, s is the sample standard deviation, n is the sample size, and 1.96 is the critical value corresponding to a 95% confidence level.

Table A.2.8 Common Parameter Estimation Methods in Statistics.

Name	Detailed Explanation
Maximum Likelihood Estimation (MLE)	MLE is a commonly used parameter estimation method. It seeks the parameter value that is most likely to produce the observed sample data. This method is based on the likelihood function of the sample data and aims to maximize the function to obtain the optimal parameter estimate.
Least Squares Estimation	Least squares estimation is commonly used for linear regression problems. It seeks model parameters that minimize the sum of squared errors between the predicted values and the observed values.
Bayesian Estimation	Bayesian estimation is based on Bayesian theory. It uses prior distributions and sample data to estimate the posterior distribution of parameters. This method is particularly suitable when uncertainty needs to be taken into account.
Method of Moments	The method of moments estimates parameters by equating sample moments to theoretical moments. It is a straightforward method and is especially useful when sample sizes are small.
Method of Conditional Moments	This method uses conditional moments or conditional distributions to estimate parameters. It is frequently used in economics and econometrics.
Maximum a Posteriori Estimation (MAP)	MAP estimation combines MLE and Bayesian estimation. It seeks parameter values that maximize the product of the likelihood function and the prior distribution.
Direct Value Method	In some cases, direct value methods can be used, such as Monte Carlo simulation or optimization methods, to estimate parameters—especially when a closed-form solution cannot be obtained through analytical methods.

Table A.2.9 An Introduction to Key Concepts in Statistical Hypothesis Testing.

Term	Explanation
Null Hypothesis (H0)	A default assumption, opposing the alternative hypothesis (H1). It typically indicates no difference or no effect. It is a statement researchers attempt to disprove through experimentation. For example, if a drug is believed to have no effect, then the null hypothesis would assert that the drug has no therapeutic effect.
Alternative Hypothesis (H1)	The hypothesis that contradicts the null hypothesis, usually representing the presence of an effect or difference. The goal of hypothesis testing is to provide evidence to support the alternative hypothesis. For instance, an alternative hypothesis might claim that a drug has a significant therapeutic effect.
Significance Level (α)	The significance level is a threshold value used to determine whether a hypothesis should be rejected. Common values include 0.05, 0.01, or 0.001, representing a 5%, 1%, or 0.1% chance of making a Type I error (rejecting a true null hypothesis).
Test Statistic	A statistic computed from sample data, used to assess whether the null hypothesis should be rejected. Different hypothesis tests use different test statistics, such as the t-statistic or z-statistic.
p-value	The p-value is the probability of obtaining test results at least as extreme as the results observed, assuming the null hypothesis is true. A small p-value indicates that the observed data is unlikely under the null hypothesis, suggesting it should be rejected.
Rejection Region	The range of test statistic values for which the null hypothesis is rejected. If the test statistic falls within this region, the null hypothesis is rejected. The boundary of this region is determined by the significance level.
Acceptance Region	The range of test statistic values for which the null hypothesis is not rejected. If the test result falls within this region, the null hypothesis is accepted (not rejected).
Type I Error	Occurs when the null hypothesis is true but is incorrectly rejected. This error is denoted by α.
Type II Error	Occurs when the null hypothesis is false but is incorrectly accepted. This error is denoted by β, and its probability depends on sample size and effect size.
Statistical Significance	When a result is statistically significant, it means that the p-value is below the significance level, and the null hypothesis is rejected. Statistical significance indicates the presence of an effect, but it does not imply practical or substantive importance.
Effect Size	In addition to statistical significance, effect size quantifies the magnitude of the effect. A small p-value may not correspond to a large effect, especially in large samples. Effect size is context-dependent and important for understanding the practical significance of findings.

Table A.2.10 Common Hypothesis Testing Methods in Statistics.

Name	Explanation
t-Test	Used to compare whether the means of two groups are significantly different. The *t*-test has two types: independent samples *t*-test and paired samples *t*-test.
Analysis of Variance (ANOVA)	Used to compare whether the means of three or more groups are significantly different. ANOVA includes one-way ANOVA (single factor) and factorial ANOVA (multiple factors).
Chi-square Test	Used to test the association or independence between two or more categorical variables. Chi-square tests include the chi-square test of independence and chi-square goodness-of-fit test.
Correlation Analysis	Used to test the correlation between two continuous variables. Pearson correlation coefficient is used for linear correlation, and Spearman rank correlation coefficient is used for nonlinear correlation.

Chapter A.3

Table A.3.1 Major Methods in the Development of Causal Inference.

Type	Concept	Features
Randomized Controlled Trials (RCTs)	Through random assignment of subjects to treatment and control groups, RCTs eliminate confounding variables. A/B testing is a widely used real-world implementation of RCTs.	Advantage: Considered the gold standard for establishing causal relationships. Randomization balances both known and unknown confounders across groups. Limitation: In some cases, RCTs may be infeasible due to ethical or practical constraints, such as assigning people to smoke or not.
Observational Studies	By matching treatment and control groups based on their propensity scores, PSM reduces selection bias. It attempts to replicate the balance of covariates achieved through randomization. Matching is done based on the probability of receiving treatment given the observed covariates. As long as the matched samples have similar scores, they can be directly compared. The method requires two key assumptions. (1) All variables that influence both treatment and outcome must be observed and measured. (2) All covariates must be accurately included in the score model.	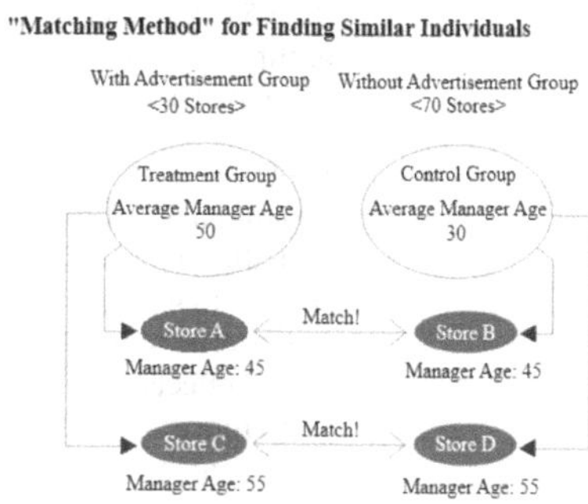
	Uses variables that are correlated with the treatment but uncorrelated with the error term in the outcome equation to identify causal effects. The diagram illustrates the IV affecting the treatment, but not directly the outcome. For a variable to be a valid instrument, it must satisfy the following. (1) Relevance: The IV must be correlated with the treatment. (2) Exclusion Restriction: The IV must not affect the outcome directly, only through the treatment. Additionally, there must be no "fourth variable" that simultaneously affects both the instrument and the outcome.	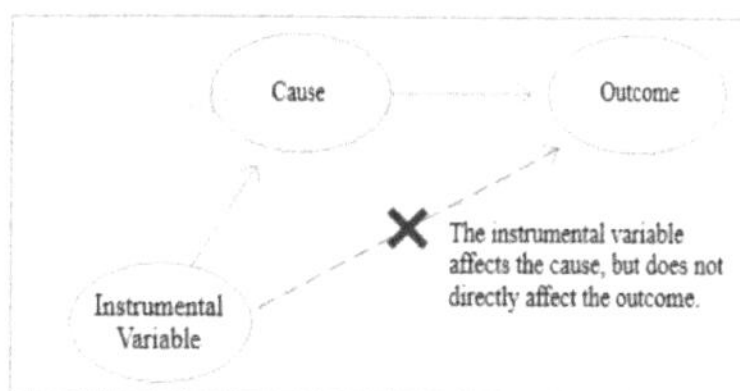 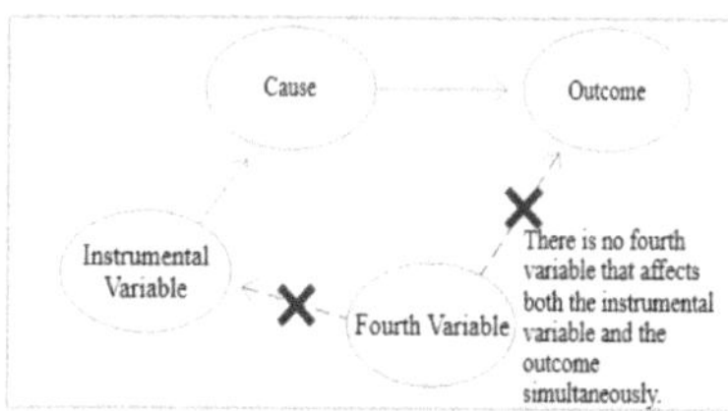

(Continued)

Table A.3.1 (Continued)

<table>
<tr><th>Type</th><th>Concept</th><th>Features</th></tr>
<tr><td></td><td>Leverages a cutoff point in a continuous variable (forcing variable) to infer causal effects (a form of counterfactual reasoning). RDD assumes that observations just above and below the cutoff are comparable, except for the treatment assignment. For example, suppose stores with more than 50 employees are eligible to run advertisements (treatment group), while stores with fewer than 50 employees are not (control group). The causal effect can be estimated around this cutoff.</td><td>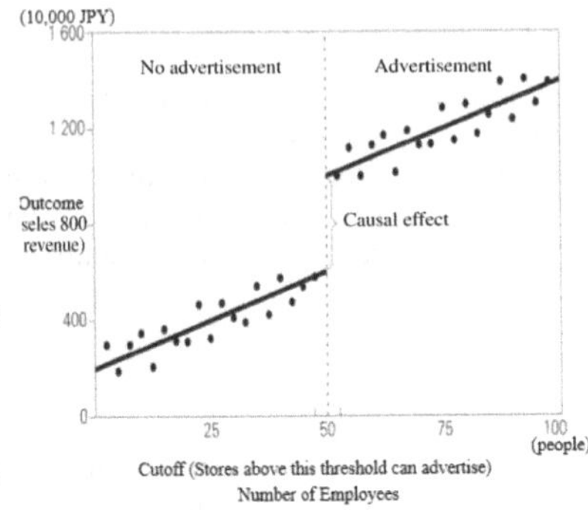
</td></tr>
<tr><td></td><td>Difference-in-differences (DiD):

By comparing the change before and after treatment in the treatment group and control group (counterfactual), DiD identifies causal effects. One dimension is the "difference" before and after the treatment, and the other is the "difference" in the control group over the same period. The difference between these two "differences" gives the treatment effect—this is called "difference-in-differences." This method helps control for natural changes over time (i.e., "trends").</td><td>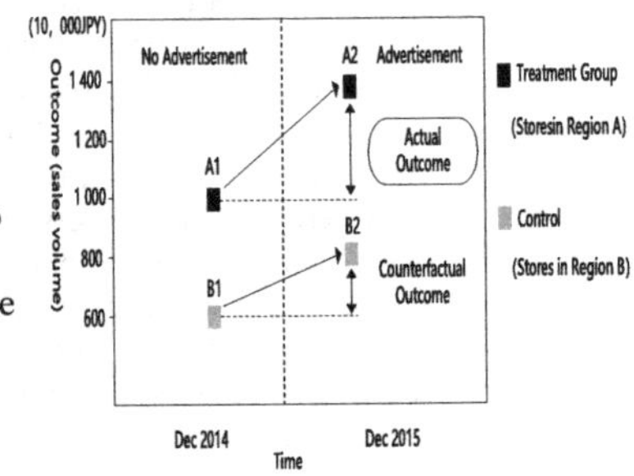
</td></tr>
</table>

Table A.3.1 (Continued)

Type	Concept	Features
	This method follows the logic of "finding a reference" (counterfactual) by constructing a hypothetical control group. The method simulates the outcome that would have occurred in the absence of the treatment and constructs a synthetic control group that approximates the treated unit. For example, if a certain store was affected by a pandemic, we could use a weighted combination of similar stores in other unaffected regions (e.g., Store A, Store B, Store C) to form a synthetic control unit. The difference between the actual store's sales during the pandemic and the synthetic control's predicted sales gives an estimate of the causal impact of the pandemic.	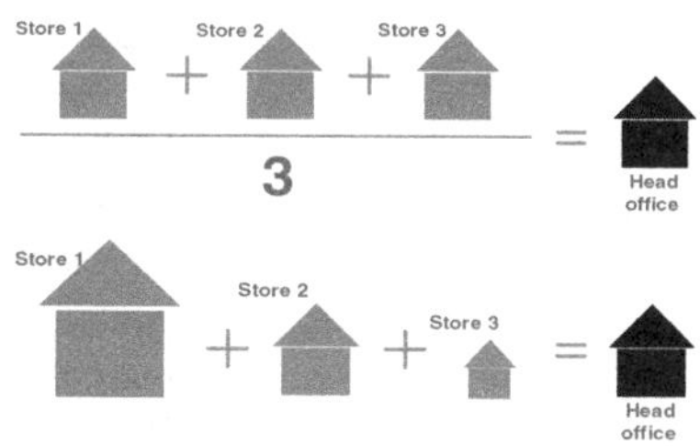
Structural Equation Modeling (SEM)	SEM constructs models that include multiple variables and their causal relationships. These models estimate both direct and indirect causal mechanisms.	Advantage: Capable of modeling complex causal relationships and multiple latent variables. Limitation: Requires strong modeling assumptions, and results depend heavily on model specification and variable measurement.

Chapter A.5

Table A.5.1 Characteristic Attributes of Manual Vehicle Agent.

System Variables	Experimental Settings
ID	Each manual vehicle agent has a unique ID.
OperationCost	Daily operation cost (fund consumption per cycle), with a cost per cycle ratio set to 2.
$DisCost_R$	Order acquisition cost (assets consumed when searching for orders), calculated as $DisCost_R = k * x$, (x: *agent's moving distance*, $x > 0$; k: *cost per step, set to* 2).
Vision	Local search range (larger range → higher probability of finding orders/partners)
Speed	Set to 1 (faster speed → easier to obtain orders).
OrderAbility	Order processing capability, ranging from [2, 10].
M_t	Memory sequence (clears part of memory via forgetting vector within a cycle; forgetting vector f_t: [0.3, 0.7]; update vector i_t: [0.5, 0.9]).
ThreBirth	Reproduction threshold (varies by agent; assets exceeding this threshold → parent agent generates offspring), set to 50.
$P_{Mutation}$	Mutation threshold, set to 0.7.
Interval	Replication interval (time between two consecutive replications), set to 20.

Table A.5.2 State Attributes of Manual Vehicle Agent.

System Variables	Experimental Settings
$Location_S(t)$	Agent's position, represented by a two-dimensional coordinate; agents continuously change position to find on-demand resources. Each agent has a unique two-dimensional coordinate as its position, adjusted dynamically to search for on-demand resources.
$Capital_S(t)$	Determines the agent's survival (exit if assets < death threshold) and reproduction (generate offspring if assets > reproduction threshold). $Capital(t+1) =$ $Capital(t) + \sum_{i=1}^{Num} Profit(i) - DisCost(t) - OrderAbility_t - OperationCost$ where *Num*: Number of orders found by the agent in this cycle; *Profit*(t): Profit from processing order i; *OperationCost*: Daily operation cost; *DisCost*(t): Moving distance cost at time t; and $OrderAbility_t$: Cost of order processing at time t (0 if the agent is idle).
$\text{Inventory}_{s(t)}$	Records the number of orders within the agent's current vision, categorized by demand type (e.g., raw material demand, finished product demand).
$Defict_S(t)$	Number of consecutive cycles where the agent's assets incur losses, affecting its decision-making. Increases by 1 for each consecutive cycle of asset loss; resets to 0 if assets do not decrease; higher values enhance the agent's tendency to seek cooperation. The calculation formula is as follows: $if\ (Capital_S(t) - Capital_S(t-1) < 0)$ $Defict_S(t) = Defict_S(t-1) + 1;$ $else\ Defict_S(t) = 0$
CoPre_R	The agent's tendency to cooperate with other partners. Correlates with *Defict_(t)*: stronger loss cycles lead to a higher willingness to cooperate.

Table A.5.3 Perceived Events of Manual Vehicle Agent.

System Variables	Experimental Settings
$CustomerInView_E(t)$	Order distribution within the agent's vision directly affects its funding sources (more orders in vision → higher potential funding).
$Partner_E(t)$	Partners within the agent's vision influence its partner selection (more available partners → wider selection range).
$Supplier_E(t)$	Autonomous vehicle agents within the agent's vision are invokable, affecting order completion efficiency and cost (invoking autonomous agents → higher efficiency, lower cost).

Table A.5.4 Behaviors of Manual Vehicle Agent.

Behaviors	Experimental Settings
Movement	Searches for orders in the environment: moves toward "required resources" if detected, else toward a random point; movement consumes funds. Fund consumption is calculated based on moving distance (consistent with $DisCost_R$ in Table A.5.8).
Order Processing	After finding "demand resources," moves to the resource position and processes orders.
Reproduction and Extinction	Reproduction: Generates same-type offspring if assets reach the threshold (set to 300); offspring attributes inherit/mutate from parents; reproduction consumes assets. Extinction: Exits (dies) if assets reach 0. Assets after reproduction: $Capital_S(t+1) = Capital_S(t) - Offspring\_self.\ Capital_S(t)$. $Offspring\_self.\ Capital_S(t)$: Initial assets given to offspring.
Genetic Operations	Involves inheritance, crossover, mutation: • Inheritance: Offspring inherits parent attributes if random value (0–1) < genetic threshold. • Crossover/mutation: New agents are created via crossover/mutation if random value > genetic threshold. Changes agent attributes/capabilities; random value range: [0,1]
Collaboration	Collaborates with in-vision partners to complete orders; profit distribution depends on effective value and investment cost. Effective value: $Value(i) = OrderAbility_i * T_{Process}$ where $OrderAbility_i$ is the order processing capability of the supply-side agent, and $T_{Process}$ is the collaboration duration.

Table A.5.5 Decision-Making Mechanism of Manual Vehicle Agent.

Decisions	Experimental Settings
Core decision mode	Mainly adopts the "Observe-Act" mode, where manual vehicle agents make decisions based on their current state.
Whether to seek cooperation with peer agents?	Determined by the agent's own situation: If assets decrease for consecutive cycles and a random value (0–1) exceeds the cooperation threshold, the agent searches for potential partners within its vision.
Whether to accept cooperation requests from same-type agents?	Evaluates the received request (profit/required resources) against its own conditions (sufficient required resources): Accepts the request if conditions are met and a random probability value exceeds the cooperation threshold.
Whether to request cooperation with autonomous vehicle agents?	Compares cost–benefit of processing orders alone vs. with autonomous vehicle agents: Invokes in-vision autonomous vehicle agents for cooperation if processing alone is less cost-effective or efficient.

Table A.5.6 Main Settings of Mechanical Vehicle Agents.

System Variables	**Experimental Settings**
ID	Each mechanical vehicle agent has a unique ID.
$Location_A$	The location of the mechanical vehicle agent is set by the platform and remains unchanged.
$Vision_A$	The order scope that the mechanical vehicle agent can autonomously process without being invoked is set to 10.
Processig Power	The order processing power of the mechanical vehicle agent is set to 2.
Action	The mechanical vehicle agent perceives and processes orders within its sensing radius.

Table A.5.7 Behavioral Process of Vehicle Agents.

Input: Current agent information list;
Output: Updated agent information list.

L1 (Autonomous characteristics) considered in scenarios 1–4	**Step 1**: Manual vehicle agents acquire order and neighbor information within their vision, records the status of orders at the current location in a storage sequence, and long-unprocessed orders (within a specified time) will be lost.
	Step 2: Based on order status, manual vehicle agents select the unlocked order with the highest profit (within its economic capacity) from their vision, lock the selected order, and update their own order status simultaneously.
	Step 3: Manual vehicle agents determine their next movement location according to the locked order's location, and updates their information sequence and location information.
	Step 4: Calculate the assets of each manual vehicle agent to decide whether it reproduces or exits (dies). If it operates normally, proceed to the next cycle.
	Step 5: Manual vehicle agents update their information list, decision-making mechanism, and rule sequence in the behavior set.
L2 (Evolutionary characteristics) considered in scenarios 2–4	**Step 6**: Manual vehicle agents acquire orders within their vision, lock pending orders according to the nearest distance principle, and process orders based on their processing capacity.
	Step 7: Locations with more orders (from high-asset manual vehicle agents) are used as learning knowledge for agent location data; low-asset agents obtain location information of resource-rich areas through learning as decision-making references.
L3 (Interactive characteristics) considered in scenarios 3 and 4	**Step 8**: After locking an order, manual vehicle agents select suitable partners based on funds and collaboration costs, and consider whether there are autonomous vehicle agents nearby to decide on collaboration. If the order remains unprocessed within a certain time, it will be unlocked.
L4 (Emergent characteristics) considered in scenario 4	**Step 9**: Different tax rates affect the decision-making of manual vehicle agents, thus influencing the system. The platform induces agents to choose areas with high-order loss rates by reducing order tax rates in these areas, thereby lowering the order loss rate.
	Step 10: The platform agent updates and collects data to provide a basis for the subsequent operation of the system.

Table A.5.8 Main Settings of Platform Agents.

Setting Categories	Specific Settings
Scene and Boundary	Scene scope: Defines the system boundary; Agent activity range: Set as 140×100 coordinate units.
Entity Quantity	Initial number of manual vehicle agents: 100 (changes dynamically during system evolution).
Tax Mechanism	After manual vehicle agents complete an order, they pay taxes to the platform; Platform tax rate: 2%.
Inducement Mechanism	• Unprocessed orders (beyond a specified time) disappear. • The agent activity area is evenly divided into nine regions. • Regulation rule: If the order loss rate in a region exceeds the regulation threshold α (initially 60%) within the regulation cycle *TR* (set to 10), the platform lowers the tax rate in that region to reduce the order loss rate.
Intervention Mechanism	Adjust the number and distribution locations of autonomous vehicle agents in the system.

Table A.5.9 Demand-Side Settings.

System Variables	Experimental Settings
Initial number of orders	Initial market resource (i.e., initial order quantity), set to 200.
Order generation rule	Market trends are reflected by the order generation function: $Y = N + M \cdot \sin(t)$ (where t: time cycle; N: base number of orders = 100; M: order fluctuation range = 40).
OrderProfit	Manual vehicle agents obtain profits by processing orders; profit is randomly set within [30, 100].
OrderCost	Initial cost for vehicle agents to process an order, randomly set within [10, 30].
Life_of_Orders	Orders are withdrawn if unprocessed within the survival duration; survival duration is set to 30 (time units).

Chapter A.6

Table A.6.1 Development of Virtual Worlds in the 21st Century.

Time	Work	Impact
Virtual Worlds with Mature Economic Systems		
2003	Second Life	A fully developed virtual world with 3D creation tools and a complete virtual economic system.
2009	Blue Mars	Utilized advanced graphics technology to build a more realistic 3D world.
Decentralized Virtual Worlds (2007 to Present)		
2007	Solipsis	The first open-source decentralized virtual world.
2008	Imprudence	One of the earliest open-source third-party browsers for accessing Second Life.
2009	Open Simulator	The first protocol-compliant simulation of the Second Life virtual world.
2010	Metaverse	Established an interconnected metaverse through public virtual world protocols.

Table A.6.2 Experimental System Model Library.

Name	Agent Model
Agent Structure	The structure of an agent can be described by a set of attributes related to time: R: agent's characteristic attributes; S: agent's state attributes; P: The internal states and external events perceived by the agent; D: Agent's decision-making mechanism; A: The set of behaviors of the agent; C: Constraints on the agent, including environment, other agents, and task objectives.
Agent Behavior Rules	The behavior of a single agent is driven by a predefined time-dependent behavior log. The behavior log is an editable structure that clearly defines the set of optional behaviors for the agent in each time period. The event chain of the agent is generated based on the behavior log, which involves solving two key problems: (1) forming the sequence of agent activities based on constraints of start time, duration, and end time and (2) classifying activities by priority level, including primary and subordinate activities.
Agent Learning Rules	An agent's behavior log may change under the following two scenarios: (1) The agent adjusts its daily behaviors based on its internal state and responds to emergency events and (2) External interventions forcibly change the agent's behavior, requiring the agent to re-plan its actions.
Environment Update Rules	The parameters of agents and environmental variables are updated in real time. For example: The gas concentration at different locations in the coal mine roadway changes dynamically over time; The number of miners at different working faces also changes dynamically.
Disaster Source Agent	Features include specific scenario of the disaster occurrence, magnitude of the disaster, and spread rate of the disaster.
Miner Agent	Researchers should use a suitable algorithm to generate an artificial population that matches the overall statistical characteristics of real miners. Features include initial objective, current location data, personal attributes (e.g., health status, movement speed), psychological factors (e.g., panic, stress, herd behavior), and knowledge level (e.g., familiarity with the environment, cognitive learning ability)

Table A.6.3 Experimental System Knowledge Base.

Name	Domain Knowledge
Evacuation Path Planning Rules	Evacuation path planning rules are based on the network structure of the coal mine and related constraints to achieve multiobjective path planning. For example, routes that minimize health cost or time cost provide decision-making support for developing evacuation strategies.
Disaster Source Evolution Rules	Disaster source agents use disaster evolution rules to calculate their potential diffusion area and impact range and update environmental variables accordingly. For instance, the Gaussian plume dispersion model recommended by the International Atomic Energy Agency (IAEA) is used to simulate the spread of heavy gas clouds in coal mines.
Miner Evacuation Behavior Rules	The formulation of these rules considers various factors affecting evacuees, such as miners' judgment ability, consistency in group behavior, physical health, guidance signs, and passage blockages. These rules are based on existing theoretical knowledge or statistical facts. For example, when personnel density is low (per capita area $> 2.3\,m^2$), evacuation speed is approximately 1 m/s; when density is high (per capita area $< 0.5\,m^2$), evacuation speed drops to 0.37 m/s (see Ref. [11] in Chapter 6).
Decision Analysis Rules	In multi-attribute decision-making problems, an evaluation vector matrix is established based on multi-attribute evaluations from multiple experts regarding emergency events in coal mine scenarios. Fuzzy decision analysis rules integrate qualitative and quantitative features to address complex decision-making scenarios.
Optimal Resource Scheduling Rules	These rules are used to establish multistage optimal resource scheduling equations. They can be categorized into multiobjective resource scheduling models and decision analysis models (e.g., multi-attribute group decision analysis models, fuzzy decision analysis models).

Table A.6.4 Experimental System Database.

Category	Content Description
Disaster Source Library	It includes spatial locations of disaster sources, distribution schemes, types of disaster sources, and specialized response plans.
Resource Library	It contains information about the types, quantities, and distribution locations of rescue capsules and emergency resources.
Basic Geographic Data Library	It includes spatial distributions of key facilities, geographic data, and environmental data for the coal mine. Environmental data includes tunnel layouts, work faces, equipment, and other related equipment.
Preplanning Library	It stores pre-plans, which are emergency response plans prepared in advance for various potential safety incidents in coal mines. All pre-plans are stored in standardized digital formats.
Case Library	It includes typical cases and applicable regulations. Texts, images, and videos related to typical cases are stored in digital formats.

Chapter A.7

Table A.7.1 Classical Experiment Design Methods.

Method Type	Method Name	Method Statement
Full Factorial Design	One factor design at a time	Under the condition of keeping the other $k - 1$ factors unchanged at the initial level, each factor is adjusted within its allowable range to determine the caused change in response. However, this method is limited as it examines only one factor at a time. With a large number of factors, it becomes inefficient and incapable of detecting interactions between factors.
	N^k factorial design	The experimental model consists of k factors, and the level set of each factor is $\{N_1, N_2, \dots, N_k\}$. In each complete experiment, all possible combinations of values for the k factors are studied. This method helps researchers investigate the individual effects (or main effects) of each factor and determine if there are interactions among factors.
	Grid design	A regular grid pattern is typically used to select the experimental points, with equal sampling on each level of the factors. This is suitable for single-factor or few-factors cases where interactions among factors are not considered.
	Supersaturated design	By introducing redundancy, the coverage of observed data can be increased within a limited number of experiments. These additional data can be from higher-order interaction of the original factors or virtual factors (such as center points). Compared to traditional full factorial design, supersaturated design can provide more information with fewer experiments.
Partial Factorial Design	Uniform design	Experiment design is carried out by a two-dimensional uniform design table, where each row represents an experiment point and each column represents a factor, and the levels of each factor under the experimental point are given. With the given factors and levels, a set of experimental points is selected by a reasonable sampling method to explore and reveal the relationship among factors as much as possible within a limited number of experiments. It is mainly used to fully understand the relationship between multiple factors and factor levels and usually requires a relatively large number of experiments.

Table A.7.1 (Continued)

Method Type	Method Name	Method Statement
	Orthogonal design	By selecting a set of specific experimental points, the experiment results can be used not only to study the main effects of each factor but also to effectively separate the interactions between factors. This is mainly used for quickly screening major factors and understanding their impacts, especially in cases of limited resources or time constraints. It is suitable for optimization problems and rapid screening experiments.
	Latin square design	The randomized complete block design makes the experimental block group form a relatively homogeneous experimental unit, so as to effectively control the interference factor. Latin square design is a partitioned group design with two orthogonal block variables that can locally control two interference factors in both directions. It is mainly used in experiment design to reduce interference and control interaction. By reducing interference effectively, it is used in research requiring independence between different factors.
	Fractional factor design	Based on the number of determined factors k and the expected resolution p, fewer experimental points N^{k-p} are selected to better estimate the main effect or the interaction effect of two factors. It reduces experimental complexity by omitting certain interactions and is often used to screen and identify major factors, focusing on the study of major factors.

Table A.7.2 Commonly Used Modern Experiment Design Methods.

Method Type	Method Name	Method Statement
Group Screening Design	Sequential branching design	A large number of factors are usually involved in computational experiments, so it is very important to find important factors to avoid dimensional disasters. The sequential branching method detects whether factor groups are significant by aggregating all factors into factor groups. And the factor groups with significance are tested.
	Iterative fractional factorial design	By iteratively reducing the number of factors, the large experimental space is gradually narrowed down to a smaller and manageable experimental matrix. The limitation is that each iteration can only cover a portion of the factors, so it may miss some complex interactions or minor factors. It focuses on optimizing multi-factor systems and is suitable for problems that require a deeper understanding and improvement.
	Multistep group screening design	Multiple factors were studied and screened in stages. In each stage, only one group of factors is studied. First, a smaller subset of factors is selected for study and then whether to introduce the next group of factors for further study is based on the results of the previous stage. It focuses more on screening and excluding factors and is suitable for quickly identifying initial problems.
Space Filling Design	Latin hypercube design	Latin hypercube sampling is used to select experimental points, uniformly covering the multidimensional design space to ensure that each factor level appears in the design and to comprehensively explore the combinations of factors in the experiment. It helps to capture potential relationships and variations in the multidimensional design space, especially suitable for multi-factor, multilevel cases that require the comprehensive coverage of design space to understand the relationships among factors to the maximum extent.
	Ball packing design	By changing the levels of multiple factors systematically, the complex interactions among factors are studied to optimize the response variable. It can deeply study the complex relationships and interactions among factors and is used for optimizing and improving systems or processes. It is suitable for situations that require a deep understanding of multi-factor systems and optimizing responses, often used in product development and engineering design.

Table A.7.3 Algorithm 1: Demand Environment Modeling and Extreme Environment Generation by Environment Agent (EA).

Input: Historical dataset $\boldsymbol{D}$; Domain knowledge base $\boldsymbol{K}$; Adversarial prompt set $\boldsymbol{P}$.

Output: Upper and lower bounds of the environmental distribution.

Initialize 4 empty sets, namely the semantic feature set C, the scenario set E, the credibility score set R, and the high-risk scenario set E_{high}.

for each d **in** D **do** // Extract features

  for each p **in** P **do**

    Use LLM to extract the semantic feature c from d, and add c to the set C;

    Extract the demand fluctuation trend c from d and add c to the set C to determine Y_{target}; // Extract features

  end for

end for

Extract the event logic rule set L from the domain knowledge base K through knowledge distillation;

for each c **in** C **do**

  for each l **in** L **do**

    Annotate key features for the prompt set p;

    Generate long - tail features and save them in the set F_{tail} ;

  end for

end for

$InputPrompt = \emptyset(p, F_{tail})$, $PL(e) \sim LLM(InputPrompt; Logits_{new})$, take p and F_{tail} as the LLM's input, and take Formula (7.14) as the generation criterion to generate the demand environment distribution $PL(e)$ in time period T;

Extract N independent and identically distributed samples from $PL(e)$ based on the Monte Carlo method, and each sample is a generated demand scenario e_p;

Merge all e_p into the scenario set E ($E = E \cup e_p$);

for each e **in** E **do** // Calculate credibility scores

  Calculate the credibility score r_s of each demand scenario based on Formula (7.15), and add r_s to the credibility score set R;

end for

Determine the demand scenario credibility threshold θ_{high} based on Formula (7.18) ;

for $i \leftarrow 1$ to $|E|$ **do** // Extract features

  if $r_s[i] \geq \theta_{high}$ **then**

    Add scenario $S[i]$ to the high-credibility demand scenario set E_{high};

  end if

end for

Extract the vector set V from the relevant variables of the high-credibility scenario set E_{high}.

According to the minimum and maximum values of the variables in the data vector set V, calculate the upper bound V_{max} and the lower bound V_{min} of the environment distribution;

return The calculated upper bound V_{max} and lower bound V_{min} of the demand environment distribution.

Table A.7.4 Algorithm 2: Demand Environment Modeling and Extreme Environment Generation Algorithm Implemented by Environment Agent (EA).

Input: Historical social relationship dataset D;
Output: Supplier subject collaboration network N.
Initialize sets, including the backbone relationship network S, individual set I, group set G, individual feature set I_f, group feature set G_f, and relationship set N.
for each d **in** D **do** // Classify individuals and groups
  if d is an individual **then**
    Add it to the individual set I;
    Convert text and data into the feature vector h_i of Agent i based on Formula (7.21)
      (including individuals in the group);
  else if d is a group **then**
Add it to the group set G, and convert text and data into the feature vector ;
  end if
end for
Extract the key feature semantics containing (Q, K, V) in individuals and between individuals and groups (based on Formula (7.22)) to obtain the semantic vector *Ate*, and integrate it into the "latent feature space" of Agent i to obtain the final feature representation z_i of each individual;
Combine the internal features of each groupbased on Formula (7.25), analyze and update the feature representation and selection tendency of Agent i itself, and form the feature representation g of the implicit collaborative relationship contained in the group;
for each i **in** I **do**
  for each j **in** I **where** $i \neq j$ **do**
    Calculate the relationship strength s_{ij} of agent i and j ;
    If s_{ij} meets the set threshold $\mathcal{M}_{ij}$ then
      Add (i, j, s_{ij}) to the relationship set N;
    end if
end for
  for each g in G **do**
    Calculate the relationship strength s_{ig} between them based on the features of i and g;
    If s_{ij} meets the set threshold $\mathcal{M}_{ij}$ then
      Add (i, g, s_{ig}) to the relationship set N;
    end if
  end for
for each g_1 in G **do** // Calculate relationships between groups
  for each g_2 in G where $g_1 \neq g_2$ do
    Calculate the relationship strength $s_{g_1 g_2}$ between them based on the features of g_1 and g_2;
    if $s_{g_1 g_2}$ meets the set threshold $\mathcal{M}_{g_1 g_2}$ then
      Add (g_1, g_2, $s_{g_1 g_2}$) to the relationship set N;
    end if
  end for
end for
return Generate the backbone relationship set N of agents collaboration network over time series.

Table A.7.5 Representation of Scenario Vectors in API Service Market.

Scenario Vector ε_t	Description
$e_t = (SA, Similarity)$	**SA** is the total activity value of APIs, representing the total number of Mashup calls to APIs at different time periods, indirectly indicating demand conditions. ***Similarity*** reflects changes in the annual market demand structure based on the proportional distribution of Mashup calls to various APIs.
$a_t = \begin{pmatrix} NF, WF, WE, LCC_S, \\ Reachability \end{pmatrix}$	*NF* (node fraction) is the ratio of nodes in the generated relationship graph to the original graph, reflecting node completeness. *WF* (weight fraction) is the ratio of retained edge weights to the total weight, indicating the preservation of important relationships. *WE* (weight entropy) measures edge weight distribution diversity. LCC_S (largest connected component size) indicates network integrity. ***Reachability*** shows the proportion of node pairs with at least one existing path, reflecting network connectivity.
u_t	The total number of APIs is N_{total}. Individual efficiency values are sorted in descending order, and equal-width binning defines m ecological niches ($m \approx \sqrt{N_{total}}$). The system efficiency u_t is calculated using Formula (7.4). The individual efficiency calculation is as follows: $I\_U_i = \alpha \cdot \ln\left(\frac{C_i}{C_{max}} + 1\right) + \beta \cdot \ln\left(\frac{D_i}{D_{max}} + 1\right)$ where C_i is the number of Mashup calls for the current API node, C_{max} is the maximum calls among all API nodes, D_i is the network weighted degree centrality, D_{max} is the maximum weighted degree, and α and β adjust the contributions of market demand and network influence.

Table A.7.6 Evaluation Metrics Adopted in Experiments.

Type	Metric	Description
Quality Metrics	Comprehensive difference degree δ between scenario parameters generated by each method and baseline method (*Original*)	Calculated according to Eq. (7.6) to measure how well the generated scenario τ preserves the characteristics of real scenarios.
Efficiency Metrics	Total time of multiple experiments Sum *Sum*(*ks*)	Measure resource consumption in long-term or batch tasks.
	Average time per single experiment Mean (ks)	Reflects processing efficiency under regular scenario conditions.
	Maximum time for single experiment during the experimental process Peak (ks)	Used to evaluate the stability of experimental methods under worst-case conditions.
	Time per 1000 nodes ($/1kn$)	The experimental system simulates the evolution of 1000 nodes in one cycle, measuring the time consumption to quantify sensitivity to node scale.
	Time per 1000 edges ($/1ke$)	The system also simulates the establishment of 1000 relational connections in one cycle, measuring time consumption to quantify sensitivity to interaction scale.

Table A.7.7 Design of Environment Generation-Related Parameters.

Parameter	Description
$T_c(e_k)$	Parse scenario s through the rule engine to check for violations of hard constraints in L (e.g., economic conservation). For example, if scenario s indicates that a service's commission increases by 10%, but its customers' consumption rises by 20%, while the rule in L states that the service price is positively correlated with the churn rate ($\rho \geq 0.5$), this creates a conflict and will reduce the score of *T(ek)*.
$T_p(e_k)$	Initialize to 1, with smoothing factor $\alpha = 0.5$. Extract call rates of n types of APIs ($n = 4$) in s, and compare them with the statistical distribution in dataset D. p_k denotes the call rate of the i-th type of API in scenario s; $max(p)$ denotes the maximum value among the call rates of the i-th type of API in dataset $\boldsymbol{D}$.
r_s	Each element in e corresponds to weights w_1, w_2, with $\sum_{i=1}^{n} w_i = 1$. For each element F_i, determine the credibility score r_i based on the deviation from corresponding real-world parameters, with $w_1 = 0.5$, $w_2 = 0.5$.
θ_{high}	Evaluate the effect of θ_{high} based on the degree of matching between generated scenarios and actual situations. If the scenarios lack key information, the threshold is lowered; if excessive noise is present, the threshold is raised. Through iteration, the optimal threshold is found and set to 0.72.

Table A.7.8 Design of Environment Generation-Related Parameters.

Parameter	Description
$T_c(e_k)$	Parse scenario s through the rule engine to check for violations of hard constraints in L (e.g., economic conservation). For example, if scenario s indicates that a service's commission increases by 10%, but its customers' consumption rises by 20%, while the rule in L states that the service price is positively correlated with the churn rate ($\rho \geq 0.5$), this creates a conflict and will reduce the score of *T(ek).*
$T_p(e_k)$	Initialize to 1, with smoothing factor $\alpha = 0.5$. Extract call rates of n types of APIs ($n = 4$) in s, and compare them with the statistical distribution in dataset D. p_k denotes the call rate of the i-th type of API in scenario s; $max(p)$ denotes the maximum value among the call rates of the i-th type of API in dataset $\boldsymbol{D}$.
r_s	Each element in e corresponds to weights w_1, w_2, with $\sum_{i=1}^{n} w_i = 1$. For each element F_i, determine the credibility score r_i based on the deviation from corresponding real-world parameters, with $w_1 = 0.5$, $w_2 = 0.5$.
θ_{high}	Evaluate the effect of θ_{high} based on the degree of matching between generated scenarios and actual situations. If the scenarios lack key information, the threshold is lowered; if excessive noise is present, the threshold is raised. Through iteration, the optimal threshold is found and set to 0.72.

Table A.7.9 Design of Parameters Related to the Scenario Generation.

Parameters	Description
η	Randomly selected between 0.001 and 0.01 based on dataset size or model complexity. Set to 0.0055 based on the model's convergence and performance.
Gradient norm convergence threshold ε	Initially varied between [0.01, 0.1]. As the threshold decreased, convergence slowed and performance gains became negligible, suggesting the optimal value lies near the current threshold. Set to 0.9 after tuning.

Chapter A.8

Table A.8.1 Mapping Between the Real World and the Computational Experiment System.

Model	Element	Real World	Computational System
Agent Model	Rider agent	Rider agents have indicators such as salary, speed, and unit consumption. They can independently choose or modify their operational area, within which they accept orders, and they also have the autonomy to determine their working hours.	The individual utility of a rider is defined as follows: $Utility_{T,i} = crra\left(\sum_{t=0}^{t=T} Reward_{t,i}\right) - \sum_{t=0}^{t=T} Cost_{t,i}$ $crra(z) = \frac{z^{1-\eta}-1}{1-\eta}\ \eta > 0$ where Reward refers to the income earned by the rider, and Cost represents expenditures, including time and distance. The CRRA model (constant relative risk aversion) is used to capture the cumulative payoff of an agent's actions up to time T, reflecting the principle of diminishing marginal utility.
	Platform agent	During the rider's labor process, the platform system is responsible for guiding their work, including order dispatching, rider matching, delivery pricing, and route planning. The platform also administers rewards and penalties to workers based on performance. The platform's strategy optimization is driven by the objective of maximizing the social welfare function Swf, which requires balancing income equality and system productivity.	The system utility function is defined as: $Swf_T = eq_t\left(a_t^c\right) * prod_t\left(a_t^c\right)$ where a_t^c denotes the total income of riders in the system at time step t; N is the number of riders in the system; eq_t represents income equality; and $prod_t$ denotes system productivity. $eq_t\left(a_t^c\right) = 1 - gini\left(a_t^c\right)\frac{N}{N-1}, 0 \le eq\left(a_t^c\right) \le 1$ where 1 indicates perfect equality and 0 indicates complete inequality. $prod_t\left(a_t^c\right) = \sum_{i=1}^{N} a_{i,t}^c$ where $prod_t$ indicates the total income of all riders.

(Continued)

Table A.8.1 (Continued)

Model	Element	Real World	Computational System
Rules Model	Rider agent learning and evolution	If the price per delivery order decreases, riders must work longer hours, reduce the delivery time per order, and complete more deliveries in order to maintain their income.	The objective of Rider agent learning and optimization is to maximize individual utility per unit of time. Depending on the scenario, various machine learning approaches can be adopted, including large language model (LLM), evolutionary learning, and reinforcement learning (RL). On the one hand, agents can utilize imitation learning to dynamically adjust the weights of their utility functions based on environmental cues and the behavior of neighboring agents. On the other hand, agents can adopt RL- or LLM-based policies during decision-making and continuously improve their strategies through interaction with the environment.
	Platform agent learning and evolution	The platform's strategy is continuously fine-tuned based on current system performance metrics to ensure its effectiveness. It updates delivery routes by analyzing rider trajectories and average delivery times, progressively shortening the guidance time allocated for each route. This process effectively constitutes a continuous test of user endurance and adaptability.	The degree of involution is measured by comparing system-level performance with individual-level utility. $Involution_t = \frac{Swf_T}{ave(Utility_{T,i})}$ $Involution_t$: A larger value indicates more severe involution, meaning that as individual utility decreases while system-level performance remains stable, the degree of involution intensifies.
Environment Model	Order distribution	The city serves as the operational area for the riders. For the purposes of administrative management and statistical analysis, the entire city is divided into distinct zones.	A discrete grid map is used to represent the spatial environment of the riders, while orders are generated through Monte Carlo simulation. The generation formula is: $f(x) = \sum_{i=1}^{5} a_i e^{\left(-\left(\frac{x-b_i}{c_i}\right)^2\right)}$

Index

b

c

d

t

u

Printed in the USA/Agawam, MA
July 23, 2026

908085.015